Non-Classical Problems in the Theory of Elastic Stability

When a structure is put under an increasing compressive load, it may become unstable and buckle. Buckling is a particularly significant concern in designing shell structures such as aircraft, automobiles, ships, or bridges. This book discusses stability analysis and buckling problems and offers practical tools for dealing with uncertainties that inherently exist in real systems. The techniques are based on two competing yet complementary theories, which are developed in the text. First, the authors present the probabilistic theory of stability, with particular emphasis on reliability. Both theoretical and computational issues are discussed. Second, they present the alternative to probability based on the notion of "anti-optimization," a theory that is valid when the necessary information for probabilistic analysis is absent, that is, when only scant data are available. Design engineers, researchers, and graduate students in aerospace, mechanical, marine, and civil engineering and theoretical and applied mechanics who are concerned with issues of structural reliability and integrity will find this book a particularly useful reference.

Isaac Elishakoff is a professor in the Department of Mechanical Engineering at Florida Atlantic University.

Yiwei Li is an engineering software developer at Alpine Engineered Products, Inc.

James H. Starnes, Jr., is the head of the Structural Mechanics Branch at NASA Langley Research Center.

Non-Classical Problems in the Theory of Elastic Stability

Deterministic, Probabilistic and Anti-Optimization Approaches

ISAAC ELISHAKOFF
Florida Atlantic University

YIWEI LI
Alpine Engineered Products, Inc.

JAMES H. STARNES, JR.
NASA Langley Research Center

CAMBRIDGE
UNIVERSITY PRESS

Dedicated to the memory
of Professor Dr. Ir. Warner Tjardus Koiter
– colleague, co-author, mentor, and friend

CAMBRIDGE UNIVERSITY PRESS
Cambridge, New York, Melbourne, Madrid, Cape Town, Singapore, São Paulo

Cambridge University Press
The Edinburgh Building, Cambridge CB2 2RU, UK

Published in the United States of America by Cambridge University Press, New York

www.cambridge.org
Information on this title: www.cambridge.org/9780521782104

First published 2001
This digitally printed first paperback version 2005

A catalogue record for this publication is available from the British Library

Library of Congress Cataloguing in Publication data

Elishakoff, Isaac. 1944–
Non-classical problems in the theory of elastic stability / Isaac Elishakoff, Yiwei Li, James H. Starnes, Jr.
p. cm.
Includes bibliographical references and index.
ISBN 0-521-78210-4
1. Structural stability – Statistical methods. 2. Probabilities. 3. Elastic analysis (Engineering) 4. Buckling (Mechanics) I. Li, Yiwei. II. Starnes, James H. III. Title.
TA656 .E45 2000
624.1′7 – dc21

00-023667

ISBN-13 978-0-521-78210-4 hardback
ISBN-10 0-521-78210-4 hardback

ISBN-13 978-0-521-02010-7 paperback
ISBN-10 0-521-02010-7 paperback

Contents

Preface: Why Still Another Book on Stability?

> The preface is the most important part of the book. Even reviewers read a preface.
> Philip Guedalla

There are at present numerous books available on the theory of stability and its applications to structures. One author even remarked sarcastically that if they were put in a single bookcase, it would buckle under their weight. We do not complain that "... of the making of many books there is no end" (Ecclesiastes 12:12), rather we ask a natural question: Is there a legitimate place for a new book in this field?

The answer to this question is affirmative, if a book has its unique, distinct characteristics. We have chosen to deal with non-classical problems. To the best of our knowledge, none of the subjects, touched upon in this monograph, have been discussed exclusively in the existing books on buckling analysis. Thus we feel that this book will not be *just* another new book on buckling. Indeed, most existing books may be classified as belonging to one of the following two categories: textbooks – which often look very much alike, maybe not without reason, since the subject is the same – and monographs – which have an encyclopedic nature, trying to comprise an uncomprisable – to cover all or nearly all pertinent topics. This latter task of listing all the results (even only those of major importance) appears to be impossible indeed.

The purpose of this book is to present two competing theories, which incorporate ever-present uncertainty in the stability applications of the real world. These uncertainties are first and foremost due to unavoidable initial imperfections, deviations of the structure from its intended, nominal, ideal shapes. Other uncertainties are the material characteristics and/or realizations of the boundary conditions. These topics are almost never touched upon in the texts or monographs. Here we first present the probabilistic theory of stability. The bridge is made between a description of random imperfections as random fields, its description as random vectors, and the Monte Carlo methods. Special emphasis is devoted to evaluation of reliability, the probability that the structure will not fail prior to preselected load level; reliability concept is the powerful tool that needs to be introduced in the practical design of uncertain structures, if the probabilistic paradigm is adopted. The book presents a unified probabilistic theory of stability. It elucidates both the theoretical and computational aspects in a single package.

Special emphasis is placed on asymptotic evaluations and non–Monte Carlo analytical methods.

Along with the probabilistic theory of stability, we expose the alternative to probability, based on the notion of "anti-optimization"; this theory is valid when the necessary information for the probabilistic analysis is absent and only the scant data are available. Such a theory appears to be of prime importance for estimating the worst case scenario with limited data present. Thus, these two theories complement each other. Ideally, if sufficient data are available, one would resort to the probabilistic theory; however, realistically, in most circumstances its alternative, anti-optimization, should be preferred.

The main objective of this book is, therefore, to provide a strong impetus to both theorists and practitioners so that they will become acquainted with the probabilistic and anti-optimization theories as the most practical tools for dealing with the uncertainties that are present whether theorists or practitioners want to acknowledge them or not and whether they know the appropriate analyses or not. Uncertainty analysis is thus not a luxury but rather a mere necessity to rigorously reflect the working conditions of the real systems. This book, therefore, fills the gap that exists in the present-day literature and practice. In fact, engineers and researchers can no longer pretend that their deterministic approaches represent the truth, that directly, without invoking the uncertainty aspects, could be utilized by the practicing engineers. Recognition of this fact is extremely important not only to the practitioners but also for the theorists for they would try, it is hoped, in their future analyses, to devote additional effort to the rigor of their models, especially when the overlooked uncertainty may drastically change their results obtained for ideal, maybe even never existing, situations.

From the engineering point of view, the developments in stability of structures can be divided into two periods. The first period dates from 1744, when Leonhard Euler communicated to the world his famous formula for the buckling load of a column. The second dates from 1945, when Warner Koiter submitted his PhD thesis to the Delft University of Technology. He uncovered the disastrous effect of initial geometric imperfections on the load-carrying capacity of shell structures. During the two centuries of the post-Euler, pre-Koiter era, engineers and scientists made outstanding contributions in shedding light on the buckling of structures. New problems arose with the technological revolution that took place in the twentieth century. With the modern need to develop lightweight vehicles, the buckling concept has become even more eminent. Lorenz, Southwell, and Timoshenko generalized Euler's classical formula for columns to the case of thin cylindrical shells. Fortunately, buckling specialists were interested in experimental validation of their findings. This led to a surprising and rather disappointing observation: The experimental results did not vindicate the theoretical investigations, most of the data were much lower than what the linear analyses predicted. Koiter's teachings appeared then to explain that the unavoidable imperfections, deviations from the nominally ideal shape, play a dominant role in the drastic reduction of the load-carrying capacity of cylindrical shells and some other structures. Thus, the notion of *imperfection sensitivity* came into being. This is how the buckling topic emerged into healthy adolescence after two hundred years of childhood. The theoretical works of Budiansky and Hutchinson at Harvard University, of Thompson at

University College, London, and of G. W. Hunt of University of Bath, experimental studies by Singer at the Technion–Israel Institute of Technology, combined theoretical-numerical-experimental investigations by Arbocz at the Delft University of Technology, to name just a few, have been spectacularly instrumental in providing this old yet still mysterious field with new impetus and ideas.

Practical and theoretical people in the aeronautics and aerospace fields, who must design real shells and who have invested considerable funds in the research on shell structures and their stability problems, still have not fully adopted the concept of imperfection sensitivity but are using the so-called *knockdown factor*. This appears to be the case since basically engineering scientists did not present to them a means of incorporating theoretical findings into practice. Knockdown factor, which is defined so that its product with the classical buckling load yields a lower bound to all existing experimental data, was compiled primarily by Weingarten, Seide, and Peterson. This is both welcome and disappointing. It is welcome given that many of the shell structures we employ perform extremely responsible functions; it is better to overdesign than underdesign, if we must choose between these two alternatives and do not want to take excessive risks. On the other hand, it is disappointing that the results of several decades of research are mostly ignored and unreflected in engineering practice and that His Excellency the Knockdown Factor is reigning in the kingdom of the designers.

It has been recognized since 1958, when V. V. Bolotin of the Moscow Power Engineering Institute introduced the notion of *randomness* into the theory of elastic stability, that for the buckling theories to become practical, they must be combined with analysis of the uncertain initial imperfections. We are aware that there are no two identical structures, even if they are produced by the same manufacturing process. The apparent reluctance of the designers to take advantage of theoretical findings stems from the fact that most theoretical and/or numerical imperfection studies are conditional on detailed a priori knowledge of the geometric imperfections of the particular structure that is available. In an ideal case, the imperfections can be measured and incorporated into the analyses to predict the buckling loads. For example, Horton of Georgia Institute of Technology tested a large-scale shell with a diameter of 60 ft, and Arbocz and Williams measured the imperfection profile of 10-ft diameter stiffened shells at the NASA Langley Research Center. More recently, Ariane interstage shells (some of them 80 ft in diameter) were measured in the Fokker Aircraft Company. This approach, which is justified for single prototype-like structures, is impractical as a general method for behavior prediction. Information on the type and magnitude of imperfections of a particular structure would be too specific and hardly applicable to other realizations of the same structure, even if they were produced by the same manufacturing process.

With these considerations in view and also bearing in mind the large scatter of the experimental results, it becomes clear that practical applications of the imperfection-sensitivity theories are conditional on their being fused with a probabilistic analysis of the imperfections and buckling loads. This is because engineers do not want to overdesign or underdesign the structures. Apparently, the knockdown factor approach penalizes ingeniously designed shells, those with fewer imperfections.

However, appreciation of the probabilistic approach is not sufficient to solve the problem. Indeed, the early studies on random imperfections fell into two classes. The first class considered overly simplified models of single-degree-of-freedom finite structures with the initial imperfection amplitude as a random variable and with attendant straightforward analysis. The second class treated the initial imperfections as random fields (i.e., random functions of both the axial and circumferential coordinates). To facilitate the purely (and often restrictive) analytical treatment, researchers adopted far-reaching assumptions regarding the probabilistic nature of initial imperfection (statistical homogeneity and ergodicity); they also treated infinitely long structures rather than ones of finite length. Thus a new thinking was necessary to discard negative characteristics of existing probabilistic treatments, while retaining and reinforcing their positive characteristics. We take the liberty of quoting Johann Arbocz's extensive review (1981) on the past, present, and future of shell stability analysis: "... it was not until 1979, when Elishakoff published his reliability study on the buckling of a stochastically imperfect finite column on a nonlinear elastic foundation, that a method has been proposed, which made it possible to introduce the results of the initial imperfection surveys routinely into the analysis." Basically, the idea was to utilize the Monte Carlo method for solution of the stability problem of axially compressed cylindrical shells with random initial imperfections. The latter are expanded in terms of the buckling modes of the perfect structure, and then the Fourier coefficients are treated as random variables. Next, using a special numerical procedure, the Fourier coefficients of the desired large sample of random initial imperfection shapes are simulated after a sufficiently large sample of initial imperfection measurements become available. This is followed by a deterministic buckling load of each of the simulated shells, with subsequent statistical analysis.

Meaningful probabilistic analysis is conditional on probabilistic characterization of the input variables and functions. This can be performed only when appropriate measurement data are available. Fortunately, Babcock and Arbocz at the California Institute of Technology, as well as some other investigators realized the need for large-scale experiments, as a result of which several *initial-imperfection data banks* have been developed (those at the Delft University of Technology and the Technion–Israel Institute of Technology being the most notable). Instead of making unnecessary and often restrictive assumptions regarding the properties of the initial imperfections, these data banks are amenable to direct statistical analysis, since they provide indispensable tools for further probabilistic analysis.

One must recognize that, at first glance, the Monte Carlo method may not appear to be attractive analytically; it may turn out to be numerically time-consuming especially for some highly complex structures. The answer to such a possible reservation is that the Monte Carlo method is the only *universal* technique for probabilistic analysis of structures and that its use does not entail some of the heavily simplified solution procedures that are often required for idealized structural models.

Complex multi-mode non-linear deterministic analysis should be balanced by maximum resemblance to real-life situations. As Timothy Ferris instructively put it, "science is said to proceed on two legs, on the theory (or, loosely, of deduction) and the other of observation and experiment (or induction). Its progress, however, is less often a commanding stride than kind of halting stagger – more like the path of a wandering minstrel than the straight-ruled trajectory of a military marching band." We trust that the direct introduction of the results of experiments into the probabilistic analysis puts us on the right track, combining both the deductive and the inductive facets of engineering. This became feasible by employing a numerical tool closely resembling the experiments themselves, namely by the Monte Carlo method. At the same time, we do not advocate an abandonment of analytical techniques that, although applicable, may complement the Monte Carlo method especially where small imperfections are involved. The Monte Carlo method needs very accurate numerical techniques to evaluate buckling loads of each shell in the ensemble of shells; careful numerical techniques like STAGS, BOSOR, PANDA, and those based on multi-mode imperfection methods are consistent with the reliability analysis.

Can one use probabilistic modeling if the data are extremely limited? This is an intriguing question. The answer is affirmative for those who make a fetish of the probabilistic models, but we feel that it should be negative. Indeed, the central premise of this book is the need for sound theoretical, experimental, and numerical analyses. Rigorous deterministic analysis is a cornerstone of every meaningful probabilistic treatment of a problem at hand. Theoretical analysis elucidates the physical phenomenon but may not be feasible without some idealizations. Experimental analysis then provides the data used by the numerical analyst to create statistical "brothers and sisters" of the measured shells. According to the *John Wiley Dictionary of Scientific Terms*, *reliability* is "the probability that a component part, equipment, or system will satisfactorily perform its intended function under given circumstances, such as environmental conditions, limitations as the operating time, and frequency and thoroughness of maintenance, for a specified period of time." In the structural-stability context, reliability is the probability that the structure will sustain loads in excess of the specified level. Here either extremely high reliabilities or equivalently extremely low probabilities of failure are needed. To perform such a refined analysis, refined data and authentic analytical and numerical tools are called for. In the absence of experimental data (as may well be the case with variation of the elastic moduli), probabilistic methods cannot be recommended for design purposes. Fortunately, there exists a new discipline, called *convex modeling of uncertainty* (or, in a more general context, *set-theoretical modeling of uncertainty*), developed by Ben-Haim and Elishakoff for applied-mechanics problems. This discipline does not seek to make something out of nothing (i.e., it does not create the probabilistic model out of extremely limited or absent data), but it does represent an alternative technique for such special yet often encountered situations. Set-theoretical uncertainty in effect represents *anti-optimization* under uncertainty: It operates with the least favorable scenarios based on the limited experimental data describing uncertain variables or functions.

As we saw earlier, accurate determination of the reliability or the least favorable buckling loads calls for a threefold effort in: (a) deepening theoretical insight into the phenomenon, (b) collecting experimental information, and (c) conducting proper numerical analysis. Choice of a suitable model of uncertainty is also a prominent decision to be made, based on the amount and character of the information.

This monograph reflects this philosophy. Neither a textbook nor an encyclopedia, it covers the deterministic, probabilistic, and set-theoretical approaches for buckling of structures. Chapters 1, 2 (except Sections 2.1 and 2.2), and 5 are written by three of us; the first two sections of Chapter 2 are based on our joint papers with Professor Koiter of the Delft University of Technology on the effect of thickness variation in perfect or imperfect isotropic shells. The last two sections of Chapter 2 represent a generalization for the case of composite shells. Chapters 3 (except Section 3.2), 4, 6, and 7 are written by the first author; Section 3.2 is based on our joint paper with Professor Masanobu Shinozuka of the University of Southern California. Chapters 3 and 4 are based on the work of the first author (Section 3.1), and his joint studies with Professor Johann Arbocz of the Delft University of Technology (Sections 3.3, 3.4, and 4.2), Professor Koyohiro Ikeda of Tohoku University and Professor Kazuo Murota of the Kyoto University (Section 4.1) and other investigators. Contents of works of Professor Wei-Chau Xie of the University of Waterloo and of Dr. Jun Zhang and Professor Bruce Ellingwood of Johns Hopkins University constitute the central part of the last two sections of Chapter 4. Section 3.5 is based on our joint investigation with Dr. David Bushnell of Lockheed Company. Section 5.3 is based on the joint paper by the first and third authors with Professor Guoqiang Cai of the Florida Atlantic University.

The organization of the present book is as follows: The first two chapters are devoted to some new deterministic problems. Chapter 1 discusses the mode localization in the deterministic setting, an extremely recent topic in the buckling context. In particular, we consider multi-span columns and plates, with unavoidable misplacements of the stiffeners location. Generally, a misplacement can be regarded as one type of imperfection, although it does not lead to as drastic a change in the buckling load as geometric imperfections do in shell buckling. As a matter of fact, in many experimental studies (e.g., of Singer or of Tenerelli and Horton) of the shell buckling, only localized buckling modes were observed. In strong ring-stiffened shells, two or three buckles emerged when the critical load was reached and were confined to a single span. In weak ring-stiffened shells, buckles appeared around the whole circumference but extended only through part of the spans. Chapter 1 advocates that, if each span is treated as an element of a periodic structure, the localization phenomena can be explained by using the analytical finite difference calculus. Chapter 2 is devoted to the influence of thickness variation on the buckling of perfect or imperfect, isotropic or composite, circular cylindrical shells.

Chapters 3 and 4 deal with stochastic buckling of structures with random imperfections. Chapter 3 focuses on the Monte Carlo method, whereas Chapter 4 discusses approximate analytical and numerical techniques, including the asymptotic analysis, the first-order second-moment method, the mode localization due to random misplacements, and the finite-element method for structures with random material properties.

Convex modeling of uncertainty in buckling problems is the focal point of Chapter 5. Here the realistic situation of data scarcity is considered, and the minimum buckling loads in an ensemble of plates and shells with uncertain material properties are derived. Chapter 6 discusses the Godunov-Conte shooting method (representing an extended version of the paper by Elishakoff and Charmats), whereas Chapter 7 deals with the application of computerized symbolic algebra – an able and obedient "servant" of the present-day researchers. It constitutes an extended version of the paper by Elishakoff and Tang.

We hope that the monograph will prove useful for researchers, engineers, and senior graduate students specializing in aeronautical, aerospace, mechanical, civil, nuclear, and marine engineering. We hope that our "pluralistic" philosophy will have a definite impact on the entire engineering profession. As Abraham Maslow remarks, "If the only tool you have is a hammer, you tend to treat everything as if it were a nail." We strongly trust that the modern analyst cannot afford to confine himself/herself to only one of the trio, namely, (a) deterministic, (b) probabilistic, and (c) set-theoretical *Weltanschauung*. Instead, we, as engineers must be pragmatic and flexible in choosing the most appropriate methods, consistent with available experimental information.

Some pertinent questions remain still to be tackled. Two of the central questions were posed by Professor Bernard Budiansky in his correspondence with Professor Johann Arbocz (Arbocz and Singer, 2000): "Are we necessarily doomed to accept forever the unhappy coexistence of efficient shell design and imperfection-sensitivity? Or are there some structural design secrets to be discovered that will retain minimum weight and rid us of the curse of imperfection-sensitivity?"

This book deals with how to live with the above curse most efficiently, combining deterministic probabilistic and anti-optimization "cures," in the hope that some "genes" will be uncovered that will make the multiplicity of studies on the "curse" unnecessary even if still instructive.

We thank (a) Elsevier Science Publishers for allowing us to reproduce Figures 3.13–3.29 from the article by I. Elishakoff and J. Arbocz, "Reliability of Axially Compressed Cylindrical Shells with Random Axisymmetric Imperfections," *International Journal of Solids and Structures*, Vol. 18, 563–85, 1982; Figures 4.1–4.4 from the article by K. Ikeda, K. Murota, and I. Elishakoff, "Reliability of Structures Subject to Normally Distributed Initial Imperfections," *Computers and Structures*, Vol. 59, 463–9, 1995; Figures 5.10–5.14 from the study by I. Elishakoff, G. Q. Cai, and J. H. Starnes, Jr., "Nonlinear Buckling of a Column with Initial Imperfection via Stochastic and Non-Stochastic Convex Models," *International Journal of Nonlinear Mechanics*, Vol. 29, 71–82, 1994; Figure 3.34 from the book *Buckling of Structures – Theory and Experiment* (Elsevier, 1988), edited by I. Elishakoff, J. Arbocz, C. D. Babcock, Jr., and A. Libai; (b) AIAA Press for permission to reproduce Figures 4.5–4.6 from the article by I. Elishakoff, S. Van Manen, P. G. Vermeulen, and J. Arbocz, "First-Order Second-Moment Analysis of the Buckling of Shells with Random Initial Imperfections," *AIAA Journal*, Vol. 25, 1113–17, 1987; Figures 4.7–4.11 from the paper by W.-C. Xie, "Buckling Mode Localization in Randomly Disordered Multispan Continuous Beam," *AIAA Journal*, Vol. 33, 1142–9, 1995; (c) American Society of Civil Engineering for permission to reproduce Figures 4.12–4.14 from the paper by J. Zhang and B. Ellingwood, "Effects of Uncertain Material Properties on Structural Stability," *Journal of Structural Engineering*, Vol. 121, 705–16, 1995; and (d) American Society of Mechanical Engineering for the permission to reproduce Figures 3.30–3.33 and 3.35–3.37 from the paper by I. Elishakoff and J. Arbocz, "Reliability of Axially Compressed Cylindrical Shells with General Nonsymmetric Imperfections," *Journal of Applied Mechanics*, Vol. 52, 122–8, 1985.

The authors gratefully acknowledge scientific cooperation, over the last two decades, on various buckling problems with the following colleagues: Professor J. Arbocz, Ir. J. van Geer, Professor J. Kalker, Professor W. T. Koiter, Ir. A. Scheurkogel, Professor W. D. Verduyn, and Ir. P. G. Vermeulen of Delft University of Technology; Professor J. Ari-Gur of Western Michigan University; Professor M. Baruch, Professor

Y. Ben-Haim, Mr. M. Charmatz, Mr. E. Itzhak, Dr. S. Marcus, Professor J. Singer, and Professor Y. Stavsky of the Technion–Israel Institute of Technology; Professor C. W. Bert of the University of Oklahoma; Professor V. Birman of the University of Missouri–Rolla; Dr. D. Bushnell of Lockheed Missiles and Space Company, Inc.; Professor G. Q. Cai, Dr. L. P. Zhu, and Dr. M. Zingales of Florida Atlantic University; Professor K. Ikeda of Tohoku University, Japan; Dr. S. van Manen of TNO (the Netherlands); Professor K. Murota of Kyoto University, Japan; Dr. W. J. Stroud of the NASA Langley Research Center; Professor M. Shinozuka of the University of Southern California; and Professor W. C. Xie of the University of Waterloo.

Over the last two decades, helpful discussions on aspects in uncertain buckling with Professor J. Amazigo of the University of Nigeria; Professor S. T. Ariaratnam of the University of Waterloo; Professor G. Augusti of the University of Rome; Professor V. V. Bolotin, Professor B. P. Marakov, and Dr. V. M. Leizerakh of the Moscow Power Engineering Institute–State University; Professor C. D. Babcock, Jr. of Caltech; Professor B. Budiansky of Harvard University; Professor B. Frazer of the University of New South Wales; Professor S. H. Crandall of M.I.T; Professor N. J. Hoff of Stanford University; Professor I. Kaplyevatsky and Professor I. Sheinman of the Technion–Israel Institute of Technology; Professor D. Shilkrut of the University of Negev; Professor S. Krenk of Lund University; Professor G. Palassoupulos of the Military Academy of Greece; Dr. M. Stein of the NASA Langley Research Center; Professor R. C. Tennyson and Professor J. Hansen of the University of Toronto; and others (who may have been unintentionally overlooked) are most appreciated. Most work reported in this book was made possible through a grant from NASA Langley Research Center. This help is gratefully acknowledged. First author appreciates spending July 2000 at the Delft University of Technology as the W. T. Koiter Visiting Chair Professor, and especially the warm and kind hospitality of Professor Rene de Borst and Professor Fred van Keulen; during this stay the final touches of the manuscript were completed.

The responsibility for any imperfections in this monograph, be they random or otherwise, rests solely upon the authors. We would appreciate hearing your comments via electronic mail (*ielishak@me.fau.edu*), fax (561-297-2825), or regular correspondence.

CHAPTER ONE

Mode Localization in Buckling of Structures

Think globally, act locally.
Anonymous

The genuine spirit of localism . . .
George Borrow

This chapter investigates the buckling mode localization in the periodic multi-span beams and plates. We start our discussion with disorder in two- and three-span elastic plates; then we focus our attention on the multi-span beams and plates with a disorder occurring in an arbitrary single span. The analytical finite difference calculus is employed to derive the transcendental equations from which buckling load is calculated. The underlying treatment is general, and the solution thus obtained is exact within the theory used. Numerical results show that the buckling mode is highly localized in the vicinity of the disordered span of the beam or the plate. In the multi-span, elastic plates are considered with transverse stiffeners, and the discreteness of the stiffeners is accounted for. The torsional rigidity of the stiffener is found to play an important role in the buckling mode pattern. When the torsional rigidity is properly adjusted, the stiffener can be used in passive control; that is, it can serve as an isolator of deformation for the structure at buckling so that the deflection is limited to only a small area.

1.1 Localization in Elastic Plates Due to Misplacement in the Stiffener Location

Traditionally, the stability of the stiffened plate has been studied following three different settings. One approach consists of replacing the stiffened plate by an "equivalent" orthotropic plate after the stiffeners are smeared out, in an energetic sense, over the entire surface of the plate (Brush and Almroth, 1975; McFarland, Smith, and Bernhart, 1972). This approach appears reasonable for plates with many closely spaced stiffeners but is invalid for plates with very few stiffeners. The second approach is based on energy consideration and treats the contributions of the plate and the stiffeners separately; the Rayleigh-Ritz method has been utilized widely to estimate the buckling load of the stiffened plate structures (Bulson, 1969; Liew and Wang, 1990). This method may predict well the global buckling but fails to detect the possible localization of buckling mode due to the small changes in the location of the stiffeners. The third approach is the method applicable to equally spaced stiffeners by the *analytical* finite difference calculus (Bleich, 1952; Wah and Calcote, 1970). The latter method, though powerful for studying plates with periodically spaced stiffeners or supports, is inapplicable if the periodicity is disturbed as is usually the case when misplacement in the location of the stiffener or support is present through imprecision of construction or assembly. Despite their usefulness and simplicity, the previously mentioned methods can only be employed to investigate the global buckling of the structure and appear incapable of revealing the localization phenomena when the structure is sparsely or irregularly stiffened and the buckling mode is likely to be localized.

Localization phenomenon was first uncovered by the Nobel laureate in physics P. W. Anderson (1958). Its occurrence in structures has recently attracted much attention. Much research has been conducted in recent years to study the localization phenomenon in vibration of structures (for example, Hodges, 1982; Hodges and Woodhouse, 1983; Pierre and Dowell, 1987; Pierre and Chat, 1989; Pierre, 1990) and in acoustics (De Jong, 1994; Maidanik and Dickey, 1996). Cai and Lin (1991) studied the localization of wave propagation in randomly disordered periodic structures. Apparently, Pierre and Plaut (1989) were the first investigators to consider the localization in buckling; they studied the simplest two-span column with a deterministic disorder. A more general case, the multi-span column, was recently treated by Nayfeh and Hawwa (1994a, 1994b) using the transfer matrix method. Ariaratnam and Xie (1996) investigated the localization in the buckling of a system of rigid bars connected with springs in the stochastic setting. Xie (1995) studied the buckling mode localization in randomly disordered multi-span beams by the finite-element method. Tvergaard and Needleman (1983) discussed the development of localized patterns in the elastic-plastic and thermal buckling problems. The deterministic buckling localization in cylindrical shells was investigated by El Naschie (1975a, 1975b, 1977, 1990).

In this section, we investigate the effect of small structural irregularity, due to the misplacement of stiffeners or interior supports, on both the buckling load and the buckling mode of the rib-stiffened plate. Since the buckling mode shape is of main interest, the interaction between the plate and stiffeners should be properly taken into account. Here, the integration of the general governing differential equation is performed for the stiffened elastic plate. By considering the rib-stiffened plate as a physically continuous plate with as many spans as the number of ribs, the stiffeners are accounted for through the conditions of continuity. Two cases commonly encountered in practice are considered: one with simple support under the ribs and one without. It is found that in the presence of small misplacement of stiffeners or interior supports, the buckling mode shapes experience dramatic changes to become strongly localized. We will first deal with the localization phenomenon in the buckling of a two-span plate with a single rib using different parameters for the stiffener. Furthermore, a stiffened three-span plate will be investigated, and the optimal configuration of stiffener placement, which yields the highest buckling strength, will be discussed along with the attendant localization sensitivity to deterministic misplacement.

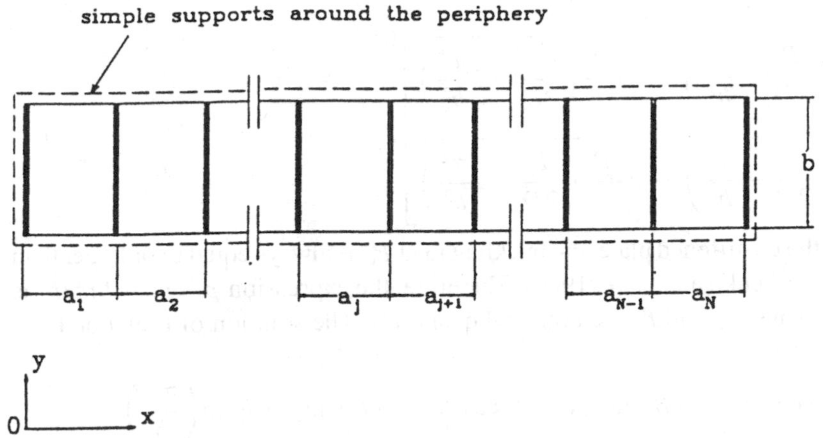

Figure 1.1 A simply supported rectangular plate reinforced with transverse stiffeners.

We first consider a rib-stiffened rectangular plate subjected in its mid-plane to uniform compression P in the x direction (Figure 1.1). The differential equation of the deflection surface of the plate under consideration is

$$D\left(\frac{\partial^4 w}{\partial x^4} + 2\frac{\partial^4 w}{\partial x^2 \partial y^2} + \frac{\partial^4 w}{\partial y^4}\right) + P\frac{\partial^2 w}{\partial x^2} = 0 \tag{1.1}$$

where w is the transverse displacement, downward positive; D is the flexural rigidity of the plate. The solution of Equation (1.1) can be represented in the following form:

$$w(x, y) = X(x)\sin\left(\frac{\pi y}{b}\right) \tag{1.2}$$

Substitution of Equation (1.2) into Equation (1.1) results in

$$\frac{d^4 X}{dx^4} + \left(\frac{P}{D} - 2\frac{\pi^2}{b^2}\right)\frac{d^2 X}{dx^2} + \frac{\pi^4}{b^4}X = 0, \qquad X = Ae^{sx} \tag{1.3}$$

The corresponding characteristic equation reads

$$s^4 + \left(\frac{P}{D} - 2\frac{\pi^2}{b^2}\right)s^2 + \frac{\pi^4}{b^4} = 0 \tag{1.4}$$

or

$$s^2 = -\left(\frac{P}{2D} - \frac{\pi^2}{b^2}\right) \pm \sqrt{\frac{P}{2D}\left(\frac{P}{2D} - 2\frac{\pi^2}{b^2}\right)} \tag{1.5}$$

For rib-stiffened, or intermediately supported, plates, we have roots s_i $(i = 1, 2, 3, 4)$ as follows:

$$s_1 = i\beta_1, \qquad s_2 = -i\beta_1, \qquad s_3 = i\beta_2, \qquad s_4 = -i\beta_2 \tag{1.6}$$

where

$$\beta_1 = \left[\left(\frac{P}{2D} - \frac{\pi^2}{b^2}\right) + \sqrt{\frac{P}{2D}\left(\frac{P}{2D} - 2\frac{\pi^2}{b^2}\right)}\right]^{1/2},$$
$$\beta_2 = \left[\left(\frac{P}{2D} - \frac{\pi^2}{b^2}\right) - \sqrt{\frac{P}{2D}\left(\frac{P}{2D} - 2\frac{\pi^2}{b^2}\right)}\right]^{1/2} \tag{1.7}$$

Even for the unstiffened plate, the buckling load P_{cr} is always equal to or larger than $4\pi^2 D/b^2$ (Timoshenko and Gere, 1961). Therefore, the expression $p/2D - 2\pi^2/b^2$ is non-negative. Thus, β_1 and β_2 are both real quantities. The solution of Equation (1.1) thus can be written as

$$w(x) = [A\cos(\beta_1 x) + B\sin(\beta_1 x) + C\cos(\beta_2 x) + D\sin(\beta_2 x)]\sin\left(\frac{\pi y}{b}\right) \tag{1.8}$$

where A, B, C, and D are constants of integration, which are to be determined using the continuity and boundary conditions. For the arbitrary, jth span, the solution can be written as

$$w_j(x_j) = [A_j\cos(\beta_1 x_j) + B_j\sin(\beta_1 x_j) + C_j\cos(\beta_2 x_j) + D_j\sin(\beta_2 x_j)]\sin\left(\frac{\pi y}{b}\right), \qquad 0 \le x_j \le a_j \tag{1.9}$$

where a_j is the length of the jth span, and j ranges from 1 to N for an N-span plate. We consider the plate simply supported along its periphery. The boundary conditions read

$$\begin{aligned} w_1|_{x_1=0} &= 0 \\ M_x^{(1)}\big|_{x_1=0} &= -D\left(\frac{\partial^2 w_1}{\partial x_1^2} + \nu\frac{\partial^2 w_1}{\partial y^2}\right)\bigg|_{x_1=0} = 0 \\ M_x^{(N)}\big|_{x_N=a_N} &= -D\left(\frac{\partial^2 w_N}{\partial x_N^2} + \nu\frac{\partial^2 w_N}{\partial y^2}\right)\bigg|_{x_N=a_N} = 0 \\ w_N|_{x_N=a_N} &= 0 \end{aligned} \tag{1.10}$$

where $M_x^{(1)}$ and $M_x^{(N)}$ are the bending moments in the first and last spans of the continuous plate; ν is the Poisson's ratio. In view of (1.9), the preceding boundary conditions become

$$A_1 + C_1 = 0 \tag{1.11}$$

$$\beta_1^2 A_1 + \beta_2^2 C_1 = 0 \tag{1.12}$$

$$\beta_1^2\cos(\beta_1 a_N)A_N + \beta_1^2\sin(\beta_1 a_N)B_N + \beta_2^2\cos(\beta_2 a_N)C_N + \beta_2^2\sin(\beta_2 a_N)D_N = 0 \tag{1.13}$$

$$\cos(\beta_1 a_N)A_N + \sin(\beta_1 a_N)B_N + \cos(\beta_2 a_N)C_N + \sin(\beta_2 a_N)D_N = 0 \tag{1.14}$$

Regarding the continuity conditions between two successive spans, two cases of practical interest deserve consideration.

Case A: Simple support under the rib. In some applications, the flexural rigidity of the stiffener is not large enough, and a vertical support is installed under the stiffener to suppress the transverse displacement. In this case, the continuity conditions between the two typical neighboring spans numbered j and $j+1$ are

$$
\begin{aligned}
&w_{j+1}|_{x_{j+1}=0} = 0\\
&w_j|_{x_j=a_j} = 0\\
&\left.\frac{\partial w_j}{\partial x_j}\right|_{x_j=a_j} = \left.\frac{\partial w_{j+1}}{\partial x_{j+1}}\right|_{x_{j+1}=0}\\
&M_x^{(j+1)}\big|_{x_{j+1}=0} - M_x^{(j)}\big|_{x_j=a_j} = (GJ)_j \left.\frac{\partial^3 w_{j+1}}{\partial x_{j+1}\partial y^2}\right|_{x_{j+1}=0}
\end{aligned}
$$

or

$$
\begin{aligned}
&-D\left(\frac{\partial^2 w_{j+1}}{\partial x_{j+1}^2} + \nu\frac{\partial^2 w_{j+1}}{\partial y^2}\right)\bigg|_{x_{j+1}=0} + D\left(\frac{\partial^2 w_j}{\partial x_j^2} + \nu\frac{\partial^2 w_j}{\partial y^2}\right)\bigg|_{x_j=a_j}\\
&\quad = (GJ)_j \left.\frac{\partial^3 w_{j+1}}{\partial x_{j+1}\partial y^2}\right|_{x_{j+1}=0} \qquad (1.15)
\end{aligned}
$$

where $(GJ)_j$ denotes the torsional rigidity of the jth rib.

Substituting Equations (1.9) into the preceding conditions of continuity leads to the following four equations:

$$A_{j+1} + C_{j+1} = 0 \qquad (1.16)$$

$$\cos(\beta_1 a_j)A_j + \sin(\beta_1 a_j)B_j + \cos(\beta_2 a_j)C_j + \sin(\beta_2 a_j)D_j = 0 \qquad (1.17)$$

$$
\begin{aligned}
&-\beta_1\sin(\beta_1 a_j)A_j + \beta_1\cos(\beta_1 a_j)B_j - \beta_2\sin(\beta_2 a_j)C_j\\
&\quad + \beta_2\cos(\beta_2 a_j)D_j - \beta_1 B_{j+1} - \beta_2 D_{j+1} = 0 \qquad (1.18)
\end{aligned}
$$

$$
\begin{aligned}
&-\beta_1^2\cos(\beta_1 a_j)A_j - \beta_1^2\sin(\beta_1 a_j)B_j - \beta_2^2\cos(\beta_2 a_j)C_j - \beta_2^2\sin(\beta_2 a_j)D_j\\
&\quad + \beta_1^2 A_{j+1} + \frac{(GJ)_j}{D}\frac{\pi^2}{b^2}\beta_1 B_j + \beta_2^2 C_{j+1} + \frac{(GJ)_j}{D}\frac{\pi^2}{b^2}\beta_2 D_{j+1} = 0 \qquad (1.19)
\end{aligned}
$$

Case B: No support under the rib. In this case, the bending, in addition to the torsion, of ribs should be taken into account. The conditions of continuity between two consecutive spans j and $j+1$ read

$$
\begin{aligned}
&w_j|_{x_j=a_j} = w_{j+1}|_{x_{j+1}=0}\\
&\left.\frac{\partial w_j}{\partial x_j}\right|_{x_j=a_j} = \left.\frac{\partial w_{j+1}}{\partial x_{j+1}}\right|_{x_{j+1}=0}\\
&M_x^{(j+1)}\big|_{x_{j+1}=0} - M_x^{(j)}\big|_{x_j=a_j} = (GJ)_j \left.\frac{\partial^3 w_{j+1}}{\partial x_{j+1}\partial y^2}\right|_{x_{j+1}=0}
\end{aligned}
$$

or

$$
\begin{aligned}
&\left(\frac{\partial^2 w_j}{\partial x_j^2} + \nu\frac{\partial^2 w_j}{\partial y^2}\right)\bigg|_{x_j=a_j} - \left(\frac{\partial^2 w_{j+1}}{\partial x_{j+1}^2} + \nu\frac{\partial^2 w_{j+1}}{\partial y^2}\right)\bigg|_{x_{j+1}=0}\\
&\quad = (GJ)_j \left.\frac{\partial^3 w_{j+1}}{\partial x_{j+1}\partial y^2}\right|_{x_{j+1}=0}\\
&V_x^{(j+1)}\big|_{x_{j+1}=0} - V_x^{(j)}\big|_{x_j=a_j} = (EI)_j \left.\frac{\partial^4 w_{j+1}}{\partial y^4}\right|_{x_{j+1}=0}
\end{aligned}
$$

or

$$\left[\frac{\partial^3 w_j}{\partial x_j^3} + (2-\nu)\frac{\partial^2 w_j}{\partial x_j \partial y^2}\right]_{x_j=a_j} - \left[\frac{\partial^3 w_{j+1}}{\partial x_{j+1}^3} + (2-\nu)\frac{\partial^2 w_{j+1}}{\partial x_{j+1}\partial y^2}\right]_{x_{j+1}=0}$$
$$= (EI)_j \frac{\partial^4 w_{j+1}}{\partial y^4}\bigg|_{x_{j+1}=0} \tag{1.20}$$

where $V_x^{(j)}$ and $V_x^{(j+1)}$ are the shearing forces in the jth and $(j+1)$th spans of the plate; $(EI)_j$ is the flexural rigidity of the jth rib.

These conditions of continuity can, in turn, be expressed by the following equations in terms of the constants of integration:

$$\cos(\beta_1 a_j)A_j + \sin(\beta_1 a_j)B_j + \cos(\beta_2 a_j)C_j + \sin(\beta_2 a_j)D_j - A_{j+1} - C_{j+1} = 0 \tag{1.21}$$

$$-\beta_1 \sin(\beta_1 a_j)A_j + \beta_1 \cos(\beta_1 a_j)B_j - \beta_2 \sin(\beta_2 a_j)C_j + \beta_2 \cos(\beta_2 a_j)D_j$$
$$- \beta_1 B_{j+1} - \beta_2 D_{j+1} = 0 \tag{1.22}$$

$$-\beta_1^2 \cos(\beta_1 a_j)A_j - \beta_1^2 \sin(\beta_1 a_j)B_j - \beta_2^2 \cos(\beta_2 a_j)C_j - \beta_2^2 \sin(\beta_2 a_j)D_j$$
$$+ \beta_1^2 A_{j+1} + \frac{(GJ)_j}{D}\frac{\pi^2}{b^2}\beta_1 B_{j+1} + \beta_2^2 C_{j+1} + \frac{(GJ)_j}{D}\frac{\pi^2}{b^2}\beta_2 D_{j+1} = 0 \tag{1.23}$$

$$\beta_1^3 \sin(\beta_1 a_j)A_j - \beta_1^3 \cos(\beta_1 a_j)B_j + \beta_2^3 \sin(\beta_2 a_j)C_j - \beta_2^3 \cos(\beta_2 a_j)D_j$$
$$- \frac{(EI)_j}{D}\left(\frac{\pi}{b}\right)^4 A_{j+1} + \beta_1^3 B_{j+1} - \frac{(EI)_j}{D}\left(\frac{\pi}{b}\right)^4 C_{j+1} + \beta_2^3 D_{j+1} = 0 \tag{1.24}$$

By introducing the following non-dimensional quantities

$$\lambda = \frac{Pb^2}{\pi^2 D}, \qquad r_j = \frac{a_j}{b}, \qquad \tau_j = \frac{(GJ)_j}{bD}, \qquad \omega_j = \frac{(EI)_j}{bD} \qquad (j = 1 \sim N)$$
$$\bar{\beta}_1 = \left[\frac{\lambda}{2} - 1 + \sqrt{\frac{\lambda}{2}\left(\frac{\lambda}{2} - 2\right)}\right]^{1/2}, \qquad \bar{\beta}_2 = \left[\frac{\lambda}{2} - 1 - \sqrt{\frac{\lambda}{2}\left(\frac{\lambda}{2} - 2\right)}\right]^{1/2} \tag{1.25}$$

the boundary conditions for the simply supported continuous plate, Equations (1.11)–(1.14), can be written as

$$A_1 + C_1 = 0 \tag{1.26}$$

$$\bar{\beta}_1^2 A_1 + \bar{\beta}_2^2 C_1 = 0 \tag{1.27}$$

$$\bar{\beta}_1^2 \cos(\bar{\beta}_1 r_N \pi)A_N + \bar{\beta}_1^2 \sin(\bar{\beta}_1 r_N \pi)B_N + \bar{\beta}_2^2 \cos(\bar{\beta}_2 r_N \pi)C_2$$
$$+ \bar{\beta}_2^2 \sin(\bar{\beta}_2 r_N \pi)D_N = 0 \tag{1.28}$$

$$\cos(\bar{\beta}_1 r_N \pi)A_N + \sin(\bar{\beta}_1 r_N \pi)B_N + \cos(\bar{\beta}_2 r_N \pi)C_N + \sin(\bar{\beta}_2 r_N \pi)D_N = 0 \tag{1.29}$$

The conditions of continuity for Case A, Equations (1.16)–(1.19), are transformed into the following equations:

$$A_{j+1} + C_{j+1} = 0 \tag{1.30}$$

$$\cos(\bar{\beta}_1 r_j \pi)A_j + \sin(\bar{\beta}_1 r_j \pi)B_j + \cos(\bar{\beta}_2 r_j \pi)C_j + \sin(\bar{\beta}_2 r_j \pi)D_j = 0 \tag{1.31}$$

$$-\bar{\beta}_1 \sin(\bar{\beta}_1 r_j \pi) A_j + \bar{\beta}_1 \cos(\bar{\beta}_1 r_j \pi) B_j - \bar{\beta}_2 \sin(\bar{\beta}_2 r_j \pi) C_j + \bar{\beta}_2 \cos(\bar{\beta}_2 r_j \pi) D_j - \bar{\beta}_1 B_{j+1} - \bar{\beta}_2 D_{j+1} = 0 \tag{1.32}$$

$$-\bar{\beta}_1^2 \cos(\bar{\beta}_1 r_j \pi) A_j - \bar{\beta}_1^2 \sin(\bar{\beta}_1 r_j \pi) B_j - \bar{\beta}_2^2 \cos(\bar{\beta}_2 r_j \pi) C_j - \bar{\beta}_2^2 \sin(\bar{\beta}_2 r_j \pi) D_j + \bar{\beta}_1^2 A_{j+1} + \tau_j \pi \bar{\beta}_1 B_{j+1} + \bar{\beta}_2^2 C_{j+1} + \tau_j \pi \bar{\beta}_2 D_{j+1} = 0 \tag{1.33}$$

For Case B, Equations (1.21)–(1.24) are rendered into the following form:

$$\cos(\bar{\beta}_1 r_j \pi) A_j + \sin(\bar{\beta}_1 r_j \pi) B_j + \cos(\bar{\beta}_2 r_j \pi) C_j + \sin(\bar{\beta}_2 r_j \pi) D_j - A_{j+1} - C_{j+1} = 0 \tag{1.34}$$

$$-\bar{\beta}_1 \sin(\bar{\beta}_1 r_j \pi) A_j + \bar{\beta}_1 \cos(\bar{\beta}_1 r_j \pi) B_j - \bar{\beta}_2 \sin(\bar{\beta}_2 r_j \pi) C_j + \bar{\beta}_2 \cos(\bar{\beta}_2 r_j \pi) D_j - \bar{\beta}_1 B_{j+1} - \bar{\beta}_2 D_{j+1} = 0 \tag{1.35}$$

$$-\bar{\beta}_1^2 \cos(\bar{\beta}_1 r_j \pi) A_j - \bar{\beta}_1^2 \sin(\bar{\beta}_1 r_j \pi) B_j - \bar{\beta}_2^2 \cos(\bar{\beta}_2 r_j \pi) C_j - \bar{\beta}_2^2 \sin(\bar{\beta}_2 r_j \pi) D_j + \bar{\beta}_1^2 A_{j+1} + \tau_j \pi \bar{\beta}_1 B_{j+1} + \bar{\beta}_2^2 C_{j+1} + \tau_j \pi \bar{\beta}_2 D_{j+1} = 0 \tag{1.36}$$

$$\bar{\beta}_1^3 \sin(\bar{\beta}_1 r_j \pi) A_j - \bar{\beta}_1^3 \cos(\bar{\beta}_1 r_j \pi) B_j + \bar{\beta}_2^3 \sin(\bar{\beta}_2 r_j \pi) C_j - \bar{\beta}_2^3 \cos(\bar{\beta}_2 r_j \pi) D_j - \omega_j \pi A_{j+1} + \bar{\beta}_1^3 B_{j+1} - \omega_j \pi C_{j+1} + \bar{\beta}_2^3 D_{j+1} = 0 \tag{1.37}$$

For a general N-span continuous plate, we have four equations for boundary conditions in the form of Equations (1.26)–(1.29). Since there are four equations for each rib or interior support such as Equations (1.30)–(1.33) or Equations (1.34)–(1.37), $4 \times (N-1)$ equations can be established from the continuity considerations. Altogether, there are $4N$ algebraic equations for the same number of unknown coefficients A_j, B_j, C_j, and D_j ($j = 1 \sim N$),

$$[\boldsymbol{F}(\lambda)]_{4\times N}\{\boldsymbol{\Delta}\}_{N\times 1} = \boldsymbol{0} \tag{1.38}$$

where elements of matrix $[F(\lambda)]$ are composed of such parameters as those denoted in (1.25) and $\{\Delta\}$ is a column containing A_j, B_j, C_j, and D_j. These equations are linear and homogeneous. A non-trivial solution is obtained by setting the determinant of the matrix $\boldsymbol{F}(\lambda)$ equal to zero, which yields a transcendental equation whose smallest root is the critical buckling load λ. Once we evaluate the buckling load λ, Equation (1.38) is used to determine, to an arbitrary constant multiple, the coefficients A_j, B_j, C_j, and D_j, which can then be substituted back into Equation (1.9) to obtain the buckling mode shape of the entire plate. Note that, in the special case of plates with equally spaced stiffeners, the *analytical* finite difference calculus discussed by Wah and Calcote (1970) can be used. In this section, however, since we will concentrate on the two- or three-span plates with stiffeners that are not necessarily uniformly spaced, the use of the previously mentioned method is not viable.

To investigate the variation of the buckling mode of the stiffened plate due to a small structural irregularity, we study the simplest case where there is a single stiffener that is slightly misplaced from the mid-span (Figure 1.2).

Let us consider a square plate. Intuitively, we know that, to produce the highest reinforcement on the plate, the single stiffener should be placed equidistantly from the plate edge parallel to it. We will therefore study the effect of misplacement from such an idealized situation. Hence, we use the following non-dimensional notations for

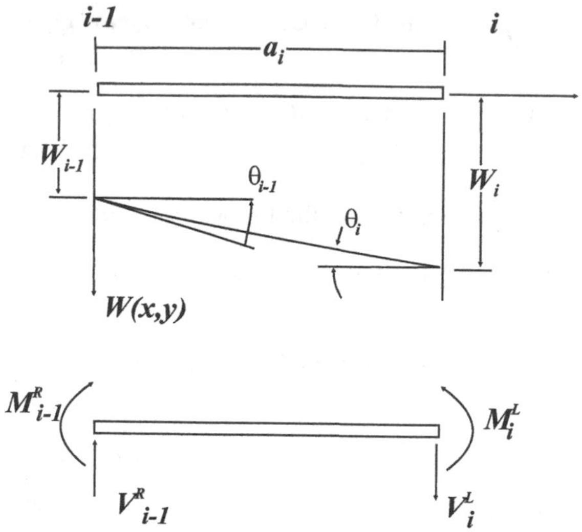

Figure 1.2 Notations and positive directions.

specifying the positions of the stiffeners:

$$r_1 = \frac{1}{2} + \delta, \qquad r_2 = \frac{1}{2} - \delta, \qquad \delta = \frac{d}{a} \tag{1.39}$$

where d denotes the misplacement of the stiffener from the middle, and δ is its non-dimensional counterpart. Positiveness of d (or δ) indicates that the stiffener is shifted to the right of its designed position; when d (or δ) is negative, the stiffener is located to the left of its designed position. $\boldsymbol{F}(\lambda)$ in Equation (1.38) is now an 8×8 matrix with elements as follows:

Case A:

$$\begin{aligned}
&F_{1,1}=1, \quad F_{1,3}=1, \quad F_{2,1}=\bar{\beta}_1^2, \quad F_{2,3}=\bar{\beta}_2^2, \quad F_{3,5}=\bar{\beta}_1^2\cos(\bar{\beta}_1 r_2\pi)\\
&F_{3,6}=\bar{\beta}_1^2\sin(\bar{\beta}_1 r_2\pi), \quad F_{3,7}=\bar{\beta}_2^2\cos(\bar{\beta}_2 r_2\pi), \quad F_{3,8}=\bar{\beta}_2^2\sin(\bar{\beta}_2 r_2\pi)\\
&F_{4,5}=\cos(\bar{\beta}_1 r_2\pi), \quad F_{4,6}=\sin(\bar{\beta}_1 r_2\pi), \quad F_{4,7}=\cos(\bar{\beta}_2 r_2\pi)\\
&F_{4,8}=\sin(\bar{\beta}_2 r_2\pi), \quad F_{5,5}=1, \quad F_{5,7}=1, \quad F_{6,1}=\cos(\bar{\beta}_1 r_1\pi)\\
&F_{6,2}=\sin(\bar{\beta}_1 r_1\pi), \quad F_{6,3}=\cos(\bar{\beta}_2 r_1\pi), \quad F_{6,4}=\sin(\bar{\beta}_2 r_1\pi)\\
&F_{7,1}=-\bar{\beta}_1\sin(\bar{\beta}_1 r_1\pi), \quad F_{7,2}=\bar{\beta}_1\cos(\bar{\beta}_1 r_1\pi), \quad F_{7,3}=\bar{\beta}_2\cos(\bar{\beta}_2 r_1\pi)\\
&F_{7,4}=\bar{\beta}_2\cos(\bar{\beta}_2 r_1\pi), \quad F_{7,6}=-\bar{\beta}_1, \quad F_{7,8}=-\bar{\beta}_2\\
&F_{8,1}=-\bar{\beta}_1^2\cos(\bar{\beta}_1 r_1\pi), \quad F_{8,2}=-\bar{\beta}_1^2\sin(\bar{\beta}_2 r_1\pi), \quad F_{8,3}=\bar{\beta}_2^2\cos(\bar{\beta}_2 r_1\pi)\\
&F_{8,4}=-\bar{\beta}_2^2\sin(\bar{\beta}_2 r_1\pi), \quad F_{8,5}=-\bar{\beta}_1^2, \quad F_{8,6}=\tau_1\pi\bar{\beta}_1\\
&F_{8,7}=\bar{\beta}_2^2, \quad F_{8,8}=\tau_1\pi\bar{\beta}_1
\end{aligned} \tag{1.40}$$

The remaining elements are zero.

Case B:

$$\begin{aligned}
&F_{1,1}=1, \quad F_{1,3}=1, \quad F_{2,1}=\bar{\beta}_1^2, \quad F_{2,3}=\bar{\beta}_2^2, \quad F_{3,5}=\bar{\beta}_1^2\cos(\bar{\beta}_1 r_2\pi)\\
&F_{3,6}=\bar{\beta}_1^2\sin(\bar{\beta}_1 r_2\pi), \quad F_{3,7}=\bar{\beta}_2^2\cos(\bar{\beta}_2 r_2\pi), \quad F_{3,8}=\bar{\beta}_2^2\sin(\bar{\beta}_2 r_2\pi)
\end{aligned}$$

$$
\begin{aligned}
&F_{4,5} = \cos(\bar{\beta}_1 r_2 \pi), \quad F_{4,6} = \sin(\bar{\beta}_1 r_2 \pi), \quad F_{4,7} = \cos(\bar{\beta}_2 r_2 \pi)\\
&F_{4,8} = \sin(\bar{\beta}_2 r_2 \pi), \quad F_{5,1} = \cos(\bar{\beta}_1 r_1 \pi), \quad F_{5,2} = \sin(\bar{\beta}_1 r_1 \pi)\\
&F_{5,3} = \cos(\bar{\beta}_2 r_1 \pi), \quad F_{5,4} = \sin(\bar{\beta}_{12} r_1 \pi), \quad F_{5,5} = -1, \quad F_{5,7} = -1\\
&F_{6,1} = -\bar{\beta}_1 \cos(\bar{\beta}_1 r_1 \pi), \quad F_{6,2} = \bar{\beta}_1 \sin(\bar{\beta}_1 r_1 \pi), \quad F_{6,3} = -\bar{\beta}_2 \cos(\bar{\beta}_2 r_1 \pi)\\
&F_{6,4} = \bar{\beta}_2 \sin(\bar{\beta}_2 r_1 \pi), \quad F_{6,6} = -\bar{\beta}_1, \quad F_{6,8} = -\bar{\beta}_2\\
&F_{7,1} = -\bar{\beta}_1^2 \cos(\bar{\beta}_1 r_1 \pi), \quad F_{7,2} = -\bar{\beta}_1^2 \sin(\bar{\beta}_1 r_1 \pi), \quad F_{7,3} = -\bar{\beta}_2^2 \cos(\bar{\beta}_2 r_1 \pi)\\
&F_{7,4} = -\bar{\beta}_2^2 \sin(\bar{\beta}_2 r_1 \pi), \quad F_{7,5} = -\bar{\beta}_1^2, \quad F_{7,6} = \tau_1 \pi \bar{\beta}_1, \quad F_{7,7} = \bar{\beta}_2^2\\
&F_{7,8} = \tau_1 \pi \bar{\beta}_2, \quad F_{8,1} = \bar{\beta}_1^3 \sin(\bar{\beta}_1 r_1 \pi), \quad F_{8,2} = -\bar{\beta}_1^3 \cos(\bar{\beta}_1 r_1 \pi)\\
&F_{8,3} = \bar{\beta}_2^3 \sin(\bar{\beta}_2 r_1 \pi), \quad F_{8,4} = -\bar{\beta}_2^3 \cos(\bar{\beta}_2 r_1 \pi), \quad F_{8,5} = -\omega_1 \pi\\
&F_{8,6} = \bar{\beta}_1^3, \quad F_{8,7} = -\omega\pi, \quad F_{8,8} = \bar{\beta}_2^3
\end{aligned}
\tag{1.41}
$$

The rest of the elements equal zero.

The column $\{\Delta\}$ is now defined as

$$\{\Delta\} = \{A_1, B_1, C_1, D_1, A_2, B_2, C_2, D_2\}^T \tag{1.42}$$

Setting the determinant of $[\boldsymbol{F}(\lambda)]$ to zero and using the quasi-Newton root-searching method, one can find the smallest root of $\det[\boldsymbol{F}(\lambda)] = 0$, which corresponds to the buckling load of the structure. Then, substituting λ back into Equation (1.38) and taking any seven equations out of the eight equations (1.38), we can solve for $\{\Delta\}$. The buckling mode reads

$$
\begin{aligned}
&w_1(x_1) = [A_1 \cos(\beta_1 x_1) + B_1 \sin(\beta_1 x_1) + C_1 \cos(\beta_2 x_1) + D_1 \sin(\beta_2 x_1)] \sin\left(\frac{\pi y}{b}\right),\\
&0 \le x_1 \le \frac{a}{2} - d\\
&w_2(x_2) = [A_2 \cos(\beta_1 x_2) + B_1 \sin(\beta_1 x_2) + C_2 \cos(\beta_2 x_2) + D_2 \sin(\beta_2 x_2)] \sin\left(\frac{\pi y}{b}\right),\\
&0 \le x_2 \le \frac{a}{2} + d
\end{aligned}
\tag{1.43}
$$

Consider now a three-span continuous plate with stiffeners. The plate is all-round simply supported and subjected to the uni-axial uniform compression in the direction perpendicular to the stiffeners (Figure 1.3). For this specific problem, a set of 12 algebraic equations can be established in the form of Equation (1.38).

Suppose that the two stiffeners are located at distances ξ_1 and ξ_2 from the left edge, respectively. Following the same procedure as discussed earlier, the buckling mode for the three spans are expressed as

$$
\begin{aligned}
&w_1(x_1) = [A_1 \cos(\beta_1 x_1) + B_1 \sin(\beta_1 x_1) + C_1 \cos(\beta_2 x_1) + D_1 \sin(\beta_2 x_1)] \sin\left(\frac{\pi y}{b}\right),\\
&0 \le x_1 \le \xi_1\\
&w_2(x_2) = [A_2 \cos(\beta_1 x_2) + B_2 \sin(\beta_1 x_2) + C_2 \cos(\beta_2 x_2) + D_2 \sin(\beta_2 x_2)] \sin\left(\frac{\pi y}{b}\right),\\
&0 \le x_2 \le \xi_2 - \xi_1
\end{aligned}
$$

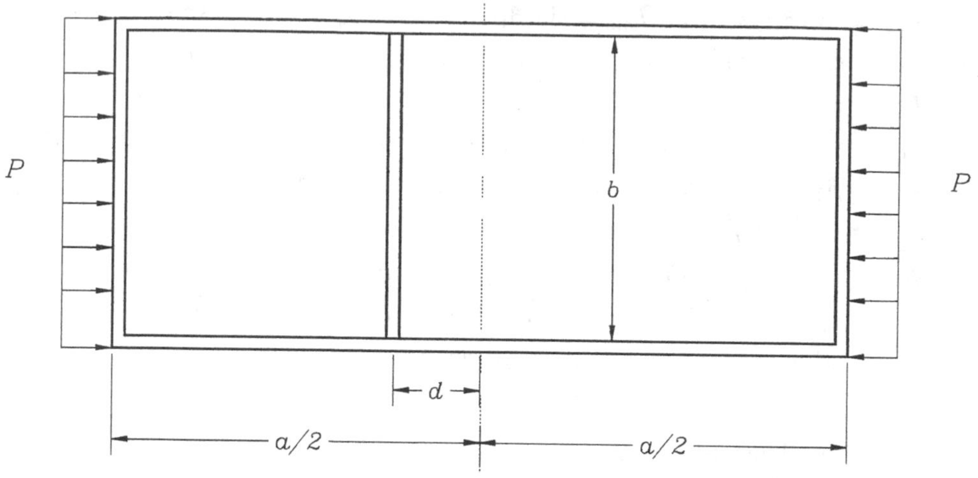

Figure 1.3 Uni-axially compressed rectangular plate stiffened with a single misplaced rib.

$$w_3(x) = [A_3 \cos(\beta_1 x_3) + B_3 \sin(\beta_2 x_3) + C_3 \cos(\beta_2 x_3) + D_3 \sin(\beta_2 x_3)] \sin\left(\frac{\pi y}{b}\right),$$
$$0 \le x_3 \le a - \xi_2 \tag{1.44}$$

We are interested in the variation of the buckling load and the buckling mode with the small misplacement of the stiffeners. In addition, the optimal position of the stiffeners, which yields the highest buckling strength, also appears to be of interest. Numerical calculations are performed for both the single rib-stiffened plate and the stiffened three-span continuous plate. Structures with different parameters for the torsional and flexural rigidities are also investigated. For the plate with a single stiffener, attachment of the stiffener to the middle location between the two parallel edges of the plate provides the structure with the most favorable load-carrying capacity. This conclusion holds true for Case A and, as will be seen later, also for Case B. It is found that for Case A, where there is a support that prevents the vertical displacement of the plate, the magnitude of the non-dimensional torsional rigidity τ has only a moderate effect on the buckling load when τ is larger than 10 (Figure 1.4). Deviation of the stiffener from the mid-span position reduces the buckling strength and, more importantly, changes the buckling mode from an overall pattern to the local pattern in the plate segment with longer span. The more misplaced the stiffener from the mid-span, the greater the reduction in buckling load is, and the more localized the buckling mode becomes. That implies that the deflection of the plate on one side of the stiffener is much greater than that on the other side. For example, for Case A with a torsional rigidity of $\tau = 20$, the ratio of the maximum deflection in the left segment to that in the right segment is about 4.5 when $\delta = 0.01$. If a bigger misplacement is involved, say $\delta = 0.02$, then the ratio of the maximum deflections in the two segments increases to 7. However, for the stiffener with non-dimensional torsional rigidity $\tau < 5$, small misplacement does not significantly affect the buckling load; for instance, when $\tau = 2$, a deviation of magnitude $\delta = 0.05$ produces only 4% reduction in buckling load.

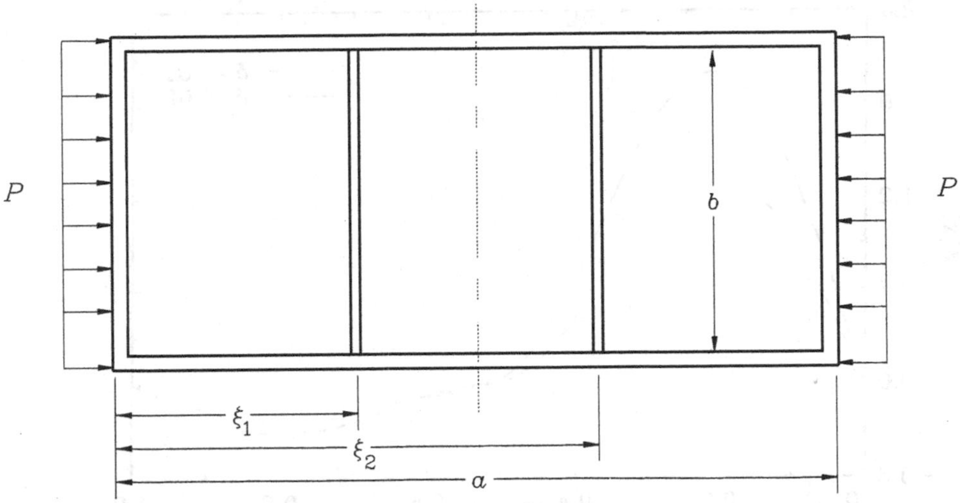

Figure 1.4 Uni-axially compressed three-span stiffened plate.

Figures 1.5 and 1.6 show the buckling mode shape of the plate in Case A for different values of τ. With a stiffener of τ larger than 30, the shorter segment of the plate is almost undeflected as buckling mode is localized in the longer segment. For Case B, the flexural rigidity of the stiffener plays a more important role in the buckling strength than the torsional rigidity, although the influence of torsional rigidity is still remarkable on the buckling mode shape. For example, it can be seen from Figure 1.7 that only stiffener with flexural rigidity $\omega \geq 5$ has noticeable strengthening effect, and that when ω falls below two, the position of the stiffener becomes almost irrelevant for the magnitude of the buckling load. When a plate is reinforced with a rib of moderate flexural stiffness,

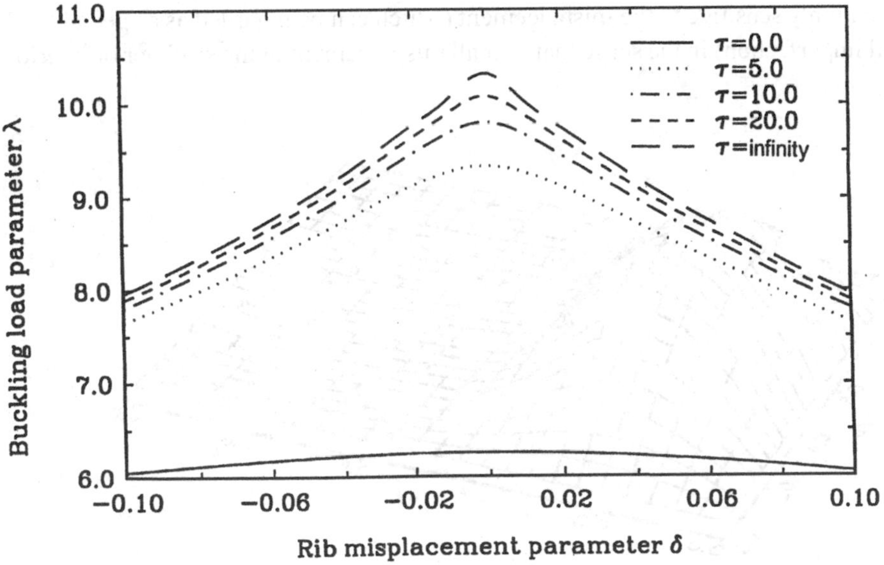

Figure 1.5 Loci of buckling loads for a plate stiffened by a single rib with different values of τ (Case A).

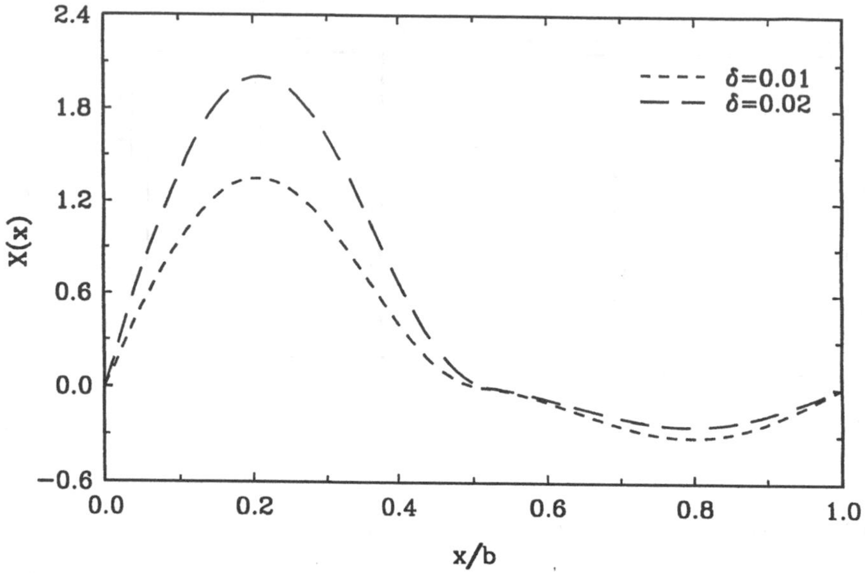

Figure 1.6 Buckling mode localization for a plate stiffened with a single rib of $\tau = 20.0$ (Case A).

the longer segment of the plate is severely deflected at the onset of buckling while the short segment experiences only a scant deformation. So the buckling is still fairly localized, as can be seen from the mode shapes depicted in Figures 1.7, 1.8, and 1.9. It is interesting to note that the reduction in the overall strength of the plate by mispositioning a stronger stiffener can be greater. For instance, when $\tau = 30$ and $\omega = 20$, a 5% deviation from the mid-point produces 13% decrease in the buckling strength. Thus, we can see that a unilateral increase in the stiffener's strength may make the whole structure highly sensitive to the misplacement (which can be regarded as a special kind of initial imperfection) in the sense that a small misplacement of the stiffener or interior

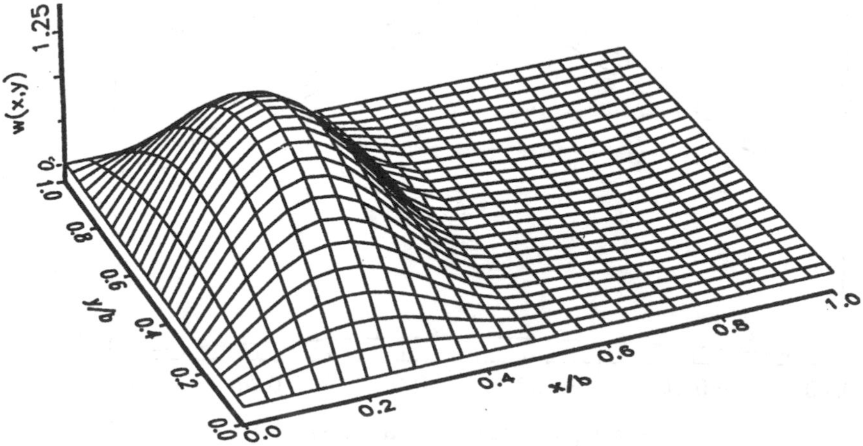

Figure 1.7 Buckling mode shape for a plate stiffened by a single rib with misplacement $\delta = 0.02$ (Case A).

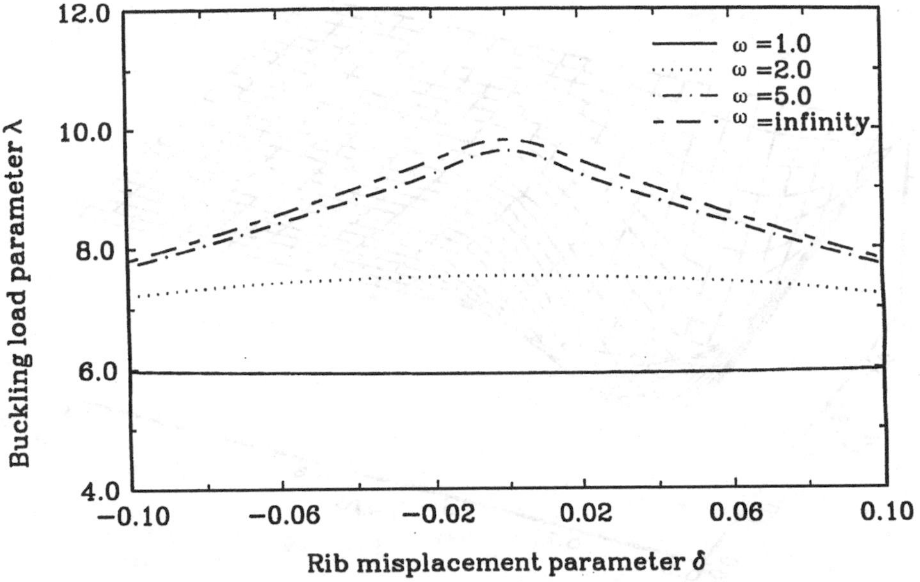

Figure 1.8 Loci of buckling loads for a plate stiffened by a single rib with different values of ω ($\tau = 10$, Case B).

support can lower the buckling load of the plate, and more importantly, localize the buckling mode shape. Figure 1.10 shows the buckling mode for the plate stiffened by a single rib with misplacement $\delta = 0.01$ (Case B).

As compared with the single rib-stiffened plate, the three-span plate is even more sensitive to the misplacement of the stiffener. For example, if one stiffener is precisely

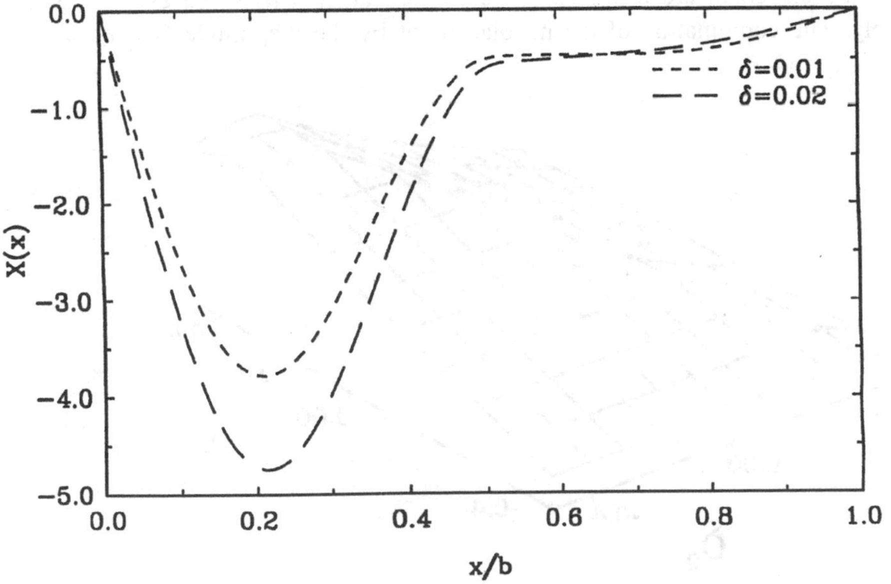

Figure 1.9 Buckling mode localization for a plate stiffened by a single rib of $\tau = 20.0$ and $\omega = 10.0$ (Case B).

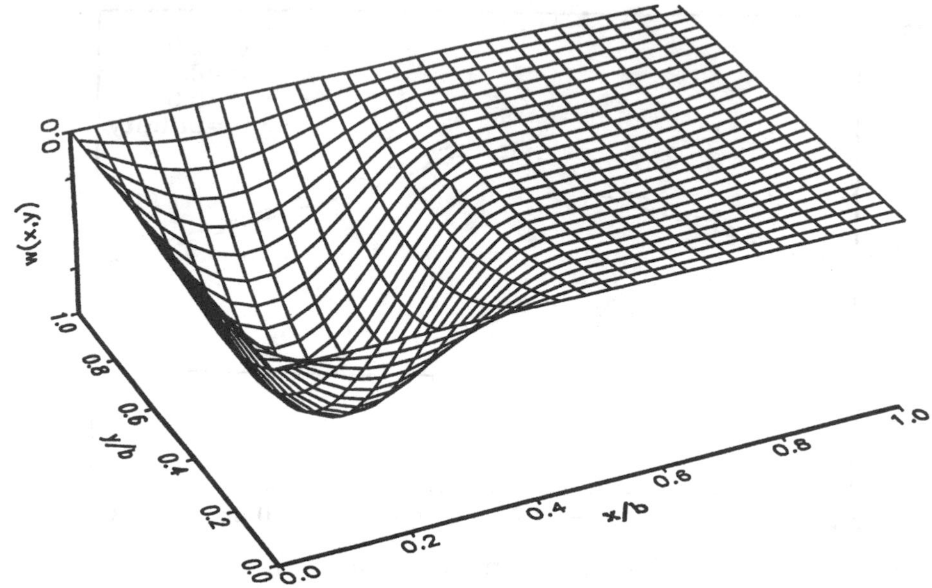

Figure 1.10 Buckling mode shape for a plate stiffened by a single rib with misplacement $\delta = 0.01$ (Case B).

fixed at $\xi_1 = a/3$, and the other stiffener is supposed to be located at $\xi_2 = 2a/3$ but somehow was misplaced from this position by, say, $\delta_2 = -0.02$ (the negative sign represents the misplacement is in the negative x direction), the buckling load is decreased by 9.5% from its counterpart without misplacement. Interestingly enough, some patterns of the misplacement are detrimental, while others can be helpful. For instance, suppose the stiffeners are designed to be located at $\xi_1 = a/3$ and $\xi_2 = 2a/3$, respectively. The combination of the misplacement by the magnitude $\delta_1 = \delta_2 = 0.02$

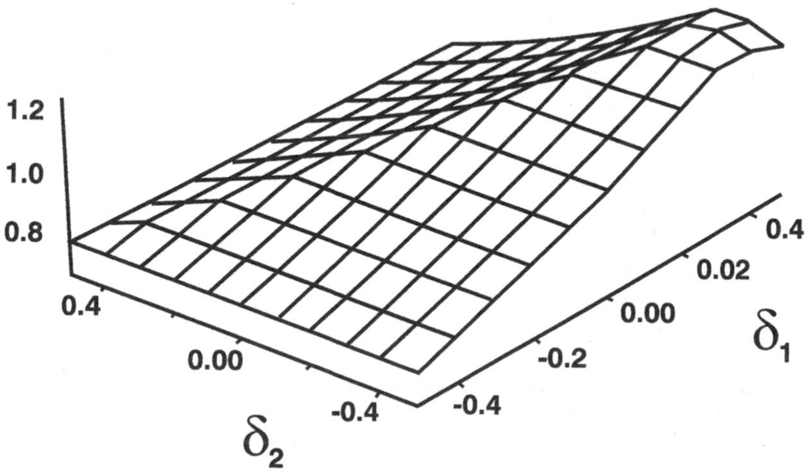

Figure 1.11 Effect of misplacement on the buckling load of a stiffened, three-span plate ($\tau_1 = \tau_2 = 20.0$).

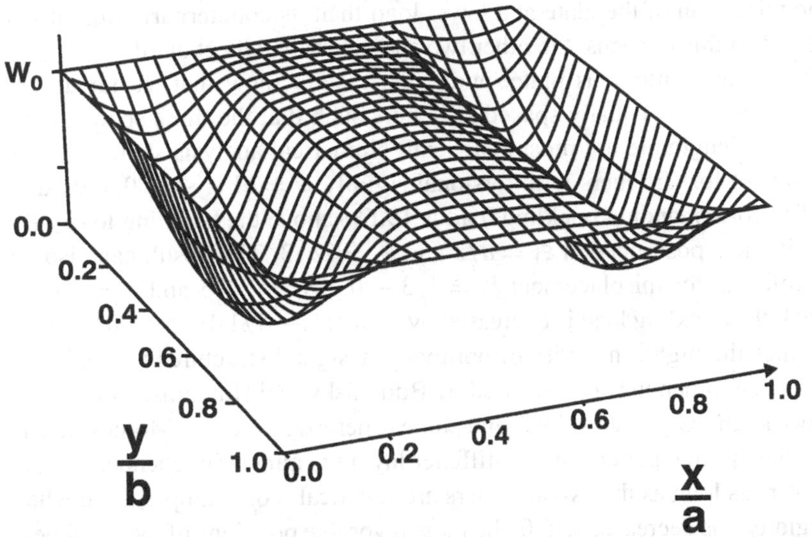

Figure 1.12 Buckling mode shape for a stiffened, three-span plate without stiffener misplacement ($\tau_1 = \tau_2 = 20.0$, $\delta_1 = \delta_2 = 0.0$).

(meaning the misplacements are both in the positive x direction) from $\xi_1 = a/3$ and $\xi_2 = 2a/3$, respectively, cuts down the buckling load by 9.6%. However, misplacement of $\delta_1 = -\delta_2 = -0.02$ (meaning the left stiffener has been moved slightly to the left and the right stiffener to the right) boosts the buckling strength by 11% (Figure 1.11). As for the buckling mode, in the majority of the situations, the buckling patter is still severely localized (Figure 1.12). This is shown by Figure 1.13 where a small misplacement of $\delta_2 = -0.01$ of the right stiffener triggers the onset of the local

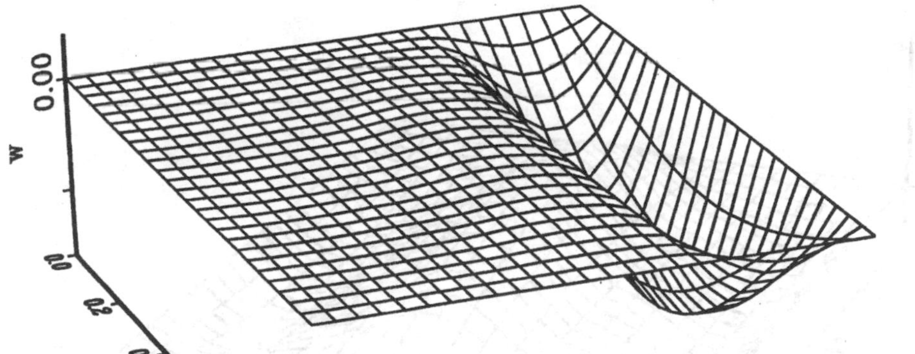

Figure 1.13 Buckling mode shape for a stiffened, three-span plate with slight misplacement ($\tau_1 = \tau_2 = 20.0$, $\delta_1 = 0.0$, $\delta_2 = 0.01$).

buckling in the third span of the plate at a lower load than its counterpart without the misplacement. When this happens, the other two spans hardly deflect at all.

For the three-span continuous plate, numerical results show that, as far as the buckling load is concerned, equally spaced stiffeners such that the three spans of the plate have the same length are not most beneficial. Numerical analysis shows the optimal stiffener layout for stiffeners with torsional rigidity $\tau = 20$ is $\xi_1 = 0.329a$ and $\xi_2 = 0.671a$. The corresponding buckling load is 19.6% above the buckling load with two identical stiffeners positioned at $\xi_1 = a/3$ and $\xi_2 = 2a/3$. This result can also be interpreted as follows: for misplacement $\delta_1 = 1/3 - 0.329 \approx 0.005$ and $\delta_2 = 2/3 - 0.671 \approx -0.004$, the buckling load is decreased by 16% $((1. - 1/1.19) \times 100\%)$. This again demonstrates the high sensitivity of optimally designed structures to small imperfections. This phenomenon was discussed by Budiansky and Hutchinson (1979) as well as Zyczkowski and Gajewski (1983) and some other investigators. Moreover, it is found here that the optimal pattern of the stiffener layout is almost independent of the specific value of τ, as long as the two stiffeners are identical. For example, even when the torsional rigidity τ is decreased to 5.0, the most favorable positions of two stiffeners are hardly changed, as they are now $\xi_1 = 0.329a$ and $\xi_2 = 0.671a$. Figure 1.14 depicts the buckling mode for such a situation, from which we can observe that with the optimal stiffener layout the buckling mode is a global one (that is, all parts of the place deflect to a comparable degree, and the potential capability of the structure is fully tapped).

As shown, the buckling mode localization phenomenon resulting from small misplacement in the stiffened or continuous plates should not be overlooked, especially in those applications where the mode shape is a significant concern. Because of the imprecision in the fabrication, the misplacement is always present and can affect the buckling characteristics of the structure to a large extent. When the structure is designed in terms of the optimal stiffener layout, misplacement may reduce the buckling load

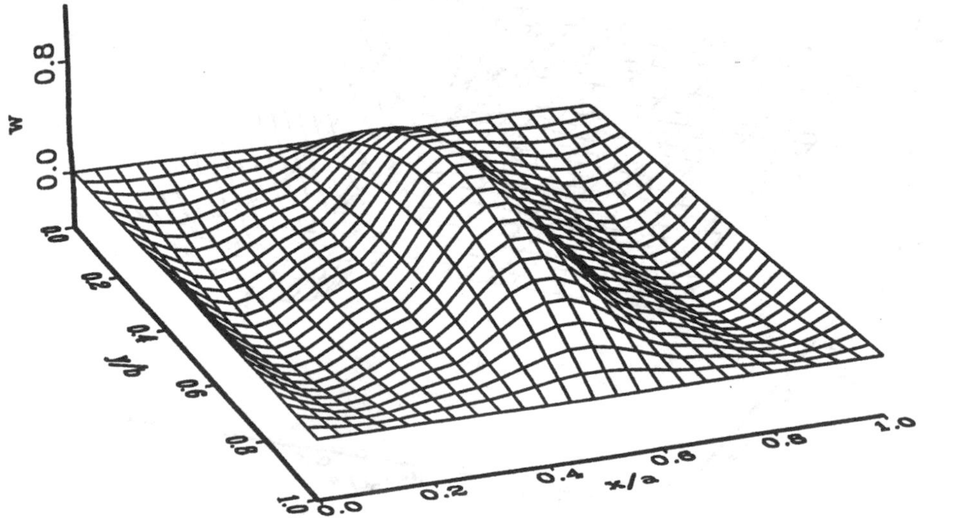

Figure 1.14 Buckling mode shape for a stiffened, three-span plate with optimal stiffener placement ($\tau_1 = \rho_2 = 20.0$, $\xi_1 = 0.329a$, $\xi_2 = 0.671a$).

and may cause the buckling mode to become highly localized in a manner that one segment of the structure deflects appreciably while the capability of the other parts of the structure has not been brought into full play.

1.2 Localization in a Multi-Span Periodic Column with a Disorder in a Single Span

In Section 1.1, we considered two- and three-span plates. We are also interested in studying the buckling mode localization in multi-span plates. Generally, such an analysis, though numerically cumbersome, is quite straightforward. In a multi-span plate, one should write down expressions for the buckling mode for each span, as we did in the case of the three-span plate [for example, Equation (1.44)]. Satisfaction of the boundary and continuity conditions leads to an eigenvalue problem; the lowest eigenvalue represents the buckling load, and its corresponding eigenvector depicts the buckling mode. Extensive numerical effort is usually required to investigate the localization phenomenon. This is especially true if the number of spans is large. One might be, however, interested in the situation where numerical analysis is less demanding. This is where the case of the so-called *perfect misplacement* comes in. Perfect misplacement refers to a set of misplacements appearing in an extremely specialized manner. Indeed, Koiter's classical work (1963) where he studied the effect of initial imperfections on the buckling of cylindrical shells was of this kind: he studied the effect of specialized initial imperfections varying sinusoidally (see p. 107). In this and the following section, we address the issue of a multi-span structure, where ideally stiffeners are spaced periodically but where, due to a production process, the periodicity breaks down in one single span of the multi-span structure. This case may, at first glance, appear rather artificial, but we can envision that this idealized situation arises in the following way. Suppose that the large periodic structure must be assembled; however, from the point of view of transporting the structure, it must be assembled from several sub-structures. Each sub-structure is a periodic structure. We can visualize that the exact periodicity was destroyed in the assembling process. Thus, we may end up with a structure where misplacement in one single span destroys the overall periodicity of the entire construction. Because we will use mathematical methods pertinent to ideal periodic structures (that is, without any misplacement) it appears instructive to start our discussion with an extremely brief overview of the analysis of periodic structures.

Large-scale periodic structures are often encountered in engineering practice. They appear in different forms, such as beams on equidistant supports, gridwork structures with equally spaced and geometrically similar stiffening components, and plates and shells with uniformly distributed stiffeners. Periodic structures have drawn substantial interest among researchers, and many mathematical methods have been developed, starting with the pioneering work by Brillouin (1953). The reader may also consult several studies (for example, Krein, 1933; Miles, 1956; Lin and McDaniel, 1969; Sen-Gupta, 1970; Mead, 1971; Abramovich and Elishakoff, 1987; Zhu, Elishakoff, and Lin, 1994; and Elishakoff, Lin, and Zhu, 1994). To analyze perfectly periodic structures, the method of the *analytical* finite difference calculus (Bleich, 1952; Wah and Calcote, 1970) appears to be very instrumental. For the treatment of periodic structures, this method has decisive advantage over the conventional matrix methods used in the structural analysis. The method usually leads to a determinative matrix that can be orders of magnitude smaller than those necessary in either the force or displacement method or the conventional finite element method. With the size of the matrix reduced, the numerical accuracy is improved, and the computational effort is cut down dramatically. However, the applicability of this method is confined to only those structures that are uniform in the spacing and stiffness characteristics of the constituent elements. When the periodicity is disturbed, this method can no longer be applied. The deviation from complete periodicity is commonly known as disorder or irregularity. Disorders may arise from various imprecisions in the fabrication process or from geometric and material variations in the different parts of the structures. From the standpoint of structural analysis, if the constituent units of the structure are different from one to another, we need to analyze each of them separately, with attendant satisfaction of continuity conditions from one unit to another, as was done in Section 1.1 for the two- and three-span plates. This procedure usually results in matrices of high order if the structure is composed of a large number of units. Because inversion and other operations on such matrices may be involved in the calculations, numerical errors are unavoidable. Therefore, it is often desirable to reduce the order of the matrices as far as possible. Fortunately, many large-scale engineering structures are essentially periodic, and disorders often take place in some localized areas of the structure.

Here, we discuss the general N-span beam with torsional springs at supports, which is structurally periodic except that one of the spans of the beam contains a disorder. By combining the finite difference calculus with the conventional displacement method, we present the exact solution for the buckling of a large-scale, multi-span periodic beam having disorder in an *arbitrary single* span of the beam. Section 1.3 will discuss the buckling of multi-span plates with a single disorder. It is shown that even a single disorder could be responsible for the highly localized pattern of buckling modes.

The governing differential equation for the typical ith span of axially compressed, continuous beam with uniform cross section reads:

$$EI\frac{d^2w}{dx_i^2} + Pw = M_{i-1}^R\left(\frac{x_i}{a_i} - 1\right) - M_i^L\frac{x_i}{a_i} \tag{1.45}$$

or

$$\frac{d^2w}{dx_i^2} + k^2w = \frac{M_{i-1}^R}{EI}\left(\frac{x_i}{a_i} - 1\right) - \frac{M_i^L}{EI}\frac{x_i}{a_i} \tag{1.46}$$

where $k = \sqrt{P/EI}$, $P =$ the axial load on the beam, $E =$ the Young's modulus, and $I =$ the moment of inertia of the cross-section of the beam; $w =$ the deflection of the beam and $a_i =$ the length of the ith span (Figure 1.14); M_{i-1}^R, M_i^L are the bending moments at two supports of that span, respectively. The superscript $R(L)$ indicates that span of the beam is to the *right* (*left*) of the support in question.

The general solution to Equation (1.46) is

$$w = C_i \sin(kx_i) + D_i \cos(kx_i) + \frac{M_{i-1}^R}{P}\left(\frac{x_i}{a_i} - 1\right) - \frac{M_i^L}{P}\frac{x_i}{a_i} \tag{1.47}$$

where C_i and D_i are arbitrary constants that are determined by the use of boundary conditions. Here we discuss the case of transversely rigid supports. Thus, the boundary conditions at the supports are

$$(1)\quad w = 0 \quad \text{at} \quad x_i = 0; \qquad (2)\quad w = 0 \quad \text{at} \quad x_i = a_i \tag{1.48}$$

from which C_i and D_i can be evaluated as

$$C_i = -\frac{M_{i-1}^R \cos(ka_i)}{P\sin(ka_i)} + \frac{M_i^L}{P\sin(ka_i)}, \qquad D_i = \frac{M_{i-1}^R}{P} \tag{1.49}$$

Substitution into Equation (1.47) results in

$$w = \frac{M_{i-1}^R}{P}\left[-\frac{\cos(ka_i)}{\sin(ka_i)}\sin(kx_i) + \cos(kx_i) + \frac{x_i}{a_i} - 1\right] + \frac{M_i^L}{P}\left[-\frac{\sin(kx_i)}{\sin(ka_i)} + \frac{x_i}{a_i}\right] \tag{1.50}$$

from which

$$w' = \frac{M_{i-1}^R}{P}\left[\frac{1}{a_i} - k\frac{\sin(ka_i)\sin(kx_i) + \cos(ka_i)\cos(kx_i)}{\sin(ka_i)}\right]$$
$$-\frac{M_i^L}{P}\left[\frac{1}{a_i} - \frac{k\cos(kx_i)}{\sin(ka_i)}\right] \tag{1.51}$$

In the following, we will use the analytical finite difference calculus (see, for example, Wah and Calcote, 1970). We will use the angles of rotation at supports as principal variables, since the deflections at all the supports are zero. The angles of rotation at supports $i-1$ and i are obtained from Equation (1.51) by setting x_i equal to 0 and a_i, respectively,

$$\theta_{i-1} = \frac{M_{i-1}^R}{k^2 a_i EI}\left[1 - \frac{ka_i\cos(ka_i)}{\sin(ka_i)}\right] - \frac{M_i^L}{k^2 a_i EI}\left[1 - \frac{ka_i}{\sin(ka_i)}\right] \tag{1.52}$$

$$\theta_i = \frac{M_{i-1}^R}{k^2 a_i EI}\left[1 - \frac{ka_i}{\sin(ka_i)}\right] - \frac{M_i^L}{k^2 a_i EI}\left[1 - \frac{ka_i\cos(ka_i)}{\sin(ka_i)}\right] \tag{1.53}$$

We may also express M_{i-1}^R and M_i^L, the bending moments at supports $i-1$ and i, respectively, in terms of the rotational angles θ_i and θ_{i+1} as follows:

$$M_{i-1}^R = \frac{2EI}{a_i}[2c_1\theta_{i-1} + c_2\theta_i]$$
$$M_i^L = -\frac{2EI}{a_i}[2c_1\theta_i + c_2\theta_{i-1}] \tag{1.54}$$

where

$$c_1 = \frac{ka_i[\sin(ka_i) - ka_i\cos(ka_i)]}{4[2 - 2\cos(ka_i) - ka_i\sin(ka_i)]}$$
$$c_2 = \frac{ka_i[ka_i - \sin(ka_i)]}{2[2 - 2\cos(ka_i) - ka_i\sin(ka_i)]} \tag{1.55}$$

Note that the preceding derivation could also be performed straightforwardly by using the stability functions (Horne and Merchant, 1965). Equilibrium at a typical ith support requires that

$$M_i^L - M_i^R = J\theta_i \tag{1.56}$$

where J is the torsional modulus of the spring (Figure 1.14), θ_i is the rotational angle at support i.

Using Equation (1.54), we obtain

$$c_2(\theta_{i+1} + \theta_{i-1}) + 4(c_1 + \Psi)\theta_i = 0 \tag{1.57}$$

where $\Psi = Ja/8EI$.

Introducing the shifting operator $\boldsymbol{E}$ (Wah and Calcote, 1970) which is defined as $\boldsymbol{E}\theta_i = \theta_{i+1}$, Equation (1.47) can be rewritten as

$$[c_2(\boldsymbol{E} + \boldsymbol{E}^{-1}) + 4(c_1 + \Psi)]\theta_i = 0 \tag{1.58}$$

Equation (1.48) is a second-order finite difference equation, whose solution is obtained by letting.

$$\theta_i = \exp(i\phi) \tag{1.59}$$

A direct substitution of Equation (1.59) into Equation (1.58) yields

$$\cosh\phi = -\frac{2(c_1 + \Psi)}{c_2} \tag{1.60}$$

Three different cases must be considered:

Case A: $-2(c_1 + \Psi)/c_2 \geq 1$

The solution of Equation (1.60) has the following form

$$\phi_{1,2} = \pm\phi; \qquad \phi = \cosh^{-1}\left[-\frac{2(c_1 + \Psi)}{c_2}\right] \tag{1.61}$$

The attendant solution for θ_i reads

$$\theta_i = Ae^{\phi i} + Be^{-\phi i} \tag{1.62}$$

where A and B are arbitrary constants.

Case B: $-2(c_1 + \Psi)/c_2 \leq -1$

The solution to Equation (1.60) takes the form

$$\phi_{1,2} = \pm(\alpha + j\pi); \qquad \alpha = \cosh^{-1}\left[\frac{2(c_1 + \Psi)}{c_2}\right], \qquad j = \sqrt{-1} \tag{1.63}$$

and the solution for θ_i is

$$\theta_i = [A\cosh(i\alpha) + B\sinh(i\alpha)]\cos(i\pi) \tag{1.64}$$

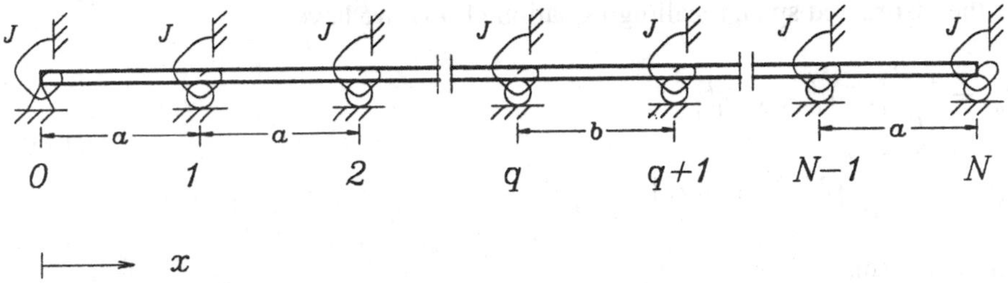

Figure 1.15 A multi-span beam with a single disorder in the $(q+1)$th span.

Case C: $-1 \leq -2(c_1+\Psi)/c_2 \leq 1$
The solution to Equation (1.60) now becomes

$$\phi_{1,2} = \pm\beta; \qquad \beta = \cos^{-1}\left[-\frac{2(c_1+\Psi)}{c_2}\right] \tag{1.65}$$

and, for θ_i, we have

$$\theta_i = A\cos(i\beta) + B\sin(i\beta) \tag{1.66}$$

Suppose that we have an N-span beam (Figure 1.15), which is periodic except that its $(q+1)$th span contains a span length imperfection that makes the span slightly longer or shorter than the other spans of the structure, that is,

$$a_i = a \quad \text{for } i = 1, \ldots, q, q+2, \ldots, N; \qquad a_{q+1} = b \quad (a \neq b) \tag{1.67}$$

As a particular case $b = a$, we recover the perfectly periodic structure.

To facilitate the solution of the problem, we can treat the entire beam as being composed of three segments. The first q-span periodic beam constitutes segment I; the $(q+1)$th span, namely, the single disordered span, constitutes segment II; and the last $(N-q-1)$ spans of periodic beam represent segment III. Assume first that both segments I and III themselves contain a large number of spans. For segments I and III per se, the finite difference calculus is applicable due to their structural periodicity. As to the disordered span, segment II, a separate consideration should be made, and the conventional displacement method is used here. By following this procedure, we construct a solution composed of three parts with each part corresponding to a specific segment of the beam. Continuity conditions between those different segments are utilized in combination with boundary conditions at the ends of the beam to establish an eigenvalue problem.

For the first q spans of periodic beam, we perform the analytical finite difference analysis

$$\theta_r = \theta_r^I; \qquad (r = 0, 1, \ldots, q) \tag{1.68}$$

where the superscript denotes the sequence number of the segment in question; θ_r^I takes one of the three forms represented by Equations (1.62), (1.64), and (1.66), depending on the physical and geometrical conditions of the segment.

For the disordered span, recalling Equation (1.44), we have

$$\begin{aligned} M_q^R &= \frac{2EI}{b}\left[2\bar{c}_1\theta_0^{II} + \bar{c}_2\theta_1^{II}\right] \\ M_{q+1}^L &= -\frac{2EI}{b}\left[2\bar{c}_1\theta_1^{II} + \bar{c}_2\theta_0^{II}\right] \end{aligned} \tag{1.69}$$

or, in another form,

$$\begin{aligned} \theta_0^{II} &= \frac{b}{2EI}\frac{2\bar{c}_1 M_q^R + \bar{c}_2 M_{q+1}^L}{4\bar{c}_1^2 - \bar{c}_2^2} \\ \theta_1^{II} &= -\frac{b}{2EI}\frac{2\bar{c}_1 M_{q+1}^L + \bar{c}_2 M_q^R}{4\bar{c}_1^2 - \bar{c}_2^2} \end{aligned} \tag{1.70}$$

where

$$\begin{aligned} \bar{c}_1 &= \frac{kb[\sin(kb) - kb\cos(kb)]}{4[2 - 2\cos(kb) - kb\sin(kb)]} \\ \bar{c}_2 &= \frac{kb[kb - \sin(kb)]}{2[2 - 2\cos(kb) - kb\sin(kb)]} \end{aligned} \tag{1.71}$$

Note that $\bar{c}_1$ and $\bar{c}_2$ are obtained from Equation (1.55) by formally replacing a_i with b.

The treatment of the last $N - q - 1$ spans of periodic beam is similar to that of segment I,

$$\theta_s = \theta_{s-q-1}^{III}; \qquad (s = q+1, q+2, \ldots, N) \tag{1.72}$$

where θ_{s-q-1}^{III} again adopts one of the three forms denoted by Equations (1.62), (1.64), and (1.66).

Consider now a beam simply supported at its two ends (other boundary conditions can be treated in a similar manner). Then the boundary condition at the left end of the beam can be represented as

$$-M_0^R = J\theta_0^I \qquad \text{or} \qquad (2c_1 + 4\Psi)\theta_0^I + c_2\theta_1^I = 0 \tag{1.73}$$

while the boundary condition at the right end of the beam reads

$$M_N^L = J\theta_{N-q-1}^{III} \qquad \text{or} \qquad (2c_1 + 4\Psi)\theta_{N-q-1}^{III} + c_2\theta_{N-q-2}^{III} = 0 \tag{1.74}$$

Conditions of continuity between the periodic spans and the disordered span of the beam, namely, between segment I and segment II, are

$$\begin{aligned} M_q^L - M_q^R = J\theta_q \qquad &\text{or} \qquad (2c_1 + 4\Psi)\theta_q^I + c_2\theta_{q-1}^I + \frac{2}{2EI}M_q^R = 0 \\ \theta_q^I = \theta_0^{II} \qquad &\text{or} \qquad \theta_q^I - \frac{b}{a\left(4\bar{c}_1^2 - \bar{c}_2^2\right)}\left(2\bar{c}_1\frac{a}{2EI}M_q^R + \bar{c}_2\frac{a}{2EI}M_{q+1}^L\right) = 0 \end{aligned} \tag{1.75}$$

Analogously, the continuity conditions between the second and the third segments are

$$M_{q+1}^L - M_{q+1}^R = J\theta_0^{III} \qquad \text{or} \qquad (2c_1 + 4\Psi)\theta_0^{III} + c_2\theta_1^{III} - \frac{a}{2EI}M_{q+1}^L = 0$$

$$\theta_1^{II} = \theta_0^{III} \qquad \text{or} \qquad \theta_0^{III} + \frac{b}{a(4\bar{c}_1^2 - \bar{c}_2^2)}\left(\bar{c}_2\frac{a}{2EI}M_q^R + 2\bar{c}_1\frac{a}{2EI}M_{q+1}^L\right) = 0 \tag{1.76}$$

Equations (1.75) and (1.76) should be formulated in terms of the three different cases because, in each case, the resulting expressions are different.

For Case A, the rotation angles in the first and third segments can be expressed as

$$\begin{aligned} &\theta_r^I = A_1 e^{\phi r} + B_1 e^{-\phi r} \qquad (r = 0, 1, \ldots, q) \\ &\theta_{s-q-1}^{III} = A_2 e^{\phi(s-q-1)} + B_2 e^{-\phi(s-q-1)} \qquad (s = q+1, \ldots, N) \end{aligned} \tag{1.77}$$

Substituting the preceding expressions in boundary conditions (1.73) and (1.74) and continuity conditions (1.75) and (1.76), we obtain six homogeneous algebraic equations:

$$(2c_1 + 4\Psi + c_2 e^{\phi})A_1 + (2c_1 + 4\Psi + c_2 e^{-\phi})B_1 = 0 \tag{1.78}$$

$$\left[(2c_1 + 4\Psi)e^{\phi q} + c_2 e^{\phi(q-1)}\right]A_1 + \left[(2c_1 + 4\Psi)e^{-\phi q} + c_2 e^{-\phi(q-1)}\right]B_1 + \bar{M}_q^R = 0 \tag{1.79}$$

$$e^{\phi q}A_1 + e^{-\phi q}B_1 - \frac{2\bar{c}_1}{4\bar{c}_1^2 - \bar{c}_2^2}\left(\frac{b}{a}\right)\bar{M}_q^R - \frac{\bar{c}_2}{4c_1^2 - \bar{c}_2^2}\left(\frac{b}{a}\right)\bar{M}_{q+1}^L = 0 \tag{1.80}$$

$$(2c_1 + 4\Psi + c_2 e^{\phi})A_2 + (2c_1 + 4\Psi + c_2 e^{-\phi})B_2 - \bar{M}_{q+1}^L = 0 \tag{1.81}$$

$$A_2 + B_2 + \frac{\bar{c}_2}{4\bar{c}_1^2 + \bar{c}_2^2}\left(\frac{b}{a}\right)\bar{M}_q^R + \frac{2\bar{c}_1}{4c_1^2 + \bar{c}_2^2}\left(\frac{b}{a}\right)\bar{M}_{q+1}^L = 0 \tag{1.82}$$

$$\begin{aligned} &\left[(2c_1 + 4\Psi)e^{\phi(N-q-1)} + c_2 e^{\phi(N-q-2)}\right]A_2 \\ &\quad + \left[(2c_1 + 4\Psi)e^{-\phi(N-q-1)} + c_2 e^{-\phi(N-q-1)}\right]B_2 = 0 \end{aligned} \tag{1.83}$$

where

$$\bar{M}_q^R = \frac{M_q^R a}{2EI}; \qquad \bar{M}_{q+1}^L = \frac{\bar{M}_{q+1}^L a}{2EI} \tag{1.84}$$

For Case B, the solutions for the first and third segments are as follows:

$$\begin{aligned} &\theta_r^I = A_1\cosh(\alpha r)\cos(\pi r) + B_1\sinh(\alpha r)\cos(\pi r) \qquad (r = 0, 1, \ldots, q) \\ &\theta_{s-q-1}^{III} = A_2\cosh[\alpha(s-q-1)]\cos[\pi(s-q-1)] \\ &\qquad + B_2\sinh[\alpha(s-q-1)]\cos[\pi(s-q-1)] \qquad (s = q+1, \ldots, N) \end{aligned} \tag{1.85}$$

and performing the substitution similar to that in Case A, we arrive at the following six homogeneous equations:

$$[2c_1 + 4\Psi - c_2\cosh(\alpha)]A_1 - c_2\sinh(\alpha)B_1 = 0 \tag{1.86}$$

$$\{(2c_1+4\Psi)\cosh(\alpha q)\cos(\pi q)+c_2\cosh[\alpha(q-1)]\cos[\pi(q-1)]\}A_1$$
$$\{(2c_1+4\Psi)\sinh(\alpha q)\cos(\pi q)+c_2\sinh[\alpha(q-1)]\cos[\pi(q-1)]\}B_1+\bar{M}_q^R=0 \tag{1.87}$$

$$\cosh(\alpha q)\cos(\pi q)A_1+\sinh(\alpha q)\cos(\pi q)B_1 - \frac{2\bar{c}_1}{4\bar{c}_1^2-\bar{c}_2^2}\left(\frac{b}{a}\right)\bar{M}_q^R-\frac{\bar{c}_2}{4\bar{c}_1^2-\bar{c}_2^2}\left(\frac{b}{a}\right)\bar{M}_{q+1}^L=0 \tag{1.88}$$

$$[2c_1+4\Psi-c_2\cosh(\alpha)]A_2-c_2\sinh(\alpha)B_2-M_{q+1}^L=0 \tag{1.89}$$

$$A_2+\frac{\bar{c}_2}{4\bar{c}_1^2-\bar{c}_2^2}\left(\frac{b}{a}\right)\bar{M}_q^R+\frac{2\bar{c}_1}{4\bar{c}_1^2-\bar{c}_2^2}\left(\frac{b}{a}\right)\bar{M}_{q+1}^L=0 \tag{1.90}$$

$$\begin{aligned}&\{(2c_1+4\Psi)\cosh[\alpha(N-q-1)]\cos[\pi(N-q-1)]\\&\quad+c_2\cosh[\alpha(N-q-2)]\cos[\pi(N-q-2)]\}A_2\\&\{(2c_1+4\Psi)\sinh[\alpha(N-q-1)]\cos[\pi(N-q-1)]\\&\quad+c_2\sinh[\alpha(N-q-2)]\cos[\pi(N-q-2)]\}B_2=0\end{aligned} \tag{1.91}$$

For Case C, the solutions are in the following form:

$$\theta_r^I=A_2\cos(\beta r)+B_1\sin(\beta r)\qquad (r=0,1,\ldots,q)$$
$$\theta_{s-q-1}^{III}=A_2\cos[\beta(s-q-1)]+B_2\sin[\beta(s-q-1)]\qquad (s=q+1,\ldots,N) \tag{1.92}$$

and the corresponding equations are

$$[2c_1+4\Psi+c_2\cos(\beta)]A_1+c_2\sin(\beta)B_1=0 \tag{1.93}$$

$$\begin{aligned}&\{(2c_1+4\Psi)\cos(\beta q)+c_2\cos[\beta(q-1)]\}A_1+\{(2c_1+4\Psi)\sin(\beta q)\\&\quad+c_2\sin[\beta(q-1)]\}B_1+\bar{M}_q^R=0\end{aligned} \tag{1.94}$$

$$\cos(\beta q)A_1+\sin(\beta q)B_1-\frac{2\bar{c}_1}{4\bar{c}_1^2-\bar{c}_2^2}\left(\frac{b}{a}\right)\bar{M}_q^R-\frac{\bar{c}_2}{4\bar{c}_1^2-\bar{c}_2^2}\left(\frac{b}{a}\right)\bar{M}_{q+1}^L=0 \tag{1.95}$$

$$[2c_1+4\Psi+c_2\cos(\beta)]A_2+c_2\sin(\beta)B_2-\bar{M}_{q+1}^L=0 \tag{1.96}$$

$$A_2+\frac{\bar{c}_2}{4\bar{c}_1^2-\bar{c}_2^2}\left(\frac{b}{a}\right)\bar{M}_q^R+\frac{2\bar{c}_1}{4\bar{c}_1^2-\bar{c}_2^2}\left(\frac{b}{a}\right)\bar{M}_{q+1}^L=0 \tag{1.97}$$

$$\begin{aligned}&\{(2c_1+4\Psi)\cos[\beta(N-q-1)]+c_2\cos[\beta(N-q-2)]\}A_2\\&\quad+\{(2c_1+4\Psi)\sin[\beta(N-q-1)]+c_2\sin[\beta(N-q-2)]\}B_2=0\end{aligned} \tag{1.98}$$

Thus, for each case, we have six homogeneous algebraic equations, which can be expressed in a matrix form as follows:

$$[F(K)]_{6\times 6}\{\delta\}_{6\times 1}=0,\qquad K=ka \tag{1.99}$$

where $[F(K)]$ is the matrix, and $\{\delta\}^T=\{A_1,B_1,\bar{M}_q^R,\bar{M}_{q+1}^L,A_2,B_2\}$.

Non-triviality of $\{\delta\}$ requires that the determinant of the coefficient matrix vanish,

$$\det[F(K)]=0 \tag{1.100}$$

which constitutes a transcendental equation for the non-dimensional buckling load parameter K. After the buckling load parameter K is determined, we can use Equation (1.50) to calculate, span by span, the buckling mode shapes for the entire structure.

It is worth mentioning that, if the disorder occurs in the first or last span of the beam, the whole beam can be partitioned into two segments: one is the disordered span and the other is the $N-1$ spans of periodic beam. If this happens, only four equations are needed to characterize the problem so that instead of having a 6×6 matrix for $[F(K)]$, we will have a 4×4 determinant matrix.

In the following numerical examples, the non-dimensional spring constant Ψ is fixed at 0.3 because this particular case for perfectly periodic beam was considered by Wah and Calcote (1970), and a comparison can be made with their results.

As a first example, we discuss a simply supported, 100-span continuous beam. The disorder arises from a span length imperfection characterized by $b/a = 1.1$, where a and b are the lengths of periodic spans and the disordered span, respectively. Because $b/a > 1$, the disordered span is longer than other spans of the structure. The disorder may appear in any span of the beam. Figure 1.16 depicts the results of the buckling load parameter K for the beams with or without torsional springs. The most critical situation for the beam without torsional springs occurs when the disorder occurs at either the first or the last span, for which the buckling load parameter K equals 3.05. (If there is no disorder, the buckling load parameter is π. Thus, the effect on the buckling load itself is somewhat detrimental with buckling load reduced by 3%.) If the disorder appears in one

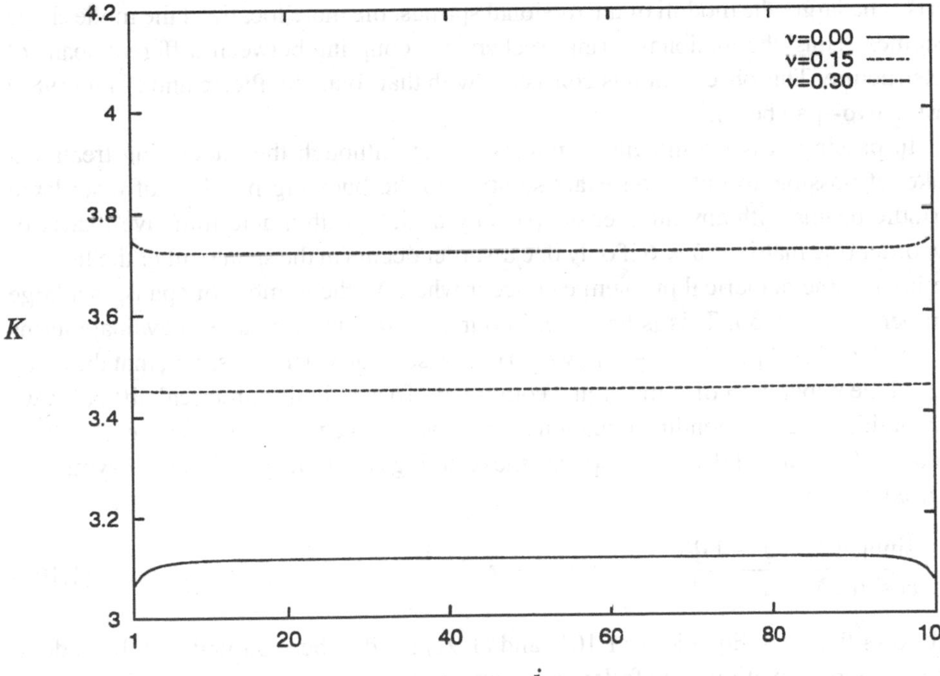

Figure 1.16 Variation of buckling load with the location of the disorder (i, sequence number of the span where the disorder occurs).

of the spans close to the center of the beam, the buckling load increases. However, for beams with torsional springs, numerical results display a quite different picture. For this case, the occurrence of disorder near the boundaries may be more advantageous. The buckling load is maximum for this configuration, it decreases as the disorder moves away from the boundaries. Nevertheless, for both cases, the buckling load remains almost unchanged with the location of disorder, once the disorder is 10 spans away from boundaries. This mean that the effect of the boundary dies out if the disorder is sufficiently far from the boundary. The location of the disorder has almost no effect on the buckling load if the spring modulus $\Psi = 0.15$. If we refer to the buckling load in the absence of disorder as the classical buckling load, then the classical buckling load parameter K equals π for the case without torsional springs and has a value of 3.760 for the beam with torsional springs of $\Psi = 0.03$ (Wah and Calcote, 1970). With the disorder present, numerical results show that the buckling load parameter K is below the corresponding classical value. For instance, K is less than 3.12 for the beam without torsional springs and is no more than 3.72 for the beam with torsional springs of $\Psi = 0.3$. Thus, we can see that such a disorder, namely, the span length imperfection, may have a degrading effect on the load-carrying capacity of the structure.

The second example is mainly devoted to the discussion of the buckling mode shapes for disordered periodic beam. Figures 1.17 and 1.18 depict the buckling mode shapes for an 11-span beam with the disorder appearing at various locations of the beam. Again, the disorder is introduced by a span length imperfection specified by $b/a = 1.1$. A significant phenomenon is that the buckling mode shape exhibits a strong localization around the disordered span, when the torsional springs are used as supports. The larger the moduli of the torsional springs, the more localized the mode shape becomes. Thus, the torsional spring weakens the coupling between different spans of the structure. This observation is consistent with that found by Pierre and Plaut (1984) for the two-span beam.

In passing, it is worthwhile to point out that, although the underlying treatment makes it possible to obtain an exact solution to the buckling problem of disordered periodic beams with any number of spans by dealing with a determinative matrix of low order (the matrix is 6×6 if only one disorder occurs in the span neither the first nor the last), some numerical problem can occur when N, the number of spans, is a large number, say $N \geq 50$. This is because, in our calculations, we have to evaluate terms $e^{\phi(N-q-1)}$ and $\cosh[\alpha(N-q-1)]$, which can be so large when q is small, that they may exceed the upper limit of some digital computers. To avoid this numerical difficulty, we may divide the corresponding equation by a relevant large term, for example, $e^{\phi(N-q-1)}$ or $\cosh[\alpha(N-q-1)]$, and manipulate the resulting equation by making an asymptotic approximation

$$\frac{\sinh[\alpha(N-q-1)]}{\cosh[\alpha(N-q-1)]} \to 1 \quad \text{for } q \ll N \tag{1.101}$$

It follows that only Equations (1.101) and (1.91) need to be modified and they adopt, after the approximations, the following forms, respectively,

$$[(2c_1 + 4\Psi) + c_2 e^{-\phi}]A_2 = 0 \tag{1.102}$$

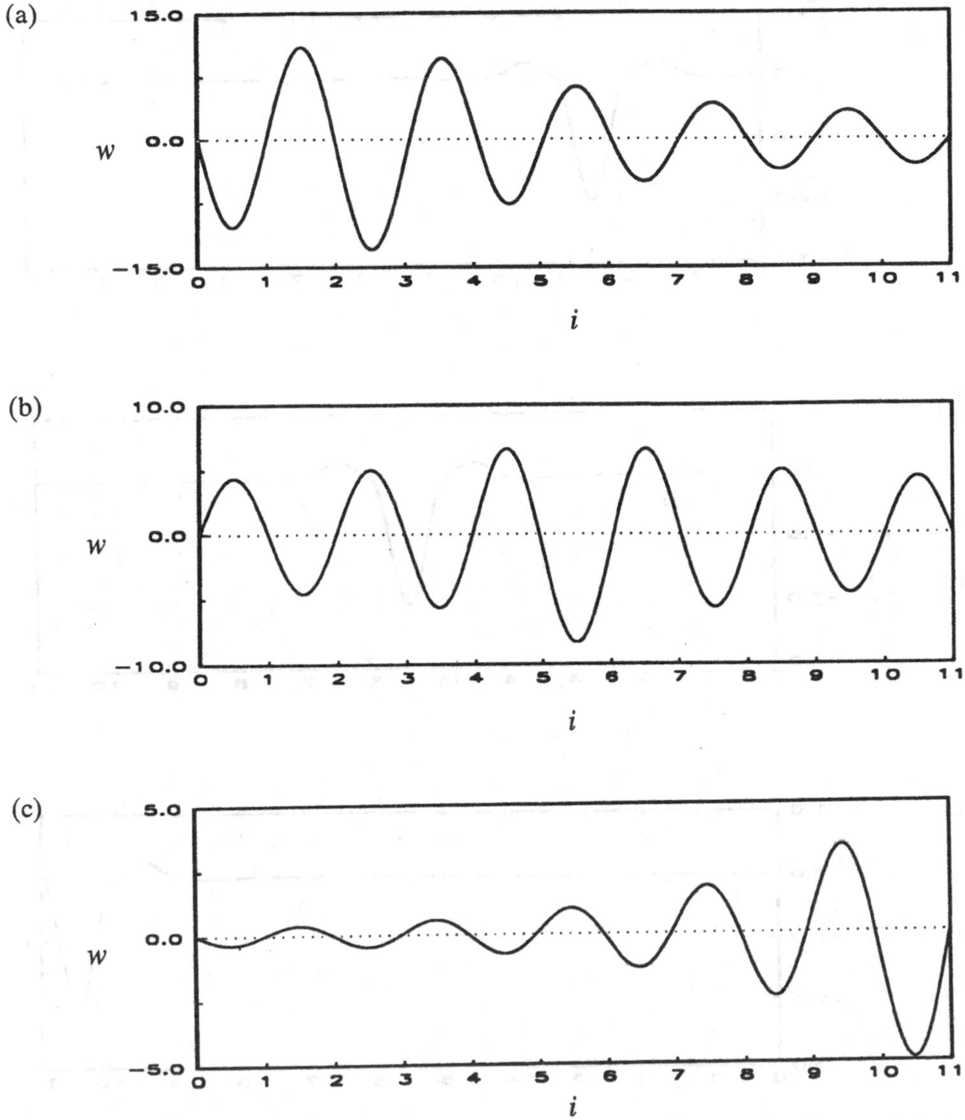

Figure 1.17 Buckling mode shapes for a disordered 11-span beam without torsional spring (i, support sequence number; w, deflection). (a) Disorder occurs in the third span; (b) disorder occurs in the mid-span (sixth span); (c) disorder occurs in the last span.

and

$$\begin{aligned}\{(2c_1 + 4\Psi)\cos[\pi(N-q-1)] + c_2 e^{-\alpha}\cos[\pi(N-q-2)]\}A_2 \\ + \{(2c_1 + 4\Psi)\cos[\pi(N-q-1)] + c_2 e^{-\alpha}\cos[\pi(N-q-2)]\}B_2 = 0\end{aligned} \tag{1.103}$$

Figure 1.19 depicts the buckling modes of a 100-span beam and a 400-span beam with torsional springs of $\Psi = 0.3$; both beams contain a disorder in the 40th span. From

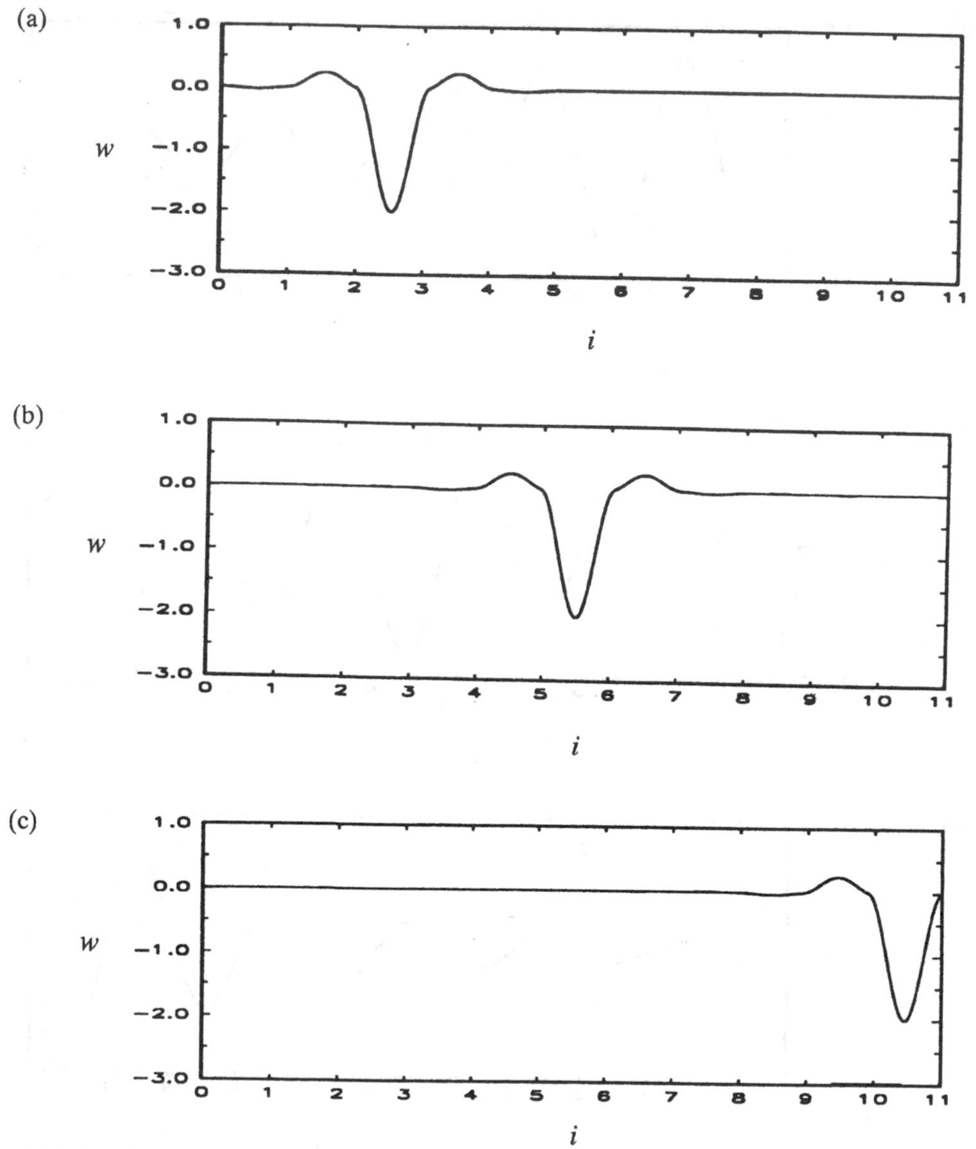

Figure 1.18 Buckling mode shapes for a disordered 11-span beam with torsional spring of $\nu = 10$ (i, support sequence number; w, deflection). (a) Disorder occurs in the third span; (b) disorder occurs in the mid-span (sixth span); (c) disorder occurs in the last span.

Figure 1.19, it is clear that the deflection of the beam at buckling dies out quickly as the distance from the disordered span increases. The envelope of the buckling mode is depicted in Figure 1.19. If we take a logarithm of the function, we obtain Figure 1.20, which displays a nearly straight line. Thus, we establish that the deflection at buckling decays exponentially. The exponential decay constant (Pierre, 1990) is usually referred to as the *Lyapunov exponent* (Arnold, Crauel, and Eckmann, 1991; Ariaratnam and Xie, 1995) in the literature [note that its counterpart in vibration problems is commonly known as the *logarithmic decrement* (Thomson, 1981)] and in our case equals −0.260.

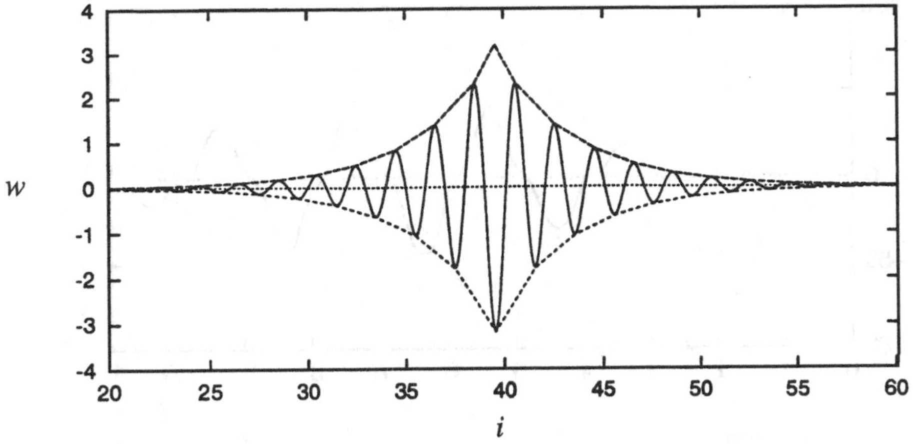

Figure 1.19 Envelope of the buckling mode shape for a 100-span beam.

In the preceding examples, we dealt with the disorder of "detrimental" character; that is, the disordered span is longer than the periodic spans, and such a disorder leads to a reduction in the buckling load of the structure. As a third example, we consider also another possibility with attendant opposite effect; that is, the disordered span is shorter than the other spans ($b/a < 1$). In this case, the disorder turns out to be "beneficial" because it results in slightly higher buckling loads. For example, for a perfectly periodic 11-span beam with torsional springs of non-dimensional modulus $\Psi = 0.3$, the buckling load parameter K is, as mentioned earlier, 3.76. For the same structure but with disorder, K varies in the range from 3.79 to 3.82, depending on the location of the disorder. Thus, we can see that the "beneficial" effect of such a disorder on the buckling load is very small – the increase in buckling load is less than 2%. However, the impact of disorder on the buckling mode shape is noteworthy. Figure 1.21 portrays the buckling mode shapes of 11-span beams containing such a disorder, from which it can be seen that the change in buckling mode is significant, especially when the

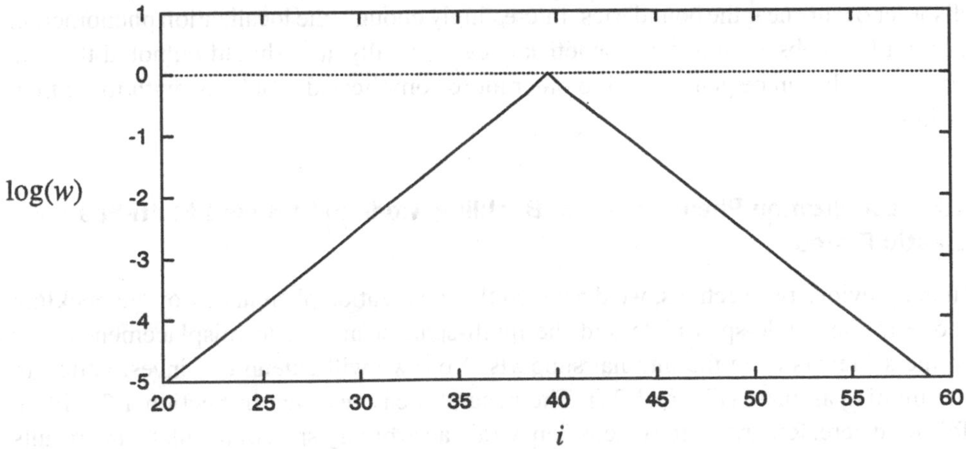

Figure 1.20 Logarithmic plot of the envelope function.

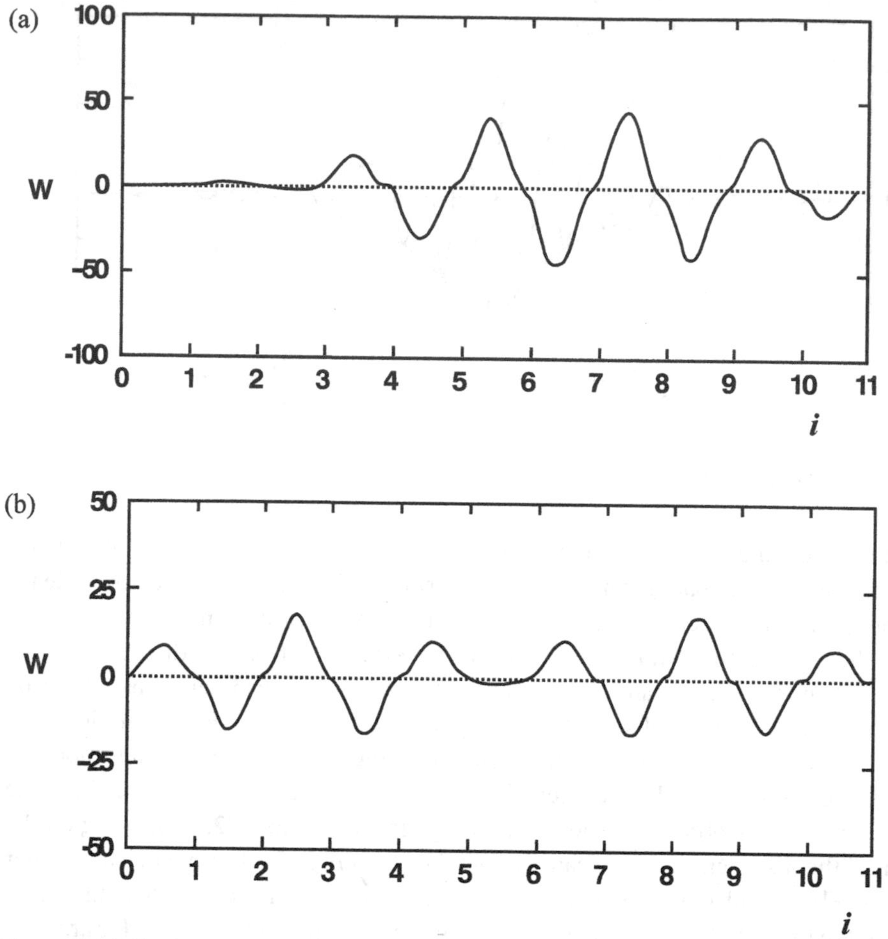

Figure 1.21 Buckling mode shape for a disordered 11-span beam (disordered span is shorter than the periodic span). (a) Disorder occurs in the third span; (b) disorder occurs in the mid-span (sixth span).

disorder occurs near the boundaries. Interestingly enough, the localization phenomenon has not been observed in this "beneficial" case. Finally, it is should be noted that the present study can be generalized to other more complicated problems, with more than a single disorder.

1.3 Localization Phenomenon of Buckling Mode in Stiffened Multi-Span Elastic Plates

In the previous two sections, we discussed the localization phenomena of the buckling mode in the single-span plate and the multi-span beam due to misplacement either in the stiffeners or in the internal supports. Now we will extend our investigation to the multi-span plate (Figure 1.22). The general idea presented in Section 1.2 will be followed here. Regarding the discussion within an arbitrary span of the plate, the results obtained in Section 1.1 will be employed.

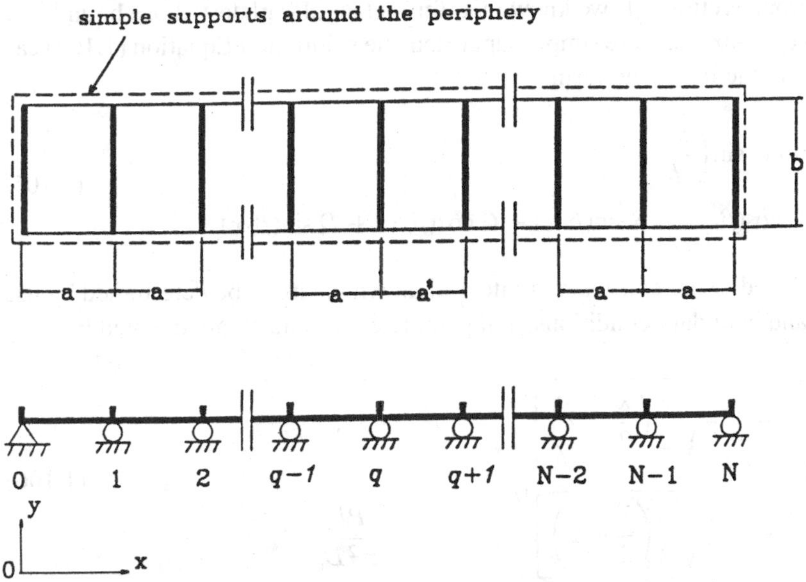

Figure 1.22 An N-span continuous plate with a single disorder in the $(q+1)$th span.

Again, the differential equation of the deflection surface of the plate subjected to a uniform compression P in the x direction (Figure 1.23) is

$$D_p\left(\frac{\partial^4 w}{\partial x^4}+2\frac{\partial^4 w}{\partial x^2\partial y^2}+\frac{\partial^4 w}{\partial y^4}\right)+P\frac{\partial^2 w}{\partial x^2}=0 \tag{1.104}$$

where w is the transverse displacement, downward positive; D_p is the flexural rigidity

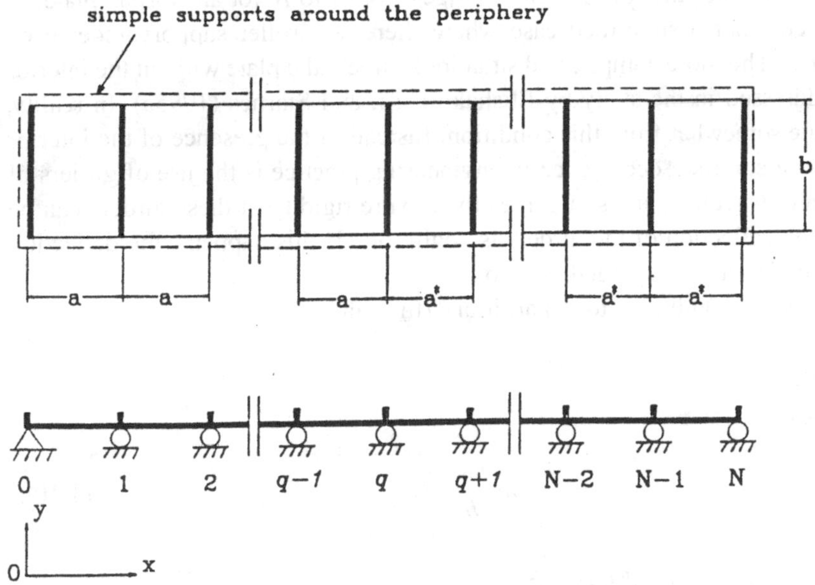

Figure 1.23 A piecewise periodic continuous plate.

of the plate; From Section 1.1, we know that, for rectangular plates whose boundaries are parallel to the x-axis and are simply supported, the solution of Equation (1.104) can be represented in the following form:

$$\begin{aligned} w(x) &= W(x)\sin\left(\frac{\pi y}{b}\right) \\ W(x) &= A\cos(\beta_1 x) + B\sin(\beta_1 x) + C\cos(\beta_2 x) + D\sin(\beta_2 x) \end{aligned} \tag{1.105}$$

where A, B, C, and D are constants of integration, which are to be determined by use of continuity and boundary conditions; and parameters β_1 and β_2 are denoted by

$$\begin{aligned} \beta_1 &= \frac{\pi}{b}\left[\frac{\lambda}{2} - 1 + \sqrt{\frac{\lambda}{2}\left(\frac{\lambda}{2} - 2\right)}\right]^{1/2} \\ \beta_2 &= \frac{\pi}{b}\left[\frac{\lambda}{2} - 1 - \sqrt{\frac{\lambda}{2}\left(\frac{\lambda}{2} - 2\right)}\right]^{1/2}, \qquad \lambda = \frac{Pb^2}{\pi^2 D_c} \end{aligned} \tag{1.106}$$

Here b stands for the width of the plate.

Catering to a single, arbitrary ith span, the solution can be written as

$$\begin{aligned} w_i &= W_i \sin\left(\frac{\pi y}{b}\right) \\ W_i &= A_i\cos(\beta_1 x_i) + B_i\cos(\beta_1 x_j) + C_i\cos(\beta_2 x_j) + D_i\sin(\beta_2 x_i), \\ &\quad 0 \le x_i \le a_i \end{aligned} \tag{1.107}$$

where a_i is the length of the ith span and i ranges from 1 to N for an N-span plate.

Here we consider a simplified case where there is a roller support under each interior stiffener. The more complicated situation, namely the plate without the interior support, is addressed in the study by Elishakoff, Li, and Starnes (1995a). In reality, we may deviate somewhat from this condition. Instead of the presence of the interior supports, a more common occurrence in engineering practice is the use of girders of joists with plate structures, and sometimes the flexure rigidity of these girders can be so large that the deflection of the girders is negligible. If this happens, the deflection along the stiffeners can be regarded as zero.

Using boundary conditions for an arbitrary ith span

$$\begin{aligned} & w_i|_{x_i=0} = 0 \\ & w_i|_{x_i=a_j} = 0 \\ & \left.\frac{dw_i}{dx_i}\right|_{x_i=0} = \theta_{i-1}; \qquad \theta_{i-1} = \Theta_{i-1}\sin\frac{\pi y}{b} \\ & \left.\frac{dw_i}{dx_i}\right|_{x_i=a_i} = \theta_i; \qquad \theta_i = \Theta_i \sin\frac{\pi y}{b} \end{aligned} \tag{1.108}$$

coefficients A_i, B_i, C_i, and D_i can be determined with the aid of *Mathematica* (Wolfram, 1991)

$$\begin{aligned}
A_i &= \frac{1}{S_i}\{[\beta_2 \sin(\beta_1 a_i) - \beta_1 \sin(\beta_2 a_i)]\Theta_i \\
&\quad + [-\beta_2 \cos(\beta_2 a_i)\sin(\beta_1 a_i) + \beta_1 \cos(\beta_1 a_i)\sin(\beta_2 a_i)]\Theta_{i-1}\} \\
B_i &= \frac{1}{S_i}\{\beta_2[-\cos(\beta_1 a_i) + \cos(\beta_2 a_i)]\Theta_i \\
&\quad + [\beta_2 \cos(\beta_1 a_i)\cos(\beta_2 a_i) - \beta_2 + \beta_1 \sin(\beta_1 a_i)\sin(\beta_2 a_i)]\Theta_{i-1}\} \\
C_i &= \frac{1}{S_i}\{[-\beta_2 \sin(\beta_1 a_i) + \beta_1 \sin(\beta_2 a_i)]\Theta_i \\
&\quad + [\beta_2 \cos(\beta_2 a_i)\sin(\beta_1 a_i) - \beta_1 \cos(\beta_1 a_i)\sin(\beta_2 a_i)]\Theta_{i-1}\} \\
D_i &= \frac{1}{S_i}\{\beta_1[\cos(\beta_1 a_i) - \cos(\beta_2 a_i)]\Theta_i \\
&\quad + [-\beta_1 + \beta_1 \cos(\beta_1 a_i)\cos(\beta_2 a_i) + \beta_2 \sin(\beta_1 a_i)\sin(\beta_2 a_i)]\Theta_{i-1}\} \\
S_i &= 2\beta_1\beta_2[-1 + \cos(\beta_1 a_i)\cos(\beta_2 a_i)] + (\beta_1^2 + \beta_2^2)\sin(\beta_1 a_i)\sin(\beta_2 a_i)
\end{aligned} \tag{1.109}$$

The expression for the bending moment is

$$M_x = -D_p\left(\frac{\partial^2 w_i}{\partial x_i^2} + \nu\frac{\partial^2 w_i}{\partial y^2}\right) \tag{1.110}$$

Using Equation (1.107), a moment-slope relationship can be established as follows:

$$\begin{aligned}
M_{i-1}^R &= M_x|_{x_i=0} = m_{i-1}^R \sin\frac{\pi y}{b}; \qquad m_{i-1}^R = \frac{D_p}{a_i}[c_1\Theta_{i-1} + c_2\Theta_i] \\
M_i^L &= M_x|_{x_i=a_i} = m_i^L \sin\frac{\pi y}{b}; \qquad m_i^L = \frac{D_p}{a_i}[c_1\Theta_i + c_2\Theta_{i-1}]
\end{aligned} \tag{1.111}$$

where M_{j-1}^R and M_j^L are the bending moments at the two supports of the span, respectively. The superscript $R(L)$ indicates that span of the plate is to the *right* (*left*) of the support in question (Figure 1.24). The coefficients c_1 and c_2 are defined as

$$\begin{aligned}
c_1 &= \frac{a_i}{S_i}(\beta_1^2 - \beta_2^2)[-\beta_2 \cos(\beta_2 a_i)\sin(\beta_1 a_i) + \beta_1 \cos(\beta_1 a_i)\sin(\beta_2 a_i)] \\
c_2 &= \frac{a_i}{S_i}(\beta_1^2 - \beta_2^2)[\beta_2 \sin(\beta_1 a_i) - \beta_1 \sin(\beta_2 a_i)]
\end{aligned} \tag{1.112}$$

If a number of spans of the plate have the common length $a_i = a$ and are made of the same material, then the analytical finite difference calculus may be applied in the discussion of that part of the plate.

Equilibrium at a typical support r reads

$$M_r^R - M_r^L = GJ\frac{\partial^2\theta_r}{\partial y^2} \tag{1.113}$$

where GJ is the torsional rigidity of the transverse stiffener.

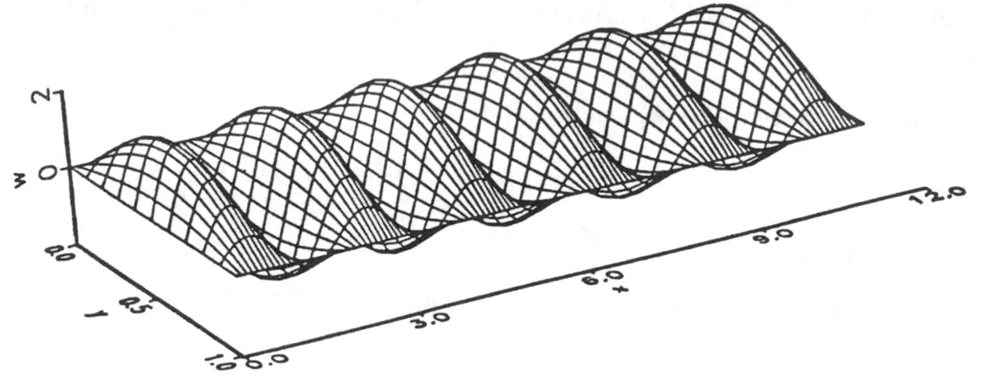

Figure 1.24 Buckling mode for an 11-span periodic plate.

Substituting Equation (1.111) into Equation (1.113), we have

$$(2c_1 + k)\Theta_r + c_2(\Theta_{r+1} + \Theta_{r-1}) = 0 \tag{1.114}$$

where $k = \pi^2 aGJ/b^2 D_c$.

Introducing the shifting operator $\boldsymbol{E}$ which is defined as $\boldsymbol{E}\theta_i = \theta_{i+1}$, Equation (1.114) can be rewritten as

$$\left[c_2(\boldsymbol{E} + \boldsymbol{E}^{-1}) + (2c_1 + k)\right]\Theta_r = 0 \tag{1.115}$$

Equation (1.115) is a second-order finite difference equation, whose solution is obtained by letting

$$\Theta_r = e^{\phi r} \tag{1.116}$$

Substitution into Equation (1.115) results in

$$\cosh(\phi) = -\frac{2c_1 + k}{2c_2} \tag{1.117}$$

Three different cases may arise and deserve separate considerations.

Case 1: $-(2c_1 + k)/2c_2 \geq 1$
The solution of Equation (1.117) has the following form:

$$\phi_{1,2} = \pm\alpha; \qquad \alpha = \cosh^{-1}\left[-\frac{2c_1 + k}{2c_2}\right] \tag{1.118}$$

The attendant solution for Θ_i reads

$$\Theta_i = Ae^{\alpha i} + Be^{-\alpha i} \tag{1.119}$$

where A and B are arbitrary constants.

Case 2: $-(2c_1 + k)/2c_2 \leq -1$
The solution to Equation (1.117) takes the form

$$\phi_{1,2} = \pm(\alpha + j\pi); \qquad \alpha = \cosh^{-1}\left[\frac{2c_1 + k}{2c_2}\right], \qquad j = \sqrt{-1} \tag{1.120}$$

and the solution for Θ_i is

$$\Theta_i = [A\cosh(\alpha i) + B\sinh(\alpha i)]\cos(\pi i) \tag{1.121}$$

Case 3: $-1 \leq -(2c_1 + k)/2c_2 \leq 1$
The solution to Equation (1.117) now becomes

$$\phi_{1,2} = \pm\alpha; \qquad \alpha = \cos^{-1}\left[-\frac{2c_1 + k}{2c_2}\right] \tag{1.122}$$

and, for Θ_i, we obtain

$$\Theta_i = A\cos(\alpha i) + B\sin(\alpha i) \tag{1.123}$$

Upon introducing the following notations

$$\begin{aligned} &f_1(\alpha, r) = e^{\alpha r}, \qquad &&f_2(\alpha, r) = \cosh(\alpha r)\cos(\pi r), \qquad &&f_3(\alpha, r) = \cos(\alpha r) \\ &g_1(\alpha, r) = e^{-\alpha r}, \qquad &&g_2(\alpha, r) = \sinh(\alpha r)\cos(\pi r), \qquad &&g_3(\alpha, r) = \sin(\alpha r) \end{aligned} \tag{1.124}$$

the three cases have a unified expression

$$\Theta_{r,i} = A_1 f_i(\alpha, r) + B_1 g_i(\alpha, r), \qquad i = 1, 2, 3 \tag{1.125}$$

where $\Theta_{r,i}$ corresponds to the three different cases when the subscript i varies from 1 to 3.

We will consider two different kinds of continuous plates. The first kind of the multi-span plate is structurally periodic except for a single disordered span that contains an imperfection. The second kind is a two-piecewise-periodic plate, which means that its first qth spans and the rest of the $N - q$ spans are periodic per se, but they do not have the same periodicity.

Suppose that we have an N-span plate (Figure 1.25), which is periodic except that its $(q + 1)$th span contains a span length imperfection that makes that span slightly longer or shorter than the other spans of the structure, that is,

$$a_i = a \quad \text{for } i = 1, \ldots, q, q + 2, \ldots, N; \qquad a_{q+1} = a^* \quad (a \neq a^*) \tag{1.126}$$

As a particular case ($a^* = a$), we recover the original, perfectly periodic structure.

To facilitate the solution of the problem, we can treat the entire continuous plate as being composed of three segments. The first q-span periodic plate constitutes segment I; the $(q + 1)$th span, namely, the disordered span, constitutes segment II; and the last $(N - q - 1)$ spans of periodic plate represent segment III. Assume first that both segment I and III themselves contain a large number of spans. For segments I and III, the finite difference calculus is applicable because of their structural periodicity. With respect to the disordered span, segment II, a separate consideration should be made. By following this procedure, we construct a solution composed of three parts with each part corresponding to a specific segment of the plate. Continuity conditions between those different segments are utilized in combination with boundary conditions at the ends of the plate to establish an eigenvalue problem.

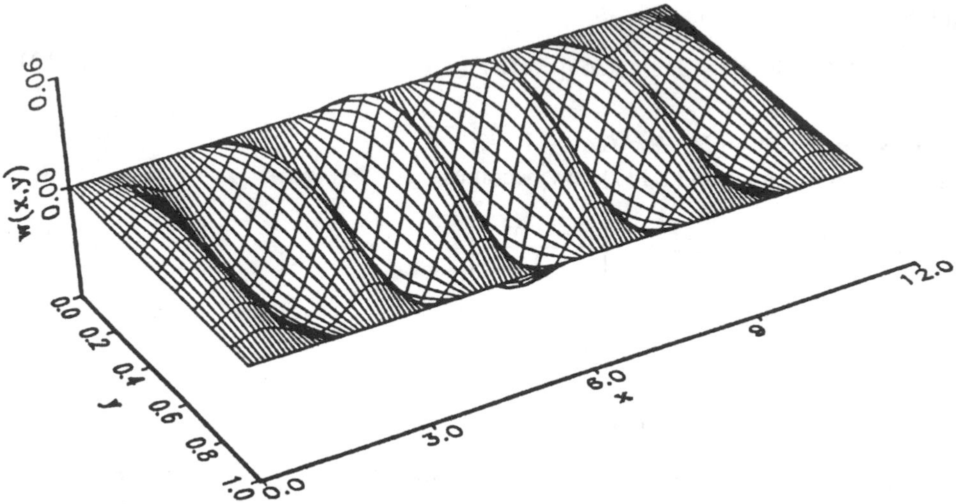

Figure 1.25 Buckling mode for a disordered 11-span plate ($k = 5$).

For the first q spans of periodic plate, we perform the finite difference calculus analysis

$$\theta_r = \theta_r^I = \Theta_r^I \sin\frac{\pi y}{b}; \qquad (r = 0, 1, \ldots, q) \tag{1.127}$$

where the superscript denotes the sequence number of the segment in question; Θ_r^I takes on one of the three forms represented by Equations (1.119), (1.121), and (1.123), depending on the physical and geometrical conditions of the segment.

For the disordered span, recalling Equation (1.111), we have

$$\begin{aligned} M_q^R &= \frac{D_c}{a^*}\left[\bar{c}_1\theta_0^{II} + \bar{c}_2\theta_1^{II}\right] \\ M_{q+1}^L &= -\frac{D_c}{a^*}\left[\bar{c}_1\theta_1^{II} + \bar{c}_2\theta_0^{II}\right] \end{aligned} \tag{1.128}$$

or, in another form,

$$\begin{aligned} \theta_0^{II} &= \frac{a^*}{D_c}\frac{\bar{c}_1 M_q^R + \bar{c}_2 M_{q+1}^L}{\bar{c}_1^2 - \bar{c}_2^2} \\ \theta_1^{II} &= -\frac{a^*}{D_c}\frac{\bar{c}_1 M_{q+1}^L + \bar{c}_2 M_q^R}{\bar{c}_1^2 - \bar{c}_2^2} \end{aligned} \tag{1.129}$$

where $\bar{c}_1$ and $\bar{c}_2$ are obtained from the expressions for c_1 and c_2 by formally replacing a_i in Equation (1.112) with a^*.

The treatment of the last $N - q - 1$ spans of periodic beam is similar to that of segment I,

$$\theta_s = \theta_{s-q-1}^{III}; \qquad (s = q + 1, q + 2, \ldots, N) \tag{1.130}$$

Consider now a plate simply supported at its two ends (other boundary conditions can be treated in a similar manner). Then the boundary condition at the left end of the plate can be represented as

$$M_0^R = GJ_1 \frac{\partial^2 \theta_0^I}{\partial y^2} \qquad \text{or} \qquad (c_1 + k_1)\Theta_0^I + C_2\Theta_1^I = 0 \tag{1.131}$$

while the boundary condition at the right end of the plate reads

$$-M_N^L = GJ_1 \frac{\partial^2 \theta_{N-q-1}^{III}}{\partial y^2} \qquad \text{or} \qquad (c_1 + k_1)\Theta_{N-q-1}^{III} + c_2\Theta_{N-q-2}^{III} = 0 \tag{1.132}$$

where GJ_1 is the torsional rigidity of the transverse stiffeners at the two boundaries. In order to have a purely periodic structure, the stiffeners at the boundaries should have half the stiffness of those interior stiffeners (that is, $GJ_1 = GJ/2$ or $k_1 = k/2$). This "half-stiffener" concept has been used widely in the monograph (Wah and Calcote, 1970).

Conditions of continuity between the periodic spans and the disordered span of the beam, namely, between segment I and segment II, are

$$\begin{aligned} -M_q^L + M_q^R &= GJ\frac{\partial^2 \theta_q}{\partial^2 y} \qquad \text{or} \qquad (c_1 + k)\Theta_q^I + c_2\Theta_{q-1}^I + \frac{a}{D_c} m_q^R = 0 \\ \theta_q^I &= \theta_0^{II} \qquad \text{or} \qquad \Theta_q^I - \frac{a^*}{a(\bar{c}_1^2 - \bar{c}_2^2)}\left(\bar{c}_1 \frac{a}{D_c} m_q^R + \bar{c}_2 \frac{a}{D_c} m_{q+1}^L\right) = 0 \end{aligned} \tag{1.133}$$

Analogously, the continuity conditions between the second and the third segments are

$$\begin{aligned} -M_{q+1}^L + M_{q+1}^R &= GJ\frac{\partial^2 \theta_0^{III}}{\partial y^2} \qquad \text{or} \qquad (c_1 + k)\Theta_0^{III} + c_2\Theta_1^{III} - \frac{a}{D} m_{q+1}^L = 0 \\ \theta_1^{II} &= \theta_0^{III} \qquad \text{or} \qquad \Theta_0^{III} + \frac{a^*}{a(\bar{c}_1^2 - \bar{c}_2^2)}\left(\bar{c}_2 \frac{a}{D} m_q^R + \bar{c}_1 \frac{a}{D} m_{q+1}^L\right) = 0 \end{aligned} \tag{1.134}$$

Using the unified expression (1.125), the rotation angles in the first and third segments can be expressed as

$$\begin{aligned} \Theta_r^I &= A_1 f_i(\alpha, r) + B_1 g_i(\alpha, r) \qquad (r = 0, 1, \ldots, q) \\ \Theta_{s-q-1}^{III} &= A_2 f_i(\alpha, s-q-1) + B_2 g_i(\alpha, s-q-1) \qquad (s = q+1, \ldots, N) \end{aligned} \tag{1.135}$$

Substituting expressions (1.135) in the boundary conditions (1.131) and (1.132) and the continuity conditions (1.133) and (1.134), we obtain six homogeneous algebraic equations,

$$[(c_1 + k_1)f_i(\alpha, 0) + c_2 f_i(\alpha, 1)]A_1 + [(c_1 + k_1)g_i(\alpha, 0) + c_2 g_i(\alpha, 1)]B_1 = 0 \tag{1.136}$$

$$\begin{aligned} &[(c_1 + k)f_i(\alpha, q) + c_2 f_i(\alpha, q-1)]A_1 \\ &\quad + [(c_1 + k)g_i(\alpha, q) + c_2 g_i(\alpha, q-1)]B_1 + \bar{m}_q^R = 0 \end{aligned} \tag{1.137}$$

$$f_i(\alpha, q)A_1 + g_i(\alpha, q)B_1 - \frac{\bar{c}_1}{\bar{c}_1^2 - \bar{c}_2^2}\left(\frac{a^*}{a}\right)\bar{m}_q^R - \frac{\bar{c}_2}{c_1^2 - \bar{c}_2^2}\left(\frac{a^*}{a}\right)\bar{m}_{q+1}^L = 0 \tag{1.138}$$

$$[(c_1 + k)f_i(\alpha, 0) + c_2 f_i(\alpha, 1)]A_2 + [(c_1 + k)g_i(\alpha, 0) + c_2 g_i(\alpha, 1)]B_2 - \bar{m}^L_{q+1} = 0 \tag{1.139}$$

$$A_2 + B_2 + \frac{\bar{c}_2}{\bar{c}_1^2 + \bar{c}_2^2}\left(\frac{a^*}{a}\right)\bar{m}^R_q + \frac{\bar{c}_1}{c_1^2 - \bar{c}_2^2}\left(\frac{a^*}{a}\right)\bar{m}^L_{q+1} = 0 \tag{1.140}$$

$$[(c_1 + k_1)f_i(\alpha, N - q - 1) + c_2 f_i(\alpha, N - q - 2)]A_2 + [(c_1 + k_1)g_i(\alpha, N - q - 1) + c_2 g_i(\alpha, N - q - 2)]B_2 = 0 \tag{1.141}$$

where

$$\bar{m}^R_q = \frac{m^R_q a}{D_c}; \qquad \bar{m}^L_{q+1} = \frac{m^L_{q+1} a}{D_c} \tag{1.142}$$

Thus, we have six homogeneous algebraic equations, which can be expressed in a matrix form as follows:

$$[F(\lambda)]_{6\times 6}\{\delta\}_{6\times 1} = 0 \tag{1.143}$$

where $[F(\lambda)]$ is the coefficient matrix, and $\{\delta\}^T = \{A_1, B_1, \bar{m}^R_q, \bar{m}^L_{q+1}, A_2, B_2\}$.

Another kind of the continuous plate is characterized by

$$a_i = \begin{cases} a & \text{for } i = 1, \ldots, q \\ a^* & \text{for } i = q + 1, \ldots, N \end{cases} \qquad (a \neq a^*) \tag{1.144}$$

For this problem, the first periodic segment, which consists of the first q spans of the plate, may fall into one of the three different cases; the second periodic segment, which is comprised of the remaining $N - q$ spans may present itself if another three different cases. Therefore, there might be nine separate cases altogether, which makes the search for a solution rather complicated.

Recalling Equation (1.124), a unified solution for this piecewise periodic plate can be written as

$$\Theta_r = \begin{cases} A_1 f_i(\alpha_1, r) + B_1 g_i(\alpha_1, r), & i = 1, 2, 3; \quad r = 0, \ldots, q \\ A_2 f_j(\alpha_2, r) + B_2 g_j(\alpha_2, r), & j = 1, 2, 3; \quad r = q + 1, \ldots, N \end{cases} \tag{1.145}$$

Using the boundary conditions (simple supports at two ends)

$$(1) \quad \left(M^R_0\right)^I = GJ_1 \frac{\partial^2 \theta^I_0}{\partial y^2}; \qquad (2) \quad -\left(M^L_N\right)^{II} = GJ_1 \frac{\partial^2 \theta^{II}_{N-q}}{\partial y^2} \tag{1.146}$$

and the conditions of continuity

$$(1) \quad \theta^I_q = \theta^{II}_0; \qquad (2) \quad \left(M^L_q\right)^I = \left(M^R_0\right)^{II} \tag{1.147}$$

the following homogeneous equations are established

$$[(k_1 + c_1)f_i(\alpha_1, 0) + c_2 f_i(\alpha_1, 1)]A_1 + [(k_1 + c_1)g_i(\alpha_1, 0) + c_2 g_i(\alpha_1, 1)]B_1 = 0 \tag{1.148}$$

$$[(k + \bar{c}_1)f_j(\alpha_2, N - q) + \bar{c}_2 f_j(\alpha_2, N - q - 1)]A_2 + [(k + \bar{c}_1)g_j(\bar{\alpha}_2, N - q) + c_2 g_j(\alpha_2, N - q - 1)]B_2 = 0 \tag{1.149}$$

$$f_i(\alpha_1, q)A_1 + g_i(\alpha_1, q - 1)B_1 - f_j(\alpha_2, 0)A_2 - g_j(\alpha_2, 0)B_2 = 0 \tag{1.150}$$

where the sub-indices i and j take on the value of 1, 2, or 3, depending upon which particular case the segments fall into.

$$[c_1 f_i(\alpha_1, q) + c_2 f_i(\alpha_1, q-1)]A_1 + [c_1 g_i(\alpha_1, q-1) + c_2 g_i(\alpha_1, q-1)]B_1$$
$$- \left(\frac{a}{a^*}\right)[\bar{c}_1 f_j(\alpha_2, 0) + \bar{c}_2 f_j(\alpha_2, 1)]A_2$$
$$+ \left(\frac{a}{a^*}\right)[\bar{c}_1 g_j(\alpha_2, 0) + c_2 g_j(\alpha_2, 1)]B_2 = 0 \quad (1.151)$$

Again, Equations (1.148) to (1.151) can be written in matrix form

$$[F(\lambda)]_{4\times 4}\{\delta\}_{4\times 1} = 0 \quad (1.152)$$

where $[F(\lambda)]$ is the coefficient matrix, and $\{\delta\}^T = \{A_1, B_1, A_2, B_2\}$.

Both Equations (1.143) and (1.152) are homogeneous algebraic equations. As such, non-triviality of $\{\delta\}$ requires that the determinant of the coefficient matrix vanish

$$\text{Det}[F(\lambda)] = 0 \quad (1.153)$$

which constitutes a transcendental equation from which the non-dimensional buckling load parameter λ can be solved in terms of other geometric and material properties of the structure in question. After determining the buckling load parameter λ, we can use Equation (1.107) to calculate, span by span, the buckling mode shape for the entire structure, after the type of case has been ascertained.

In this section, we discuss the buckling load and mode shapes of the two different types of multi-span plates described in Section 1.2.

As the first example, consider the following case:

$$\frac{a}{b} = 1, \qquad \frac{a^*}{a} = 1.1, \qquad N = 11, \qquad q = 5 \quad (1.154)$$

As is shown in the preceding data, the plate consists of 11 spans, of which the sixth span contains a length imperfection that makes that span a bit longer than the other spans. Numerical results show that such an imperfection has a slight degrading effect on the buckling load. For instance, when k, the parameter characterizing the torsional rigidity of the stiffener, equals 5, buckling load parameter λ is 5.06. Compared with its counterpart of the periodic plate, which is $\lambda = 5.26$, the reduction rate is only 4%. The buckling load reduction remains almost unchanged with the torsional rigidity of the stiffeners. Even when the torsional rigidity doubles, the reduction rate only amounts to 5%. So, the buckling load decrease induced by the presence of the imperfection is not significant. However, the buckling modes are appreciably different for the plates with and without the imperfection (Figures 1.26–1.28). Moreover, as k increases, the buckling mode of the disordered plate becomes increasingly localized (Figures 1.26 and 1.27). The overall behavior of such plates is very similar to that of the continuous beams with torsional springs discussed in Section 1.2, despite the difference of structural dimensionality between beams and plates.

The second example is a 10-span, piecewise periodic plate whose first five spans have the length of a and the other five spans have the length of a^* (assuming $a^* > a$),

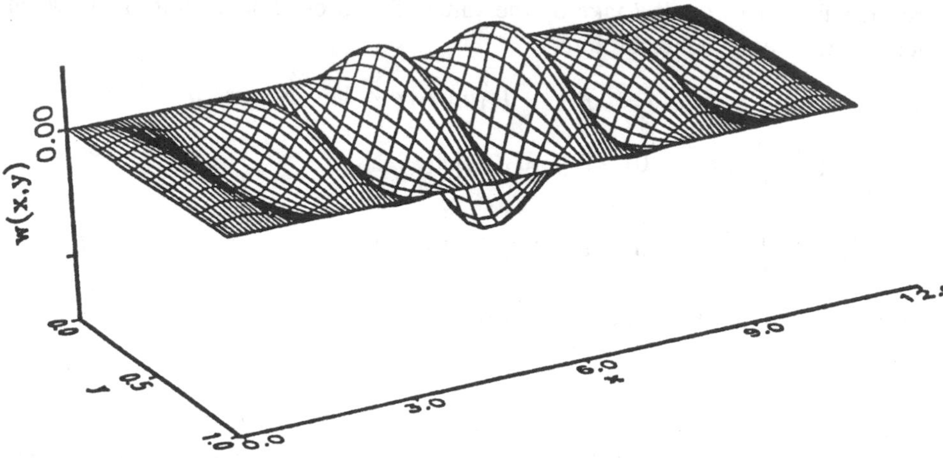

Figure 1.26 Buckling mode for a disordered 11-span plate ($k = 10$).

and the plate is reinforced by transverse stiffeners with torsional rigidity of $k = 10$. When the ratio of a^* to a equals unity, the plate reduces to a purely periodic plate. Here we discuss the case where the ratio is different from unity, say $a^*/a = 1.1$. As far as the buckling load is concerned, the difference between such a structure and the purely periodic plate is relatively minor. For the preceding structure, the buckling load parameter λ equals 5.35; for the corresponding, exactly periodic plate, λ is 5.77. So the difference in the buckling load between the two is only 7%. More significant, however, is the difference in the buckling mode. For instance, when $a^*/a = 1.1$ and $k = 10$, the deflection of the structure at buckling is largely confined to the left end, whereas those spans of the plate near the other end hardly experience any deformation (Figure 1.28).

From the examples already shown, it is well demonstrated that the torsional stiffness of the stiffeners should not be ignored in the investigation, just for the sake of

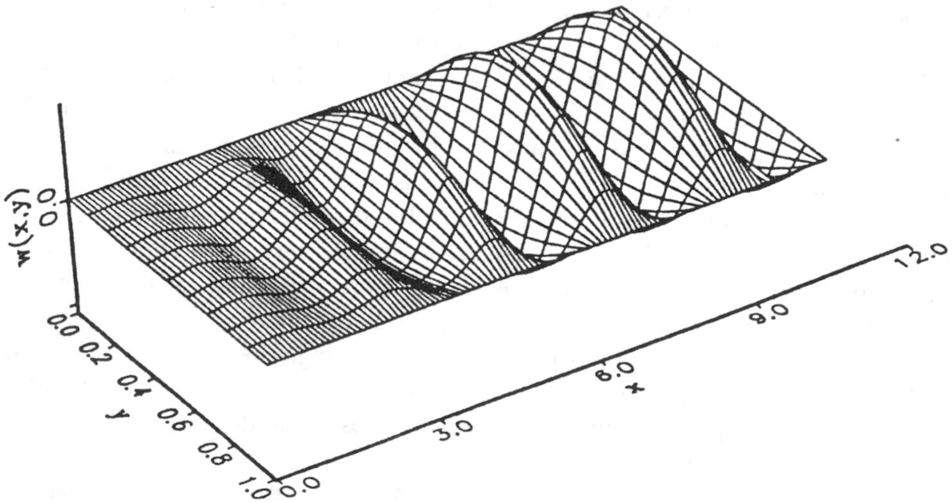

Figure 1.27 Buckling mode for a piecewise periodic plate ($k = 5$).

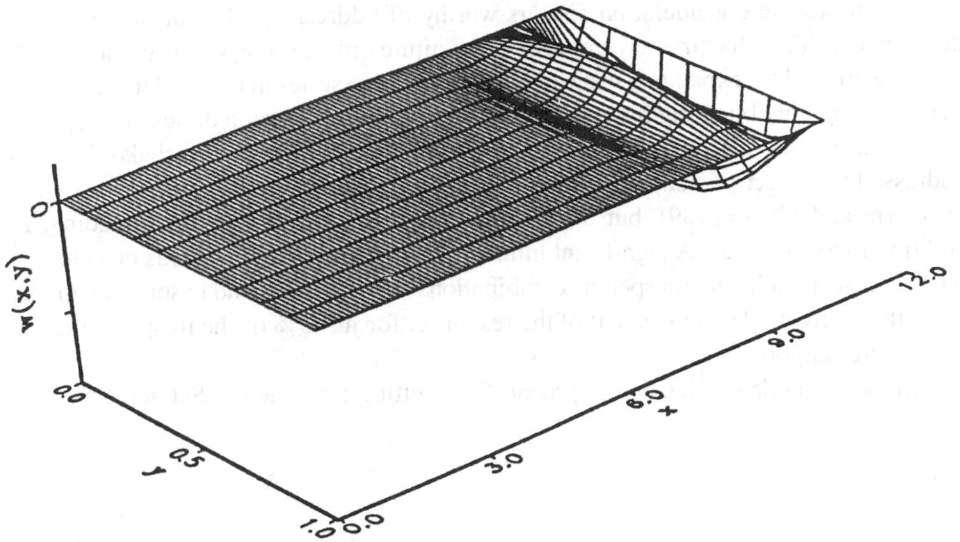

Figure 1.28 Buckling mode for a piecewise periodic plate ($k = 10$).

simplification in analysis. As it turns out, the torsional stiffness plays quite an important role. It not only boosts the strength of the structure but also localizes the loss of geometric rigidity of the structure at buckling to a small area so that any damages, should they occur, are kept to a minimum.

For the plates contemplated here, small structural irregularities do not significantly alter the load-carrying capacity. However, the presence of such irregularities confines the buckling pattern associated with large deflection to a limited fraction of the structure. In this regard, the effect of such irregularities on the buckling loads can be considered favorable. As Nayfeh and Hawwa (1994b) pointed out, by means of "deliberately" inducing some irregularities in the system, one may confine the structural buckling to a limited part of the system only, which can be regarded as a passive control of the buckling process.

The method used here seems to have a wide application, especially in continuous plates with a large number of spans. In fact, the larger the number of spans is, the more advantageous the present method is over the other traditional methods such as the one based on the integration of the governing differential equation because it leads to a determinative matrix that can be orders of magnitude smaller than those necessary in other methods. Besides, the solutions generated by the present method are analytic and exact and, thus, can be used as benchmarks for other numerical methods.

The method of investigation presented here can be applied to discuss essentially periodic structures where irregularities occur only in some local areas. It takes into consideration the discreteness of the stiffeners and, in particular, their torsional rigidity. As it turns out, the torsional rigidity of the stiffeners are important and should not be ignored in the discussion of the buckling mode shape. Adjusting the torsional rigidity of the stiffeners, one can achieve the goal of localizing the deflection of the structure at buckling to a small area. For further discussion, the reader is referred to the paper by Elishakoff, Li, and Starnes (1995).

The following consideration appears worthy of addressing: The buckling modes themselves, although extremely important, constitute only auxiliary information. They could be utilized for expanding the response quantities in series in terms of these modes. Yet, studying only buckling modes does not lead to the conclusion about the response quantities that the designer may be concerned with. Zingales and Elishakoff (2000) addressed the effect of small misplacements in the same two-span column considered by Pierre and Plaut (1989), but the column was subjected to transverse loading, in addition to axial forces. A significant influence of the misplacements was detected on the response; it included, for specific combinations of parameters and in some locations, more than a fivefold enhancement of the response, for just 3% of the misplacement in the middle support.

Buckling mode localization in probabilistic setting is studied in Section 4.5.

CHAPTER TWO

Deterministic Problems of Shells with Variable Thickness

Shells were first used by the Creator of the earth and its inhabitants. The list of natural shell-like structures is long, and the strength properties of some of them are remarkable. Egg shells range in size from those of the smallest insects to the large ostrich eggs, and cellular structures are the building blocks for both plants and animals. Bamboo is basically a thin-walled cylindrical structure, as is the root section of a bird's feather. The latter structural element develops remarkable load-carrying abilities.

E. E. Sechler

The degeneration of a bifurcation point into a limit point due to presence of imperfections... suggests that bifurcation is an exception rather than the rule. In spite of this the literature on elastic stability and buckling of structures deals preponderantly with bifurcation buckling problems. Human frailty is, perhaps, the main reason for this state of affairs.

W. T. Koiter

This chapter focuses on the buckling of cylindrical shells with small thickness variations. Two important cases of thickness variation pattern are considered. Asymptotic formulas up to the second order of the thickness variation parameter ϵ are derived by combining the perturbation and weighted residual methods. The expressions obtained in this study reduce to Koiter's formulas, when only the first-order term of the thickness variation parameter is retained in the analysis. Results from the asymptotic formula are compared with the those obtained through the purely numerical techniques of the finite difference method and the shooting method. We first deal with homogeneous shells; then we discuss shells made of composite materials in some detail.

2.1 Introductory Remarks

There is a vast literature devoted to buckling of cylindrical shells of constant thickness. The problem regarding the influence of thickness variation on the buckling load has not gained attention and remains open even today. To the best of our knowledge, the first work on the effect of thickness variation on the buckling of shells was undertaken by

Elishakoff, Li, and Starnes (1992). Both the thickness variation and the initial geometric imperfections were considered axisymmetrical. Solution was composed of two terms: The first term was associated with the shell of constant thickness, and the second one incorporated the effects of the thickness variation. The former coincided with Koiter's analytical investigation (Koiter, 1963) for constant thickness shells with axisymmetric imperfection, whereas the latter term was derived numerically using the shooting method. Koiter (personal communication, 1992) derived an analytical formula for the buckling load of a perfect, non-uniform cylindrical shell. The attendant derivation, obtained by using the energy method, was included in Koiter et al. (1994b). The latter study supports the central result of the combined theoretical-numerical investigation (Elishakoff et al., 1992), namely that the effect of thickness variation becomes remarkable when the thickness pattern is co-configurational with the initial imperfection. However, further investigation shows that the most detrimental effect of the axisymmetric thickness variation occurs at twice the wave number of the classical buckling mode.

This chapter examines in detail the buckling of the cylindrical shell with small thickness variations. Our analysis is based on a system of linearized governing differential equations of perfect shells with variable thickness. The asymptotic formulas in terms of ϵ (ϵ is the thickness non-uniformity parameter) are derived by a hybrid perturbation-weighted residuals methods. In comparison with formulas (Koiter, personal communication, 1992) that are linear in ϵ, these asymptotic formulas also contain the quadratic term, which results in a higher accuracy. In addition to the analytic investigation, numerical study was also performed, and the results stemming from different methods were compared and discussed.

2.1.1 Basic Equations for Homogeneous Shells

The linear equations governing the axially compressed, non-uniform cylindrical shell (Figure 2.1) are as follows:

$$
\begin{aligned}
&h^2\nabla^2\nabla^2 F + 2\left(\frac{dh}{dx}\right)^2\left(\frac{\partial^2 F}{\partial x^2} - \nu\frac{\partial^2 F}{\partial y^2}\right) - h\frac{d^2h}{dx^2}\left(\frac{\partial^2 F}{\partial x^2} - \nu\frac{\partial^2 F}{\partial y^2}\right)\\
&\quad - 2h\frac{dh}{dx}\left(\frac{\partial^3 F}{\partial x^3} - \nu\frac{\partial^3 F}{\partial x\partial y^2}\right) - 2(1+\nu)h\frac{dh}{dx}\frac{\partial^3 F}{\partial x\partial y^2} = \frac{Eh^3}{R}\frac{\partial^2 W}{\partial x^2}
\end{aligned}
\tag{2.1}
$$

$$
\begin{aligned}
&\frac{Eh^3}{12(1-\nu^2)}\nabla^2\nabla^2 W + \frac{1}{R}\frac{\partial^2 F}{\partial x^2} + \frac{3Eh^2}{12(1-\nu^2)}\frac{dh}{dx}\nabla^2 W\\
&\quad + \frac{3Eh^2}{12(1-\nu^2)}\left(\frac{\partial^3 W}{\partial x^3} + \nu\frac{\partial^3 W}{\partial x\partial y^2}\right)\frac{dh}{dx}\\
&\quad + \frac{6Eh}{12(1-\nu^2)}\left(\frac{dh}{dx}\right)^2\left(\frac{\partial^2 W}{\partial x^2} + \nu\frac{\partial^2 W}{\partial^2 y}\right)\\
&\quad + \frac{3Eh^2}{12(1-\nu^2)}\left(\frac{\partial^2 W}{\partial x^2} + \nu\frac{\partial^2 W}{\partial y^2}\right)\frac{d^2h}{dx^2}\\
&\quad + \frac{3Eh^2}{12(1+\nu)}\frac{dh}{dx}\frac{\partial^3 W}{\partial x\partial y^2} + P_0\frac{\partial^2 W}{\partial x^2} = 0
\end{aligned}
\tag{2.2}
$$

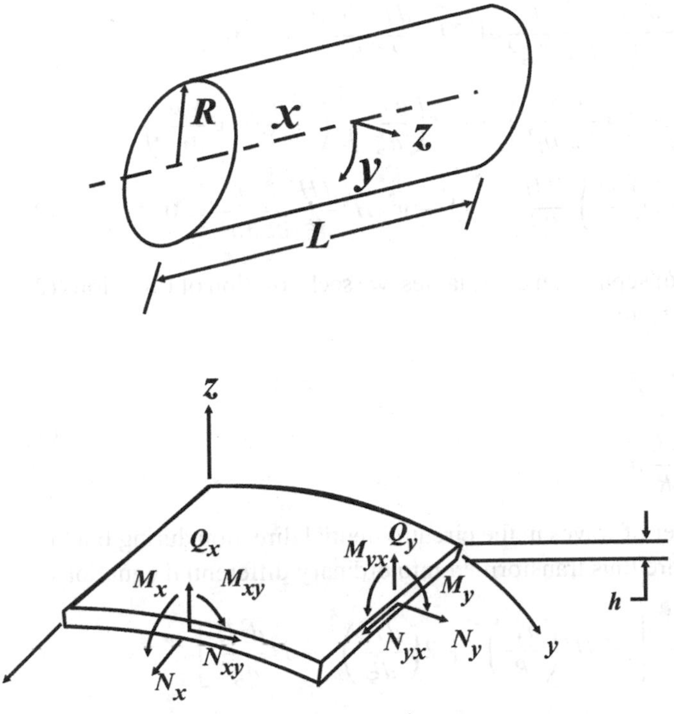

Figure 2.1 Notations and sign conventions.

where W and F represent the radial displacement (positive outward) and the Airy stress function, respectively; ν is Poisson's ratio; E the modulus of elasticity; P_0 denotes the uniform axial load at the ends of the shell; $h(x)$ is the shell thickness, assumed here varying only axisymmetrically,

$$h(x) = h_0\left(1 - \epsilon \cos \frac{2px}{R}\right) \tag{2.3}$$

where h_0 is the nominal thickness of the shell and ϵ and p are the non-dimensional parameters indicating the magnitude and wave of the thickness variation, respectively.

By introducing the following non-dimensional parameters

$$\xi = \frac{x}{L}, \qquad \eta = \frac{y}{L}, \qquad w = \frac{W}{L}, \qquad f = \frac{F}{D_0}, \qquad H = \frac{h}{h_0} \tag{2.4}$$

where $D_0 = Eh_0^3/12(1-\nu^2)$, the governing equations (2.1) and (2.2) can be rewritten into their non-dimensional form,

$$\begin{aligned} &H^2\bar{\nabla}^2\bar{\nabla}^2 f + 2\left(\frac{dH}{d\xi}\right)^2\left(\frac{\partial^2 f}{\partial \xi^2} - \nu\frac{\partial^2 f}{\partial \eta^2}\right) - H\frac{\partial^2 H}{\partial \xi^2}\left(\frac{\partial^2 f}{\partial \xi^2} - \nu\frac{\partial^2 f}{\partial \eta^2}\right) \\ &\quad - 2H\frac{dH}{d\xi}\left(\frac{\partial^3 f}{\partial \xi^3} - \nu\frac{\partial^3 f}{\partial \xi \partial \eta^2}\right) - 2(1+\nu)H\frac{dH}{d\xi}\frac{\partial^3 f}{\partial \xi \partial \eta^2} \\ &\quad = \frac{12(1-\nu^2)L^2}{Rh_0}H^3\frac{\partial^2 w}{\partial \xi^2} \end{aligned} \tag{2.5}$$

$$
\begin{aligned}
&H^3\bar\nabla^2\bar\nabla^2 w + \frac{P_0 L^2}{D_0}\frac{\partial^2 w}{\partial \xi^2} + \frac{L^2}{Rh_0}\frac{\partial^2 f}{\partial \xi^2} + 3H^2\frac{dH}{d\xi}\frac{\partial}{\partial \xi}\bar\nabla^2 w \\
&+ 3H^2\frac{dH}{d\xi}\left(\frac{\partial^3 w}{\partial \xi^3} + \nu\frac{\partial^3 w}{\partial \xi \partial \eta^2}\right) + 6H\left(\frac{dH}{d\xi}\right)^2\left(\frac{\partial^2 w}{\partial \xi^2} + \nu\frac{\partial^2 w}{\partial \eta^2}\right) \\
&+ 3H^2\left(\frac{\partial^2 w}{\partial \xi^2} + \nu\frac{\partial^2 w}{\partial \eta^2}\right)\frac{d^2 H}{d\xi^2} + 3(1-\nu^2)H^2\frac{dH}{d\xi}\frac{\partial^3 w}{\partial \xi \partial \eta^2} = 0
\end{aligned}
\tag{2.6}
$$

Furthermore, in view of separation of variables, we seek solution of Equations (2.5) and (2.6) in the following form:

$$
\begin{aligned}
f(\xi,\eta) &= \bar f(\xi)\cos\frac{nL}{R}\eta \\
w(\xi,\eta) &= \bar w(\xi)\cos\frac{nL}{R}\eta
\end{aligned}
\tag{2.7}
$$

where n denotes the number of waves in the circumferential direction during buckling. Equations (2.5) and (2.6) are thus transformed into ordinary differential equations

$$
\begin{aligned}
&H^2\bar f^{(4)} - 2H\frac{dH}{d\xi}\bar f''' + \left[-2H^2\left(\frac{nL}{R}\right)^2 + 2\left(\frac{dH}{d\xi}\right)^2 - H\frac{d^2H}{d\xi^2}\right]\bar f'' \\
&+ 2H\frac{dH}{d\xi}\left(\frac{nL}{R}\right)^{2\bar f'} + \left[H^2\left(\frac{nL}{R}\right)^4 + 2\left(\frac{dH}{d\xi}\right)^2\nu\left(\frac{nL}{R}\right)^2\right. \\
&\left. - H\frac{d^2H}{d\xi}\nu\left(\frac{nL}{R}\right)^2\right]\bar f = \frac{12(1-\nu^2)L^2}{Rh_0}H^3\bar w''
\end{aligned}
\tag{2.8}
$$

$$
\begin{aligned}
&H^3\bar w^{(4)} + 6H^2\frac{dH}{d\xi}\bar w''' + \left[-2H^3\left(\frac{nL}{R}\right)^2 + \frac{P_0L^2}{D_0} + 6H\left(\frac{dH}{d\xi}\right)^2 + 3H^2\frac{d^2H}{d\xi^2}\right]\bar w'' \\
&- 6H^2\frac{dH}{d\xi}\left(\frac{nL}{R}\right)^2\bar w' + \left[H^3\left(\frac{nL}{R}\right)^4 - 6H\left(\frac{dH}{d\xi}\right)^2\nu\left(\frac{nL}{R}\right)^2\right. \\
&\left. - 3H^2\frac{d^2H}{d\xi^2}\nu\left(\frac{nL}{R}\right)^2\right]\bar w + \frac{L^2}{Rh_0}\bar f'' = 0
\end{aligned}
\tag{2.9}
$$

In this chapter, three different methods are used to obtain the classical buckling load P_{cl}. First, we evaluate the buckling load via an analytical technique, and then compare it with the results of purely numerical calculations.

2.1.2 Hybrid Perturbation-Weighted Residuals Method

In the hybrid perturbation-weighted residuals method (Koiter et al., 1994a), we assume $\bar w(\xi)$ in the form

$$
\bar w(\xi) = A\cos\frac{pL}{R}\xi + B\cos\frac{3pL}{R}\xi
\tag{2.10}
$$

where p is the number of half-waves along the shell length at buckling and A and B are undetermined constants. This buckling pattern satisfies the boundary conditions of the simple supports. The first term of the two-term approximation (2.10) is the exact buckling mode for the shell of constant thickness, and the second term is introduced to account for the thickness variation.

In order to solve the compatibility equation (2.8) for $\bar{f}$, the perturbation procedure will be employed here. To this end, $\bar{f}$ is expressed in terms of the thickness variation parameter ϵ as

$$\bar{f}(\xi) = f_0(\xi) + \epsilon f_1(\xi) + \epsilon^2 f_2(\xi) + \cdots \tag{2.11}$$

Substituting (2.11) into (2.8) and keeping (2.3) in mind, one has, after collecting the like terms in ϵ,

$$\begin{aligned}
&\bar{f}^{(4)} - 2N^2 f_0'' + f_0 - 4c^2z^2\bar{w}'' + \epsilon\big[f_1^{(4)} - 2N^2 f_1'' + N^4 f_1 - 2\cos 2P\xi f_0^{(4)} \\
&\quad - 4P\sin 2P\xi f_0''' + 4N^2\cos 2P\xi f_0'' - 4P^2\cos 2P\xi f_0'' \\
&\quad - 4PN^2\sin 2P\xi f_0' - (2N^4 + 4\nu P^2N^2)\cos 2P\xi f_0 + 12c^2z^2\cos 2P\xi\bar{w}''\big] \\
&\quad + \epsilon^2\big[f_2^{(4)} - 2N^2 f_2'' + N^4 f_2 + \cos^2 2P\xi f_0^{(4)} - 2\cos 2P\xi f_1^{(4)} \\
&\quad + 4P\cos 2P\xi\sin 2P\xi f_0''' - 4P\sin 2P\xi f_1''' + (-2N^2\cos^2 2P\xi \\
&\quad + 8P^2\sin^2 2P\xi + 4P^2\cos^2 2P\xi) f_0'' - (4P^2 - 4N^2)\cos 2P\xi f_1'' \\
&\quad - 4PN^2\cos 2P\xi\sin 2P\xi f_0' + 4PN^2\sin 2P\xi f_1' \\
&\quad + (N^4\cos^2 2P\xi + 8\nu N^2P^2\sin^2 2P\xi + 4\nu P^2N^2\cos^2 2P\xi) f_0 \\
&\quad - (2N^4 + 4P^2N^2\nu)\cos 2P\xi f_1 - 12c^2z^2\cos^2 2P\xi\bar{w}''\big] + \cdots = 0
\end{aligned} \tag{2.12}$$

where

$$P = \frac{pL}{R}, \qquad N = \frac{nL}{R}, \qquad z = \frac{L}{\sqrt{Rh_0}}, \qquad c = \sqrt{3(1-\nu^2)} \tag{2.13}$$

From Equation (2.12), we obtain

$$\mathscr{L}(f_0) = 4c^2z^2\bar{w}'' \tag{2.14}$$

$$\begin{aligned}
\mathscr{L}(f_1) = {}& 2\cos 2P\xi f_0^{(4)} + 4P\sin 2P\xi f_0''' - 4N^2\cos 2P\xi f_0'' \\
&+ 4P^2\cos 2P\xi f_0'' - 4PN^2\sin 2P\xi f_0' \\
&+ (2N^4 + 4\nu P^2N^2)\cos 2P\xi f_0 - 12c^2z^2\cos 2P\xi\bar{w}''
\end{aligned} \tag{2.15}$$

$$\begin{aligned}
\mathscr{L}(f_2) = {}& -\cos^2 2P\xi f_0^{(4)} + 2\cos 2P\xi f_1^{(4)} - 4P\cos 2P\xi\sin 2P\xi f_0''' \\
&+ 4P\sin 2P\xi f_1''' - (-2N^2\cos^2 2P\xi + 8P^2\sin^2 2P\xi \\
&+ 4P^2\cos^2 2P\xi) f_0'' + (4P^2 - 4N^2)\cos 2P\xi f_1'' \\
&+ 4PN^2\cos 2P\xi\sin 2P\xi f_0' - 4PN^2\sin 2P\xi f_1' \\
&- (N^4\cos^2 2P\xi + 8\nu N^2P^2\sin^2 2P\xi + 4\nu P^2N^2\cos^2 2P\xi) f_0 \\
&+ (2N^4 + 4P^2N^2\nu)\cos 2P\xi f_1 + 12c^2z^2\cos^2 2P\xi\bar{w}''
\end{aligned} \tag{2.16}$$

where the operator $\mathscr{L}(\cdot)$ is defined as

$$\mathscr{L}(f) = f^{(4)} - 2N^2 f'' + N^4 f \tag{2.17}$$

Equations (2.14)–(2.16) can be solved analytically with the aid of the computerized symbolic algebra *Mathematica* (Wolfram, 1991) for f_0, f_1, and f_2 to yield

$$\begin{aligned} f_0 &= a_1 \cos P\xi + a_2 \cos 3P\xi \\ f_1 &= a_3 \cos P\xi + a_4 \cos 3P\xi + a_5 \cos 5P\xi \\ f_2 &= a_6 \cos P\xi + a_7 \cos 3P\xi + a_8 \cos 5P\xi + a_9 \cos 7P\xi \end{aligned} \tag{2.18}$$

where $a_1, a_2, \ldots, a_9$ are coefficients depending on A and B, and are given in the paper by Koiter et al. (1994a). Consult Chapter 7 for more information on the use of computerized symbolic algebra in the buckling analysis.

Applying the weighted residuals method, namely, in our case the Boobnov-Galerkin procedure, to the equilibrium equation (2.9), we arrive at

$$\begin{aligned} \int_{-1/2}^{1/2} \Bigg\{ & H^3 \bar{w}^{(4)} + 6H^2 \frac{dH}{d\xi} \bar{w}''' + \left[-2H^3 N^2 + 4\alpha c z^2 + 6H \left(\frac{dH}{d\xi} \right)^2 + 3H^2 \frac{d^2 H}{d\xi^2} \right] \bar{w}'' \\ & - 6H^2 \frac{dH}{d\xi} N^2 \bar{w}' + \left[H^3 N^4 - 6H \left(\frac{dH}{d\xi} \right)^2 \nu N^2 - 3H^2 \frac{d^2 H}{d\xi^2} \nu N^2 \right] \bar{w} \\ & + z^2 (f_0'' + \epsilon f_1'' + \epsilon^2 f_2 + \cdots) \Bigg\} \begin{Bmatrix} \cos P\xi \\ \cos 3P\xi \end{Bmatrix} d\xi = 0 \end{aligned} \tag{2.19}$$

where α is the buckling load reduction factor due to the thickness variation defined as

$$\alpha = \frac{P_0}{P_{0,\text{const}}}, \qquad P_{0,\text{const}} = \frac{E h_0^2}{R\sqrt{3(1-\nu^2)}} \tag{2.20}$$

and $P_{0,\text{const}}$ is the classical buckling load of the uniform shell with constant thickness h_0.

Case A: We evaluate the classical buckling load corresponding to the buckling mode at the top of the Koiter's semi-circle (Koiter, 1980) (Figure 2.2). In this case, the buckling mode has the same wave numbers in both the axial and the circumferential directions, and the buckling wave numbers p and n can be expressed as follows:

$$p = n = \frac{p_0}{2}, \qquad p_0^2 = 2c \frac{R}{h_0} \tag{2.21}$$

then the thickness variation pattern (2.3) becomes

$$h = h_0 \left(1 - \epsilon \cos \frac{p_0 x}{R} \right) \tag{2.22}$$

With this assumption, substituting (2.11) and (2.18) into (2.19) and making some algebraic manipulations lead, when retaining only the terms up to ϵ^2, to the following

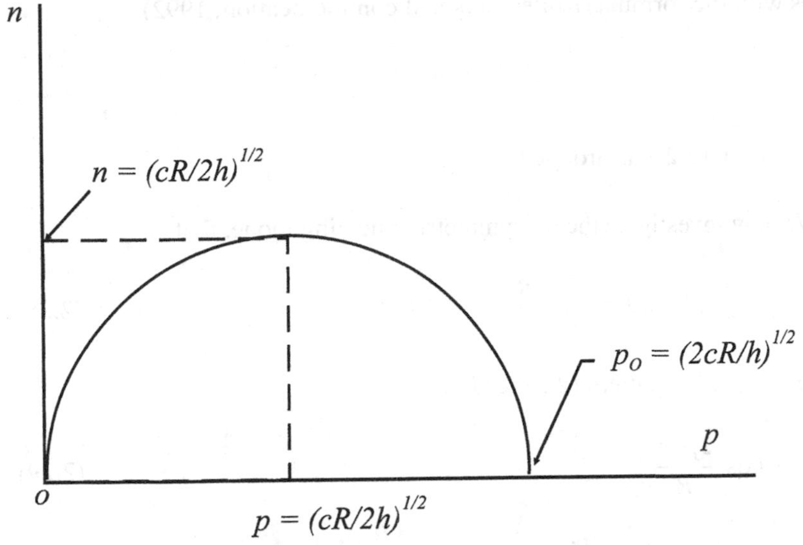

Figure 2.2 Koiter's semi-circle.

eigenvalue problem:

$$[C(\epsilon,\alpha)]_{2\times 2}\begin{Bmatrix} A \\ B \end{Bmatrix} = 0 \tag{2.23}$$

where $[C(\epsilon,\alpha)]$ is the coefficient matrix containing the thickness variation parameter ϵ and the buckling load reduction factor α. The elements of matrix $[C(\epsilon,\alpha)]$ are as follows:

$$C_{11} = P^4\left[4 - 4\alpha - 2\epsilon\nu + \frac{58 - 4\nu + 13\nu^2}{25}\epsilon^2\right]$$

$$C_{12} = C_{21} = P^4\left[-\frac{336 + 66\nu}{25}\epsilon + \frac{66 + 300\nu + 9\nu^2}{50}\epsilon^2\right] \tag{2.24}$$

$$C_{22} = P^4\left[\frac{1412 - 900\alpha}{25} + \frac{1571010 - 11988\nu + 1377\nu^2}{21125}\epsilon^2\right]$$

The requirement of vanishing of the determinant matrix $[C(\epsilon,\alpha)]$ results in the following equation, when the terms higher than ϵ^2 are neglected:

$$144\alpha^2 + \left(-\frac{9248}{25} + 72\nu\epsilon + \frac{-8048400 + 169632\nu - 400968\nu^2}{21125}\epsilon^2\right)\alpha$$

$$+\frac{5648}{25} - \frac{2824}{24}\nu\epsilon - \frac{1737952\nu - 478708\nu^2}{21125}\epsilon^2 = 0 \tag{2.25}$$

From Equation (2.25) an asymptotic expression can be obtained for the buckling load reduction factor due to the thickness variation

$$\alpha = 1 - \frac{1}{2}\nu\epsilon - \frac{(832 + 464\nu - 23\nu^2)}{512}\epsilon^2 \tag{2.26}$$

which coincides with the formula (Koiter, personal communication, 1992)

$$\alpha = 1 - \frac{1}{2}\nu\epsilon \tag{2.27}$$

if the quadratic term in (2.26) is dropped.

Case B: We now investigate the axisymmetric buckling mode, that is,

$$n = 0, \qquad p = p_0, \qquad P_0^2 = 2c\frac{R}{h_0} \tag{2.28}$$

then the thickness variation pattern (2.3) reads

$$h = h_0\left(1 - \epsilon \cos \frac{2p_0 x}{R}\right) \tag{2.29}$$

For this case, we obtain, by retaining the terms up to ϵ^2, the following asymptotic expression for the buckling load reduction factor:

$$\alpha = 1 - \epsilon - \frac{25}{32}\epsilon^2 \tag{2.30}$$

which again coincides with Koiter's linear formula (Koiter, personal communication, 1992)

$$\alpha = 1 - \epsilon \tag{2.31}$$

if the quadratic term is ignored.

2.1.3 Solution by Finite Difference Method

The finite difference method, which is particularly useful for the buckling problems of structures of complicated geometry or varying flexural rigidity, is used here. This method is based on the use of approximate algebraic expressions for the derivatives of unknown variables that appear in the fundamental governing equations. The following expressions of the central difference are used to approximate the corresponding derivatives:

$$\begin{aligned}
\Delta f_i &= \frac{f_{i+1} - f_{i-1}}{2d}, \qquad \Delta^2 f_i = \frac{f_{i+1} - 2f_i + f_{i-1}}{d^2} \\
\Delta^3 f_i &= \frac{f_{i+2} - 2f_{i+1} - 2f_{i-1} - f_{i-2}}{2d^3}, \\
\Delta^4 f_i &= \frac{f_{i+2} - 4f_{i+1} + 6f_i - 4f_{i-1} + f_{i-2}}{d^4}
\end{aligned} \tag{2.32}$$

where d is the distance between neighboring nodal points.

Using Equation (2.32), the differential equations (2.8) and (2.9) are approximated by the finite difference equations,

$$\left(\frac{G_{1i}}{d^4} - \frac{G_{2i}}{2d^3}\right) f_{i-2} + \left(-4\frac{G_{1i}}{d^4} + \frac{G_{2i}}{d^3} + \frac{G_{3i}}{d^2} - \frac{G_{4i}}{2d}\right) f_{i-1}$$
$$+ \left(6\frac{G_{1i}}{d^4} - 2\frac{G_{3i}}{d^2} + G_{5i}\right) f_i + \left(-4\frac{G_{1i}}{d^4} - \frac{G_{2i}}{d^3} + \frac{G_{3i}}{d^2} + \frac{G_{4i}}{2d}\right) f_{i-1}$$
$$+ \left(\frac{G_{1i}}{d^4} + \frac{G_{2i}}{2d^3}\right) f_{i+2} + \frac{G_{6i}}{d^2} w_{i-1} - 2\frac{G_{6i}}{d^2} w_i + \frac{G_{6i}}{d^2} w_{i+1} = 0 \tag{2.33}$$

$$\frac{G_{7i}}{d^2} f_{i-1} - 2\frac{G_{7i}}{d^2} f_i + \frac{G_{7i}}{d^2} f_{i+1} + \left(\frac{G_{8i}}{d^4} - \frac{G_{9i}}{2d^3}\right) w_{i-2}$$
$$+ \left(-4\frac{G_{8i}}{d^4} + \frac{G_{9i}}{d^3} + \frac{G_{10i}}{d^2} - \frac{G_{11i}}{2d}\right) w_{i-1} + \left(6\frac{G_{8i}}{d^4} - 2\frac{G_{10i}}{d^2} + G_{12i}\right) w_i$$
$$+ \left(-4\frac{G_{8i}}{d^4} - \frac{G_{9i}}{d^3} + \frac{G_{10i}}{d^2} + \frac{G_{11i}}{2d}\right) w_{i+1} + \left(\frac{G_{8i}}{d^4} + \frac{G_{9i}}{2d^3}\right) w_{i+2} = 0 \tag{2.34}$$

Here the derivatives $H'(\xi_i)$ and $H''(\xi_i)$ are evaluated analytically. By subdividing the shell length domain $(-L/2, L/2)$ into M equal segments and applying (2.33) and (2.34) to each nodal point, points near the ends of the shell are influenced by the boundary conditions. Here we consider the case of simply supported boundary conditions, namely,

$$w_0 = w_0'' = f_0 = f_0'' = w_M = w_M'' = f_M = f_M'' = 0 \tag{2.35}$$

or in view of (2.32),

$$\begin{aligned}
G_{1i} &= [H(\xi_i)]^2, \qquad G_{2i} = -2H(\xi_i)H'(\xi_i) \\
G_{3i} &= -2N^2[H(\xi_i)]^2 + 2[H'(\xi_i)]^2 - H(\xi_i)H''(\xi_i) \\
G_{4i} &= 3N^2 H(\xi_i)H'(\xi_i), \\
G_{5i} &= H^2N^4 + 2\nu N^2[H'(\xi_i)]^2 - \nu N^2 H(\xi_i)H''(\xi_i) \\
G_{6i} &= -12(1-\nu^2)z^2[H(\xi_i)]^3, \qquad G_{7i} = z^2 \\
G_{8i} &= [H(\xi_i)]^3, \qquad G_{9i} = 6[H(\xi_i)]^2 H'(\xi_i) \\
G_{10i} &= -2N^2H^3 + 4\alpha c z^2 + 6H(\xi_i)[H'(\xi_i)]^2 + 3[H(\xi_i)]^2 H''(\xi_i) \\
G_{11i} &= -6N^2[H(\xi_i)]^2 H'(\xi_i), \\
G_{12i} &= N^4[H(\xi_i)]^3 - 6\nu N^2 H[H'(\xi_i)]^2 - 3\nu N^2[H(\xi_i)]^2
\end{aligned} \tag{2.36}$$

$$w_{-1} = w_1, \qquad f_{-1} = f_1, \qquad w_{M+1} = w_{M-1}, \qquad f_{M+1} = f_{M-1} \tag{2.37}$$

Thus, we establish a system of simultaneous algebraic equations,

$$[C(\xi_i, \alpha)]_{(2M+2)\times(2M+2)}\{\delta\}_{(2M+2)\times(2M+2)} = 0 \tag{2.38}$$

where $[C(\xi_i, \alpha)]$ is the coefficient matrix, whose elements depend on the shell geometry, nodal point coordinates, and elastic constants as well as the unknown buckling load reduction factor α; $\{\delta\}$ represents a column vector containing the unknown values of functions of w and f at the nodal points.

Setting the determinant of $[C(\xi_i, \alpha)]$ equal to zero gives the approximate value for the classical buckling load reduction factor α or the classical buckling load that improves in accuracy with an increase in the number of subdivided segments. In the implementation of this process, the classical buckling load reduction rate α is sought through iterations.

2.1.4 Solution by Godunov-Conte Shooting Method

The differential equations (2.8) and (2.9), together with the boundary conditions of simple supports ($\bar{f} = \bar{f}'' = \bar{w} = \bar{w}'' = 0$) at the ends of the shell, can be solved for the classical buckling load by use of the shooting method. However, as pointed out by Grigolyuk et al. (1971) in the problem of buckling of cylindrical shells, when the non-dimensional parameter $L(1-\nu^2)^{0.25}(Rh)^{-0.5}$ exceeds 10, the coefficient matrix of the algebraic equations, from which the missing initial conditions are solved, becomes too ill-conditioned, which may lead to the loss of accuracy or even completely incorrect results. Here the modified version of the shooting method, known as the Godunov-Conte method (Elishakoff and Charmats, 1977), is employed. It utilizes the Gram-Schmidt orthogonalization procedure during the integration steps to prevent the ill-conditioning problem so that more accurate results can be obtained than those furnished by the ordinary shooting method. For the detailed description of the Godunov-Conte method, consult Chapter 5.

2.1.5 Numerical Results and Discussion

The results for the classical buckling load reduction α from the preceding three methods are listed in Tables 2.1 and 2.2 for different values of the thickness variation parameter ϵ.

A very good match between the results from different methods is shown up to the value $\epsilon = 0.05$. The increasingly bigger difference is observed between the results of the first-order approximation given by Equation (2.27) [or Equation (2.31)] and those

Table 2.1. Comparison of buckling loads derived via different methods for Case A ($\nu = 0.3$)

	Asymptotic formula			
ϵ	Koiter's formula, Eq. (2.27)	Second-order approximation, Eq. (2.26)	Shooting method	Finite difference
0.0	1.0	1.0	1.0	1.0
0.01	0.999	0.998	0.999	0.998
0.05	0.993	0.988	0.989	0.988
0.10	0.985	0.966	0.967	0.966
0.15	0.975	0.935	0.939	0.938

Table 2.2. Comparison of buckling loads derived via different methods for Case B ($\nu = 0.3$)

	Asymptotic formula			
ϵ	Koiter's formula, Eq. (2.31)	Second-order approximation, Eq. (2.30)	Shooting method	Finite difference
0.0	1.0	1.0	1.0	1.0
0.01	0.990	0.990	0.990	0.990
0.05	0.950	0.948	0.949	0.948
0.10	0.900	0.892	0.895	0.894
0.15	0.850	0.832	0.837	0.836

of numerical solutions as ϵ becomes larger. Even though the first-order asymptotic approximate formula may not be sufficiently accurate as ϵ reaches 0.1, the second-order asymptotic formula (2.26) [or (2.30)] retains a good accuracy even for ϵ as big as 0.15. Thus, owing to its higher accuracy, Equations (2.26) and (2.30) can be used to obtain a sufficiently good estimate of the buckling load reduction factor due to the thickness variation.

These results also show that the effect of certain types of thickness variation on buckling load deserves special attention. Although the thickness variation pattern akin to the classical buckling mode (Case A) may have some effect on the classical buckling load (the classical buckling load is decreased by over 6% when $\epsilon = 0.15$), the most detrimental effect of thickness variation occurs when the wave number of axisymmetric thickness variation is twice that of the classical buckling mode (Case B). In this situation, even if the amplitude of the thickness variation is as small as 0.1, the thickness variation reduces the buckling load by 10% from its counterpart on the shell with constant thickness. When $\epsilon = 0.15$, the classical buckling load is decreased by over 15%. Thus, in the absence of initial geometric imperfection, this particular kind of thickness variation may constitute the most important factor in the buckling load reduction.

2.2 Buckling of an Axially Compressed Imperfect Cylindrical Shell of Variable Thickness

In 1963, Koiter investigated the effect of axisymmetric imperfections in the shape of the axisymmetric buckling mode on the buckling of cylindrical shells. No similar information seems to be available for the effect of axisymmetric thickness variations of type $h = h_0[1 - \epsilon \cos(px/R)]$, where the wave number p was expected (naively) to be most critical when it coincides with the wave number $p_0 = [2cR/h_0]^{1/2}$ of the axisymmetric buckling mode of the shell of uniform thickness h_0. Here we aim at providing the additional information on the effect of such thickness variations. In this section, we will follow the work by Koiter et al. (1994b).

The simplest approach to the problem is a direct discussion of the energy criterion of stability by means of the second variation of the potential energy. Comparable result are obtained by means of the linear equations of neutral equilibrium in terms of Airy stress functions F and the normal deflection w. The end conditions are largely ignored in the energy approach, but the conditions of simple support are fully taken into account in the solution of the equations of neutral equilibrium. Therefore, the latter results should be more accurate, even though the energy approach does not claim to be valid beyond the first order. Results from the asymptotic formulas are compared with those obtained through the purely numerical technique of the shooting method.

2.2.1 Direct Discussion of Energy Criterion

As discussed in Koiter et al. (1994b), the linear prebuckling state of axial compression for a cylindrical shell of constant thickness h_0 is characterized by a uniform axial stress resultant $N_x = -\lambda E h_0^2/cR$, where E is the Young's modulus and R is the radius of the shell; $c = [3(1-\nu^2)]^{1/2}$ (ν is the Poisson ratio), and the critical value of load factor λ is unity. The associated uniform outward radial deflection is $\nu\lambda h_0/c$. A deflection from the fundamental prebuckling state is described by the axial, circumferential, and radial components (u, v, w), the latter positive outward. We employ Flügge's notation of primes and dots for derivatives with respect to the axial and circumferential non-dimensional coordinates x/R and y/R. The second variation of the energy for shallow buckling modes is now given by (Koiter, 1945, 1980)

$$P_2[\underline{u}] = \int \frac{Eh_0 dS}{2(1-\nu^2)R^2}\Big[u'^2 + (v^{\cdot} + w)^2 + 2\nu u'(v^{\cdot} + w) + \frac{1}{2}(1-\nu)(u^{\cdot} + v')^2$$
$$+ \frac{h_0^2}{12R^2}\{(w'' + w^{\cdot\cdot})^2 + 2(1-\nu)[(w'^{\cdot})^2 - w''w^{\cdot\cdot}]\} - (1-\nu^2)\frac{\lambda h_0}{cR}(w')^2\Big] \quad (2.39)$$

where the integration is extended over the entire shell area $2\pi RL$. The equations of neutral equilibrium are obtained from the second variation by putting its first variation equal to zero. The resulting linear equations in u, v, and w are completely equivalent to the equations in terms of the Airy stress function and the normal deflection as discussed by Koiter (1963). To include the effect of mid-surface geometric imperfections of type $w_0 = -\mu h_0 \cos(p_0 x/R)$, where $p_0 = [2cR/h_0]^{1/2}$ is the wave number of the axisymmetric buckling mode, we supplement the second variation with the additional bilinear term

$$\lambda P'_{11}[\underline{u_0}, \underline{u}] = -\lambda \frac{Eh_0^2}{cR^3}\int w'_0 w' dS \quad (2.40)$$

We shall also need the third variation of the potential energy

$$P_3[\underline{u}] = \int \frac{Eh_0 dS}{2(1-\upsilon^2)R^3}\big[\{u' + \nu(v^{\cdot} + w)\}(w')^2$$
$$+ \{\nu u' + v^{\cdot} + w\}(w^{\cdot})^2 + (1-\nu)(\dot{u} + v')w'w^{\cdot}\big] \quad (2.41)$$

Cylindrical shells of adequate length L (say, $L/R > 1$) and with reasonably rigid support at the ends (say, simply supported or clamped ends) exhibit many simultaneous buckling modes in the interior at the critical load factor $\lambda = 1$. These modes are all sinusoidal in both directions with wave numbers p and q, connected by the equation $p^2 + q^2 = pp_0$, where $p_0 = [2cR/h_0]^{1/2}$ is a large number. In order to obtain a non-zero result for the third variation, we must retain the axisymmetric mode with wave number p_0 and one or more asymmetric modes such that the sum of positive or negative wave numbers is zero, both in the axial and circumferential directions. We select the asymmetric mode $p = q = m = \frac{1}{2}p_0$ at the top of the Koiter's semi-circle, resulting in the formulas (Koiter, 1980)

$$\begin{aligned} u/h_0 &= -\frac{\nu}{2m} b_0 \sin p_0 x/R + \frac{1-\nu}{4m} C_m \sin(mx/R)\cos(my/R), \\ v/h_0 &= -\frac{3+\nu}{4m} C_m \cos(mx/R)\sin(my/R) \\ w/h_0 &= b_0 \cos(p_0 x/R) + C_m \cos(mx/R)\cos(my/R) \end{aligned} \tag{2.42}$$

where C_m and b_0 are constants and m is the wave number of the buckling mode in the circumferential direction of the shell.

Substituting from u, v, and w into the sum of the second and third variations for the cylindrical shell of constant thickness and length L, the result is Equation (3.15) of Koiter (1980)

$$P_2[\underline{u}] + P_3[\underline{u}] = \frac{\pi}{8}\frac{ELh_0^3}{R}\left[(1-\lambda)\left(8b_0^2 + C_m^2\right) + \frac{3c}{2} b_0 C_m^2\right] \tag{2.43}$$

The additional term to allow for axisymmetric imperfections $w_0 = -\mu h_0 \cos(p_0 x/R)$ is given by

$$P_{11}'[\underline{u_0}, \underline{u}] = -\lambda \frac{Eh_0^2}{cR^3}\int w_0' w' dS = \frac{\pi}{8}\frac{ELh_0^3}{R} 16\lambda\mu b_0 \tag{2.44}$$

We turn now to the effect of small axisymmetric thickness variations described by

$$h = h_0[1 - \epsilon \cos(px/R)] \tag{2.45}$$

where $\epsilon > 0$ in order to achieve a detrimental effect by a "thinning" of the wall thickness in the region around $x = 0$ where the flexural energy dominates. The single essential change in the second variation of the energy is that the extensional rigidity now contains a factor $1 - \epsilon \cos(px/R)$, whereas the flexural rigidity contains a factor $[1 - \epsilon \cos(px/R)]^3 \approx 1 - 3\epsilon \cos(px/R)$, if we ignore higher than first-order corrections in ϵ. No change occurs in the last term of the second variation or in the additional term to allow for axisymmetric geometric imperfections of the mid-surface.

We now assume that the buckling modes of the shell with a uniform thickness remain a good approximation for the buckling modes of the shell with thickness variations. We are at least ensured that the critical load factor λ, obtained this way, is by the energy principle an upper bound for the actual critical load factor. The integrand of the second variation for shells of uniform thickness contains terms with a factor $\cos^2(mx/R)$ or $\sin^2(mx/R)$ as well as those with a factor $\cos^2(p_0/R)$. The additional terms due to the thickness variation all have an additional factor $\cos(px/R)$. For sufficiently long shells, say $L/R > 1$, the integrals of type

$$\int \cos^2(mx/R)\cos(px/R)\,dx, \qquad \int \cos^2(p_0x/R)\cos(px/R)\,dx$$

are all approximately zero, except for the cases $p = 2m = p_0$ and $p = 2p_0$, where the integrals have the value $L/4$. It is now a simple matter to evaluate the formulae for the second variation of shells with thickness variations of wave numbers $p = p_0$ or $p = 2p_0$

$$\text{Case } p = p_0: \qquad P_2[\underline{u}] = \frac{\pi}{8}\frac{ELh_0^3}{R}\left[8(1-\lambda)b_0^2 + \left(1 - \lambda - \frac{1}{2}\nu\epsilon\right)C_m^2\right] \tag{2.46}$$

$$\text{Case } p = 2p_0: \qquad P_2[\underline{u}] = \frac{\pi}{8}\frac{ELh_0^3}{R}[8(1-\lambda-\epsilon)b_0^2 + (1-\lambda)C_m^2] \tag{2.47}$$

In the absence of initial geometric imperfection, with the partial derivative of $P_2[\underline{u}]$ with respect to C_m equal to zero, we obtain

$$\text{Case } p = p_0: \qquad \lambda = 1 - \frac{1}{2}\nu\epsilon \tag{2.48}$$

$$\text{Case } p = 2p_0: \qquad \lambda = 1 - \epsilon \tag{2.49}$$

Thus, an important result of the present analysis is that axisymmetric thickness variations of wave numbers p_0 or $2p_0$ entail a *reduction of the critical load factor below unity* by $\frac{1}{2}\nu\epsilon$ in the case $p = p_0$ and by ϵ in the case $p = 2p_0$. The associated buckling modes are the non-symmetric mode $w = \cos(mx/R)\cos(my/R)$ in the case $p = p_0$ and the axisymmetric mode $w(x) = \cos(p_0x/R)$ in the case $p = 2p_0$.

It is now also a simple matter to discuss the combined effect of thickness variations of type, $h = h_0[1 - \epsilon\cos(px/R)]$, where $p = p_0$ or $p = 2p_0$, and the most critical type of asymmetrical geometric imperfections $w_0 = -\mu h_0 \cos(p_0x/R)$. For this purpose, we need the evaluation of the third variation $P_3[u]$, that is, Equation (2.41) with h_0 replaced by $h = h_0[1 - \epsilon\cos(px/R)]$. The result reads:

$$\text{Case } p = p_0: \qquad P_3[\underline{u}] = \frac{3\pi c}{16}\frac{ELh_0^3}{R}\left(1 - \frac{1}{3}\epsilon\right)b_0C_m^2 \tag{2.50}$$

$$\text{Case } p = 2p_0: \qquad P_3[\underline{u}] = \frac{3\pi c}{16}\frac{ELh_0^3}{R}\left(1 - \frac{1}{6}\epsilon\right)b_0C_m^2 \tag{2.51}$$

To a first approximation, we may ignore the factors $(1 - \epsilon/3)$ and $(1 + \epsilon/6)$ in the cubic terms. Leaving aside the dimensional factor $\pi ELh_0^3/8R$, the energy expression to be discussed is:

$$\text{Case } p = p_0: \qquad 8(1-\lambda)b_0^2 + \left(1 - \frac{1}{2}\nu\epsilon - \lambda\right)C_m^2 + 16\lambda\mu b_0 + \frac{3c}{2}b_0C_m^2 \tag{2.52}$$

$$\text{Case } p = 2p_0: \qquad 8(1-\epsilon-\lambda)b_0^2 + (1-\lambda)C_m^2 + 16\lambda\mu b_0 + \frac{3c}{2}b_0C_m^2 \tag{2.53}$$

The equations of equilibrium are obtained by putting the partial derivatives with respect to b_0 and C_m equal to zero. In case $p = p_0$, we have the equations

$$16(1-\lambda)b_0 + 16\lambda\mu + \frac{3c}{2}C_m^2 = 0 \tag{2.54}$$

$$\left(1 - \frac{1}{2}\nu\epsilon - \lambda\right)C_m + \frac{3c}{2}b_0 C_m = 0 \tag{2.55}$$

The solution $C_m = 0$ of the second equation leads to $b_0 = -\lambda\mu/(1-\lambda)$ from the first equation, and bifurcation buckling with respect to the asymmetric mode with amplitude C_m occurs at the value $b_0 = -\frac{2}{3}c(1 - \frac{1}{2}\nu\epsilon - \lambda)$.

The equation for the critical load factor is thus

$$(1-\lambda)\left(1 - \frac{1}{2}\nu\epsilon - \lambda\right) - \frac{3c}{2}\lambda\mu = 0 \tag{2.56}$$

In case $p = 2p_0$, we have the conditions of equilibrium obtained from (2.51)

$$16(1-\epsilon-\lambda)b_0 + 16\lambda\mu + \frac{3c}{2}C_m^2 = 0 \tag{2.57}$$

$$(1-\lambda)C_m + \frac{3c}{2}b_0 C_m = 0 \tag{2.58}$$

and the equation for the critical load factor is in this case

$$(1-\lambda)(1-\epsilon-\lambda) - \frac{3c}{2}\lambda\mu = 0 \tag{2.59}$$

The present analysis by means of the energy criterion already permits some significant conclusions. Periodic axisymmetric thickness variations may result in a fractional decrease of the critical load factor under axial compression, up to a fraction equal to ϵ defined by (2.45), achieved for a wave number $2p_0$, twice the wave number of the axisymmetric buckling mode of the uniform shell. It is a more or less unexpected result that the fractional decrease of the critical load factor in the case of thickness variations with the wave number of axisymmetric buckling mode is much smaller, namely $\nu\epsilon/2$. We are not surprised, however, that the reduction of the critical load factor of the order ϵ, due to thickness variations is far less detrimental than the similar reduction of order $\mu^{1/2}$, due to geometric imperfections of the mid-surface. This is also the justification a posteriori for our ignoring the factors $(1-\epsilon/3)$ and $(1+\epsilon/6)$ in Equations (2.50) and (2.51).

2.2.2 Numerical Technique

Now we will approach the problem from a different angle. Here we propose a combined analytical-numerical technique and closely follow the procedure by Koiter (1963), which deals with shells of constant thickness.

The governing equations for the buckling of the cylindrical shell with variable thickness and axisymmetric initial imperfections, under axial compression read

$$h^2\nabla^2\nabla^2 F - Eh^3\left[\left(\frac{\partial^2 W}{\partial x\partial y}\right)^2 - \frac{\partial^2 W}{\partial x^2}\frac{\partial^2 W}{\partial y^2} - \frac{d^2 W_0}{dx^2}\frac{\partial^2 W}{\partial y^2} + \frac{1}{R}\frac{\partial^2 W}{\partial x^2}\right]$$
$$+2\left(\frac{dh}{dx}\right)^2\left(\frac{\partial^2 F}{\partial x^2} - \nu\frac{\partial^2 F}{\partial y^2}\right) - h\frac{d^2 h}{dx^2}\left(\frac{\partial^2 F}{\partial x^2} - \nu\frac{\partial^2 F}{\partial y^2}\right)$$
$$-2h\frac{dh}{dx}\left(\frac{\partial^3 F}{\partial x^3} - \nu\frac{\partial^3 F}{\partial x\partial y^2}\right) - 2(1+\nu)h\frac{dh}{dx}\frac{\partial^3 F}{\partial x\partial y^2} = 0 \qquad (2.60)$$

$$\frac{Eh^3}{12(1-\nu^2)}\nabla^2\nabla^2 W - \frac{\partial^2 F}{\partial y^2}\frac{\partial^2 W}{\partial x^2} - \frac{\partial^2 F}{\partial x^2}\frac{\partial^2 W}{\partial y^2} + 2\frac{\partial^2 F}{\partial x\partial y}\frac{\partial^2 W}{\partial x\partial y} + \frac{1}{R}\frac{\partial^2 F}{\partial x^2}$$
$$-\frac{d^2 W_0}{dx^2}\frac{\partial^2 F}{\partial y^2} + \frac{3Eh^2}{12(1-\nu^2)}\frac{dh}{dx}\frac{\partial}{\partial x}\nabla^2 W + \frac{3Eh^2}{12(1+\nu)}\frac{dh}{dx}\frac{\partial^3 W}{\partial x\partial y^2}$$
$$+\frac{3Eh^2}{12(1-\nu^2)}\frac{dh}{dx}\left(\frac{\partial^3 W}{\partial x^3} + \nu\frac{\partial^3 W}{\partial x\partial y^2}\right) + \frac{6Eh}{12(1-\nu^2)}\left(\frac{dh}{dx}\right)^2\left(\frac{\partial^2 W}{\partial x^2} + \nu\frac{\partial^2 W}{\partial y^2}\right)$$
$$+\frac{3Eh^2}{12(1-\nu^2)}\frac{d^2 h}{dx^2}\left(\frac{\partial^2 W}{\partial x^2} + \nu\frac{\partial^2 W}{\partial y^2}\right) = 0 \qquad (2.61)$$

where F is the Airy stress function defined as

$$N_x = \frac{\partial^2 F}{\partial y^2}; \qquad N_y = \frac{\partial^2 F}{\partial x^2}; \qquad N_{xy} = -\frac{\partial^2 F}{\partial x\partial y} \qquad (2.62)$$

With the non-dimensional notations

$$\xi = \frac{x}{L}, \qquad \eta = \frac{y}{L}, \qquad w = \frac{W}{h_0}, \qquad f = \frac{F}{D_0}, \qquad H = \frac{h}{h_0} \qquad (2.63)$$

where $D_0 = Eh_0^3/[12(1-\nu^2)]$ is the flexural rigidity of the shell with nominal thickness h_0, the governing equations can be expressed in non-dimensional forms as

$$H^3\bar{\nabla}^2\bar{\nabla}^2 W - \frac{\partial^2 F}{\partial \eta^2}\frac{\partial^2 W}{\partial \xi^2} - \frac{\partial^2 F}{\partial \xi^2}\frac{\partial^2 W}{\partial \eta^2} + 2\frac{\partial^2 F}{\partial \xi\partial \eta}\frac{\partial^2 W}{\partial \xi\partial \eta} + \frac{L^2}{Rh_0}\frac{\partial^2 F}{\partial \xi^2}$$
$$-\frac{d^2 W_0}{\partial \xi^2}\frac{\partial^2 F}{\partial \eta^2} + 3H^2\frac{dH}{d\xi}\frac{\partial}{\partial \xi}\bar{\nabla}^2 W + 3(1-\nu)H^2\frac{dH}{d\xi}\frac{\partial^3 W}{\partial \xi\partial \eta^2}$$
$$+3H^2\frac{dH}{d\xi}\left(\frac{\partial^3 W}{\partial \xi^3} + \nu\frac{\partial^3 W}{\partial \xi\partial \eta^2}\right) + 6H\left(\frac{dH}{d\xi}\right)^2\left(\frac{\partial^2 W}{\partial \xi^2} + \nu\frac{\partial^2 W}{\partial \eta^2}\right)$$
$$+3H^2\frac{d^2 H}{d\xi^2}\left(\frac{\partial^2 W}{\partial \xi^2} + \nu\frac{\partial^2 W}{\partial \eta^2}\right) = 0 \qquad (2.64)$$

$$
\begin{aligned}
&H^2\bar{\nabla}^2\bar{\nabla}^2 F + 2\left(\frac{dH}{d\xi}\right)^2\left(\frac{\partial^2 F}{d\xi^2} - \nu\frac{\partial^2 F}{\partial\eta^2}\right) - H\frac{d^2 H}{d\xi^2}\left(\frac{\partial^2 F}{\partial\xi^2} - \nu\frac{\partial^2 F}{\partial\eta^2}\right)\\
&\quad - 2H\frac{dH}{d\xi}\left(\frac{\partial^2 F}{\partial\xi^3} - \nu\frac{\partial^3 F}{\partial\xi\partial\eta^2}\right) - 2(1+\nu)H\frac{dH}{d\xi}\frac{\partial^3 F}{\partial\xi\partial\eta^2}\\
&\quad - 12(1-\nu^2)H^3\left[\left(\frac{\partial^2 W}{\partial\xi\partial\eta}\right)^2 - \frac{\partial^2 W}{\partial\xi^2}\frac{\partial^2 W}{\partial\eta^2} - \frac{d^2 W_0}{d\xi^2}\frac{\partial^2 W}{\partial\eta^2} + \frac{L^2}{Rh_0}\frac{\partial^2 W}{\partial\xi^2}\right] = 0
\end{aligned}
\tag{2.65}
$$

where the non-dimensional Laplace operator reads

$$
\bar{\nabla}^2 = \frac{\partial^2}{\partial\xi^2} + \frac{\partial^2}{\partial\eta^2}
$$

We let

$$
W = W_I + W_{II}, \qquad F = F_I + F_{II} \tag{2.66}
$$

where W_I, F_I represent the non-dimensional prebuckling solutions, and W_{II} and F_{II} represent non-dimensional small increments at buckling. A direct substitution into Equations (2.64) and (2.65) and deletion of products of the small increments yields a set of non-linear governing equations for the prebuckling quantities

$$
\begin{aligned}
&H^3\bar{\nabla}^2\bar{\nabla}^2 W_I - \frac{\partial^2 F_I}{\partial\eta^2}\frac{\partial^2 W_I}{\partial\xi^2} - \frac{\partial^2 F_I}{\partial\xi^2}\frac{\partial^2 W_I}{\partial\eta^2} + 2\frac{\partial^2 F_I}{\partial\xi\partial\eta}\frac{\partial^2 W_I}{\partial\xi\partial\eta} + \frac{L^2}{Rh_0}\frac{\partial^2 F_I}{\partial\xi^2}\\
&\quad - \frac{d^2 W_0}{d\xi^2}\frac{\partial^2 F_I}{\partial\eta^2} + 3H^2\frac{dH}{d\xi}\frac{\partial}{\partial\xi}\bar{\nabla}^2 W_I + 3(1-\nu)H^2\frac{dH}{d\xi}\frac{\partial^3 W_I}{\partial\xi\partial\eta^2}\\
&\quad + 3H^2\frac{dH}{d\xi}\left(\frac{\partial^3 W_I}{\partial\xi^3} + \nu\frac{\partial^3 W_I}{\partial\xi\partial\eta^2}\right) + 6H\left(\frac{dH}{\partial\xi}\right)^2\left(\frac{\partial^2 W_I}{\partial\xi^2} + \nu\frac{\partial^2 W_I}{\partial\eta^2}\right)\\
&\quad + 3H^2\frac{d^2 H}{d\xi^2}\left(\frac{\partial^2 W_I}{\partial\xi^2} + \nu\frac{\partial^2 W_I}{\partial\eta^2}\right) = 0
\end{aligned}
\tag{2.67}
$$

$$
\begin{aligned}
&H^2\bar{\nabla}^2\bar{\nabla}^2 F_I + 2\left(\frac{dH}{d\xi}\right)^2\left(\frac{\partial^2 F_I}{\partial\xi^2} - \nu\frac{\partial^2 F_I}{\partial\eta^2}\right) - H\frac{d^2 H}{d\xi^2}\left(\frac{\partial^2 F_I}{d\xi^2} - \nu\frac{\partial^2 F_I}{\partial\eta^2}\right)\\
&\quad - 2H\frac{dH}{d\xi}\left(\frac{\partial^3 F_I}{\partial\xi^3} - \nu\frac{\partial^3 F_I}{\partial\xi\partial\eta^2}\right) - 2(1+\nu)H\frac{dH}{d\xi}\frac{\partial^3 F_I}{\partial\xi\partial\eta^2}\\
&\quad - 12(1-\nu^2)H^3\left[\left(\frac{\partial^2 W_I}{\partial\xi\partial\eta}\right)^2 - \frac{\partial^2 W_I}{\partial\xi^2}\frac{\partial^2 W_I}{\partial\eta^2} - \frac{d^2 W_0}{\partial\xi^2}\frac{\partial^2 W_I}{\partial\eta^2} + \frac{L^2}{Rh_0}\frac{\partial^2 W_I}{\partial\xi^2}\right] = 0
\end{aligned}
\tag{2.68}
$$

and a set of linearized equations governing the increments at buckling

$$\begin{aligned}
&H^3\bar{\nabla}^2\bar{\nabla}^2 W_{II} - \frac{\partial^2 F_I}{\partial\eta^2}\frac{\partial^2 W_{II}}{\partial\xi^2} - \frac{\partial^2 W_I}{\partial\xi^2}\frac{\partial^2 F_{II}}{\partial\eta^2} - \frac{\partial^2 F_I}{\partial\xi^2}\frac{\partial^2 W_{II}}{\partial\eta^2} - \frac{\partial^2 W_I}{\partial\eta^2}\frac{\partial^2 F_{II}}{\partial\xi^2}\\
&+ 2\frac{\partial^2 F_I}{\partial\xi\partial\eta}\frac{\partial^2 W_{II}}{\partial\xi\partial\eta} + 2\frac{\partial^2 W_I}{\partial\xi\partial\eta}\frac{\partial^2 F_{II}}{\partial\xi\partial\eta} + \frac{L^2}{Rh_0}\frac{\partial^2 F_{II}}{\partial\xi^2} - \frac{d^2 W_0}{d\xi^2}\frac{\partial^2 F_{II}}{\partial\eta^2}\\
&+ 3H^2\frac{dH}{d\xi}\frac{\partial}{\partial\xi}\bar{\nabla}^2 W_{II} + 3(1-\nu)H^2\frac{dH}{d\xi}\frac{\partial^3 W_{II}}{\partial\xi\partial\eta^2}\\
&+ 3H^2\frac{dH}{d\xi}\left(\frac{\partial^3 W_{II}}{\partial\xi^3} + \nu\frac{\partial^3 W_{II}}{\partial\xi\partial\eta^2}\right) + 6H\left(\frac{dH}{d\xi}\right)^2\left(\frac{\partial^2 W_{II}}{\partial\xi^2} + \nu\frac{\partial^2 W_{II}}{\partial\eta^2}\right)\\
&+ 3H^2\frac{d^2 H}{d\xi^2}\left(\frac{\partial^2 W_{II}}{\partial\xi^2} + \nu\frac{\partial^2 W_{II}}{\partial\eta^2}\right) = 0
\end{aligned} \tag{2.69}$$

$$\begin{aligned}
&H^2\bar{\nabla}^2\bar{\nabla}^2 F_{II} + 2\left(\frac{dH}{d\xi}\right)^2\left(\frac{\partial^2 F_{II}}{\partial\xi^2} - \nu\frac{\partial^2 F_{II}}{\partial\eta^2}\right) - H\frac{d^2 H}{d\xi^2}\left(\frac{\partial^2 F_{II}}{\partial\xi^2} - \nu\frac{\partial^2 F_{II}}{\partial\eta^2}\right)\\
&- 2H\frac{dH}{d\xi}\left(\frac{\partial^3 F_{II}}{\partial\xi^3} - \nu\frac{\partial^3 F_{II}}{\partial\xi\partial\eta^2}\right) - 2(1+\nu)H\frac{dH}{d\xi}\frac{\partial^3 F_{II}}{\partial\xi\partial\eta^2}\\
&- 12(1-\nu^2)H^3\left[2\frac{\partial^2 W_I}{\partial\xi\partial\eta}\frac{\partial^2 W_{II}}{\partial\xi\partial\eta} - \frac{\partial^2 W_I}{\partial\xi^2}\frac{\partial^2 W_{II}}{\partial\eta^2} - \frac{\partial^2 W_I}{\partial\eta^2}\frac{\partial^2 W_{II}}{\partial\xi^2}\right.\\
&\left. - \frac{d^2 W_0}{d\xi^2}\frac{\partial^2 W_{II}}{\partial\eta^2} + \frac{L^2}{Rh_0}\frac{\partial^2 W_{II}}{\partial\xi^2}\right] = 0
\end{aligned} \tag{2.70}$$

Let us first perform the prebuckling analysis. When the shell is subjected to a uniform axial compression and both the thickness variation and the initial geometric imperfection are axisymmetric, there exists an axisymmetric prebuckling state. This axisymmetric prebuckling state can be described as

$$F_I(\xi,\eta) = -\frac{1}{2}\bar{N}_0\eta^2 + F_I^*(\xi), \qquad W_I(\xi,\eta) = W_I^*(\xi) \tag{2.71}$$

where

$$\bar{N}_0 = \frac{N_0 L^2}{D_0} \tag{2.72}$$

Here N_0 is the axial compressive stress resultant.

Substitution of Equation (2.71) into the prebuckling governing equations (2.67) and (2.68) yields

$$\begin{aligned}
&H^3\frac{d^4 W_I^*}{d\xi^4} + \bar{N}_0\frac{d^2 W_I^*}{d\xi^2} + \frac{L^2}{Rh_0}\frac{d^2 F_I^*}{d\xi^2} + \bar{N}_0\frac{d^2 W_0}{d\xi^2} + 6H^2\frac{dH}{d\xi}\frac{d^3 W_I^*}{d\xi^3}\\
&+ 6H\left(\frac{dH}{d\xi}\right)^2\frac{d^2 W_I^*}{d\xi^2} + 3H^2\frac{d^2 H}{d\xi^2}\frac{d^2 W_I^*}{d\xi^2} = 0
\end{aligned} \tag{2.73}$$

$$\begin{aligned}
&H^2\frac{d^4 F_I^*}{d\xi^4} + 2\left(\frac{dH}{d\xi}\right)^2\left(\frac{d^2 F_I^*}{d\xi^2} + \nu\bar{N}_0\right) - H\frac{d^2 H}{d\xi^2}\left(\frac{d^2 F_I^*}{d\xi^2} + \nu\bar{N}_0\right)\\
&- 2H\frac{dH}{d\xi}\frac{d^3 F_I^*}{d\xi^3} - \frac{12(1-\nu^2)L^2}{Rh_0}H^3\frac{d^2 W_I^*}{d\xi^2} = 0
\end{aligned} \tag{2.74}$$

In view of the thickness variation, the prebuckling terms W_I^* and F_I^* are further assumed to have the expression as follows:

$$W_I^* = W_{I,K}^* + \epsilon W_{I,T}^*$$
$$F_I^* = F_{I,K}^* + \epsilon F_{I,T}^* \tag{2.75}$$

The parameter ϵ is exactly the same one appearing in the thickness variation pattern of Equation (2.45). This means that, when thickness is a constant (i.e., $\epsilon = 0$), the second terms in Equation (2.75) vanish. Notice that the terms with subscript 'K' describe the situation of shells of constant thickness (dealt with by Koiter, 1963) and the terms with subscript 'T' are the additional terms due to the thickness variation. Substituting Equation (2.75) into Equations (2.73) and (2.74), and regrouping in terms of ϵ, we obtain the differential equations for the terms with subscripts 'K' and 'T', respectively,

$$\frac{d^4 W_{I,K}^*}{d\xi^4} + \frac{L^2}{Rh_0}\frac{d^2 F_{I,K}^*}{d\xi^2} + \bar{N}_0\left(\frac{d^2 W_{I,K}^*}{d\xi^2} + \frac{d^2 W_0}{d\xi^2}\right) = 0 \tag{2.76}$$

$$\frac{d^4 F_{I,K}^*}{d\xi^4} - \frac{12(1-\nu^2)L^2}{Rh_0}\frac{d^2 W_{I,K}^*}{d\xi^2} = 0 \tag{2.77}$$

and

$$\begin{aligned}
&\epsilon H^3 \frac{d^4 W_{I,T}^*}{d\xi^4} + 6\epsilon H^2 \frac{dH}{d\xi}\frac{d^3 W_{I,T}^*}{d\xi^3} + \epsilon\left[6H\left(\frac{dH}{d\xi}\right)^2 + 3H^2\frac{d^2 H}{d\xi^2} + \bar{N}_0\right]\frac{d^2 W_{I,T}^*}{d\xi^2} \\
&+ \epsilon\frac{L^2}{Rh_0}\frac{d^2 F_{I,T}^*}{d\xi^2} - \left(3\epsilon\cos\frac{pL\xi}{R} - 3\epsilon^2\cos^2\frac{pL\xi}{R} + \epsilon^3\cos^3\frac{pL\xi}{R}\right)\frac{d^4 W_{I,K}^*}{d\xi^4} \\
&+ 6H^2\frac{dH}{d\xi}\frac{d^3 W_{I,K}^*}{d\xi^3} + \left[6H\left(\frac{dH}{d\xi}\right)^2 + 3H^2\frac{d^2 H}{d\xi^2}\right]\frac{d^2 W_{I,K}^*}{d\xi^2} = 0
\end{aligned} \tag{2.78}$$

$$\begin{aligned}
&\epsilon H^2\frac{d^4 F_{I,T}^*}{d\xi^4} - 2\epsilon H\frac{dH}{d\xi}\frac{d^3 F_{I,T}^*}{d\xi^3} - \epsilon\left[H\frac{d^2 H}{d\xi^2} - 2\left(\frac{dH}{d\xi}\right)^2\right]\frac{d^2 F_{I,T}^*}{d\xi^2} \\
&- \epsilon\frac{12(1-\nu^2)L^2}{Rh_0}H^3\frac{d^2 W_{I,T}^*}{d\xi^2} + \frac{12(1-\nu^2)L^2}{Rh_0}\left(3\epsilon\cos\frac{pL\xi}{R} - 3\epsilon^2\cos^2\frac{pL\xi}{R}\right. \\
&\left. + \epsilon^3\cos^3\frac{pL\xi}{R}\right)\frac{d^2 W_{I,K}^*}{d\xi^2} + \left(-2\epsilon\cos\frac{pL\xi}{R} + \epsilon^2\cos^2\frac{pL\xi}{R}\right)\frac{d^4 F_{I,K}^*}{d\xi^4} \\
&- 2H\frac{dH}{d\xi}\frac{d^3 F_{I,K}^*}{d\xi^3} - \left[H\frac{d^2 H}{d\xi^2} - 2\left(\frac{dH}{d\xi}\right)^2\right]\frac{d^2 F_{I,K}^*}{d\xi^2} \\
&- \nu\bar{N}_0\left[H\frac{d^2 H}{d\xi^2} - 2\left(\frac{dH}{d\xi}\right)^2\right] = 0
\end{aligned} \tag{2.79}$$

Equations (2.76) and (2.77) admit the following solution:

$$F^*_{I,K} = 4\mu c \frac{\lambda}{4\rho^4 + 1 - 4\lambda\rho^2} \cos \frac{p_0 L \xi}{R} \tag{2.80}$$

$$W_{I,K} = \frac{\nu}{12(1-\nu^2)} \frac{Rh_0}{L^2} \bar{N}_0 - \mu \frac{4\lambda\rho^2}{4\rho^4 + 1 - 4\lambda\rho^2} \cos \frac{p_0 L \xi}{R} \tag{2.81}$$

where $\lambda = \bar{N}_0 R h_0 / 4cL^2$ and $\rho^2 = h_0 p_0^2 / 4cR$.

Solutions given by Equations (2.81) and (2.82) are identical to the prebuckling solutions given by Koiter (1963). Here they are presented in a non-dimensional form.

The terms associated with the thickness variation are obtained by solving numerically Equations (2.78) and (2.79) with proper boundary conditions. Here the Godunov-Conte method (Godunov, 1961; Roberts and Shipman, 1972; Conte, 1966; Elishakoff and Charmats, 1977) is employed. The Godunov-Conte method is described in detail in Chapter 6. The Godunov-Conte method is a modified shooting method, using the Gram-Schmidt orthogonalization procedure during the integration steps to prevent the ill-conditioning of the coefficient matrix of the algebraic equation, from which the missing initial conditions are solved. It is worth mentioning that the Gram-Schmidt orthogonalization procedure is essential for many shells encountered in engineering practice. According to Conte (1966), when the nondimensional parameter $L(1-\nu^2)^{0.25}(Rh)^{-0.5}$ exceeds 10, the problem of ill-conditioning is bound to occur in the buckling analysis of cylindrical shells.

The boundary conditions for the terms with subscript T are taken as follows:

$$\begin{aligned} &F^*_{I,T} = \frac{d^2 F^*_{I,T}}{d\xi^2} = W^*_{I,T} = \frac{d^2 W^*_{I,T}}{d\xi^2} = 0 \\ &\xi = \pm\frac{1}{2} \end{aligned} \tag{2.82}$$

which satisfy the simply supported boundary conditions at the ends of the shell.

Now, let us perform the buckling analysis. As in the prebuckling state treatment, the terms of F^*_{II} and W^*_{II} are assumed to be composed of two parts, respectively,

$$\begin{aligned} W^*_{II} &= W^*_{II,K} + \epsilon W^*_{II,T} \\ F^*_{II} &= F^*_{II,K} + \epsilon F^*_{II,T} \end{aligned} \tag{2.83}$$

Substitution of Equation (1.83) into Equation (1.70) and again regrouping according to the parameter ϵ leads to

$$\begin{aligned} &\bar{\nabla}^2\bar{\nabla}^2 F^*_{II,K} + 12(1-\nu^2)\frac{\partial^2 W^*_{I,K}}{\partial\xi^2}\frac{\partial^2 W^*_{II,K}}{\partial\eta^2} + 12(1-\nu^2)\frac{d^2 W_0}{d\xi^2}\frac{d^2 W^*_{II,K}}{\partial\eta^2} \\ &\quad - \frac{12(1-\nu^2)L^2}{Rh_0}\frac{\partial^2 W^*_{II,K}}{\partial\xi^2} = 0 \end{aligned} \tag{2.84}$$

and

$$\epsilon H^2\bar{\nabla}^2\bar{\nabla}^2 F^*_{II,T} + 2\left(\frac{dH}{d\xi}\right)^2\left(\frac{\partial^2 F^*_{II,K}}{d\xi^2} + \epsilon\frac{\partial^2 F^*_{II,T}}{\partial^2\xi^2} - \nu\frac{\partial^2 F^*_{II,K}}{\partial\eta^2} - \nu\epsilon\frac{\partial^2 F^*_{II,T}}{\partial\eta^2}\right)$$

$$- H\frac{d^2 H}{d\xi^2}\left(\frac{\partial^2 F^*_{II,K}}{\partial\xi^2} + \epsilon\frac{\partial^2 F^*_{II,T}}{\partial\xi^2} - \nu\frac{\partial^2 F^*_{II,K}}{\partial\eta^2} - \nu\epsilon\frac{\partial^2 F^*_{II,T}}{\partial\eta^2}\right)$$

$$- 2H\frac{dH}{d\xi}\left(\frac{\partial^3 F^*_{II,K}}{d\xi^3} + \epsilon\frac{\partial^3 F^*_{II,T}}{\partial\xi^3} - \nu\frac{\partial^3 F^*_{II,K}}{\partial\xi\partial\eta^2} - \epsilon\nu\frac{\partial^3 F^*_{II,T}}{\partial\xi\partial\eta^2}\right)$$

$$- 2(1+\nu)H\frac{dH}{d\xi}\left(\frac{\partial^3 F^*_{II,K}}{\partial\xi\partial\eta^2} + \epsilon\frac{\partial^3 F^*_{II,T}}{\partial\xi\partial\eta^2}\right)$$

$$+ 12(1-\nu^2)\epsilon H^3\left(\frac{\partial^2 W^*_{I,K}}{\partial\xi^2}\frac{\partial^2 W^*_{II,T}}{\partial\eta^2} + \frac{\partial^2 W^*_{I,T}}{\partial\xi^2}\frac{\partial^2 W^*_{II,K}}{\partial\eta^2}\right.$$

$$\left. + \epsilon\frac{\partial^2 W^*_{I,T}}{\partial\xi^2}\frac{\partial^2 W^*_{II,T}}{\partial\eta^2} + \frac{d^2 W_0}{d\xi^2}\frac{\partial^2 W^*_{II,T}}{\partial\eta^2} - \frac{L^2}{Rh_0}\frac{\partial^2 W^*_{II,T}}{\partial\xi^2}\right)$$

$$- 12(1-\nu^2)\epsilon H^2\cos\frac{pL\xi}{R}\left(\frac{\partial^2 W^*_{I,K}}{\partial\xi^2}\frac{\partial^2 W^*_{II,K}}{\partial\eta^2}\right.$$

$$\left. + \frac{d^2 W_0}{d\xi^2}\frac{\partial^2 W^*_{II,K}}{\partial\eta^2} - \frac{L^2}{Rh_0}\frac{\partial^2 W^*_{II,K}}{\partial\xi^2}\right) = 0 \tag{2.85}$$

Again, as in the prebuckling analysis, the terms $W^*_{II,K}$ and $F^*_{II,K}$ follow from the analysis (Koiter, 1963) for imperfect shells of constant thickness, whereas terms $W^*_{II,T}$ and $F^*_{II,T}$ are associated solely with thickness variation.

Equation (2.84) admits a solution of the following form:

$$W^*_{II,K} = C_1\cos\frac{pL\xi}{R}\cos\frac{mL\eta}{R}$$

$$F^*_{II,K} = C_1\left(\frac{8\mu\tau^2\rho^2 tc^2}{(9\rho^2+\tau^2)^2}\cos\frac{3pL\xi}{R} + \frac{8\mu c^2\tau^2\rho^2 t - 4\rho^2 c}{(\rho^2+\tau^2)^2}\cos\frac{pL\xi}{R}\right)\cos\frac{mL\eta}{R} \tag{2.86}$$

where C_1 is an arbitrary constant, and $\tau^2 = h_0 n^2/Rc$, $t = (1-\lambda)^{-1}$.

The solution defined by Equation (2.86) satisfies the boundary conditions for simply supported edges at the shell ends $\xi = \pm\frac{1}{2}$.

Moreover, an examination of Equation (2.85) reveals that it is subject to separation of variables of ξ and η. Thus, we assume

$$W^*_{II,T} = C_1\cos\frac{pL\xi}{R}\cos\frac{mL\eta}{R}$$
$$F^*_{II,T} = C_1\Phi(\xi)\cos\frac{mL\eta}{R} \tag{2.87}$$

Here $\Phi(\xi)$ is an undetermined function of ξ only.

Substitution of Equation (2.87) into Equation (2.85) leads to

$$
\begin{aligned}
&\epsilon H^2\Phi^{(iv)} - 2\epsilon H\frac{dH}{d\xi}\Phi''' + \epsilon H^2\left[-2\frac{n^2L^2}{R^2} + \frac{2}{H^2}\left(\frac{dH}{d\xi}\right)^2 - \frac{1}{H}\frac{d^2H}{d\xi^2}\right]\Phi'' \\
&+ 2\epsilon H\frac{dH}{d\xi}\frac{n^2L^2}{R^2}\Phi' + \epsilon H^2\left[\frac{n^4L^4}{R^4} + \frac{2\nu}{H^2}\left(\frac{dH}{d\xi}\right)^2\frac{n^2L^2}{R^2} - \frac{\nu}{H}\frac{d^2H}{d\xi^2}\frac{n^2L^2}{R^2}\right]\Phi \\
&+ \left[2\left(\frac{dH}{d\xi}\right)^2 - H\frac{d^2H}{d\xi^2}\right]\left[\frac{8\mu\tau^2\rho^2tc^2}{(9\rho^2+\tau^2)^2}\left(\frac{-9p^2L^2}{R^2} + \nu\frac{n^2L^2}{R^2}\right)\cos\frac{3pL\xi}{R}\right. \\
&\left.+ \frac{8\mu c^2\tau^2\rho^2t - 4c\rho^2}{(\rho^2+\tau^2)^2}\left(-\frac{p^2L^2}{R^2} + \nu\frac{n^2L^2}{R^2}\right)\cos\frac{pL\xi}{R}\right] \\
&- 2H\frac{dH}{d\xi}\left[\frac{8\mu\tau^2\rho^2\tau c^2}{(9\rho^2+\tau^2)^2}\left(\frac{27p^3L^3}{R^3} - \nu\frac{3pn^2L^3}{R^3}\right)\sin\frac{3pL\xi}{R}\right. \\
&\left.+ \frac{8\mu c^2\tau^2\rho^2t - 4c\rho^2}{(\rho^2+\tau^2)^2}\left(\frac{p^3L^3}{R^3} - \nu\frac{pn^2L^3}{R^3}\right)\sin\frac{pL\xi}{R}\right] - 2(1+\nu)H\frac{dH}{d\xi} \\
&\times\left[\frac{24\mu c^2\tau^2\rho^2t}{(9\rho^2+\tau^2)^2}\sin\frac{3pL\xi}{R} + \frac{8\mu c^2\tau^2\rho^2t - 4c\rho^2}{(\rho^2+\tau^2)^2}\sin\frac{pL\xi}{R}\right]\frac{pn^2L^3}{R^3} \\
&+ 12(1-\nu^2)\epsilon H^3\left[-\mu(t-1)\frac{4p^2n^2L^4}{R^4}\cos\frac{2pL\xi}{R}\cos\frac{pL\xi}{R}\right. \\
&- W_{I,T}^{*\prime\prime}\frac{n^2L^2}{R^2}\cos\frac{pL\xi}{R} - \epsilon W_{I,T}^{*\prime\prime}\frac{n^2L^2}{R^2}\cos\frac{pL}{R}\xi - \frac{d^2W_0}{d\xi^2}\frac{n^2L^2}{R^2}\cos\frac{pL\xi}{R} \\
&\left.+ \frac{L^2}{Rh_0}\frac{p^2L^2}{R^2}\cos\frac{pL\xi}{R}\right] + 12(1-\nu^2)\epsilon H^2\cos\frac{pL\xi}{R} \\
&\times\left[-\mu(t-1)\frac{4p^2n^2L^4}{R^4}\cos\frac{2pL\xi}{R}\cos\frac{pL\xi}{R}\right. \\
&\left.- \frac{d^2W_0}{d\xi^2}\frac{n^2L^2}{R^2}\cos\frac{pL\xi}{R} + \frac{L^2}{Rh_0}\frac{p^2L^2}{R^2}\cos\frac{pL\xi}{R}\right] = 0
\end{aligned}
\tag{2.88}
$$

Since the second derivative $W_{I,T}^{*\prime\prime}$ is known numerically through the prebuckling analysis, Equation (2.88), representing a fourth-order ordinary differential equation with variable coefficients, can be solved for $\Phi(\xi)$ again by the Godunov-Conte method with the boundary conditions

$$\Phi(\xi) = \Phi''(\xi) = 0 \qquad \text{at} \qquad \xi = \pm\frac{1}{2} \tag{2.89}$$

Taking into account the numerically determined Airy's stress function $\Phi(\xi)$, substituting Equations (2.86) and (2.82) into Equation (2.69) and applying the Boobnov-Galerkin procedure, namely, multiplying each term of the equation by $\cos(pL\xi/R)$ and integrating over the shell length, we arrive at

$$
\begin{aligned}
&(\rho^2+\tau^2)^2 - 4\lambda\rho^2 + 4\rho^4(\rho^2+\tau^2)^{-2} + 16(c\mu)^2t^2\rho^4\tau^4[(9\rho^2+\tau^2)^{-2} \\
&\quad + (\rho^2+\tau^2)^{-2}] - 2c\mu\tau^2[t-1+8t\rho^4(\rho^2+\tau^2)^{-2}] - \Big\{\epsilon(\rho^2+\tau^2)^2 \\
&\quad - 4\alpha\lambda\rho^2\epsilon + 8\epsilon\rho^2\tau^2\mu(t-1)I_1 + 8\epsilon\frac{Rh_0}{L^2}\tau^2\left[\frac{2c\mu\tau^2\rho^2 t}{(9\rho^2+\tau^2)^2}I_2\right. \\
&\quad \left. + \frac{2\mu c\tau^2\rho^2 t-\rho^2}{(\rho^2+\tau^2)^2}I_3\right] + \frac{2\epsilon^2 Rh_0\tau^2}{cL^2}I_4 + \frac{2\epsilon Rh_0}{c^2L^2}I_5 + 8\epsilon\mu\rho^2\tau^2 I_1 \\
&\quad + 6(1+\epsilon)\left(\frac{h_0}{cR}\right)^{1/2}(\rho^2+\tau^2)\epsilon p I_6 + 6(1-\nu)(1+\epsilon)\left(\frac{h_0}{cR}\right)^{1/2}\rho\tau^2\epsilon p I_6 \\
&\quad + 6(1+\epsilon)\left(\frac{h_0}{cR}\right)^{1/2}(\rho^2+\nu\tau^2)\rho\epsilon p I_6 + 24(1+\epsilon)\frac{h_0}{cR}(\rho^2+\nu\tau^2)p^2\epsilon^2 I_7 \\
&\quad + 2\epsilon c\mu(t-1)\tau^2 - 2\epsilon(1+\epsilon)\frac{Rh_0}{cL^2}\tau^2 I_8 + 12(1+\epsilon)\frac{h_0}{cR}(\rho^2+\nu\tau^2)p^2\epsilon^2 I_9 \\
&\quad - 2(1+\epsilon)(\rho^4+2\tau^2\rho^2+\tau^4)(3\epsilon I_{10}+3\epsilon^2 I_{11}+\epsilon^3 I_{12})\Big\} = 0 \qquad (2.90)
\end{aligned}
$$

where I_i ($i = 1 \sim 12$) are integrals listed in the paper by Koiter et al. (1994b).

When the thickness is a constant, all the terms in the curved brackets in Equation (2.90) vanish; the reduced equation coincides with Equation (5.2) in the study by Koiter (1963). The smallest solution $\bar{N}_0$ of this algebraic equation is the critical buckling load, or more precisely the upper bound of the critical buckling load, of the imperfect shell of the variable thickness.

Since the critical buckling load is in need at the very beginning for the solution of $W^*_{I,T}$, $F^*_{I,T}$, and $F^*_{II,T}$ ($W^*_{II,T}$ is assumed co-configurational to $W^*_{II,K}$ in this analysis) by the shooting method, the critical buckling load is preassigned an initial guess to start the iterative procedure, which in each step includes a prebuckling analysis and a buckling analysis as well as a computation of the residue of Equation (2.90). Based on the knowledge of this residue, the initial guess is modified and used for the next iteration, and in this way the process is continued. The approximate value of the critical buckling load is obtained when the residue of Equation (2.90) changes its sign and is less than a preassumed small quantity such as 10^{-5}. It should be emphasized that the initial guess of the critical buckling load is an important consideration for fast numerical convergence. It is observed that the whole computational process might be very sensitive to the choice of this initial guess. When ϵ is small, the solution furnished by considerations of shells of constant thickness can be utilized as the initial guess.

Let us investigate a cylindrical shell with different thickness variation and initial imperfection parameters, The radius R and the length L of this cylinder are fixed at 30 cm (11.8 in.) and 46.57 cm (18.3 in.), respectively. The nominal thickness h_0 is taken equal to 0.5 mm (0.0197 in.), Young's modulus of the material E is 205.8 Gpa (29.8 $\times$ 10^6 psi).

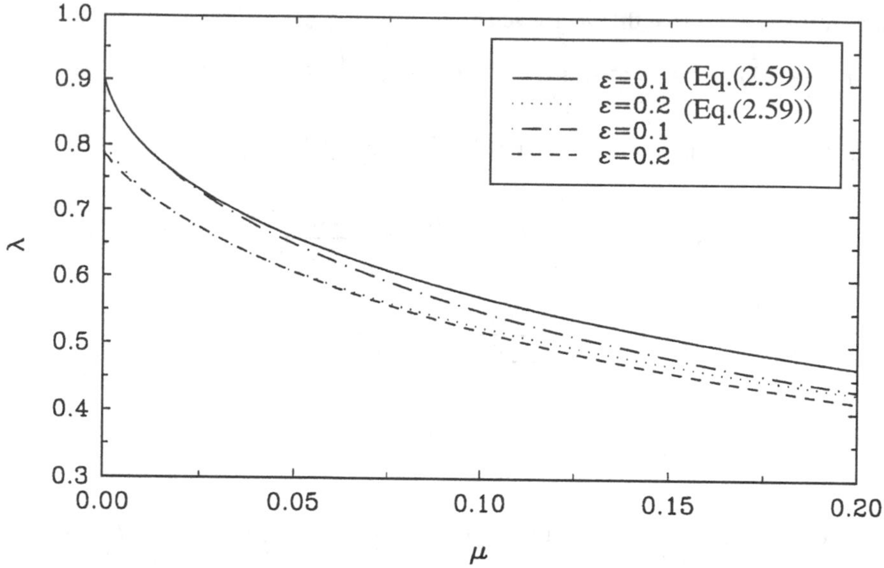

Figure 2.3 Comparison between the numerical results and those of Equation (2.59).

The Poisson ratio ν is equal to 0.3 here. The non-dimensional parameter ϵ varies from zero to 0.2.

For the shell of these dimensions, numerical calculations are performed, and results are plotted together with those of the asymptotic equation (2.59) (Figure 2.3). It appears that, for relatively large values of ϵ and μ, numerical analysis is needed; however, for small μ and ϵ, Equation (2.59) yields excellent results (Tables 2.3 and 2.4). As is revealed from Table 2.3, for small values of ϵ and μ, results provided by the two methods are nearly coincident. Comparison becomes less satisfactory for larger values of ϵ and μ. This is understandable if we recall that the asymptotic formula (2.59) represents a first-order approximation. Table 2.4 demonstrates that, for the perfect shell with the thickness variation parameter $\epsilon = 0.2$, there is a difference between the asymptotic estimate $\lambda = 0.800$ and the numerical result $\lambda = 0.787$. This difference for a relatively large value of ϵ is incompatible with our previous study (see Table 2 in Koiter et al., 1994a), where excellent agreement was documented between the results by the first-order asymptotic formula and the shooting method for values of ϵ only up to 0.05. Remarkably, however, for the same $\epsilon = 0.2$ and the imperfection amplitude $\mu = 0.01$, the agreement is extremely good: The first-order asymptotic formula yields $\lambda = 0.732$, and the numerical analysis gives $\lambda = 0.730$. Such a good agreement appears to be unexpected. To investigate this transition, additional calculations have been performed in the range of $0 \leq \mu \leq 0.01$ for ϵ fixed at 0.2. The results are listed in Table 2.5. As can be seen, the agreement between these two methods improves as the imperfection amplitude μ increases in the range under consideration. Expectedly, a bigger difference occurs for greater μ; for example, when $\mu = \epsilon = 0.2$, the first-order asymptotic formula predicts $\lambda = 0.428$ while the numerical method yields $\lambda = 0.415$. Still, in view of the complexity of the problem due to its highly nonlinear nature, the agreement between these two methods seems to be quite acceptable.

Table 2.3. Comparison of buckling loads from two different methods for $\epsilon = 0.1$ (Case $p = 2p_0$)

Imperfection amplitude, μ	Eq. (2.59)	Numerical analysis
0.00	0.900	0.894
0.01	0.801	0.800
0.05	0.660	0.649
0.10	0.571	0.549
0.15	0.511	0.482
0.20	0.467	0.432

Table 2.4. Comparison of buckling loads from two different methods for $\epsilon = 0.2$ (Case $p = 2p_0$)

Imperfection amplitude, μ	Eq. (2.59)	Numerical analysis
0.00	0.800	0.787
0.01	0.732	0.730
0.05	0.608	0.607
0.10	0.526	0.519
0.15	0.470	0.459
0.20	0.428	0.415

Table 2.5. Comparison of buckling loads from two different methods for $\epsilon = 0.2$ and $0 \leq \mu \leq 0.01$

Imperfection amplitude, μ	Eq. (2.59)	Numerical analysis
0.000	0.800	0.787
0.001	0.791	0.778
0.002	0.782	0.772
0.003	0.774	0.765
0.004	0.767	0.759
0.005	0.760	0.754
0.006	0.754	0.749
0.007	0.748	0.744
0.008	0.743	0.739
0.009	0.737	0.734
0.010	0.732	0.730

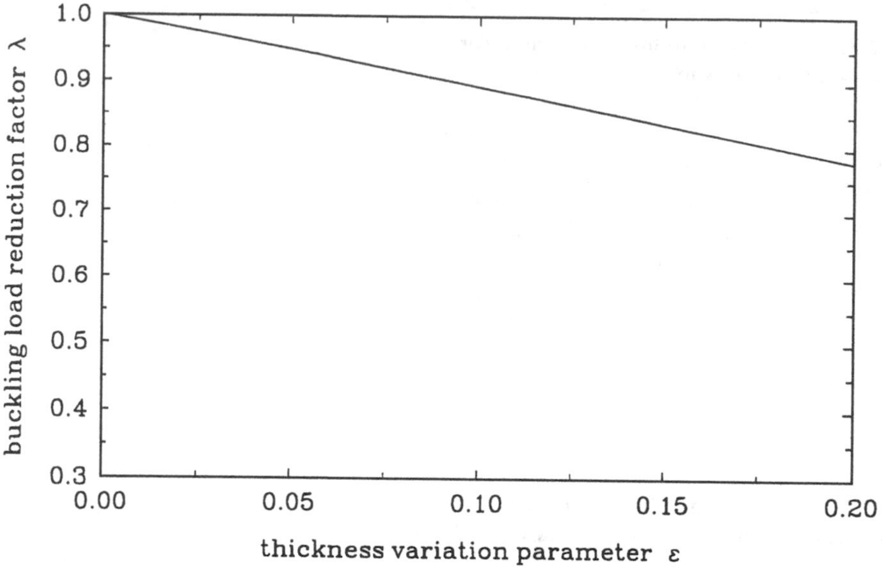

Figure 2.4 The effect of thickness variation on the buckling load of the perfect shell.

If both the initial geometrical imperfection and the thickness variation are ignored, the classical buckling load N_{cl} has a value of 1.04×10^5 N/m (592.5 lb/in.), which is calculated using the value of the nominal thickness h_0. If only the thickness variation is included, then the classical buckling load can be calculated using the methods elucidated in another study (Koiter et al., 1994a). It is found that the critical thickness variation (implying the pattern of thickness variation whose wave number is twice that of the classical buckling mode) has a quite noticeable effect on the classical buckling load. For example, even if the amplitude of the thickness variation is small, say $\epsilon = 0.1$, the thickness variation produces 10.6% reduction in the classical buckling load. When $\epsilon = 0.2$, the classical buckling load is decreased by 21.3% from its counterpart of the case without thickness variation. Figure 2.4 shows the change of classical buckling loads with the critical thickness variation when ϵ is between 0. and 0.2.

When the initial imperfection is present, the combination of the initial imperfection and thickness variation reduces the buckling load even more drastically. Here the initial imperfection amplitude μ is assumed to range from zero to 0.2. For this imperfect shell, the reduced buckling load N_0 can be readily calculated by use of Equation (5.2) by Koiter (1963) if the effect of thickness variation is not considered. Taking into account the critical thickness variation, numerical calculations shows that the influence of thickness variation is generally not as great as that of the initial imperfection for isotropic shells (Figure 2.5). In order to assess the effect of the thickness variation on the buckling load reduction, the so-called thickness variation influence factor β is introduced here, which is defined as

$$\beta = \frac{N_0 - N_{0,T}}{N_0} \times 100\% \tag{2.91}$$

where N_0 and $N_{0,T}$ are buckling loads for shells with and without thickness variation,

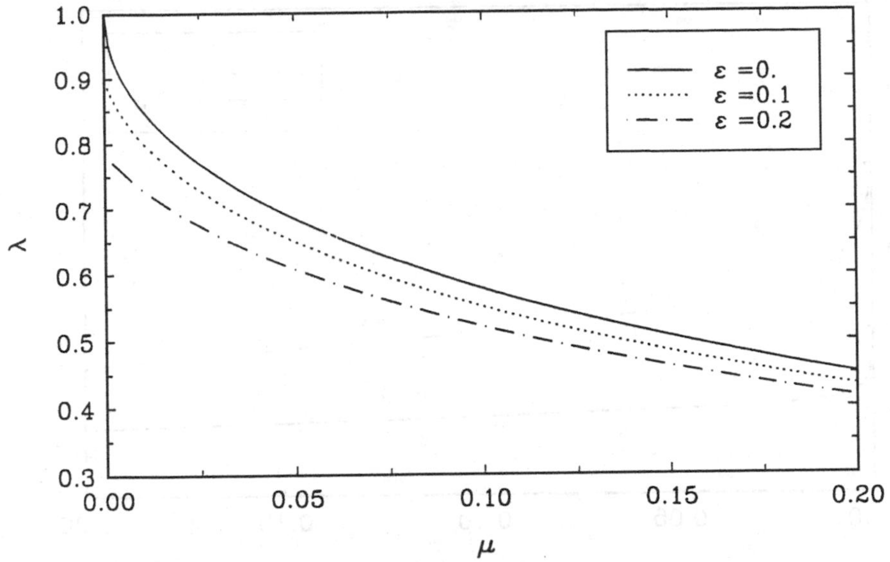

Figure 2.5 Buckling load of the imperfect shell with thickness variation.

respectively. In Figure 2.6, the thickness variation influence factor β is plotted versus the imperfection amplitude μ for $\epsilon = 0.1$ and for $\epsilon = 0.2$. The curves are seen to be similar in general form. The thickness effect is most significant in the absence of the initial imperfection. As Figure 2.5 indicates, for the shell with critical thickness variation pattern of amplitude, $\epsilon = 0.20$ and without the initial imperfection, the buckling load parameter λ is 0.787, and the thickness variation influence factor β equals 21.3%, which

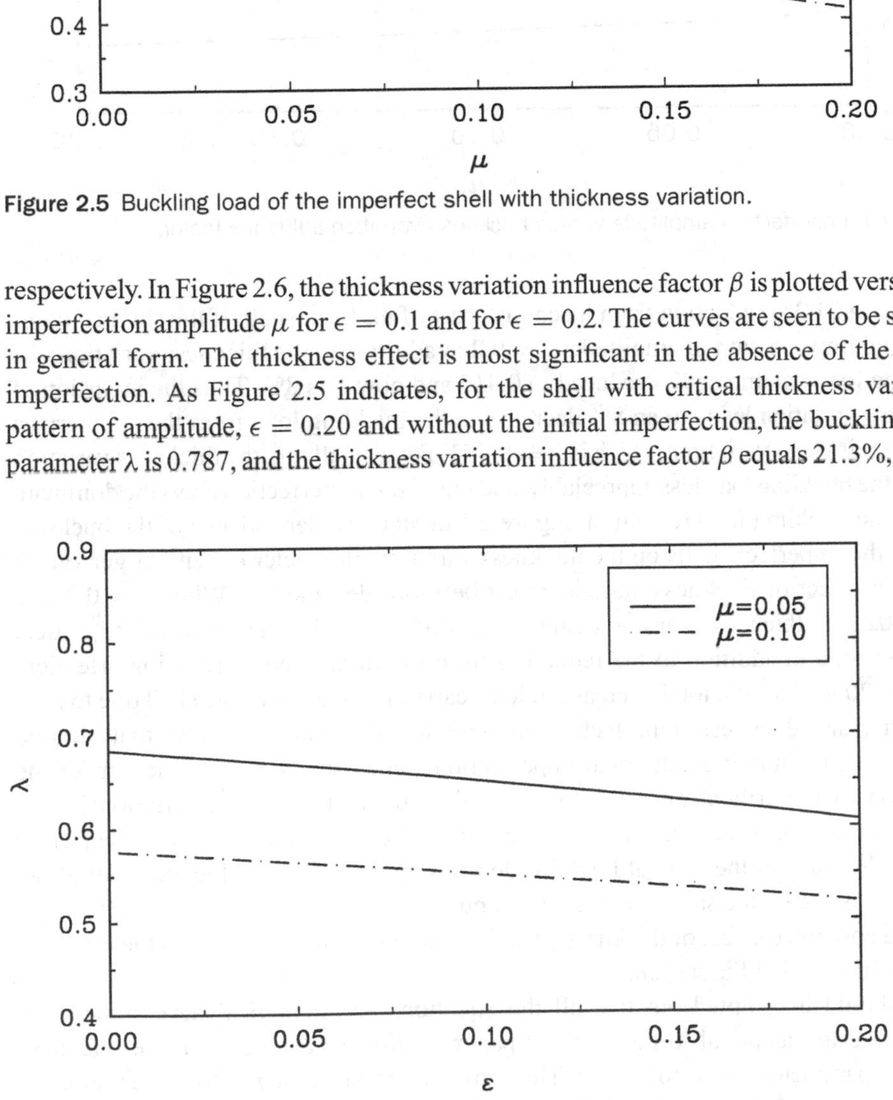

Figure 2.6 The effect of thickness variation on the buckling load of the imperfect shell.

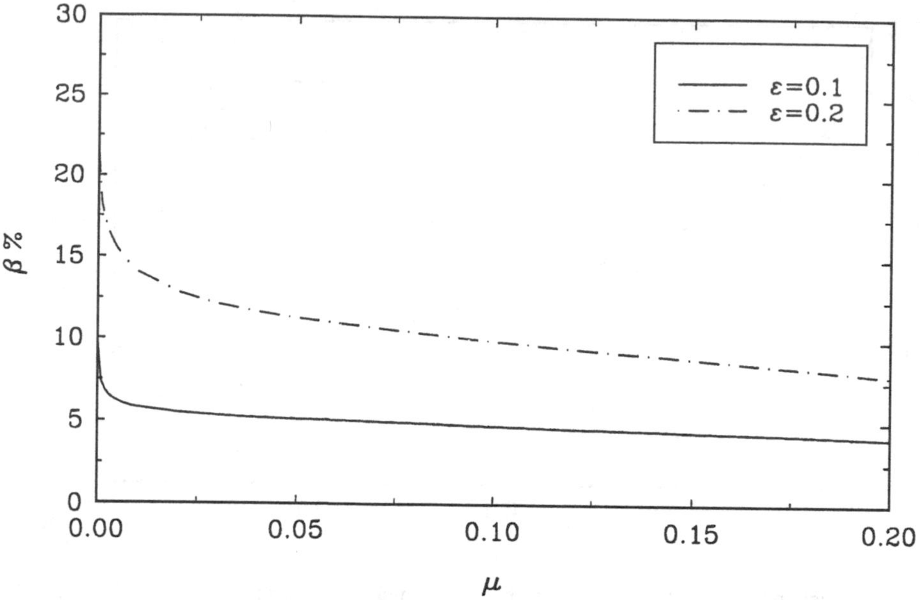

Figure 2.7 Imperfection amplitude versus thickness variation influence factor.

indicates a 21.3% reduction from its counterpart of uniform shell. If the shell contains the initial imperfection of amplitude $\mu = 0.2$ in addition to the thickness variation, the buckling load parameter λ is reduced to 0.415 and now β is 8%. Thus the inclusion of thickness variation leads to an 8% decrease in the buckling load from the counterpart without thickness variation, which is $\lambda = 0.451$. For a smaller ϵ, the thickness variation affects the buckling load less appreciably, and the initial imperfection plays the dominant role in the buckling load reduction. Figure 2.7 depicts the dependence of the buckling load of the imperfect shells on the thickness variation parameter ϵ. With larger values of ϵ, the effect of thickness variation may be more detrimental. When $\epsilon = 0.2$ and $\mu = 0.02$, the thickness variation causes 11% of the further decrease in the critical buckling load in addition to the reduction from the initial geometrical imperfection, which is 20%; thus, the total decrease in load-carrying capacity of the shell due to both geometric and thickness imperfection amounts to 31%. This illustrates that, despite the fact that the initial geometrical imperfection stands out as the main factor for the reduction of the critical buckling load and the effect of thickness variation is less significant in many cases, the thickness variations of certain patterns may cause further notable decrease in the critical buckling load. Neglect of such a thickness variation, therefore, is not on the safe side, for design purposes.

The combined effect of thickness variation and initial imperfection on the buckling load is illustrated in Figure 2.8.

It should be pointed out that all the equations and analytical developments in Section 2.2 are identical to those in the paper by Koiter (1963), when the thickness variation parameter ϵ is set to be zero. Thus, this part of the monograph can be viewed as an expansion of Koiter's work (1963), and it is intended as a contribution to the further understanding of factors leading to reduction of load-carrying capacity of shells, in addition to initial imperfections.

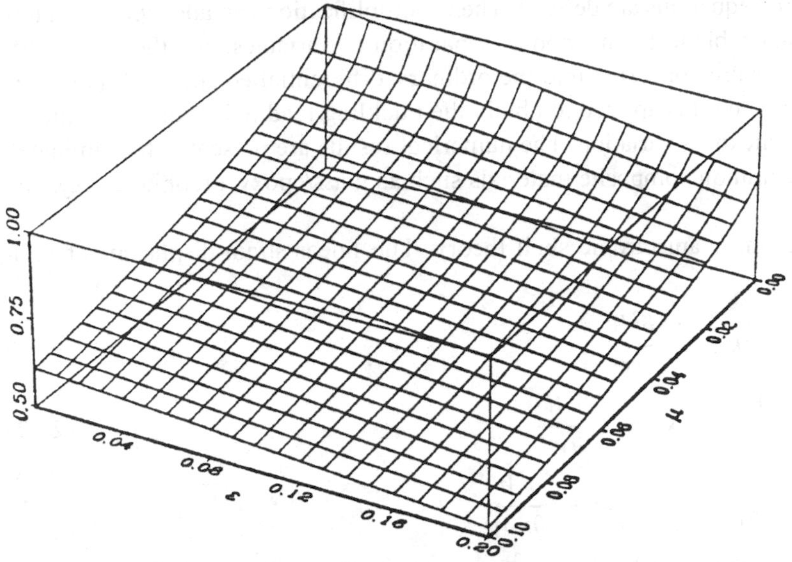

Figure 2.8 Combined effect of thickness variation and initial imperfection on the buckling load.

2.3 Axial Buckling of Composite Cylindrical Shells with Periodic Thickness Variation

Classical buckling of cylindrical shells made of composite materials was studied by several investigators (Tasi, 1966; Stavsky and Friedland, 1969; Hirano, 1979). The effect of initial geometric imperfections was pioneered by Khot (1968) and Card (1969). Further studies were performed by Khot and Venkayya (1970); Tennyson, Chan, and Muggeridge (1971); Simitses, Shaw, and Sheinman (1985); Arbocz and Hol (1991); and other investigators. The work on buckling of composite structures done before 1975 was reviewed by Tennyson (1975). Books specifically devoted to the buckling of composite shells were written by Rikards and Teters (1975) and Vanin and Semeniuk (1978). The monographs by Ambartsumian (1974), Vinson and Sierakowski (1986), Vasiliev (1993), and others also contain chapters on the buckling of composite shell structures.

Due to various factors in the manufacturing process, structures made of composite materials often exhibit certain types of variations in the wall thickness. In the context of isotropic shells, the effect of thickness variation on the buckling of axially compressed cylindrical shells has been discussed in the previous two sections. For the case of composite shells, one may conjecture that thickness variation may affect the buckling behavior of the structure as well. Despite its importance, so far, buckling of composite structures with non-uniform wall thickness has not been tackled in the literature. In the case of composite shells, with the attendant heterogeneity and anisotropy, the analysis becomes much more complicated. But results from such an analysis appear to be more important because, unlike the situation with isotropic structures, the designer may have the opportunity to reduce the deleterious effect by selecting appropriate fiber orientations.

In this section, we deal with the classical buckling load of the perfect anisotropic shell with variable thickness. First, starting from the linear shell theory, the governing partial differential equations are derived. Then a simplification is made, which renders the equations amenable to the method of separation of variables, and the partial differential equations are converted into the ordinary differential equations. Finally, the finite difference method is applied to obtain the buckling load reduction factor due to the presence of thickness variation. The numerical results are presented for laminated shells of three common composite materials such as glass/epoxy, graphite/epoxy, and boron/epoxy.

We use the linear anisotropic shell theory. The linear strain-displacement relations are

$$\begin{aligned} \epsilon_x &= \frac{\partial U}{\partial x}, & \kappa_x &= -\frac{\partial^2 W}{\partial x^2} \\ \epsilon_y &= \frac{\partial V}{\partial y} + \frac{W}{R}, & \kappa_y &= -\frac{\partial^2 W}{\partial y^2} \\ \gamma_{xy} &= \frac{\partial V}{\partial x} + \frac{\partial U}{\partial y}, & \kappa_{xy} &= -2\frac{\partial^2 W}{\partial x \partial y} \end{aligned} \tag{2.92}$$

where x and y are the axial and circumferential coordinates in the shell middle surface; U and V are the shell displacements along axial and circumferential directions, respectively; W is the radial displacement, positive outward; ϵ_x, ϵ_y, and γ_{xy} are strain components; κ_x, κ_y, and κ_{xy} are middle surface curvatures of the shell; and R is the radius of the cylindrical shell.

Thickness variation of the laminated shell invariably exists due to the imprecision involved in the fabrication process. Here, we discuss the case of idealized variation of thickness, namely, the thickness variation is axisymmetric, and in addition, of uniform nature – each lamina has the same variation pattern:

$$h_k(x) = h_{0,k}\left(1 - \epsilon \cos \frac{2px}{R}\right) = h_{0,k} H(x) \qquad (k = 1 \sim K) \tag{2.93}$$

where h_k and $h_{0,k}$ are the thickness and the nominal thickness for the kth layer, respectively; ϵ and p are the non-dimensional parameters indicating the magnitude and wave number of the thickness variation, and they are assumed to be coincident for all the constituent layers; K represents the total number of layers in the laminate. At first sight, the perfect homology of the thickness variation may appear as a restrictive assumption. Yet, if the constituent layers are produced by the same manufacturing process and if they belong to the same fleet of specimens, one can study the case of similar deviations from uniform thickness. Moreover, such a study may shed some light on the question of thickness variability and lead to relatively tractable analysis.

With the foregoing assumption, elements of the stiffness matrices $[A]$, $[B]$, and $[D]$ for the shell with variable thickness are derived as follows:

$$A_{ij} = \sum_{k=1}^{K} (\bar{Q}_{ij})_k (h_k - h_{k-1}) = H(x) \sum_{k=1}^{K} (\bar{Q}_{ij})_k (h_{0,k} - h_{0,k-1}) = H(x) a_{ij}$$

$$B_{ij} = \frac{1}{2} \sum_{k=1}^{K} (\bar{Q}_{ij})_k \left(h_k^2 - h_{k-1}^2\right) = \frac{1}{2}[H(x)]^2 \sum_{k=1}^{K} (\bar{Q}_{ij})_k \left(h_{0,k}^2 - h_{0,k-1}^2\right) = [H(x)]^2 b_{ij}$$

$$D_{ij} = \frac{1}{3}\sum_{k=1}^{K}(\bar{Q}_{ij})_k\left(k_k^3 - h_{k-1}^3\right) = \frac{1}{3}[H(x)]^3\sum_{k=1}^{K}(\bar{Q}_{ij})_k\left(h_{0,k}^3 - h_{0,k-1}^3\right) = [H(x)]^3 d_{ij}$$

$$(i, j = 1, 2, 6) \tag{2.94}$$

where a_{ij}, b_{ij}, and d_{ij} are elements of stiffness matrices for the corresponding uniform laminate; $\bar{Q}_{ij}$s are the transformed reduced stiffnesses of the individual lamina and have no bearing on the thickness. In the following, we will use the transformed stiffness matrices $[A^*]$, $[B^*]$, and $[D^*]$, which are related to the matrices in (2.94) as follows:

$$[A^*] = [A]^{-1}, \qquad [B^*] = [B][A], \qquad [D^*] = [D] - [B][A^*][B] \tag{2.95}$$

thus

$$A_{ij}^* = \frac{1}{H(x)}a_{ij}^*, \qquad B_{ij}^* = H(x)b_{ij}^*, \qquad D_{ij}^* = [H(x)]^3 d_{ij}^* \tag{2.96}$$

where $[a^*]$, $[b^*]$, and $[d^*]$ are counterparts, in the corresponding uniform laminate, of the transformed stiffness matrices $[A^*]$, $[B^*]$, and $[D^*]$, and they are derived from $[a]$, $[b]$, and $[d]$ as follows:

$$[a^*] = [a]^{-1}, \qquad [b^*] = [b][a], \qquad [d^*] = [d] - [b][a^*][b] \tag{2.97}$$

We will deal with symmetric laminates, for which there is no coupling between bending and extension. Thus, we have

$$B_{ij} = 0, \qquad B_{ij}^* = 0 \qquad (i, j = 1, 2, 6) \tag{2.98}$$

The constitutive relations for the anisotropic laminate are

$$\begin{Bmatrix} \epsilon_x \\ \epsilon_y \\ \epsilon_{xy} \end{Bmatrix} = \begin{pmatrix} A_{11}^* & A_{12}^* & A_{16}^* \\ A_{12}^* & A_{22}^* & A_{26}^* \\ A_{16}^* & A_{26}^* & A_{66}^* \end{pmatrix} \begin{Bmatrix} N_x \\ N_y \\ N_{xy} \end{Bmatrix} \tag{2.99}$$

$$\begin{Bmatrix} M_x \\ M_y \\ M_{xy} \end{Bmatrix} = \begin{pmatrix} D_{11}^* & D_{12}^* & D_{16}^* \\ D_{12}^* & D_{22}^* & D_{26}^* \\ D_{16}^* & D_{26}^* & D_{66}^* \end{pmatrix} \begin{Bmatrix} k_x \\ k_y \\ k_{xy} \end{Bmatrix} \tag{2.100}$$

where N_x, N_y, and N_{xy} are stress resultants, and M_x, M_y, and M_{xy} are bending and twisting moments, acting on the mid-surface of a laminate.

The equations of equilibrium for the cylindrical shell read:

$$\frac{\partial N_x}{\partial x} + \frac{\partial N_{xy}}{\partial y} = 0$$

$$\frac{\partial N_{xy}}{\partial x} + \frac{\partial N_y}{\partial y} = 0$$

$$\frac{\partial^2 M_x}{\partial x^2} + 2\frac{\partial^2 M_{xy}}{\partial x \partial y} + \frac{\partial^2 M_y}{\partial y^2} + N_x\frac{\partial^2 W}{\partial x^2} + 2N_{xy}\frac{\partial^2 W}{\partial x \partial y} + N_y\left(\frac{\partial^2 W}{\partial y^2} - \frac{1}{R}\right) = 0 \tag{2.101}$$

Introducing the Airy stress function F

$$N_x = \frac{\partial^2 F}{\partial y^2}, \qquad N_y = \frac{\partial^2 F}{\partial x^2}, \qquad N_{xy} = -\frac{\partial^2 F}{\partial x \partial y} \tag{2.102}$$

the first two equations of equilibrium (101) are identically satisfied. Substituting (1.99) into the third equation of equilibrium yields

$$\begin{aligned} &H^3 \nabla^4_{d*} W + 3H^2 \frac{dH}{dx}\left(2d^*_{11}\frac{\partial^3 W}{\partial x^3} + 2d^*_{12}\frac{\partial^3 W}{\partial x \partial y^2} + 6d^*_{16}\frac{\partial^3 W}{\partial x^2 \partial y} + 2d^*_{26}\frac{\partial^3 W}{\partial y^3}\right. \\ &\left. + 4d^*_{66}\frac{\partial^3 W}{\partial x \partial y^2}\right) + \left[6H\left(\frac{dH}{dx}\right)^2 + 3H^2\frac{d^2 H}{dx^2}\right]\left(d^*_{11}\frac{\partial^2 W}{\partial x^2} + d^*_{12}\frac{\partial^2 W}{\partial y^2}\right. \\ &\left. + 2d^*_{66}\frac{\partial^2 W}{\partial x \partial y}\right) + N_x\frac{\partial^2 W}{\partial x^2} + N_y\left(\frac{\partial^2 W}{\partial y^2} + \frac{1}{R}\right) + 2N_{xy}\frac{\partial^2 W}{\partial x \partial y} + z^2\frac{\partial^2 F}{\partial x^2} = 0 \end{aligned} \tag{2.103}$$

where the differential operator ∇^4_{d*} is defined as

$$\nabla^4_{d*} = d^*_{11}\frac{\partial^4}{\partial x^4} + 4d^*_{16}\frac{\partial^4}{\partial x^3 \partial y} + 2(d^*_{12} + 2d^*_{66})\frac{\partial^4}{\partial x^2 \partial y^2} + 4d^*_{26}\frac{\partial^4}{\partial x \partial y^3} + d^*_{22}\frac{\partial^4}{\partial y^4} \tag{2.104}$$

Elimination of U and V from (1.92) leads to the compatibility equation,

$$9\frac{\partial^2 \epsilon_x}{\partial y^2} + \frac{\partial^2 \epsilon_y}{\partial x^2} - \frac{\partial^2 \epsilon_{xy}}{\partial x \partial y} - \frac{1}{R}\frac{\partial^2 W}{\partial x^2} = 0 \tag{2.105}$$

Substituting (1.98) into Equation (2.105) and using the Airy stress function F, the equation of compatibility can be written as

$$\begin{aligned} &H^2 \nabla^4_{a*} F + H\frac{dH}{dx}\left(a^*_{16}\frac{\partial^3 F}{\partial y^3} + a^*_{26}\frac{\partial^3 F}{\partial x^2 \partial y} - a^*_{66}\frac{\partial^3 F}{\partial x \partial y^2} - 2a^*_{12}\frac{\partial^3 F}{\partial x \partial y^2}\right. \\ &\left. - 2a^*_{22}\frac{\partial^3 F}{\partial x^3} + 2a^*_{26}\frac{\partial^3 F}{\partial x^2 \partial y}\right) + \left[2\left(\frac{dH}{dx}\right)^2 - H\frac{d^2 H}{dx^2}\right] \\ &\times \left(a^*_{12}\frac{\partial^2 F}{\partial y^2} + a^*_{22}\frac{\partial^2 F}{\partial x^2} - a^*_{26}\frac{\partial^2 F}{\partial x \partial y}\right) - H^3 z^2 \frac{\partial^2 W}{\partial x^2} = 0 \end{aligned} \tag{2.106}$$

where

$$\nabla^4_{a*} = a^*_{22}\frac{\partial^4}{\partial x^4} - 2a^*_{26}\frac{\partial^4}{\partial x^3 \partial y} + (2a^*_{12} + a^*_{66})\frac{\partial^4}{\partial x^2 \partial y^2} - 2a^*_{16}\frac{\partial^4}{\partial x \partial y^3} + a^*_{11}\frac{\partial^4}{\partial y^4} \tag{2.107}$$

Introducing the non-dimensional quantities

$$\xi = \frac{x}{L}, \qquad \eta = \frac{y}{L}, \qquad w = \frac{W}{h_0}, \qquad f = \frac{F}{d^*_{11}}, \qquad \bar{N}_x = \frac{N_x L^2}{d^*_{11}},$$

$$\bar{N}_y = \frac{N_y L^2}{d^*_{11}} \qquad \bar{N}_{xy} = \frac{N_{xy} L^2}{d_{11}}, \qquad z = \frac{L}{\sqrt{R h_0}}, \qquad \bar{a}^*_{ij} = \frac{a^*_{ij} d^*_{11}}{h_0^2},$$

$$\bar{b}^*_{ij} = \frac{b_{ij}}{h_0}, \qquad \bar{d}^*_{ij} = \frac{d^*_{ij}}{d^*_{11}} \qquad (i, j = 1, 2, 6) \tag{2.108}$$

where h_0 is the overall thickness of the shell and L is the shell length, (1.103) and (1.106) can be written in the non-dimensional form:

$$\begin{aligned} &H^2\bar{\nabla}^4_{\bar{a}^*} f + H\frac{dH}{d\xi}\left(\bar{a}^*_{16}\frac{\partial^3 f}{\partial\eta^3} + \bar{a}^*_{26}\frac{\partial^3 f}{\partial\xi^2\partial\eta} - \bar{a}^*_{66}\frac{\partial^3 f}{\partial\xi\partial\eta^2} - 2\bar{a}^*_{12}\frac{\partial^3 f}{\partial\xi\partial\eta^2}\right. \\ &\quad \left. - 2\bar{a}^*_{22}\frac{\partial^3 f}{\partial\xi^3} + 2\bar{a}^*_{26}\frac{\partial^3 f}{\partial\xi^2\partial\eta}\right) + \left[2\left(\frac{dH}{d\xi}\right)^2 - H\frac{d^2 H}{d\xi^2}\right] \\ &\quad \times\left(\bar{a}^*_{12}\frac{\partial^2 f}{\partial\eta^2} + \bar{a}^*_{22}\frac{\partial^2 f}{\partial\xi^2} - \bar{a}^*_{26}\frac{\partial^2 f}{\partial\xi\partial\eta}\right) - H^3 z^2\frac{\partial^2 w}{\partial\xi^2} = 0 \end{aligned} \tag{2.109}$$

$$\begin{aligned} &H^3\bar{\nabla}^4_{\bar{d}^*} w + 3H^2\frac{dH}{d\xi}\left(2\bar{d}^*_{11}\frac{\partial^3 w}{\partial\xi^3} + 2\bar{d}^*_{12}\frac{\partial^3 w}{\partial\xi\partial\eta^2} + 6\bar{d}^*_{16}\frac{\partial^3 w}{\partial\xi^2\partial\eta} + 2\bar{d}^*_{26}\frac{\partial^3 w}{\partial\eta^3}\right. \\ &\quad \left. + 4\bar{d}^*_{66}\frac{\partial^3 w}{\partial\xi\partial\eta^2}\right) + \left[6H\left(\frac{dH}{d\xi}\right)^2 + 3H^2\frac{d^2 H}{d\xi^2}\right]\left(\bar{d}^*_{11}\frac{\partial^2 w}{\partial\xi^2} + \bar{d}^*_{12}\frac{\partial^2 w}{\partial\eta^2}\right. \\ &\quad \left. + 2\bar{d}^*_{66}\frac{\partial^2 w}{\partial\xi\partial\eta}\right) + \bar{N}_x\frac{\partial^2 w}{\partial\xi^2} + \bar{N}_y\left(\frac{\partial^2 w}{\partial\eta^2} + z^2\right) + 2\bar{N}_{xy}\frac{\partial^2 w}{\partial\xi\partial\eta} + z^2\frac{\partial^2 f}{\partial\xi^2} = 0 \end{aligned} \tag{2.110}$$

where

$$\bar{\nabla}^4_{\bar{a}^*} = \bar{a}^*_{22}\frac{\partial^4}{\partial\xi^4} - 2\bar{a}^*_{26}\frac{\partial^4}{\partial\xi^3\partial\eta} + (2\bar{a}^*_{12} + \bar{a}^*_{66})\frac{\partial^4}{\partial\xi^2\partial\eta^2} - 2\bar{a}^*_{16}\frac{\partial^4}{\partial\xi\partial\eta^3} + \bar{a}^*_{11}\frac{\partial^4}{\partial\eta^4} \tag{2.111}$$

$$\bar{\nabla}^4_{\bar{d}^*} = \bar{d}^*_{11}\frac{\partial^4}{\partial\xi^4} + 4\bar{d}^*_{16}\frac{\partial^4}{\partial\xi^3\partial\eta} + 2(\bar{d}^*_{12} + 2\bar{d}^*_{66})\frac{\partial^4}{\partial\xi^2\partial\eta^2} + 4\bar{d}^*_{26}\frac{\partial^4}{\partial\xi\partial\eta^3} + \bar{d}^*_{22}\frac{\partial^4}{\partial\eta^4} \tag{2.112}$$

In a perfect agreement with studies of Hirano (1979) and Vinson and Sierakowski (1986), the coupling stiffnesses (A_{16}, A_{26}, B_{16}, B_{26}, D_{16}, D_{26}) are assumed to be zero. They are identically zero for the cross-ply laminates. As for symmetric angle-ply laminates, B_{16} and B_{26} are zero, and A_{16}, A_{26}, D_{16}, and D_{26} can be neglected for laminates with "many" layers. Moreover, we confine our discussion to the buckling of shells under axial compression (i.e., $N_x = P_0$, $N_{xy} = N_y = 0$).

For the shell of constant thickness, the classical axisymmetric buckling load for the symmetrically laminated shell reads (Tasi, 1966):

$$P_{\text{cl, sym}} = \frac{2}{R}\sqrt{\frac{d^*_{11}}{a^*_{22}}} \tag{2.113}$$

The critical buckling wave number p_{cl} is (Tennyson et al., 1971)

$$p_{\mathrm{cl}}^2 = \frac{R}{4\sqrt{a_{22}^* d_{11}^*}} \tag{2.114}$$

However, in many circumstances, the buckling mode is non-axisymmetric. The more general expression for the classical buckling load is (Elishakoff, Li, and Starnes, 1993)

$$P_{m,n} = \left(\frac{L}{m\pi}\right)^2 \frac{C_{11}C_{22}C_{33} + 2C_{12}C_{23}C_{13} - C_{13}^2 C_{22} - C_{23}^2 C_{11} - C_{12}^2 C_{33}}{C_{11}C_{22} - C_{12}^2} \tag{2.115}$$

where

$$\begin{aligned}
C_{11} &= A_{11}\left(\frac{m\pi}{L}\right)^2 + A_{66}\left(\frac{n}{R}\right)^2, \qquad C_{22} = A_{22}\left(\frac{n}{R}\right)^2 + A_{66}\left(\frac{m\pi}{L}\right)^2 \\
C_{33} &= D_{11}\left(\frac{m\pi}{L}\right)^4 + 2(D_{12} + 2D_{66})\left(\frac{m\pi}{L}\right)^2\left(\frac{n}{R}\right)^2 + D_{22}\left(\frac{n}{R}\right)^4 \\
&\quad + \frac{A_{22}}{R^2} + 2\frac{B_{22}}{R}\left(\frac{n}{R}\right)^2 + 2\frac{B_{12}}{R}\left(\frac{m\pi}{L}\right)^2 \\
C_{12} &= C_{21} = (A_{12} + A_{66})\left(\frac{m\pi}{L}\right)\left(\frac{n}{R}\right) \\
C_{23} &= C_{32} = (B_{12} + 2B_{66})\left(\frac{m\pi}{L}\right)^2\left(\frac{n}{R}\right) + \frac{A_{22}}{R}\left(\frac{n}{R}\right) + B_{22}\left(\frac{n}{R}\right)^3 \\
C_{13} &= C_{31} = \frac{A_{12}}{R}\left(\frac{m\pi}{L}\right) + B_{11}\left(\frac{m\pi}{L}\right)^3 + (B_{12} + 2B_{66})\left(\frac{m\pi}{L}\right)\left(\frac{n}{R}\right)^2
\end{aligned} \tag{2.116}$$

To determine the critical buckling load P_{cl} for a cylindrical shell with given dimensions and material properties, one determines those integral numbers m and n, which minimize the value of $P_{m,n}$. To encompass the general case, we introduce the following expression for the classical buckling load of shells of constant thickness:

$$P_{\mathrm{cl}}^{(0)} = \min_{m,n}\{P_{m,n}\} = P_{\mathrm{cl,\,sym}}\phi, \qquad P_{\mathrm{cl,\,sym}} = \frac{2}{R}\sqrt{\frac{d_{11}^*}{a_{22}^*}} \tag{2.117}$$

where ϕ is a control parameter accounting for the non-axisymmetric buckling cases. When the classical buckling mode is axisymmetric, ϕ takes the value of unity.

In the case of special anisotropy, namely when

$$A_{16} = A_{26} = B_{16} = B_{26} = D_{16} = D_{26} = 0$$

the governing equations allow for separation of variables.

We seek solution of Equations (1.109) and (1.110) in the following form:

$$f(\xi,\eta) = \bar{f}(\xi)\cos\frac{nL}{R}\eta$$
$$w(\xi,\eta) = \bar{w}(\xi)\cos\frac{nL}{R}\eta \tag{2.118}$$

where n denotes the number of waves in the circumferential direction during the buckling. Using (1.118), Equations (1.109) and (1.110) are transformed into ordinary differential equations,

$$\begin{aligned}
&H^2\left[\bar{a}^*_{22}\frac{d^4\bar{f}}{d\xi^4} - N_1^2(2\bar{a}_{12}+\bar{a}_{66})\frac{d^2\bar{f}}{d\xi^2} + N_1^4\bar{a}_{11}\bar{f}\right]\\
&+ H\frac{dH}{d\xi}\left(N_1^2\bar{a}^*_{66}\frac{d\bar{f}}{d\xi} + 2N_1^2\bar{a}^*_{12}\frac{d\bar{f}}{d\xi} - 2\bar{a}^*_{22}\frac{d^3\bar{f}}{d\xi^3}\right)\\
&+\left[2\left(\frac{dH}{d\xi}\right)^2 - H\frac{d^2H}{d\xi^2}\right]\left(-N_1^2\bar{a}^*_{12}\bar{f} + \bar{a}_{22}\frac{d^2\bar{f}}{d\xi^2}\right) - H^3z^2\frac{d^2\bar{w}}{d\xi^2} = 0
\end{aligned} \tag{2.119}$$

$$\begin{aligned}
&H^3\left[\bar{d}^*_{11}\frac{d^4\bar{w}}{d\xi^4} - 2N_1^2(\bar{d}^*_{12}+2\bar{d}^*_{66})\frac{d^2\bar{w}}{d\xi^2} + N_1^4\bar{d}^*_{22}\bar{w}\right]\\
&+3H^2\frac{dH}{d\xi}\left(2\bar{d}^*_{11}\frac{d^3\bar{w}}{d\xi^3} - 2N_1^2\bar{d}^*_{12}\frac{d\bar{w}}{d\xi} - 4N_1^2\bar{d}^*_{66}\frac{d\bar{w}}{d\xi}\right)\\
&+\left[3H^2\frac{d^2H}{d\xi^2} + 6H\left(\frac{dH}{d\xi}\right)^2\right]\left(\bar{d}^*_{11}\frac{d^2\bar{w}}{d\xi^2} - N_1^2\bar{d}^*_{12}\bar{w}\right)\\
&+\lambda\Omega\frac{d^2\bar{w}}{d\xi^2} + z^2\frac{d^2\bar{f}}{d\xi^2} = 0
\end{aligned} \tag{2.120}$$

where $\lambda = P^{(\epsilon)}_{\text{cl}}/P^{(0)}_{\text{cl}}$ and is referred to as the buckling load reduction factor due to the thickness variation; $P^{(\epsilon)}_{\text{cl}}$ is the buckling load of shells with variable thickness, and Ω and N_1 are non-dimensional parameters,

$$\Omega = \frac{2\phi L^2}{R\sqrt{d^*_{11}, a^*_{22}}}, \qquad N_1 = \frac{nL}{R}$$

The governing equations (1.119) and (1.120) constitute a set of fourth-order differential equations with variable coefficients. Here we employ the finite difference method, which seems to be particularly useful for the buckling problems of structures of complicated geometry and/or varying flexural rigidity. This method is based on the use of approximate algebraic expressions for the derivatives of unknown variables that appear in the fundamental governing equations. The following expressions of the central difference are used to approximate the corresponding derivatives:

$$\Delta f_i = \frac{f_{i+1} - f_{i-1}}{2d}, \qquad \Delta^2 f_i = \frac{f_{i+1} - 2f_i + f_{i-1}}{d^2}$$
$$\Delta^3 f_i = \frac{f_{i+2} - 2f_{i+1} - 2f_{i-1} - f_{i-2}}{2d^3}, \tag{2.121}$$
$$\Delta^4 f_i = \frac{f_{i+2} - 4f_{i+1} + 6f_i - 4f_{i-1} + f_{i-2}}{d^4}$$

where d is the distance between neighboring nodal points.

Using (1.121), the differential equations (1.119) and (1.120) are approximated by the following finite difference equations:

$$\begin{aligned}
&\left(\frac{G_{1i}}{d^4} - \frac{G_{2i}}{2d^3}\right)\bar{f}_{i-2} + \left(-4\frac{G_{1i}}{d^4} + \frac{G_{2i}}{d^3} + \frac{G_{3i}}{d^2} - \frac{G_{4i}}{2d}\right)\bar{f}_{i-1} \\
&\quad + \left(6\frac{G_{1i}}{d^4} - 2\frac{G_{3i}}{d^2} + G_{5i}\right)\bar{f}_i + \left(-4\frac{G_{1i}}{d^4} - \frac{G_{2i}}{d^3} + \frac{G_{3i}}{d^2} + \frac{G_{4i}}{2d}\right)\bar{f}_{i-1} \\
&\quad + \left(\frac{G_{1i}}{d^4} + \frac{G_{2i}}{2d^3}\right)\bar{f}_{i+2} + \frac{G_{6i}}{d^2}\bar{w}_{i-1} - 2\frac{G_{6i}}{d^2}\bar{w}_i + \frac{G_{6i}}{d^2}\bar{w}_{i+1} = 0
\end{aligned} \tag{2.122}$$

$$\begin{aligned}
&\frac{G_{7i}}{d^2}\bar{f}_{i-1} - 2\frac{G_{7i}}{d^2}\bar{f}_i + \frac{G_{7i}}{d^2}\bar{f}_{i+1} + \left(\frac{G_{8i}}{d^4} - \frac{G_{9i}}{2d^3}\right)\bar{w}_{i-2} \\
&\quad + \left(-4\frac{G_{8i}}{d^4} + \frac{G_{9i}}{d^3} + \frac{G_{10i}}{d^2} - \frac{G_{11i}}{2d}\right)\bar{w}_{i-1} + \left(6\frac{G_{8i}}{d^4} - 2\frac{G_{10i}}{d^2} + G_{12i}\right)\bar{w}_i \\
&\quad + \left(-4\frac{G_{8i}}{d^4} - \frac{G_{9i}}{d^3} + \frac{G_{10i}}{d^2} + \frac{G_{11i}}{2d}\right)\bar{w}_{i+1} + \left(\frac{G_{8i}}{d^4} + \frac{G_{9i}}{2d^3}\right)\bar{w}_{i+2} = 0
\end{aligned} \tag{2.123}$$

where all the coefficients G_{ji} $(j = 1 \sim 12)$ can be analytically evaluated as follows:

$$\begin{aligned}
G_{1i} &= [H(\xi_i)]^2\bar{a}_{22}^*, \qquad G_{2i} = -2H(\xi_i)H'(\xi_i)\bar{a}_{22}^* \\
G_{3i} &= -N_1^2[H(\xi_i)]^2(2\bar{a}_{12}^* + \bar{a}_{66}^*) + \bar{a}_{22}^*\{2[H'(\xi_i)]^2 - H(\xi_i)H''(\xi_i)\} \\
G_{4i} &= (\bar{a}_{66}^* + 2\bar{a}_{12}^*)N_1^2 H(\xi_i)H'(\xi_i), \\
G_{5i} &= [H(\xi)]^2 N_1^4\bar{a}_{11}^* - N_1^2\bar{a}_{12}^*\{2[H'(\xi_i)]^2 - H(\xi_i)H''(\xi_i)\} \\
G_{6i} &= -z^2[H(\xi_i)]^3, \qquad G_{7i} = z^2, \\
G_{8i} &= [H(\xi_i)]^3\bar{d}_{11}^*, \qquad G_{9i} = 6[H(\xi_i)]^2 H'(\xi_i)\bar{d}_{11}^* \\
G_{10i} &= -2N_1^2[H(\xi)]^3(\bar{d}_{12}^* + 2\bar{d}_{66}^*) + \lambda\Omega + \{6H(\xi_i)[H'(\xi_i)]^2 \\
&\qquad + 3[H(\xi_i)]^2 H''(\xi_i)\}\bar{d}_{11}^* \\
G_{11i} &= -6N_1^2[H(\xi_i)]^2 H'(\xi_i)(\bar{d}_{12}^* + 2\bar{d}_{66}^*) \\
G_{12i} &= N_1^4[H(\xi_i)]^3\bar{d}_{22}^* - N_1^2\bar{d}_{12}^*\{6H(\xi)[H'(\xi_i)]^2 + 3[H(\xi_i)]^2 H''(\xi_i)\}
\end{aligned} \tag{2.124}$$

Note that the derivatives $H'(\xi_i)$ and $H''(\xi_i)$ are also calculated analytically. By subdividing the shell length domain $(-L/2, L/2)$ into M equal segments, we apply Equations (2.122) and (2.123) to each nodal point. Here we consider the case of simply supported boundary conditions, namely

$$\bar{w}_0 = \bar{w}_0'' = \bar{f}_0 = \bar{f}_0'' = \bar{w}_M = \bar{w}_M'' = \bar{f}_M = \bar{f}_M'' = 0 \tag{2.125}$$

or in view of (2.121)

$$\bar{w}_{-1} = \bar{w}_1, \qquad \bar{f}_{-1} = \bar{f}_1, \qquad \bar{w}_{M+1} = \bar{w}_{M-1}, \qquad \bar{f}_{M+1} = \bar{f}_{M-1} \tag{2.126}$$

Thus, we establish a system of simultaneous algebraic equations,

$$[C(\xi_i, \lambda)]_{(2M+2)\times(2M+2)}\{\delta\}_{(2M+2)\times(2M+2)} = 0 \tag{2.127}$$

where $[C(\xi_i, \lambda)]$ is the coefficient matrix, whose elements depend on the shell geometry, nodal point coordinates, elastic constants as well as the bucking load reduction factor λ; $\{\delta\}$ represents a column vector containing the sought values of functions of w and f at the nodal points. Equation (2.127), the stability equation in finite-difference form, is a set of linear, homogeneous, algebraic equations. There exist non-trivial solutions of this set of equations for some values of load parameters λ. The lowest eigenvalue λ represents the classical buckling load reduction factor. Its value is determined through an iteration procedure. Usually, the load parameter λ is increased step by step, and at each step the determinant of matrix $[C(\xi, \lambda)]$ is calculated in order to find the interval where its sign first changes. Once the interval in which the determinant of coefficient matrix changes sign is located, the secant method is used to expedite the search for the classical buckling reduction factor λ, the accuracy of which is improved with the increase in the number of subdivided segments. The numerical results obtained by the finite difference method will be discussed later.

In many circumstances, the composite shell buckles axisymmetrically. For the cases where the axisymmetric buckling mode dominates, it is possible to derive an asymptotic formula relating the buckling load reduction to the thickness variation parameter. This could be accomplished in the following fashion. We assume $\bar{w}(\xi)$ in the form

$$\bar{w}(\xi) = A\cos\frac{pL}{R}\xi + B\cos\frac{3pL}{R}\xi \tag{2.128}$$

where p is the number of half-waves along the shell length at buckling; A and B are undetermined constants. The preceding buckling pattern satisfies the boundary conditions of the simple supports if p is of the form

$$p = (2k+1)\pi\frac{R}{L} \tag{2.129}$$

where k is an integer. The first term of the two-term approximation (1.128) is the exact buckling mode for the shell of constant thickness, and the second term is introduced to account for the thickness variation.

To solve the compatibility equation (1.129) for $\bar{f}$, the perturbation procedure is employed. To this end, $\bar{f}$ is expressed in terms of the thickness variation parameter ϵ

as follows:

$$\bar{f}(\xi) = f_0(\xi) + \epsilon f_1(\xi) + \epsilon^2 f_2(\xi) + \cdots \tag{2.130}$$

Substituting (1.130) into (1.129), we have after collecting the like terms in ϵ,

$$\begin{aligned}
&\bar{a}_{22}^* f_0^{(4)} - (2\bar{a}_{12}^* + \bar{a}_{66}^*)N^2 f_0'' + \bar{a}_{11}^* f_0 - z^2 \bar{w}'' + \epsilon\big[\bar{a}_{22}^* f_1^{(4)} - (2\bar{a}_{12}^* + \bar{a}_{66}^*)N^2 f_1'' \\
&+ \bar{a}_{11}^* N^4 f_1 - 2\bar{a}_{22}^* \cos(2P\xi) f_0^{(4)} - 4\bar{a}_{22}^* P \sin(2P\xi) f_0''' \\
&+ (4\bar{a}_{12}^* + 2\bar{a}_{66}^*)N^2 \cos(2P\xi) f_0'' - 4\bar{a}_{22}^* P^2 \cos(2P\xi) f_0'' \\
&+ (4\bar{a}_{12}^* + 2\bar{a}_{66}^*)PN^2 \sin(2P\xi) f_0' - (2\bar{a}_{11}^* N^4 - 4\bar{a}_{12}^* P^2 N^2)\cos(2P\xi) f_0 \\
&+ 3z^2 \cos(2P\xi)\bar{w}''\big] + \epsilon^2\big[\bar{a}_{22}^* f_2^{(4)} - (2\bar{a}_{12}^* N^2 + \bar{a}_{66}^* N^2)f_2'' + \bar{a}_{11}^* N^4 f_2 \\
&+ \bar{a}_{22}^* \cos^2(2P\xi) f_0^{(4)} - 2\bar{a}_{22}^* \cos(2P\xi) f_1^{(4)} + 4\bar{a}_{22}^* P \cos(2P\xi)\sin(2P\xi) f_0''' \\
&+ (4\bar{a}_{12}^* N^2 + 2\bar{a}_{66}^* N^2 - \bar{a}_{22}^* P^2)\sin(2P\xi) f_1'' - (2\bar{a}_{12}^* N^2 + 2\bar{a}_{66}^* N^2 + 4\bar{a}_{22}^* P^2) \\
&\times \cos^2(2P\xi) f_0'' + 8\bar{a}_{22}^* P^2 \sin^2(2P\xi) f_0'' + (4\bar{a}_{12}^* + 2\bar{a}_{66}^*)PN^2 \sin(2P\xi) f_1' \\
&- (4\bar{a}_{12}^* + 2\bar{a}_{66}^*)N^2 P \cos(2P\xi)\sin(2P\xi) f_0' + (\bar{a}_{11}^* N^4 \cos^2(2P\xi) \\
&+ 8\bar{a}_{12}^* P^2 N^2 \sin^2(2P\xi) + 4\bar{a}_{12}^* P^2 N^2 \cos^2(2P\xi) f_0 - (2\bar{a}_{11}^* N^4 \\
&+ 4\bar{a}_{12}^* P^2 N^2)\cos(2P\xi) f_1 - 3z^2 \cos^2(2P\xi)\bar{w}''\big] + \cdots = 0 \tag{2.131}
\end{aligned}$$

where

$$P = \frac{pL}{R}, \qquad N_1 = \frac{nL}{R}, \qquad z = \frac{L}{\sqrt{Rh_0}} \tag{2.132}$$

Equation (2.131) must hold for any value of the parameter ϵ, which means that the factor for each of the powers of ϵ must be zero. Thus, Equation (2.131) may be separated into the following system of equations:

$$\mathscr{L}(f_0) = z^2 \bar{w}'' \tag{2.133}$$

$$\begin{aligned}
\mathscr{L}(f_1) = {}&2\bar{a}_{22}^* \cos(2P\xi) f_0^{(4)} + 4\bar{a}_{22}^* P \sin(2P\xi) f_0''' \\
&- (4\bar{a}_{12}^* + 2\bar{a}_{66}^*)N^2 \cos(2P\xi) f_0'' \\
&+ 4\bar{a}_{22}^* P^2 \cos(2P\xi) f_0'' - (4\bar{a}_{12}^* + 2\bar{a}_{66}^*)PN^2 \sin(2P\xi) f_0' \\
&+ (2\bar{a}_{11}^* N^4 - 4\bar{a}_{12}^* P^2 N^2)\cos(2P\xi) f_0 - 3z^2 \cos(2P\xi)\bar{w}'' \tag{2.134}
\end{aligned}$$

$$\begin{aligned}
\mathscr{L}(f_2) = {}&-\bar{a}_{22}^* \cos^2(2P\xi) f_0^{(4)} + 2\bar{a}_{22}^* \cos(2P\xi) f_1^{(4)} - 4\bar{a}_{22}^* P \cos(2P\xi)\sin(2P\xi) f_0''' \\
&- (4\bar{a}_{12}^* N^2 + 2\bar{a}_{66}^* N^2 - \bar{a}_{22}^* P^2)\sin(2P\xi) f_1'' + (2\bar{a}_{12}^* N^2 + 2\bar{a}_{66}^* N^2 \\
&+ 4\bar{a}_{22}^* P^2)\cos^2(2P\xi) f_0'' - 8\bar{a}_{22}^* P^2 \sin^2(2P\xi) f_0'' - (4\bar{a}_{12}^* + 2\bar{a}_{66}^*)PN^2 \\
&\times \sin(2P\xi) f_1' + (4\bar{a}_{12}^* + 2\bar{a}_{66}^*)N^2 P \cos(2P\xi)\sin(2P\xi) f_0' \\
&- (\bar{a}_{11}^* N^4 \cos^2(2P\xi) + 8\bar{a}_{12}^* P^2 N^2 \sin^2(2P\xi) + 4\bar{a}_{12}^* P^2 N^2 \cos^2(2P\xi) f_0 \\
&+ (2\bar{a}_{11}^* N^4 + 4\bar{a}_{12}^* P^2 N^2)\cos(2P\xi) f_1 + 3z^2 \cos^2(2P\xi)\bar{w}'' \tag{2.135}
\end{aligned}$$

where the operator $\mathscr{L}(\cdot)$ is defined as

$$\mathscr{L}(f) = \bar{a}_{22}^* f^{(4)} - (2\bar{a}_{12}^* + \bar{a}_{66}^*)N^2 f'' + \bar{a}_{11}^* N^4 f \tag{2.136}$$

Equations (2.133)–(2.135) are solved analytically with the aid of the computerized symbolic algebra *Mathematica* (Wolfram, 1991) for f_0, f_1, and f_2 to yield

$$\begin{aligned} f_0 &= a_1 \cos(P\xi) + a_2 \cos(3P\xi) \\ f_1 &= a_3 \cos(P\xi) + a_4 \cos(3P\xi) + a_5 \cos(5P\xi) \\ f_2 &= a_6 \cos(P\xi) + a_7 \cos(3P\xi) + a_8 \cos(5P\xi) + a_9 \cos(7P\xi) \end{aligned} \tag{2.137}$$

where $a_1, a_2, \ldots, a_9$ are coefficients depending on A and B, and are given in the article by Koiter et al. (1994b).

Applying the weighted residuals method, namely, in our case the Boobnov-Galerkin procedure, to the equilibrium equation (2.120), we arrive at

$$\begin{aligned} \int_{-1/2}^{1/2} \Bigg\{ & H^3 \left[\bar{d}_{11}^* \frac{d^4\bar{w}}{d\xi^4} - 2N^2(\bar{d}_{12}^* + 2\bar{d}_{66}^*) \frac{d^2\bar{w}}{d\xi^2} + N^4 \bar{d}_{22}^* \bar{w} \right] \\ & + 3H^2 \frac{dH}{d\xi} \left(2\bar{d}_{11}^* \frac{d^3\bar{w}}{d\xi^3} - 2N^2 \bar{d}_{12}^* \frac{d\bar{w}}{d\xi} - 4N^2 \bar{d}_{66}^* \frac{d\bar{w}}{d\xi} \right) \\ & + \left[3H^2 \frac{d^2 H}{d\xi^2} + 6H \left(\frac{dH}{d\xi} \right)^2 \right] \left(\bar{d}_{11}^* \frac{d^2\bar{w}}{d\xi^2} - N^2 \bar{d}_{12}^* \bar{w} \right) + \lambda\Omega \frac{d^2\bar{w}}{d\xi^2} \\ & + z^2 (f_0'' + \epsilon f_1'' + \epsilon^2 f_2'' + \cdots) \Bigg\} \psi_j(\xi)\, d\xi = 0 \end{aligned} \tag{2.138}$$

where $\psi_j(\xi)$ is either $\cos(P\xi)$ or $\cos(3P\xi)$.

We now investigate the axisymmetric buckling mode,

$$n = 0, \qquad p = 2p_{\mathrm{cl}} \tag{2.139}$$

that is, the modal number of thickness variation is twice as much as that of the classical buckling mode, which is the case where the thickness variation has the most detrimental effect on the buckling behavior of the isotropic shells. For the composite shells discussed in this paper, the aforementioned case is also the most critical one, as is shown by the numerical results in Figures 2.4, 2.5, and 2.6. Thus, in the subsequent analysis, attention will be devoted to this case to establish an asymptotic relationship between the buckling load reduction factor λ and the thickness variation parameter ϵ. Substituting (2.128) and (2.137) into (2.138) and making some algebraic manipulations lead, when retaining the terms up to ϵ^2, to the following eigenvalue problem:

$$[C(\epsilon, \lambda)]_{2\times 2} \begin{Bmatrix} A \\ B \end{Bmatrix} = 0 \tag{2.140}$$

where $[C(\epsilon, \lambda)]$ is the coefficient matrix containing the thickness variation parameter ϵ and the buckling load reduction factor λ. The elements of matrix $[C(\epsilon, \lambda)]$ read

$$\begin{aligned} C_{11} &= P^4\left(1 - \lambda - \epsilon + \frac{3}{4}\epsilon^2\right) \\ C_{12} = C_{21} &= P^4\left(-7\epsilon + \frac{27}{8}\epsilon^2\right) \\ C_{22} &= P^4\left(41 - 9\lambda + \frac{243}{4}\epsilon^2\right) \end{aligned} \tag{2.141}$$

and the characteristic equation $C_{11}C_{22} - C_{12}^2 = 0$ reads

$$9\lambda^2 + \lambda\left(-50 + 9\epsilon - \frac{135}{2}\epsilon^2\right) + 41 - 41\epsilon + \frac{85}{2}\epsilon^2 = 0 \tag{2.142}$$

Thus the following asymptotic expression for the buckling load reduction factor is obtained,

$$\lambda = 1 - \epsilon - \frac{25}{32}\epsilon^2 \tag{2.143}$$

which is identical to the formula (2.30) in the isotropic case.

In this section, six-layer cylindrical shells with different thickness variation have been investigated. The radius R and the length L are fixed at 6 in. and 30 in., respectively. The nominal thickness of each lamina is taken to be equal to 0.012 in. The non-dimensional parameter ϵ varies from zero to 0.2. Three kinds of materials are considered; the material moduli are as follows:

1. Glass/epoxy:

$$E_{11} = 7.5 \times 10^6 \text{ psi}, \qquad E_{22} = 3.5 \times 10^6 \text{ psi}$$
$$\nu_{12} = 0.25, \qquad G_{12} = 1.25 \times 10^6 \text{ psi}$$

2. Graphite/epoxy:

$$E_{11} = 20 \times 10^6 \text{ psi}, \qquad E_{22} = 1 \times 10^6 \text{ psi}$$
$$\nu_{12} = 0.25, \qquad G_{12} = 0.6 \times 10^6 \text{ psi}$$

3. Boron/epoxy:

$$E_{11} = 40 \times 10^6 \text{ psi}, \qquad E_{22} = 4.5 \times 10^6 \text{ psi}$$
$$\nu_{12} = 0.25, \qquad G_{12} = 1.5 \times 10^6 \text{ psi}$$

The laminate configuration is chosen as $[\theta/-\theta/\theta]_{\text{sym}}$, with θ ranging from 0° to 90° to show the interaction between the thickness variation and the fiber orientation.

Using (2.117), $P_{\text{cl}}^{(0)}$, the classical buckling load for the shell of constant thickness h_0, can be calculated. It is seen from Figure 2.9 that in a certain fiber-angle range (e.g., θ between 21° and 69° for glass/epoxy, θ between 14° and 76° for graphite/epoxy, and θ

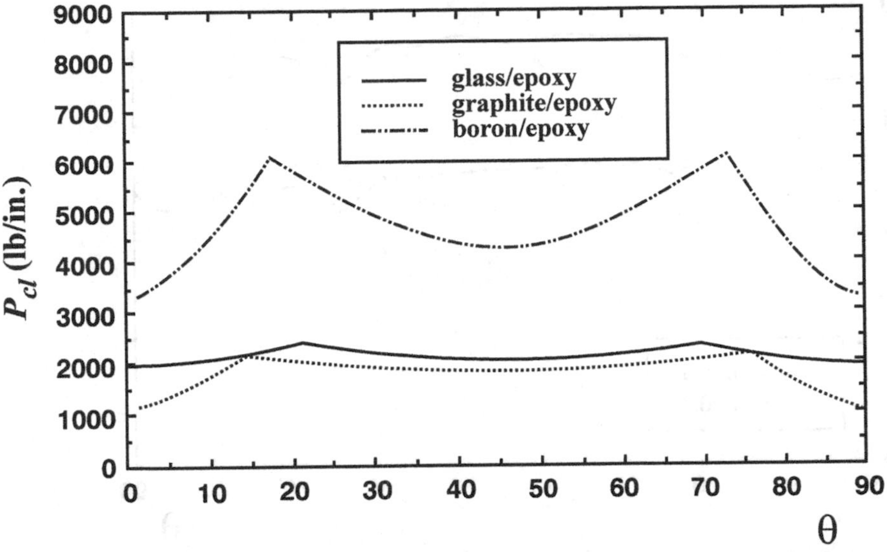

Figure 2.9 Classical buckling load $P_{cl}^{(0)}$ for shells of three different materials.

between 18° and 72° for boron/epoxy), the buckling mode is axisymmetric. Numerical results show that even though the thickness variation generally reduces the load-bearing capability of the structure, the magnitude of the reduction can vary enormously with such factors as constituent materials, fiber orientation, and thickness variation pattern. Figures 2.10, 2.11, and 2.12 illustrate that, when the thickness variation pattern is configurational to the classical axisymmetric buckling mode, the buckling load reduction

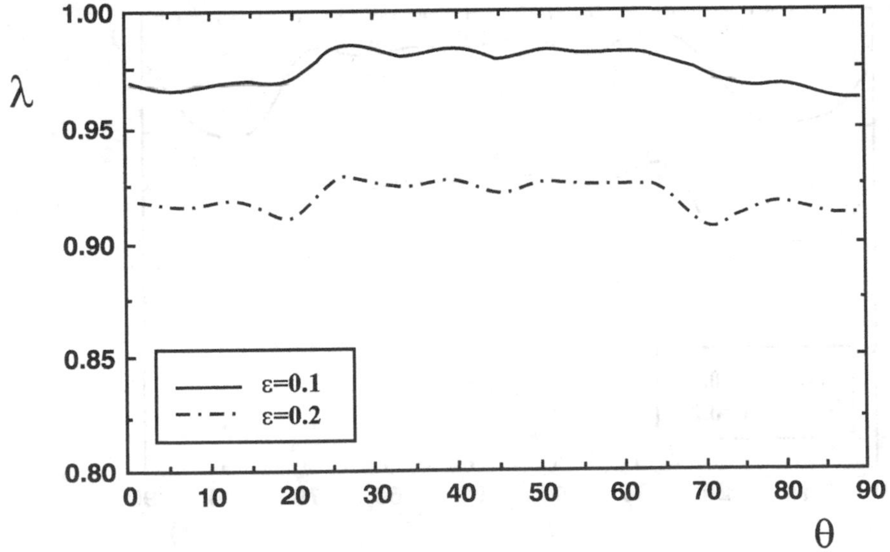

Figure 2.10 Buckling load parameter λ versus fiber angle θ for glass/epoxy shells ($[\theta/-\theta/\theta]_{sym}$, $p = p_{cl}$).

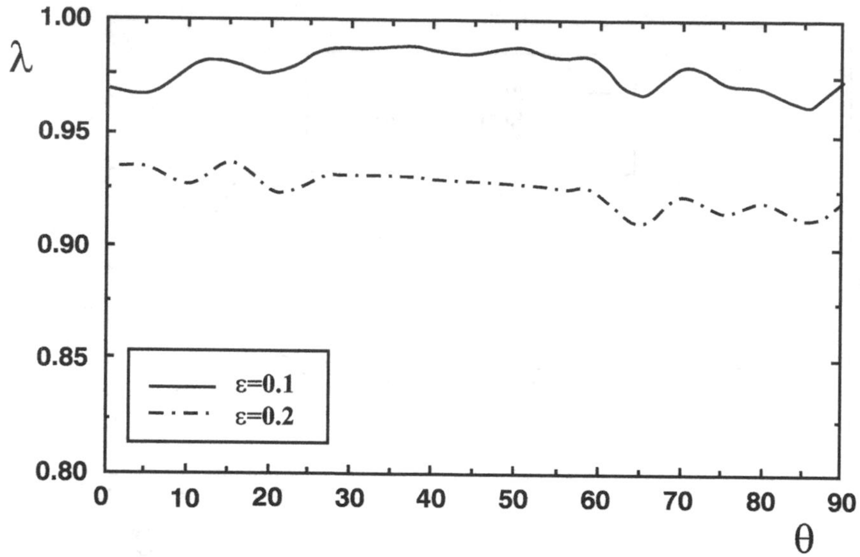

Figure 2.11 Buckling load parameter λ versus fiber angle θ for graphite/epoxy shells ($[\theta/-\theta/\theta]_{\text{sym}}$, $p = p_{\text{cl}}$).

is around 5–8% for $\epsilon = 0.1$ and around 8–12% for $\epsilon = 0.2$, the specific value varying in terms of the constituent materials and the lamination profile. Interestingly enough, in the cases where the axisymmetric buckling mode dominates, the buckling load reduction rate λ does not vary significantly in terms of the buckling pattern, whether it is axisymmetric or non-axisymmetric. From Figures 2.13, 2.14, and 2.15, the most detrimental effect of the thickness variation has been found to occur when the wave

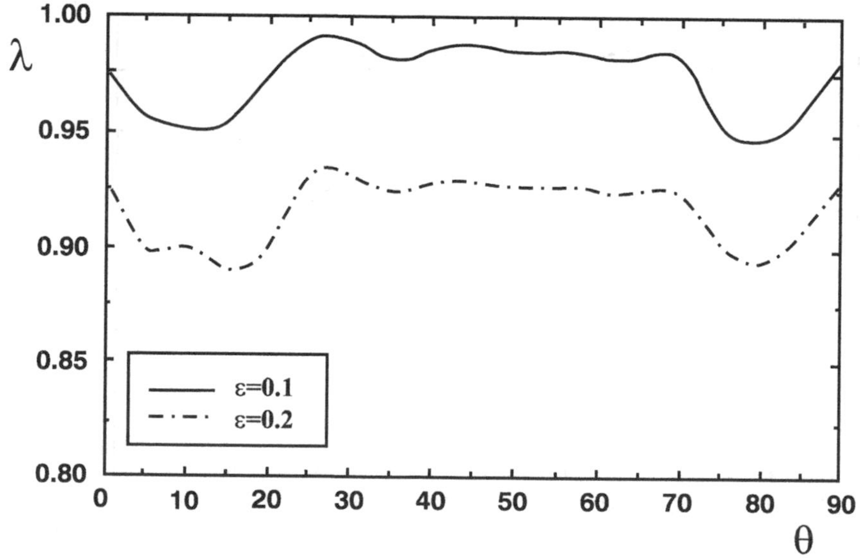

Figure 2.12 Buckling load parameter λ versus fiber angle θ for boron/epoxy shells ($[\theta/-\theta/\theta]_{\text{sym}}$, $p = p_{\text{cl}}$).

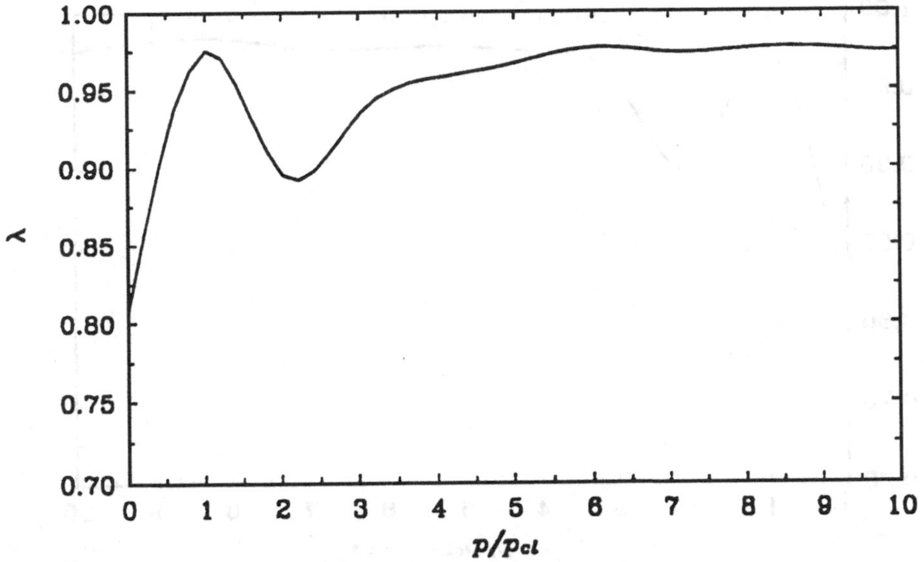

Figure 2.13 Buckling load parameter λ versus thickness variation parameter p (material: glass/epoxy; laminate profile: $[45^\circ/-45^\circ/45^\circ]_{\text{sym}}$, $\epsilon = 0.1$).

number of the axisymmetric thickness variation is twice that of the classical buckling mode. Remarkably, analogous phenomena has been analytically predicted in the isotropic case, where an asymptotic formula for the dependence of classical buckling reduction factor λ on the thickness variation is given in Equation (1.30) as

$$\lambda = 1 - \epsilon - \frac{25}{36}\epsilon^2$$

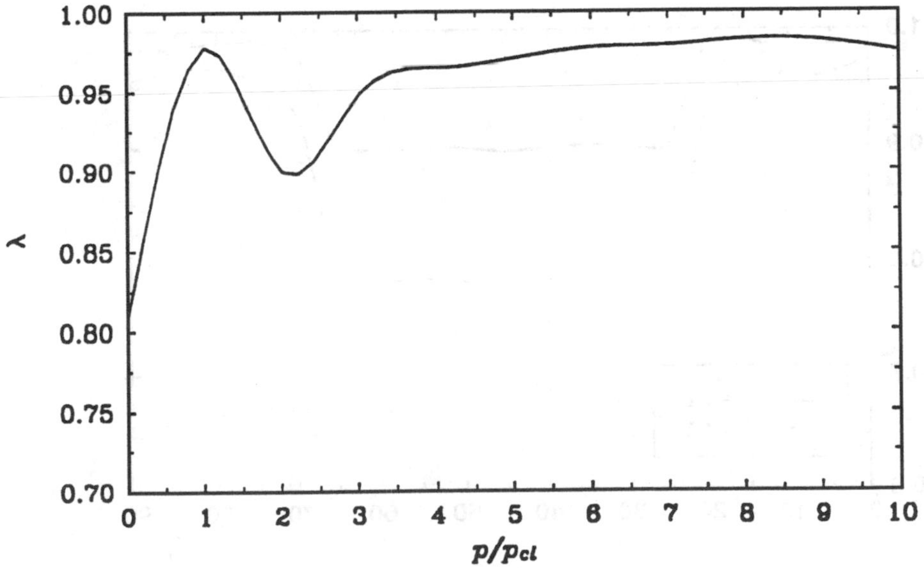

Figure 2.14 Buckling load parameter λ versus thickness variation parameter p (material: graphite/epoxy; laminate profile: $[45^\circ/-45^\circ/45^\circ]_{\text{sym}}$, $\epsilon = 0.1$).

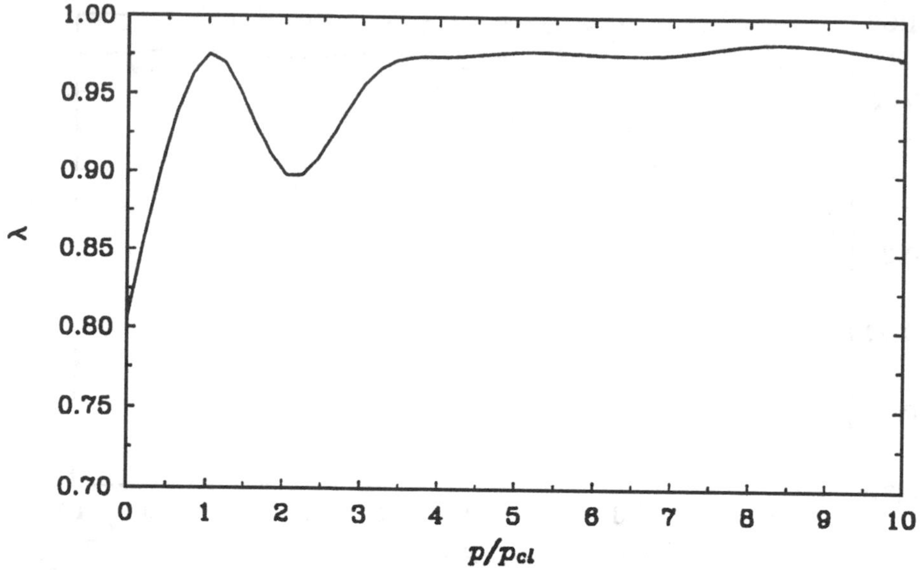

Figure 2.15 Buckling load parameter λ versus thickness variation parameter p (material: boron/epoxy; laminate profile: $[45°/-45°/45°]_{sym}$, $\epsilon = 0.1$).

However, compared with the isotropic shells, the effect of thickness variation on the buckling load is more complicated for the composite shells. From Figures 2.15, 2.16, 2.17, and 2.18, it is seen that the buckling load reduction is more remarkable in the axisymmetric buckling mode than in the asymmetric mode. For instance, if the material is glass/epoxy, the shell with the lamination profile $[\theta/-\theta/\theta]_{sym}$ buckles axisymmetrically when the fiber angle θ varies between 21° and 69°. In these cases of axisymmetric

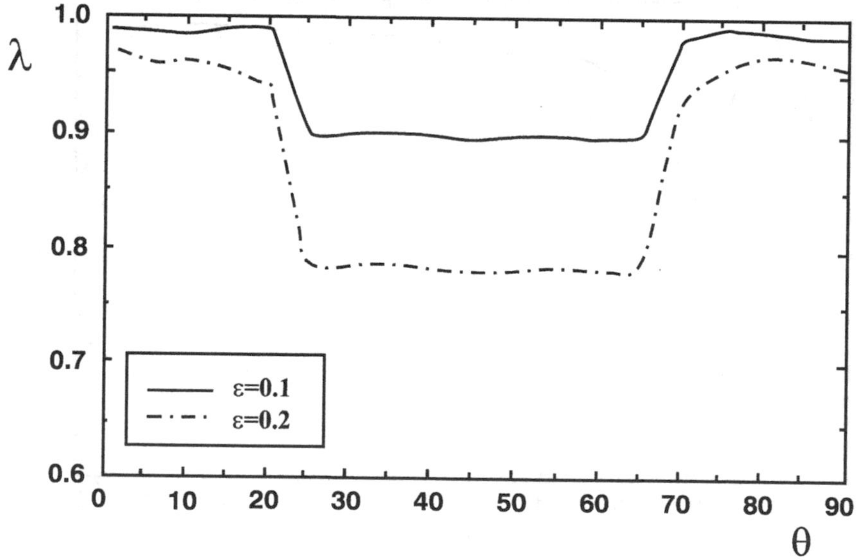

Figure 2.16 Buckling load parameter λ versus fiber angle θ for glass/epoxy shells (laminate profile: $[45°/-45°/45°]_{sym}$, $p = 2p_{cl}$).

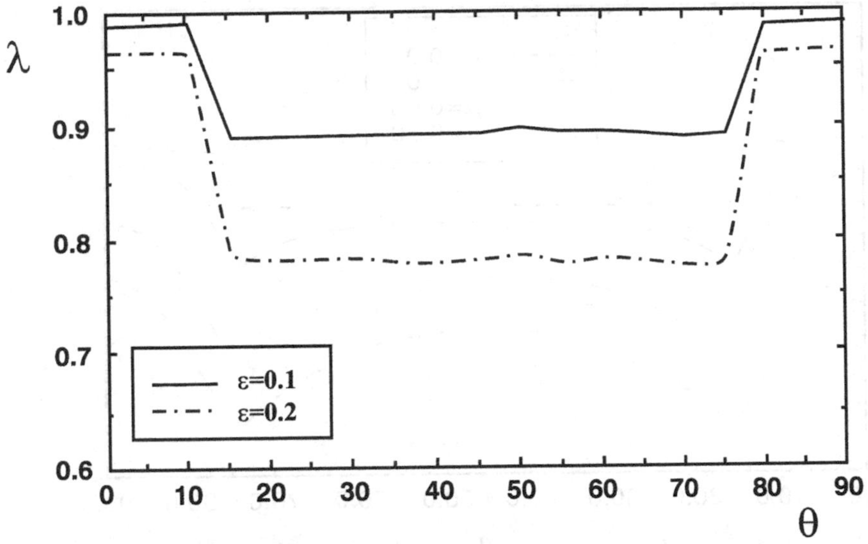

Figure 2.17 Buckling load parameter λ versus fiber angle θ for graphite/epoxy shells (laminate profile: $[45^\circ/-45^\circ/45^\circ]_{sym}$, $p = 2p_{cl}$).

buckling mode, the thickness variation of magnitude $\epsilon = 0.1$ brings about 10% buckling load reduction, whereas the buckling load reduction rate could reach 22% when ϵ is 0.2. However, the reduction rate λ remains almost the same in the entire range of axisymmetric buckling mode, regardless of what the constituent materials or the laminate fiber angles are, although these factors may affect remarkably the classical buckling load. Figures 2.17, 2.18, 2.19, and 2.20 illustrate the pattern of change of

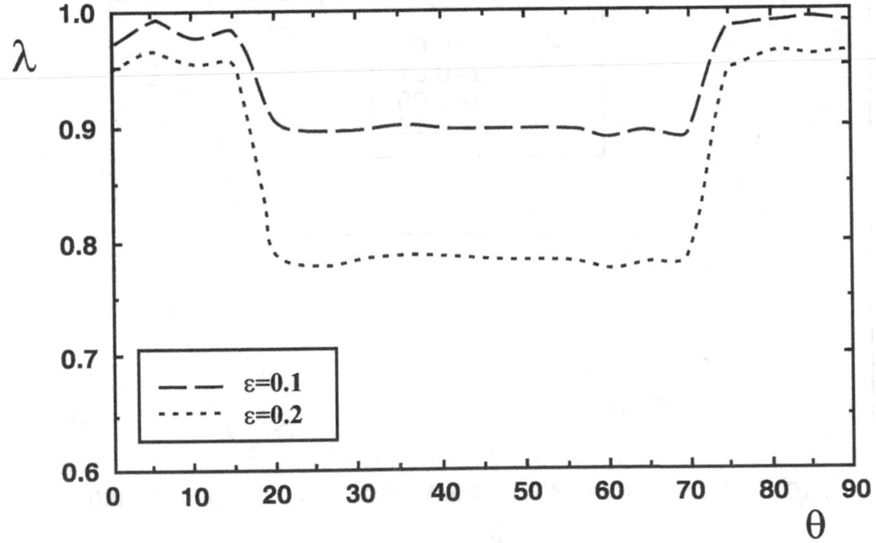

Figure 2.18 Buckling load parameter λ versus fiber angle θ for boron/epoxy shells (laminate profile: $[45^\circ/-45^\circ/45^\circ]_{sym}$, $p = 2p_{cl}$).

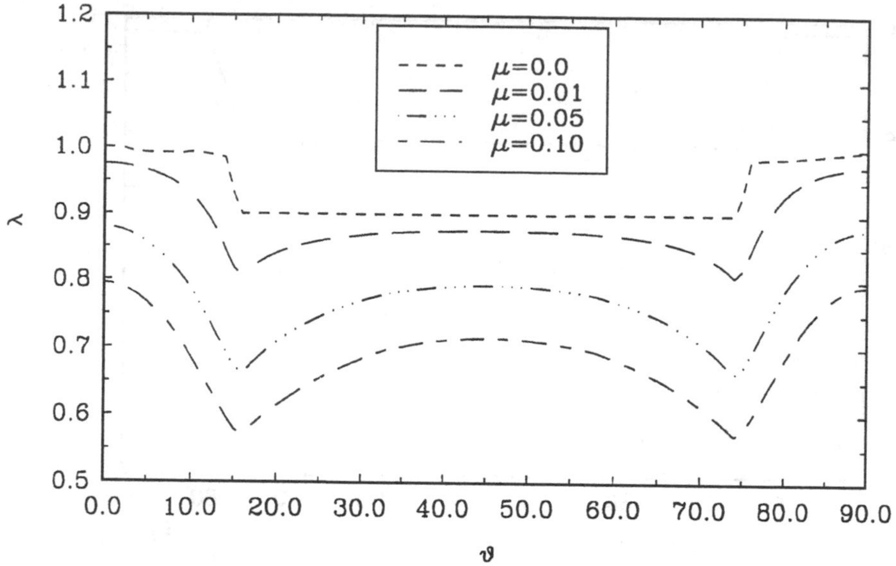

Figure 2.19 Effect of thickness variation and imperfection on the buckling load (laminate configuration $[\theta/-\theta/\theta/-\theta/\theta]_{sym}$, thickness variation parameter $\epsilon = 0.1$).

buckling load reduction factor λ with the fiber orientation when the shell contains the critical thickness variation.

As has been reported (see Khot, 1970a, 1970b), composite shells are less imperfection-sensitive than metallic shells. Nevertheless, the present investigation indicates that composite shells are as sensitive to the thickness variation as metallic shells.

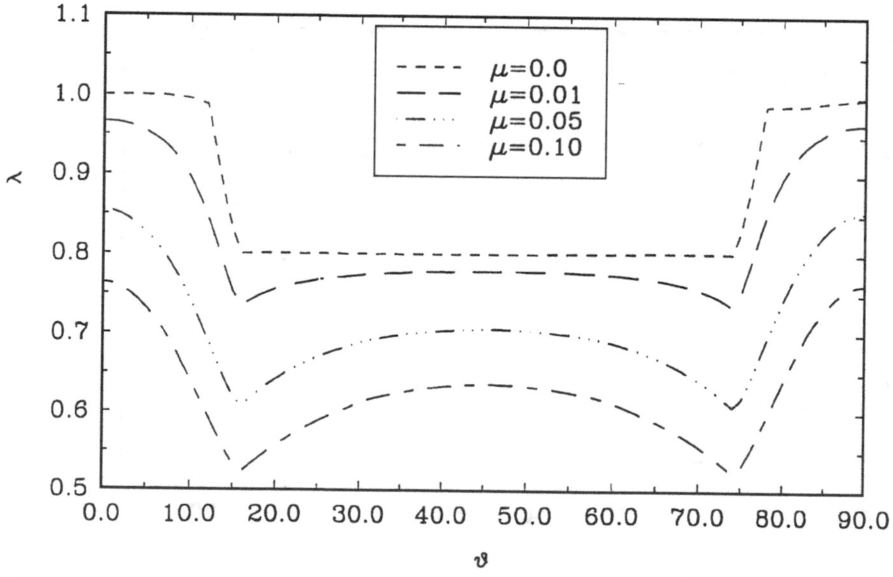

Figure 2.20 Effect of thickness variation and imperfection on the buckling load (laminate configuration $[\theta/-\theta/\theta/-\theta/\theta]_{sym}$, thickness variation parameter $\epsilon = 0.2$).

This makes the study of the effect of thickness variation more important for composite shells because they are more vulnerable to fabrication imprecision, which may directly lead to uneven wall thickness of the structure.

2.4 Effect of the Thickness Variation and Initial Imperfection on Buckling of Composite Cylindrical Shells

As in Li et al. (1997), this section aims at the combined effect of thickness variation and intial imperfection on the buckling behavior of the composite shells. We approach this problem by using Koiter's (1945, 1966b, 1980) energy criterion of elastic stability, as we have done in the isotropic case. Here, we consider the small axisymmetric thickness variation, and as a first approximation, only the terms up to the first order of thickness variation parameter are retained. The final product of this discussion is again an asymptotic formula that relates the thickness variation parameter and initial imperfection amplitude to the buckling load of the structure. Thus, the present analysis is actually a direct generalization and extension of our former investigation (Koiter et al., 1994a, 1994b) to the anisotropic case.

Membrane strain energy of a laminated cylindrical shell of length L is

$$U_m = \frac{1}{2}\int_0^{2\pi R}\int_0^L (N_x\epsilon_x + N_y\epsilon_y + N_{xy}\gamma_{xy})\,dxdy \tag{2.144}$$

Bending strain energy reads

$$U_b = \frac{1}{2}\int_0^{2\pi R}\int_0^L (M_x\kappa_x + M_y\kappa_y + M_{xy}\kappa_{xy})\,dxdy \tag{2.145}$$

For the shell under axial uniform end compression N_0, potential energy of the applied load takes the form

$$\Omega = -\frac{1}{2}\int_0^{2\pi R}\int_0^L N_0\left(\frac{\partial w}{\partial x} + \frac{\partial w_0}{\partial x}\right)^2 dxdy \tag{2.146}$$

where w_0 is the geometric initial imperfection.

Thus, the total potential energy reads

$$\Pi = U_m + U_b + \Omega \tag{2.147}$$

or, using the constitutive relations (2.99) and (2.100),

$$\Pi = \frac{1}{2}\int_0^{2\pi R}\int_0^L \Big[A_{11}\epsilon_x^2 + 2A_{12}\epsilon_x\epsilon_y + 2A_{16}\epsilon_x\gamma_{xy} + 2A_{26}\epsilon_y\gamma_{xy} + A_{22}\epsilon_y^2 + A_{66}\gamma_{xy}^2$$
$$+ D_{11}\kappa_x^2 + 2D_{12}\kappa_x\kappa_y + 2D_{16}\kappa_x\kappa_{xy} + 2D_{26}\kappa_x\kappa_{xy} + D_{22}\kappa_y^2$$
$$+ D_{66}\kappa_{xy}^2 - N_0\left(\frac{\partial w}{\partial x} + \frac{\partial w_0}{\partial x}\right)^2\Big]\,dxdy \tag{2.148}$$

Substitution of Equation (2.92) into Equation (2.148) leads to the energy expression in terms of displacements

$$\Pi = \frac{1}{2}\int_0^{2\pi R}\int_0^L \left[A_{11}\left[\frac{\partial u}{\partial x} + \frac{1}{2}\left(\frac{\partial w}{\partial x}\right)^2\right]^2 + 2A_{12}\left[\frac{\partial u}{\partial x} + \frac{1}{2}\left(\frac{\partial w}{\partial x}\right)^2\right]\right.$$
$$\times\left[\frac{\partial v}{\partial y} + \frac{w}{R} + \frac{1}{2}\left(\frac{\partial w}{\partial y}\right)^2\right] + 2A_{16}\left[\frac{\partial u}{\partial x} + \frac{1}{2}\left(\frac{\partial w}{\partial x}\right)^2\right]\left[\frac{\partial u}{\partial y} + \frac{\partial v}{\partial x} + \frac{\partial w}{\partial x}\frac{\partial w}{\partial y}\right]$$
$$+ 2A_{26}\left[\frac{\partial v}{\partial y} + \frac{w}{R} + \frac{1}{2}\left(\frac{\partial w}{\partial y}\right)^2\right]\left[\frac{\partial u}{\partial y} + \frac{\partial v}{\partial x} + \frac{\partial w}{\partial x}\frac{\partial w}{\partial y}\right]$$
$$+ A_{22}\left[\frac{\partial v}{\partial x} + \frac{w}{R} + \frac{1}{2}\left(\frac{\partial w}{\partial y}\right)^2\right]^2 + A_{66}\left[\frac{\partial u}{\partial y} + \frac{\partial v}{\partial x} + \frac{\partial w}{\partial x}\frac{\partial w}{\partial y}\right]^2$$
$$+ D_{11}\left(\frac{\partial^2 w}{\partial x^2}\right)^2 + 2D_{12}\frac{\partial^2 w}{\partial x^2}\frac{\partial^2 w}{\partial y^2} + 4D_{16}\frac{\partial^2 w}{\partial y^2}\frac{\partial^2 w}{\partial x \partial y} + 4D_{26}\frac{\partial^2 w}{\partial y^2}\frac{\partial^2 w}{\partial x \partial y}$$
$$\left. + D_{22}\left(\frac{\partial^2 w}{\partial y^2}\right)^2 + 4D_{66}\left(\frac{\partial^2 w}{\partial x \partial y}\right)^2 - N_0\left(\frac{\partial w}{\partial x} + \frac{\partial w_0}{\partial x}\right)^2\right] dxdy \qquad (2.149)$$

For the use of Koiter's energy criterion of elastic stability, variations of energy are performed at the fundamental (prebuckling) state.

The second variation of the energy for buckling modes is

$$P_2[\underline{u}] = \frac{1}{2}\int_0^{2\pi R}\int_0^L \left[A_{11}\left(\frac{\partial u}{\partial x}\right)^2 + 2A_{12}\frac{\partial u}{\partial x}\left(\frac{\partial v}{\partial y} + \frac{w}{R}\right) + 2A_{16}\frac{\partial u}{\partial x}\left(\frac{\partial u}{\partial y} + \frac{\partial v}{\partial x}\right)\right.$$
$$+ 2A_{26}\left(\frac{\partial v}{\partial y} + \frac{w}{R}\right)\left(\frac{\partial u}{\partial y} + \frac{\partial v}{\partial x}\right) + A_{22}\left(\frac{\partial v}{\partial y} + \frac{w}{R}\right)^2 + A_{66}\left(\frac{\partial u}{\partial y} + \frac{\partial v}{\partial x}\right)^2$$
$$+ D_{11}\left(\frac{\partial^2 w}{\partial x^2}\right)^2 + 2D_{12}\frac{\partial^2 w}{\partial x^2}\frac{\partial^2 w}{\partial y^2} + 4D_{16}\frac{\partial^2 w}{\partial x^2}\frac{\partial^2 w}{\partial x \partial y} + D_{22}\left(\frac{\partial^2 w}{\partial y^2}\right)^2$$
$$\left. + 4D_{26}\frac{\partial^2 w}{\partial y^2}\frac{\partial^2 w}{\partial x \partial y} + 4D_{66}\left(\frac{\partial^2 w}{\partial x \partial y}\right)^2 - N_0\left(\frac{\partial w}{\partial x}\right)^2\right] dx\, dy \qquad (2.150)$$

We will discuss the effect of the most critical type of axisymmetric geometrical imperfection $w_0(x) = -\mu h_0 \cos(2px/R)$ (Koiter, 1963; Tennyson et al., 1971), where h_0 is the nominal thickness of the shell, μ is the non-dimensional parameter describing the magnitude of the imperfection, and p is the wave number of the axisymmetric classical buckling mode, which has the following expression (Tennyson et al., 1971),

$$p^2 = \frac{R}{4\sqrt{a_{22}^* d_{11}^*}} \qquad (2.151)$$

We supplement the second variation with the additional bilinear term due to the geometric initial imperfection

$$P_{11}[\underline{u_0}, \underline{u}] = -N_0 \int_0^{2\pi R} \int_0^L \frac{\partial w}{\partial x} \frac{dw_0}{dx} \, dx \, dy \tag{2.152}$$

The third variation of the energy reads

$$\begin{aligned} P_3[\underline{u}] = \frac{1}{2} \int_0^{2\pi R} \int_0^L \Bigg\{ & A_{11} \frac{\partial u}{\partial x} \left(\frac{\partial w}{\partial x} \right)^2 + A_{12} \left[\frac{\partial u}{\partial x} \left(\frac{\partial w}{\partial y} \right)^2 + \left(\frac{\partial w}{\partial x} \right)^2 \left(\frac{\partial u}{\partial y} + \frac{w}{R} \right) \right] \\ & + 2A_{16} \left[\frac{\partial u}{\partial x} \frac{\partial w}{\partial x} \frac{\partial w}{\partial y} + \frac{1}{2} \left(\frac{\partial w}{\partial x} \right)^2 \left(\frac{\partial u}{\partial y} + \frac{\partial v}{\partial x} \right) \right] \\ & + 2A_{26} \left[\left(\frac{\partial v}{\partial y} + \frac{w}{R} \right) \frac{\partial w}{\partial x} \frac{\partial w}{\partial y} + \frac{1}{2} \left(\frac{\partial w}{\partial y} \right)^2 \left(\frac{\partial u}{\partial y} + \frac{\partial v}{\partial x} \right) \right] \\ & + A_{22} \left(\frac{\partial v}{\partial y} + \frac{w}{R} \right) \left(\frac{\partial w}{\partial y} \right)^2 + 2A_{66} \left(\frac{\partial u}{\partial y} + \frac{\partial v}{\partial x} \right) \frac{\partial w}{\partial x} \frac{\partial w}{\partial y} \Bigg\} \, dx \, dy \end{aligned} \tag{2.153}$$

We now assume that the buckling modes of the shell with a uniform thickness remain a good approximation for the buckling modes of the shell with small thickness variations. We are, again as in the isotropic case, ensured that the critical load obtained in this way is by the energy principle an upper bound for the actual critical buckling load.

According to the study of Tennyson et al. (1971), the following buckling mode expression can be adopted for the laminated cylindrical shell with the aforementioned axisymmetric initial imperfection w_0:

$$w = b \cos \frac{2px}{R} + C_n \cos \frac{px}{R} \cos \frac{ny}{R} \tag{2.154}$$

where b and C_n are constants, n is the number of waves in the circumferential direction of the shell. If we recall the shell equilibrium equations in terms of displacements u, v, and w (Vinson Sierakowski, 1986)

$$\begin{pmatrix} L_{11} & L_{12} & L_{13} \\ L_{12} & L_{22} & L_{23} \\ L_{13} & L_{23} & L_{33} \end{pmatrix} \begin{Bmatrix} u \\ v \\ w \end{Bmatrix} = \begin{Bmatrix} 0 \\ 0 \\ N_0 \dfrac{\partial^2 w}{\partial x^2} \end{Bmatrix} \tag{2.155}$$

where operators L_{ij} are

$$\begin{aligned}
L_{11} &= a_{11}\frac{\partial}{\partial x^2} + 2a_{16}\frac{\partial^2}{\partial x \partial y} + a_{66}\frac{\partial^2}{\partial y^2}, \\
L_{12} &= a_{16}\frac{\partial^2}{\partial x^2} + (a_{12} + a_{66})\frac{\partial^2}{\partial x \partial y} + a_{26}\frac{\partial^2}{\partial y^2} \\
L_{13} &= \frac{1}{R}\left(a_{12}\frac{\partial}{\partial x} + a_{26}\frac{\partial}{\partial y}\right), \quad L_{22} = a_{66}\frac{\partial^2}{\partial x^2} + 2a_{26}\frac{\partial^2}{\partial x \partial y} + a_{22}\frac{\partial^2}{\partial y^2}, \\
L_{23} &= \frac{1}{R}\left(a_{26}\frac{\partial}{\partial x} + a_{22}\frac{\partial}{\partial y}\right) \\
L_{33} &= \frac{a_{22}}{R^2} + d_{11}\frac{\partial^4}{\partial x^4} + 4d_{16}\frac{\partial^4}{\partial x^3 \partial y} + 2(d_{12} + 2d_{66})\frac{\partial^4}{\partial x^2 \partial y^2} \\
&\quad + 4d_{26}\frac{\partial^4}{\partial x \partial y^3} + d_{22}\frac{\partial^4}{\partial y^4}
\end{aligned} \tag{2.156}$$

We can obtain the expressions for u and v as follows:

$$\begin{aligned}
u &= -\frac{a_{12}}{2pa_{11}} b \sin\frac{2px}{R} + Q_n C_n \sin\frac{px}{R}\cos\frac{ny}{R} \\
v &= K_n C_n \cos\frac{px}{R}\sin\frac{ny}{R}
\end{aligned} \tag{2.157}$$

where

$$\begin{aligned}
K_n &= -\frac{n\left(p^2 a_{11}a_{22} + n^2 a_{66}a_{22} - a_{12}^2 p^2 - a_{66}a_{12}p^2\right)}{(a_{11}p^2 + a_{66}n^2)(a_{66}p^2 + a_{22}n^2) - (a_{12} + a_{66})^2 p^2 n^2} \\
Q_n &= -\frac{pa_{66}(p^2 a_{12} - n^2 a_{22})}{(a_{11}p^2 + a_{66}n^2)(a_{66}p^2 + a_{22}n^2) - (a_{12} + a_{66})^2 p^2 n^2}
\end{aligned} \tag{2.158}$$

It should be mentioned that in deriving solution (2.157), we again used an assumption in the studies of Tasi (1966) and Hirano (1979) that the coupling stiffnesses A_{16}, A_{26}, D_{16}, and D_{26} can be approximately set to zero.

In the numerical analysis of composite shells with thickness variation, we have deduced that, in the absence of the geometric imperfection, the thickness variation has the most degrading effect on the buckling load when the wave number of the thickness variation is twice that of the classical buckling mode, that is $p_1 = 2p$. This result is also observed in the isotropic shell case (Koiter et al., 1994a, 1994b). Now we are interested in the combined effect of the most critical geometric imperfection and the most detrimental thickness variation on the reduction of the buckling load.

Substituting Equations (2.154) and (2.157) into the second and third variations, we obtain, after retaining only the first-order terms in ϵ,

$$\begin{aligned}
P_2[\underline{u}] &= \frac{C_n^2 \pi L}{4R^3}\left[d_{22}n^4 + 2d_{12}n^2p^2 + 4d_{66}n^2p^2 + d_{11}p^4 + a_{22}(1 + K_n n)^2 R^2\right. \\
&\quad \left. - N_0 p^2 R^2 + 2a_{12}(1 - K_n n)Q_n p R^2 + a_{11}Q_n^2 p^2 R^2\right.
\end{aligned}$$

$$+a_{66}(K_n p + nQ_n)^2 R^2] + \frac{1}{2R^3} b^2 \left[16d_{11} p^4 \left(1 - \frac{3}{2}\epsilon\right) - \frac{a_{12}^2 R^2}{a_{11}} \left(1 - \frac{1}{2}\epsilon\right) + a_{22} R^2 \left(1 - \frac{1}{2}\epsilon\right) - 4N_0 R^2 p^2 \right] \quad (2.159)$$

$$P_{11}[\underline{u_0}, \underline{u}] = \frac{4bh_0 N_0 p^2 \mu \pi L}{R} \quad (2.160)$$

$$P_3[\underline{u}] = \frac{bC_n^2 \pi L}{8R^2} \left\{ a_{12} \left[\left(-1 + \frac{1}{2}\epsilon\right) p^2 + 4\left(1 + \frac{1}{2}\epsilon\right)(1 + K_n n) p^2 \right] + a_{22} \left(1 - \frac{1}{2}\epsilon\right) n^2 + \frac{a_{12}^2 n^2}{a_{11}} \left(-1 + \frac{1}{2}\epsilon\right) + a_{12} \left(1 - \frac{1}{2}\epsilon\right) p^2 + 4a_{11} p^3 Q_n \left(1 + \frac{1}{2}\epsilon\right) - 4a_{66} \left(1 + \frac{1}{2}\epsilon\right) np(K_n p + Q_n n) \right\} \quad (2.161)$$

Thus, the energy expression to be considered is

$$P_2[\underline{u}] + P_{11}[\underline{u_0}, \underline{u}] + P_3[\underline{u}] \quad (2.162)$$

The equations for the initial post-buckling behavior are furnished by taking the partial derivatives of the energy expression with respect to b and C_n to be zero:

$$\frac{b}{R^3} \left[16d_{11} p^4 \left(1 - \frac{3}{2}\epsilon\right) - \frac{a_{12}^2 R^2}{a_{11}} \left(1 - \frac{1}{2}\epsilon\right) + a_{22} R^2 \left(1 - \frac{1}{2}\epsilon\right) - 4N_0 R^2 p^2 \right] + \frac{4N_0 h_0 p^2 \mu}{R} + \frac{C_n^2}{8R^2} \left\{ a_{12} \left[\left(-1 + \frac{1}{2}\epsilon\right) p^2 + 4\left(1 + \frac{1}{2}\epsilon\right)(1 + K_n n) p^2 \right] + a_{22} \left(1 - \frac{1}{2}\epsilon\right) n^2 + \frac{a_{12}^2 n^2}{a_{11}} \left(-1 + \frac{1}{2}\epsilon\right) + a_{12} \left(1 - \frac{1}{2}\epsilon\right) p^2 + 4a_{11} p^3 Q_n \left(1 + \frac{1}{2}\epsilon\right) - 4a_{66} \left(1 + \frac{1}{2}\epsilon\right) np(K_n p + Q_n n) \right\} = 0 \quad (2.163)$$

and

$$\frac{C_n}{2R^3} [d_{22} n^4 + 2d_{12} n^2 p^2 + 4d_{66} n^2 p^2 + d_{11} p^4 + a_{22}(1 + K_n n)^2 R^2 - N_0 p^2 R^2 + 2a_{12}(1 + K_n n) Q_n p R^2 + a_{11} Q_n^2 p^2 R^2 + a_{66}(K_n p + nQ_n)^2 R^2] + \frac{bC_n}{4R^2} \left\{ a_{12} \left[\left(-1 + \frac{1}{2}\epsilon\right) p^2 + 4\left(1 + \frac{1}{2}\epsilon\right)(1 + K_n n) p^2 \right] + a_{22} \left(1 - \frac{1}{2}\epsilon\right) n^2 + \frac{a_{12}^2 n^2}{a_{11}} \left(-1 + \frac{1}{2}\epsilon\right) + a_{12} \left(1 - \frac{1}{2}\epsilon\right) p^2 + 4a_{11} p^3 Q_n \left(1 + \frac{1}{2}\epsilon\right) - 4a_{66} \left(1 + \frac{1}{2}\epsilon\right) np(K_n p + Q_n n) \right\} = 0 \quad (2.164)$$

The solution $C_n = 0$ of Equation (2.164) leads to

$$b = -\frac{4N_0 h_0 p^2 \mu R^2}{16d_{11}p^4\left(1-\frac{3}{2}\epsilon\right) - \frac{a_{12}^2 R^2}{a_{11}}\left(1-\frac{1}{2}\epsilon\right) + a_{22}R^2\left(1-\frac{1}{2}\epsilon\right) - 4N_0 R^2 p^2} \tag{2.165}$$

from Equation (2.163), and bifurcation buckling with respect to the asymmetric mode with amplitude C_n occurs at

$$\begin{aligned} b = &-\frac{2}{R}[d_{22}n^4 + 2d_{12}n^2p^2 + 4d_{66}n^2p^2 + d_{11}p^4 + a_{22}(1+K_n n)^2R^2 - N_0p^2R^2 \\ &+ 2a_{12}(1+K_n n)Q_n pR^2 + a_{11}Q_n^2p^2R^2 + a_{66}(K_n p + nQ_n)^2R^2] \\ &\div \left\{a_{12}\left[\left(-1+\frac{1}{2}\epsilon\right)p^2 + 4\left(1+\frac{1}{2}\epsilon\right)(1+K_n n)p^2\right] + a_{22}\left(1-\frac{1}{2}\epsilon\right)n^2\right. \\ &+ \frac{a_{12}^2n^2}{a_{11}}\left(-1+\frac{1}{2}\epsilon\right) + a_{12}\left(1-\frac{1}{2}\epsilon\right)p^2 + 4a_{11}p^3Q_n\left(1+\frac{1}{2}\epsilon\right) \\ &\left.- 4a_{66}\left(1+\frac{1}{2}\epsilon\right)np(K_n p + Q_n n)\right\} \end{aligned} \tag{2.166}$$

Equating expressions (2.165) and (2.166), we obtain the equation for the critical buckling load N_0

$$\begin{aligned} &\left[16d_{11}p^4\left(1-\frac{3}{2}\epsilon\right) - \frac{a_{12}^2R^2}{a_{11}}\left(1-\frac{1}{2}\epsilon\right) + a_{22}R^2\left(1-\frac{1}{2}\epsilon\right) - 4N_0R^2p^2\right] \\ &\quad\times [d_{22}n^4 + 2d_{12}n^2p^2 + 4d_{66}n^2p^2 + d_{11}p^4 + a_{22}(1+K_n n)^2R^2 \\ &\quad - N_0p^2R^2 + 2a_{12}(1+K_n n)Q_n pR^2 + a_{11}Q_n^2p^2R^2 + a_{66}(K_n p + nQ_n)^2R^2] \\ &\quad - 2N_0h_0p^2\mu R^3\left\{a_{12}\left[\left(-1+\frac{1}{2}\epsilon\right)p^2 + 4\left(1+\frac{1}{2}\epsilon\right)(1+K_n n)p^2\right]\right. \\ &\quad + a_{22}\left(1-\frac{1}{2}\epsilon\right)n^2 + \frac{a_{12}^2n^2}{a_{11}}\left(-1+\frac{1}{2}\epsilon\right) + a_{12}\left(1-\frac{1}{2}\epsilon\right)p^2 \\ &\quad \left.+ 4a_{11}p^3Q_n\left(1+\frac{1}{2}\epsilon\right) - 4a_{66}\left(1+\frac{1}{2}\epsilon\right)np(K_n p + Q_n n)\right\} = 0 \end{aligned} \tag{2.167}$$

It should be pointed out that, in solving Equation (2.167), integer search should be performed with respect to the circumferential wave number n to arrive at the lowest value of N_0.

We define the non-dimensional critical load parameter λ

$$\lambda = \frac{N_0}{N_{\mathrm{cl}}} \tag{2.168}$$

where N_{cl} is the classical buckling load in the absence of both initial imperfection and thickness variation and has the following expression (Vinson and Sierakowski, 1986):

$$N_{cl} = \min_{m,n}\{N_{m,n}\}$$

$$N_{m,n} = \left(\frac{L}{m\pi}\right)^2 \frac{C_{11}C_{22}C_{33} + 2C_{12}C_{23}C_{13} - C_{13}^2C_{22} - C_{23}^2C_{11} - C_{12}^2C_{33}}{C_{11}C_{22} - C_{12}^2} \tag{2.169}$$

where

$$C_{11} = A_{11}\left(\frac{m\pi}{L}\right)^2 + A_{66}\left(\frac{n}{R}\right)^2, \qquad C_{22} = A_{22}\left(\frac{n}{R}\right)^2 + A_{66}\left(\frac{m\pi}{L}\right)^2$$

$$C_{33} = D_{11}\left(\frac{m\pi}{L}\right)^4 + 2(D_{12} + 2D_{66})\left(\frac{m\pi}{L}\right)^2\left(\frac{n}{R}\right)^2 + D_{22}\left(\frac{n}{R}\right)^4 + \frac{A_{22}}{R^2}$$

$$C_{12} = (A_{12} + A_{66})\left(\frac{m\pi}{L}\right)\left(\frac{n}{R}\right), \qquad C_{13} = \frac{A_{12}}{R}\left(\frac{m\pi}{L}\right),$$

$$C_{23} = \frac{A_{22}}{R}\left(\frac{n}{R}\right) + B_{22}\left(\frac{n}{R}\right)^3 \tag{2.170}$$

Equation (2.167) is very useful in the sense that the axial buckling load can be determined from it for composite cylindrical shells containing small axisymmetric initial imperfection and thickness variation. For practical purposes, the results thus obtained should be considered conservative, since the most detrimental case of geometric imperfection and thickness variation is investigated. However, since we ignored in our derivation the higher order terms in ϵ, the results from the present study should not be deemed accurate for shells having large thickness variation.

As a numerical example, we discuss the shells made of carbon/epoxy laminae, whose elastic moduli are $E_1 = 13.75 \times 10^6$ psi, $E_2 = 1.03 \times 10^6$ psi, $\nu_{12} = 0.25$, $G_{12} = 0.42 \times 10^6$ psi. The shell is 6 in. in radius and 30 in. in length and is composed of ten equally thick layers, each being 0.012-in. thick. The laminate configuration is $[\theta/-\theta/\theta/-\theta/\theta]_{\text{sym}}$, with, the fiber angle θ varying from 0° to 90°.

Solving Equation (2.167) numerically for the critical load N_0 with integer search performed simultaneously with respect to the circumferential buckling wave number n, and then non-dimensionalizing the result in virtue of (2.168), we obtain the critical buckling load factor λ for different cases of thickness variation parameter ϵ and imperfection amplitude μ. The numerical results here obtained are in agreement with the previous first-order asymptotic formula $\lambda = 1 - \epsilon$, which holds only for the axisymmetric buckling cases for composite shells without initial imperfection. It is interesting to note that as long as the axisymmetric buckling mode dominates, the buckling load reduction factor λ remains practically constant, irrespective of the change in the laminate construction. However, after the shell involves initial imperfection, the buckling mode becomes entirely non-axisymmetric, and the buckling load reduction is strongly influenced by the stacking sequence of the laminae. Figure 2.21 depicts the results of the buckling load factor λ for shells of different laminate profiles, such as

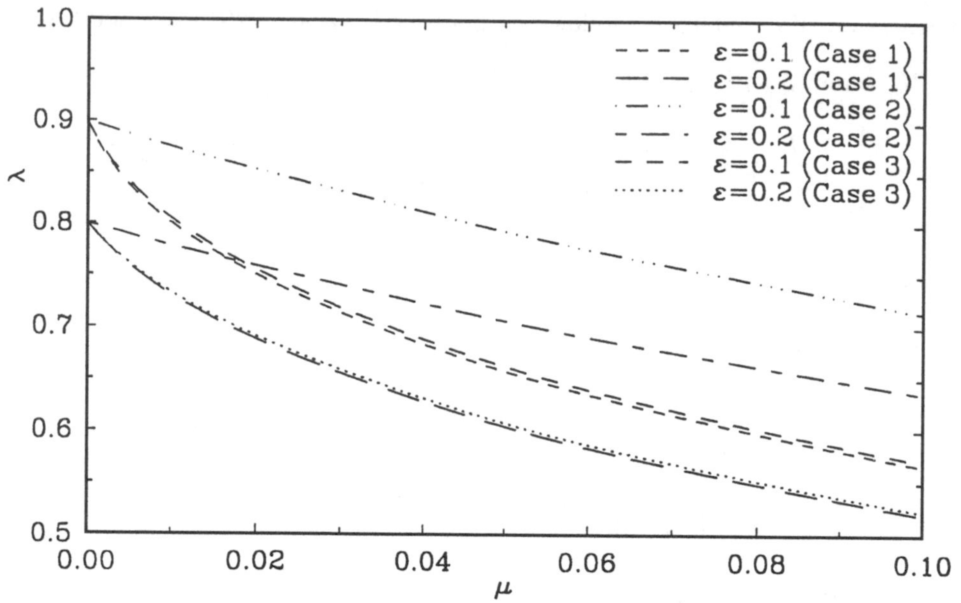

Figure 2.21 Buckling load reduction in shells with different laminate configurations (case 1: isotropic; case 2: $[45°/-45°/45°/-45°/45°]_{sym}$; case 3: $[16°/-16°/16°/-16°/16°]_{sym}$).

$[45°/-45°/45°/-45°/45°]_{sym}$ and $[16°/-16°/16°/-16°/16°]_{sym}$, together with the results for corresponding isotropic shells. It can be seen from this figure that composite shells are sensitive to thickness variation, and especially to initial imperfection, and that such a sensitivity to the thickness variation is comparable to that of the isotropic shell. Once again, we can conclude that, despite the fact that the geometric initial imperfection remains as a dominant factor for the buckling load reduction, the further degradation in the load-bearing capacity of the structure due to the effect of thickness variations should not be overlooked.

In order to investigate the accuracy of the preceding asymptotic formula, an effort has been made to use BOSOR4 (Bushnell, 1976), a well-known commercial software for buckling analysis, to generate a set of comparable data for the non-dimensional critical load parameter λ. Since the classical buckling load has been used to non-dimensionalize the critical buckling load, it is necessary to check the results from Equation (2.169) with their counterparts from the numerical software so that a common basis can be established for the follow-up comparison of results for non-dimensional critical load λ. For this purpose, software packages BOSOR4 and PANDA2 (Bushnell, 1983) (using both the linear and non-linear shell theories) were utilized for the classical buckling load and numerical results are plotted together with those from Equation (2.169) (Figure 2.22). Figure 2.22 shows that the classical buckling loads from different sources agree quite well except for the cases where θ lies between 53° and 80°. When θ is between 53° and 80°, Equation (2.169) produces a set of data that are very close to the results from PANDA2 using the shallow shell theory. However, a discrepancy is found between the results based on the shallow shell theory and those from BOSOR4 and PANDA2 using Sanders non-linear shell theory (Sanders, 1963). This indicates that, for those cases

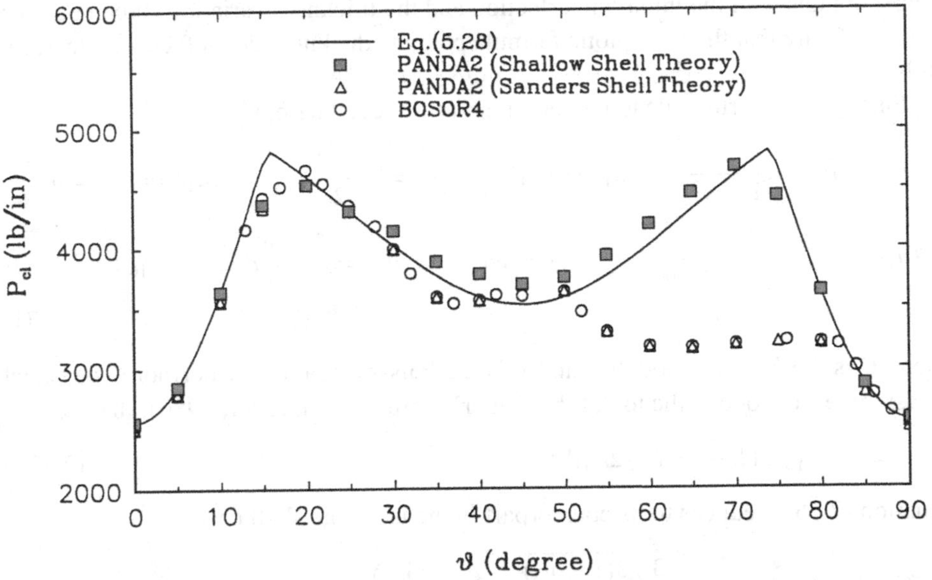

Figure 2.22 Comparison of classical buckling loads stemming from different methods.

where the lamination angle θ ranges from 53° to 80°, the shallow shell theory appears to be inadequate, and a more refined, non-linear theory may be necessary for the prediction of buckling loads. Figure 2.23 displays the results of the non-dimensional critical load parameter λ obtained from the asymptotic formula [Equation (2.167)] and BOSOR4 for a 10-layer composite shell (laminate configuration: $[16°/-16°/16°/-16°/16°]_{\text{sym}}$),

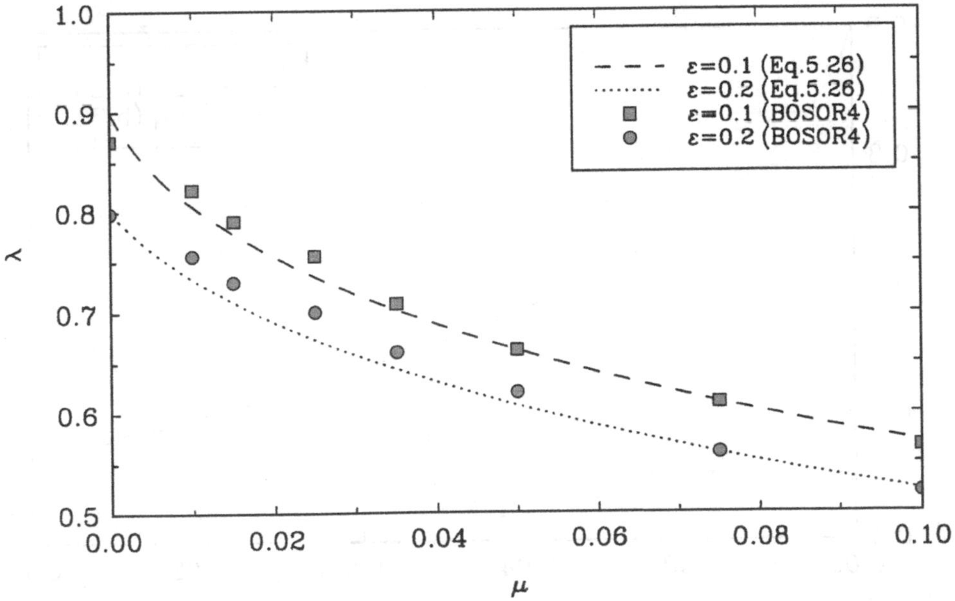

Figure 2.23 Comparison of the non-dimensional critical load λ obtained from the asymptotic formula and BOSOR4.

which contains both the initial imperfection and the thickness variation. It can be seen from this figure that the asymptotic formula predicts the knockdown factor quite accurately.

Finally, it is worthwhile to mention that, as a special case, if we let

$$a_{11} = a_{22} = \frac{Eh_0}{1-\nu^2}, \qquad a_{12} = \nu a_{11}, \qquad a_{66} = \frac{1-\nu}{2} a_{11}, \qquad a_{16} = a_{26} = 0$$

$$d_{11} = d_{22} = \frac{Eh^3}{12(1-\nu^2)}, \qquad d_{12} = \nu d_{11}, \qquad d_{66} = \frac{1-\nu}{2} d_{11}, \qquad d_{16} = d_{26} = 0 \tag{2.171}$$

where E is the Young's modulus, and ν is the Poisson's ratio; furthermore, we select the asymmetric mode at the top of the Koiter's semi-circle (Koiter, 1980); that is, let

$$p = n = [\sqrt{3(1-\nu^2)}R^2/2h_0]^{1/2} \tag{2.172}$$

Equation (2.167) reduces to its counterpart in the isotropic shell case,

$$(1-\lambda)(1-\epsilon-\lambda) - \frac{3\sqrt{3(1-\nu^2)}}{2}\lambda\mu\left(1+\frac{1}{6}\epsilon\right) = 0 \tag{2.173}$$

which is identical to Equation (2.59), if the small term $\epsilon/6$ is ignored as compared with unity (see also Koiter et al., 1994b).

Figure 2.24 shows the comparison of results in the isotropic shell case using Koiter's circle and those using integer search with respect to the circumferential wave number n, and it is seen that the agreement is excellent.

For further details, consult the study by Li et al. (1997).

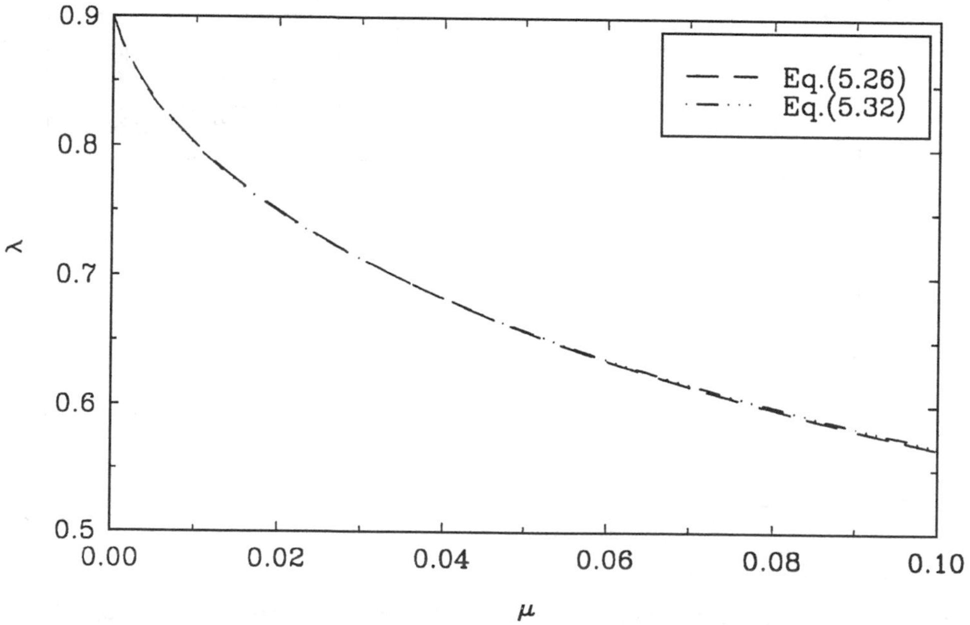

Figure 2.24 Comparison of results using Koiter's semi-circle and those using integer search for isotropic shells with thickness variation $\epsilon = 0.2$).

2.5 Effect of the Dissimilarity in Elastic Moduli on the Buckling

In this section, it is demonstrated through a simple example of the Roorda-Koiter frame that the unavoidable dissimilarity in the distribution of elastic moduli may further reduce the load-carrying capacity in addition to the well-recognized effect of initial geometric imperfections.

The works by Koiter (1945, 1963) and Budiansky and Hutchinson (1964) established that the initial geometric imperfection – deviation from the nominally ideal shape – plays a major role in the reduction of the load-carrying capacity of structures. An additional source of such a reduction was identified recently by Koiter et al. (1994b) as a non-uniform thickness of the structure. To the best of the authors' knowledge, there is only a single work, by Ikeda and Murota (1990a, 1990b), that points out that the symmetry breaking in the distribution of elastic moduli may cause additional reduction in the loads the system is able to sustain. In particular, they analyzed the von Mises truss in such a context. In this section, we perform analytical and numerical investigation on a frame structure to study the effect of dissimilarity of elastic moduli on the load-carrying capacity of the structure.

2.5.1 Analysis

In 1965, Roorda (1965) conducted a set of experiments on the two-bar frame, subjected to an eccentric load. Koiter (1967) performed an initial-imperfection analysis of this frame, referred to in the literature since then as the *Roorda-Koiter frame*. Roorda and Chilver (1970) and Bažant and Cedolin (1989) analyzed this structure from different perspectives (for additional references, see also the text of Brush and Almroth, 1975). Here, the vertical and the horizontal segments of the two-bar frame (Figure 2.25) may

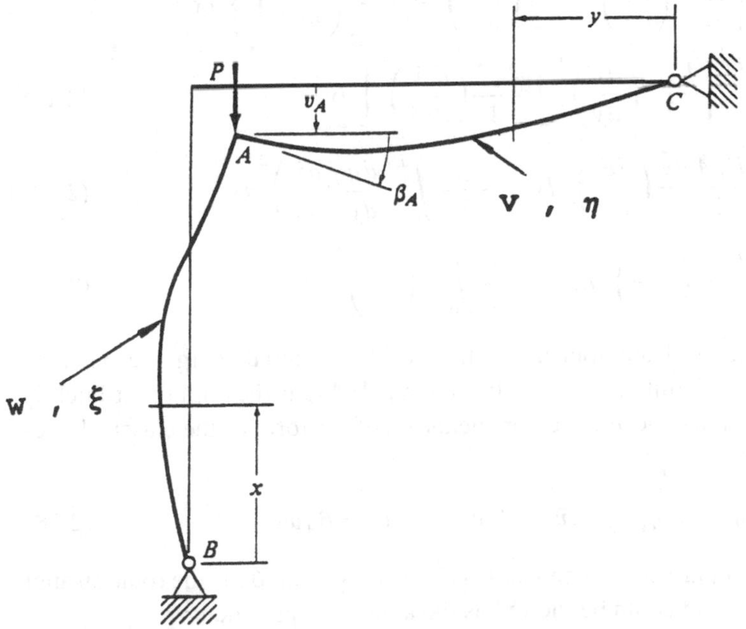

Figure 2.25 Roorda-Koiter frame.

have different lengths L_1 and L_2 and different Young's moduli E_1 and E_2. Our aim is to study the effect of small differences in the elastic moduli between the two segments, namely, the vertical and the horizontal bars of the frame structure. Such a difference may trigger the dissimilarity in the elastic moduli of the structure.

Following those previous investigators, we first analyze the perfect structure. The total potential energy of the two-bar frame is

$$\Pi = \int_0^{L_1} \left\{ \frac{E_1 A}{2} \left[\frac{d\xi}{dx} + \frac{1}{2}\left(\frac{dw}{dx}\right)^2 \right]^2 + \frac{E_1 I}{2}\left(\frac{d^2 w}{dx^2}\right)^2 \right\} dx$$
$$+ \int_0^{L_2} \left\{ \frac{E_2 A}{2} \left[\frac{d\eta}{dy} + \frac{1}{2}\left(\frac{dv}{dy}\right)^2 \right]^2 + \frac{E_2 I}{2}\left(\frac{d^2 v}{dy^2}\right)^2 \right\} dy - P v_A \tag{2.174}$$

The change in potential energy is obtained by letting

$$\xi \to \xi_0 + \bar{\xi}, \qquad \eta \to \eta_0 + \bar{\eta}, \qquad v \to v_0 + \bar{v}, \qquad w \to w_0 + \bar{w} \tag{2.175}$$

where

$$\eta_0 = v_0 = w_0 = 0, \qquad \xi_0 = -\frac{Px}{EA} \tag{2.176}$$

represents an equilibrium state on the primary equilibrium path in the neighborhood of the bifurcation point A. The energy expression may take the following form:

$$\Delta\Pi = \frac{1}{2!}\delta^2\Pi + \frac{1}{3!}\delta^3\Pi + \frac{1}{4!}\delta^4\Pi \tag{2.177}$$

where

$$\frac{1}{2!}\delta^2\Pi = \int_0^{L_1} \left\{ \frac{E_1 A}{2}\left(\frac{d\xi}{dx}\right)^2 - \frac{P}{2}\left(\frac{d\bar{w}}{dx}\right)^2 + \frac{E_1 I}{2}\left(\frac{d^2\bar{w}}{dx^2}\right)^2 \right\} dx$$
$$+ \int_0^{L_2} \left\{ \frac{E_2 A}{2}\left(\frac{d\bar{\eta}}{dy}\right)^2 + \frac{E_2 I}{2}\left(\frac{d^2\bar{v}}{dy^2}\right)^2 \right\} dy \tag{2.178}$$

$$\frac{1}{3!}\delta^3\Pi = \int_0^{L_1} \frac{E_1 A}{2}\frac{d\bar{\xi}}{dx}\left(\frac{d\bar{w}}{dx}\right)^2 dx + \frac{E_2 A}{2}\int_0^{L_2} \frac{d\bar{\eta}}{dy}\left(\frac{d\bar{v}}{dy}\right)^2 dy \tag{2.179}$$

$$\frac{1}{4!}\delta^4\Pi = \int_0^{L_1} \frac{E_1 A}{8}\left(\frac{d\bar{w}}{dx}\right)^4 dx + \frac{E_2 A}{8}\int_0^{L_2} \left(\frac{d\bar{v}}{dy}\right)^4 dy \tag{2.180}$$

correspond to second, third, and fourth variations of the potential energy, respectively.

According to Koiter's initial post-buckling theory (1945), in the initial post-buckling range, the incremental displacement components are of the form of the classical buckling mode

$$\bar{\xi} = \beta_A \xi_1, \qquad \bar{\eta} = \beta_A \eta_1, \qquad \bar{v} = \beta_A v_1, \qquad \bar{w} = \beta_A w_1 \tag{2.181}$$

where ξ_1, η_1, v_1, and w_1 are normalized classical buckling load. β_A is the rotation angle at bifurcation point A and could be viewed as the amplitude parameter.

Substituting (2.181) into the second variation leads to

$$\frac{1}{2!}\delta^2\Pi = \frac{1}{2}\int_0^{L_1}\left[E_1A\left(\frac{d\xi_1}{dx}\right)^2 - P\left(\frac{dw_1}{dx}\right)^2 + \frac{E_1I}{2}\left(\frac{d^2w_1}{dx^2}\right)^2\right]dx\beta_A^2$$
$$+\frac{1}{2}\int_0^{L_2}\left[E_2A\left(\frac{d\eta_1}{dy}\right)^2 + E_2I\left(\frac{d^2v_1}{dy^2}\right)^2\right]dy\beta_A^2 \qquad (2.182)$$

Performing variation on expression (2.182), we obtain, according to the Trefftz criterion,

$$\delta\left(\frac{1}{2!}\delta^2\Pi\right) = \int_0^{L_1}\left[E_1A\left(\frac{d\xi_1}{dx}\right)\frac{d(\delta\xi_1)}{dx} - P\left(\frac{dw_1}{dx}\right)\frac{d(\delta w_1)}{dx}\right.$$
$$\left.+\frac{E_1I}{2}\left(\frac{d^2w_1}{dx^2}\right)\frac{d^2(\delta w_1)}{dx^2}\right]dx\beta_A^2$$
$$+\int_0^{L_2}\left[E_2A\left(\frac{d\eta_1}{dy}\right)^2\frac{d(\delta\eta)}{dy} + E_2I\left(\frac{d^2v_1}{dy^2}\right)\frac{d^2(\delta v_1)}{dy^2}\right]dy\beta_A^2 = 0$$
$$(2.183)$$

Integration by parts and rearrangement gives

$$\left[E_A\frac{d\xi_1}{dx}\delta\xi_1\right]_0^{L_1} + \left[E_2A\frac{d\eta_1}{dy}\delta\eta_1\right]_0^{L_2} - \left[\left(E_1I\frac{d^3w_1}{dx^3} + P\frac{dw_1}{dx}\right)\delta w_1\right]_0^{L_1}$$
$$-\left[E_2I\frac{d^3v_1}{dy^3}\delta v_1\right]_0^{L_2} + \left[E_1I\frac{d^2w_1}{dx^2}\frac{d(\delta w_1)}{dx}\right]_0^{L_1} + \left[E_2I\frac{d^2v_1}{dy^2}\frac{d(\delta v_1)}{dy}\right]_0^{L_1}$$
$$-\int_0^{L_1}E_1A\frac{d^2\xi_1}{dx^2}\delta\xi_1\,dx - \int_0^{L_2}E_2A\frac{d^2\eta_1}{dy^2}\delta\eta_1\,dy$$
$$+\int_0^{L_1}\left(E_1I\frac{d^4w_1}{dx^4} + P\frac{d^2w_1}{dx^2}\right)\delta w_1\,dx + \int_0^{L_2}E_2I\frac{d^4v_1}{dy^4}\delta v_1\,dy = 0 \qquad (2.184)$$

The satisfaction of Equation (2.184) results in boundary conditions (using $\xi|_{x=L_1} = -v|_{y=L_2}$; $\eta|_{y=L_2} = w|_{x=L_2}$);

$$\frac{d^2w_1}{dx^2} = 0 \quad \text{at} \quad x = 0; \qquad \frac{d^2v_1}{dy^2} = 0 \quad \text{at} \quad y = 0 \qquad (2.185)$$

$$E_1A\frac{d\xi_1}{dx} + E_2I\frac{d^3v_1}{dy^3} = \frac{d^2w_1}{dx^2} + \frac{d^2v_1}{dy^2} = 0 \quad \text{at} \quad x = L_1, \qquad y = L_2 \qquad (2.186)$$

$$E_2A\frac{d\eta_1}{dy} - E_1I\frac{d^3v_1}{dx^3} - P\frac{dw_1}{dx} = 0; \qquad \text{at} \quad x = L_1, \qquad y = L_2 \qquad (2.187)$$

and governing equations:

$$\begin{aligned} &\frac{d^2\xi_1}{dx^2} = 0, \qquad E_1 I\frac{d^4 w_1}{dx^4} + P\frac{d^2 w_1}{dx^2} = 0 \\ &\frac{d^2\eta_1}{dy^2} = 0, \qquad \frac{d^4 v_1}{dy^4} = 0 \end{aligned} \tag{2.188}$$

Integration of Equations (2.188) gives

$$\begin{aligned} &\beta_A \xi_1 = C_1\frac{x}{E_1 A}, \qquad \beta_A \eta_1 = C_2\frac{y}{E_2 A} \\ &\beta_A w_1 = C_3 \sin(kx) + C_4 x, \qquad \beta_A v_1 = C_5 y + C_6 y^3 \end{aligned} \tag{2.189}$$

where $k^2 = P/E_1 I$, and C_i $(i = 1, 2, \ldots, 6)$ are integration constants.

Using boundary conditions at $x = L_1$, $y = L_2$:

$$w = \eta: \qquad C_3 \sin(kL_1) + C_4 L_1 = \frac{L_2}{E_2 A} C_2 \tag{2.190}$$

$$\xi = -v: \qquad C_5 L_2 + C_6 L_2^3 = \frac{L_1}{E_1 A} C_1 \tag{2.191}$$

$$\frac{dw}{dx} = -\beta_A: \qquad kC_3 \cos(kL_1) + C_4 = -\beta_A \tag{2.192}$$

$$\frac{dv}{dy} = -\beta_A: \qquad C_5 + 3c_6 L_2^2 = -\beta_A \tag{2.193}$$

$$E_1 A\frac{d\xi_1}{dx} + E_2 I\frac{d^3 v_1}{dy^3} = 0: \qquad C_1 + 6E_2 I C_6 = 0 \tag{2.194}$$

$$\frac{d^2 w_1}{dx^2} + \frac{d^2 v_1}{dy^2} = 0: \qquad -k^2 C_3 \sin(kL_1) + 6C_6 L_2 = 0 \tag{2.195}$$

$$E_2 A\frac{d\eta_1}{dy} - E_2 I\frac{d^3 v_1}{dx^3} - P\frac{dw_1}{dx} = 0: \qquad C_2 - P[kC_3 \cos(kL_1) + C_4] = 0 \tag{2.196}$$

Equations (2.190)–(2.196) are linear, homogeneous algebraic equations in terms of $\beta_A, C_1, \ldots, C_6$. The non-triviality condition leads to

$$\begin{aligned} &6(kL_1)\cos(kL_1) - \frac{6E_2 I}{E_1 A L_2^2}(kL_1)^2 \sin(kL_1) - 6\sin(kL_1) - 2(kL_1)^2\frac{L_2}{L_1}\sin(kL_1) \\ &\quad + \frac{P}{E_1 A}\frac{6E_1 I}{E_1 A L_1 L_2}(kL_1)^2 \sin(kL_1) + 2(kL_1)^2\frac{P}{E_2 a}\left(\frac{L_2}{L_1}\right)^2 \sin(kL_1) = 0 \end{aligned} \tag{2.197}$$

Because our discussion is limited to the elastic range, terms containing $P/E_1 A$, $6E_2 I/E_1 A L_2^2$, $P/E_2 A$, and $6E_1 I/E_1 A L_1 L_2$ are negligibly small and could be omitted in the first approximation. Thus, we have an approximate characteristic equation as the following:

$$\tan(kL_1) \approx \frac{kL_1}{1 + \frac{1}{3}(kL_1)^2 \frac{L_2}{L_1}} \tag{2.198}$$

from which, the critical buckling load parameter k_{cl} can be determined. The normalized classical buckling mode is found to be

$$\begin{aligned}
\xi_1 &= \frac{3P_{\mathrm{cl}}}{E_1 A (k_{\mathrm{cl}} L_1)^2} x \\
\eta_1 &= \frac{3P_{\mathrm{cl}}}{E_1 A (k_{\mathrm{cl}} L_2)^2} y \\
w_1 &= \frac{3}{(k_{\mathrm{cl}} L_1)^2} \left[x - L_1 \frac{\sin(k_{\mathrm{cl}} x)}{\sin(k_{\mathrm{cl}} L_1)} \right] \\
v_1 &= \frac{1}{2} y \left(1 - \frac{y^2}{L_2^2} \right), \qquad P = P_{\mathrm{cl}} + (\lambda - 1) P_{\mathrm{cl}}
\end{aligned} \tag{2.199}$$

For the case $L_1 = L_2 = L$, the buckling load parameter is found to be $k_{\mathrm{cl}} L = 3.72$. Substituting the preceding expressions back into the energy expressions, we have

$$\frac{1}{2!} \delta^2 \Pi = \left(0.478 P_{\mathrm{cl}} L + 0.109 \frac{E_2}{E_1} P_{\mathrm{cl}} L - 0.587 \lambda \right) \beta_A^2 \tag{2.200}$$

If we assume that there is a dissimilarity in the distribution of elastic moduli (i.e., generally $E_1 \neq E_2$)

$$E_2 = E_1 (1 + \epsilon) \tag{2.201}$$

then Equation (2.201) reads

$$\frac{1}{2!} \delta^2 \Pi = 0.587 (1 + 0.186 \epsilon - \lambda) P_{\mathrm{cl}} L \beta_A^2 \tag{2.202}$$

The third variation is

$$\frac{1}{3!} \delta^3 \Pi = 0.149 P_{\mathrm{cl}} L \beta_A^3 \tag{2.203}$$

The approximate expression for the potential energy increment $\Delta\Pi$ is the sum of the second and third variations of the energy expressions. For equilibrium, $d(\Delta\Pi)/d\beta_A = 0$. For $\beta_A \neq 0$, we obtain

$$\lambda = 1 + 0.186 \epsilon + 0.381 \beta_A \tag{2.204}$$

or, we can rewrite Equation (2.204) as

$$\lambda_1 = \frac{P}{P'_{\mathrm{cl}}} = 1 + \frac{0.381}{1 + 0.186 \epsilon} \beta_A; \qquad P'_{\mathrm{cl}} = P_{\mathrm{cl}} (1 + 0.186 \epsilon) \tag{2.205}$$

We now proceed with the analysis of the geometrically imperfect structure. For an *eccentric* load applied at a distance ϕL to the right of point A, the potential energy expression must be modified by adding a term (Brush and Almorth, 1975)

$$\Omega_\phi = -PL\phi\beta_A = -\lambda_1 P'_{\rm cl} L\phi\beta_A \tag{2.206}$$

and the second variation and the third variation take the form

$$\begin{aligned} \frac{1}{2}\delta^2\Pi &= A_2(1-\lambda_1)P'_{\rm cl}L\beta_A^2 \\ \frac{1}{3!}\delta^3\Pi &= A_3(1-\lambda)P'_{\rm cl}L\beta_A^3 \end{aligned} \tag{2.207}$$

where

$$A_2 = 0.587/(1+0.186\epsilon)^2, \qquad A_3 = 0.149/(1+0.186\epsilon)$$

Then

$$\begin{aligned} \Delta\Pi &= \frac{1}{2!}\delta^2\Pi + \frac{1}{3!}\delta^3\Pi + \Omega_\phi \\ &= \left[(1-\lambda_1)A_2\beta_A^2 + A_3\beta_A^3 - \lambda_1\phi\beta_A\right]P'_{\rm cl}L \end{aligned} \tag{2.208}$$

For the equilibrium, we have $d(\Delta\Pi)/d\beta_A = 0$:

$$\lambda_1 = 1 + \frac{3A_3}{2A_2}\beta_A - \frac{1}{2A_2}\frac{\phi}{\beta_A}\lambda_1 \tag{2.209}$$

or, in another form,

$$\lambda_1\beta_A = \beta_A + \frac{3A_3}{2A_2}\beta_A^2 - \frac{1}{2A_2}\phi\lambda_1 \tag{2.210}$$

Differentiating both sides of Equation (2.210) with respect to β_A, we have

$$\beta_A\frac{d\lambda_1}{d\beta_A} + \lambda_1 = 1 + \frac{3A_3}{A_2}\beta_A - \frac{1}{2A_2}\phi\frac{d\lambda_1}{d\beta_A} \tag{2.211}$$

Setting $d\lambda_1/d\beta_A$ equal to zero, we obtain the limit-point load factor λ_L for the imperfect structure

$$\lambda_L = 1 + \frac{3A_3}{A_2}\beta_A \tag{2.212}$$

or

$$\beta_A = \frac{(\lambda_L - 1)A_2}{3A_3} \tag{2.213}$$

Substituting back into Equation (2.210), we have

$$\lambda_L = 1 - \frac{(3A_3\lambda_L)^{1/2}}{A_2}(-\phi)^{1/2} \tag{2.214}$$

As a first approximation, we can set $\lambda_L = 1$ at the right-hand side of Equation (2.214), and we arrive at

$$\lambda_L \approx 1 - \frac{(3A_3)^{1/2}}{A_2}(-\phi)^{1/2} = 1 - a(-\phi)^{1/2}, \qquad a = \frac{(3A_3)^{1/2}}{A_2} \tag{2.215}$$

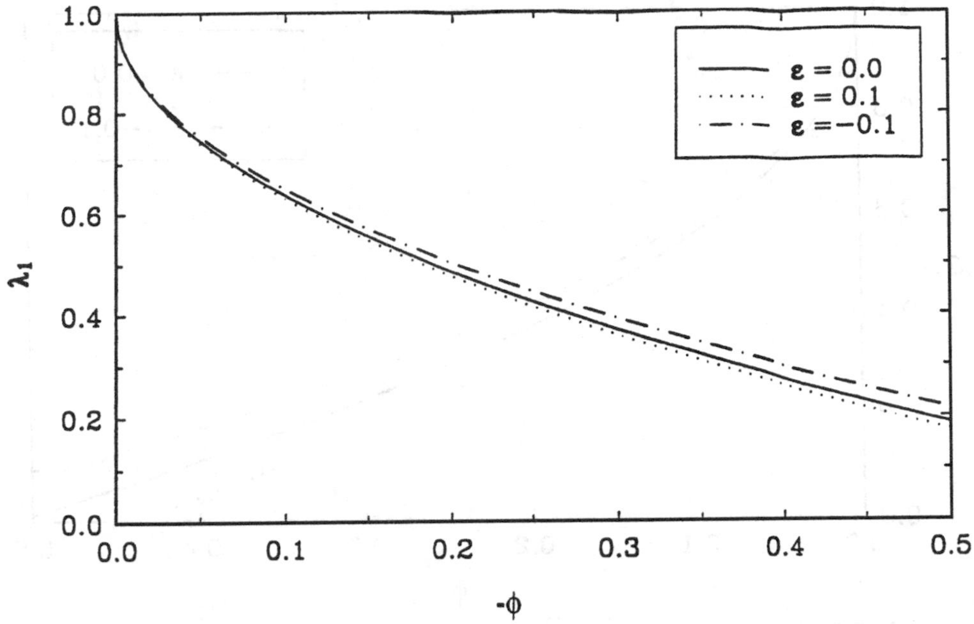

Figure 2.26 Buckling load reduction ($L_2/L_1 = 1.0$).

where a is a parameter describing the imperfection sensitivity of the structure. For $\epsilon = 0, a = 1.15$; $\epsilon = 0.1, a = 1.17$; $\epsilon = -0.1, a = 1.11$. Using these data, we can plot the relationship between the buckling load and the initial imperfection parameter. This is depicted in Figure 2.26, where the three curves correspond to three different cases of elastic moduli.

For the case where $L_2/L_1 = 0.5$, the buckling load parameter obtained from Equation (2.198) is $(k_{cl}L_1) = 3.97$. Following the same procedure as before, we again end up with Equation (42). However, A_2 and A_3 now have the following expressions:

$$A_2 = \frac{0.37}{(1 + 0.51\epsilon)^2}, \qquad A_3 = \frac{0.09}{1 + 0.51\epsilon} \tag{2.216}$$

For $\epsilon = 0$, $a = 1.40$; $\epsilon = 0.1$, $a = 1.51$; $\epsilon = -0.1$, $a = 1.30$. Having these, we can plot the curves of buckling load reduction as in Figure 2.26 and 2.27.

2.5.2 Discussion

Figures 2.26 and 2.27 demonstrate the usual pattern of the buckling load reduction due to initial imperfection. However, what is worthwhile to notice here is a further reduction in buckling load because of the presence of dissimilarity in the elastic moduli and the geometric configuration. When ϵ is positive, the horizontal bar possesses a bigger elastic modulus than the vertical bar, which results in a stiffer structure. Likewise, when L_2/L_1 is less than unity (the horizontal bar is shorter), the overall stiffness of the structure is again increased. On the one hand, a stiffer structure has a larger buckling load. Figures 2.26 and 2.27 show that the stiffer the structure is, the more sensitive it is to

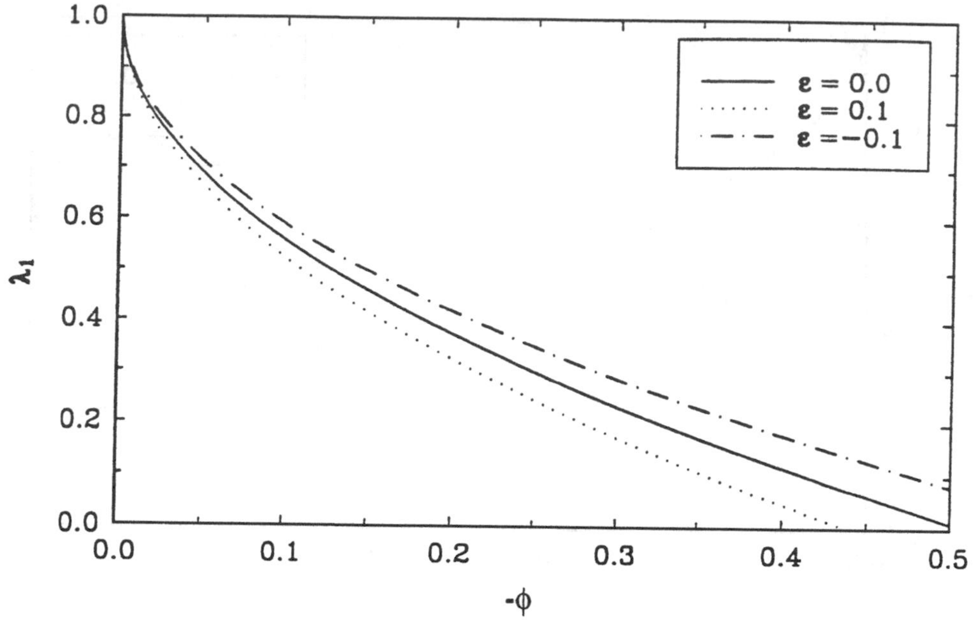

Figure 2.27 Buckling load reduction ($L_2/L_1=0.5$).

the initial imperfection. As we can see also from those two figures, when the stiffness is decreased either through a reduced elastic modulus ($\epsilon < 0$) or from an increased length ($L_2/L_1 > 1$), the structure becomes less sensitive to the initial imperfection. The change in the overall stiffness of the structure comes, as is shown here, from the dissimilarity. Thus, one may conclude that the dissimilarity in elastic moduli could contribute to the change in sensitivity of the structure to the initial imperfection. One may intentionally introduce such a dissimilarity in elastic moduli so that the structure has a decreased sensitivity to initial imperfections. In this simple two-bar structure, the decrease in sensitivity due to dissimilarity in elastic modulus is not remarkable when the two segments of the frame have the same length, namely $L_1 = L_2$. However, for other geometric configurations, the effect may become more pronounced. For $L_1 = 2L_2$ (Figure 2.27), say, at $\phi = 0.25$, $\lambda = 0.3$ for $\epsilon = 0$, and $\lambda = 0.24$ for $\epsilon = 0.1$, which indicates a 20% decrease in the limit load; for $\epsilon = -0.1$, $\lambda = 0.3$, which amounts to a 17% increase in load-carrying capacity. It appears that the subject is worth pursuing in the direction of shell structures to see how important the effect of non-homogeneity of elastic moduli is on the imperfection sensitivity.

Recently, Combescure et al. (2000) and Gusic et al. (2000) extended our results to the buckling of cylindrical shells under external pressure, under non-axisymmetric thickness imperfections. They proposed an analytical formula showing that the pressure load reduction is a linear function of thickness imperfection. The reduction factor due to the coupling of two types of imperfection (thickness imperfection and geometric imperfection) is represented as a product of each mode of reduction. The comparison with experimental results was conducted.

CHAPTER THREE

Stochastic Buckling of Structures: Monte Carlo Method

> Most calculations of imperfection-sensitivity have been carried out for simple shapes of the imperfection distribution, selected for convenience of analysis. It is generally realized, of course, that actual imperfections are unlikely to follow this regular pattern.
>
> W. T. Koiter

> There is a close connection between the concepts of stability and probability. Stable states of equilibrium or motion observed in the natural or engineering systems are the most probable ones; unstable ones are improbable and even unrealizable. The more stable a state is, the greater is the probability of its realization. Hence follows the connection between the concepts of stability and reliability.
>
> V. V. Bolotin

> But to us, probability is the very guide of life.
>
> J. Butler

In this chapter, we treat initial geometric imperfections as spacewise random functions (i.e., random fields). As a result, the buckling load – the maximum load the structure can sustain – turns out to be a random. We focus our attention on reliability of the shells, namely the probability that a structure will not fail prior to the specified load. We develop efficient techniques of simulation of random fields, based on the knowledge of the mean initial imperfection function and the auto-correlation function. This allows us to conduct, by the Monte Carlo method, an extensive analysis of the buckling of columns on non-linear elastic foundations and of circular cylindrical shells. We consider the cases of both axisymmetric and general, non-symmetric imperfections.

3.1 Introductory Remarks

It is now generally recognized that initial geometric imperfections play a dominant role in reducing the buckling load of certain structures. As is well known, an axially compressed thin shell is highly imperfection sensitive in this context (see, for example, Hutchinson and Koiter, 1970; Tvergaard, 1976; and Budiansky and Hutchinson, 1979).

This conclusion is mainly due to the work of a series of investigators, among them Koiter (1945), Donnell and Wan (1950), and Budiansky and Hutchinson (1966), who arrived at it through recourse to specialized imperfections. However, despite the accepted theoretical explanation of the buckling behavior of these structures, the use of the concept of imperfection sensitivity in engineering practice is still in the ad hoc stage, and engineers prefer to rely on the "knockdown factor" (NASA, 1968), chosen so that its product with the classical buckling load yields a lower bound to available experimental data for the configuration in question. This apparent reluctance to take advantage of theoretical findings stems from the fact that most imperfection studies are conditional on detailed advance knowledge of the geometric imperfections of the particular structure, which is rarely possible. In an ideal case, the imperfections can be measured experimentally in each shell and incorporated in the theoretical analysis to predict the buckling loads. This approach, however, while justified for single prototype-like structures, is impracticable as a general method of behavior prediction. Information on the type and magnitude of imperfections of a particular structure would be too specific and thus are not strictly valid for other realizations of the same structure even those obtained by the same manufacturing process.

With these considerations in view and also bearing in mind the scatter of the experimental results, it appears obvious that practical applications of the imperfection-sensitivity theories are conditional on their being combined with a statistical analysis of the imperfections and critical loads. The notion of randomness of the initial imperfections was given considerable attention in the literature, and a bibliography can be found in Amazigo's paper (1976). For the single-mode solutions, consult Bolotin (1969) (who pioneered the probabilistic approach to buckling problems in 1958) and Roorda (1980).

Later on, Elishakoff (1978b, 1979a, 1980b) suggested utilizing the Monte Carlo method for the solution of multi-mode problems involving random initial imperfection sensitivity. This method represents a logical remedy in view of the difficulties inherent in purely analytical approaches (based on unnecessary and often very restrictive assumptions about the properties of the initial imperfections and/or using heavily simplified solution procedures).

The first step of the Monte Carlo method consists of simulating the random initial imperfection profiles via a special numerical procedure (Elishakoff, 1979b); the second step is simply a numerical solution of the buckling problem for every realization of the initial imperfection profile; and the third and last step involves a statistical analysis of the buckling loads (for a detailed description of the Monte Carlo method, consult Elishakoff, 1983b). The reliability is determined as the fraction of an ensemble of shells of which the buckling loads exceed the specified load.

As far as we know, three works have been devoted to the analysis of cylindrical shells with general non-symmetric random imperfections. Makarov (1971), at the Moscow Power Engineering Institute and State University, carried out systematic analysis of initial imperfections. He used a series of 30 cylindrical shells made of sheets of electrical grade pressboard. The initial imperfection was represented as a double Fourier series, and the coefficients were treated as random variables. The analysis showed that the assumption of circumferential homogeneity of the initial imperfections was satisfactory (within the confidence limits used in the analysis), and the assumption of Gaussianity of their Fourier coefficients did not conflict with the experimental data. Makarov also carried out a theoretical analysis of the buckling of stochastically imperfect shells, with the experimental data obtained earlier serving as the input for the description of the imperfections. He found a theoretical mean buckling load that exceeds its experimental counterpart by a factor of 1.35.

Fersht (1974) generalized the method of truncated hierarchy, used earlier by Amazigo (1969) for axisymmetric random imperfections, to include the non-symmetric case. It turned out that for non-symmetric imperfections a closed-form expression for the buckling load is unattainable, and rather cumbersome numerical integrations have to be performed. Moreover, for the axisymmetric case, Fersht's numerical results do not agree with those of Amazigo (1969).

Hansen (1977) generalized his previous deterministic results (Hansan, 1975). The main conclusion was that the imperfection parameters associated with the non-axisymmetric modes appear only in three separate summations and that the behavior of the system is governed by the values of these summations rather than by the individual imperfection amplitudes. It was assumed that the Fourier coefficients of the initial imperfections are jointly normal random variables with zero mean and that they are statistically independent and identically distributed. Then the Monte Carlo method was applied. For each sample problem the buckling load was found via the method described by Hansen (1975), and then the mean buckling loads were calculated. The role of the non-axisymmetric imperfections turned out to be very important.

In Sections 3.2 and 3.3, we consider the buckling of beams on non-linear elastic foundations in order to elucidate the Monte Carlo methods as they are applied to stochastic buckling of structures in the first section. The other two sections (Sections 3.4 and 3.5) are devoted to discussions on cylindrical shells with random axisymmetric and non-symmetric imperfections.

3.2 Reliability Approach to the Random Imperfection Sensitivity of Columns

The reliability approach to the random imperfection sensitivity of columns is based on Elishakoff (1979a, 1983a, 1985).

3.2.1 Motivation for the Reliability Approach

The following considerations are intended as a contribution to understanding the random imperfection sensitivity of non-linear elastic structures. Specifically, we are dealing with a stochastically imperfect column on a non-linear mixed quadratic-cubic foundation. This section is a generalization of an earlier work (Elishakoff, 1979a), where buckling of a column on a purely cubic foundation was considered.

Deterministic imperfection-sensitivity studies of structures on a non-linear elastic foundation were performed by Reissner (1970, 1971) and Keener (1974). The first paper on random imperfection sensitivity of columns under such conditions was published by Fraser and Budiansky (1969), with reference to infinitely long columns on a softening (cubic) elastic foundation. They concluded that every column in the ensemble has the same buckling load, which depends on the auto-correlation function of the initial imperfections, assumed to be a random ergodic function of the axial coordinate. Finite columns on a cubic foundation were considered in several studies cited in our earlier paper (Elishakoff, 1979a), and finite columns on a quadratic-cubic foundation were considered by Hansen and Roorda (1973, 1974), who assumed that the initial imperfection function has the shape of the buckling mode of the associated linear structure. Earlier, Fraser (1965) considered a single-term approximation for finite columns on a cubic foundation. The two approaches – consideration of an infinite structure and a single-term approximation for a finite one – were bridged (Elishakoff, 1979a), the main contribution of which was a multi-mode solution of the finite structure. In that study, the initial and the additional deflection functions were expanded in terms of the buckling modes of the associated perfect linear structure. Fourier coefficients of the expansion for the initial imperfection function were simulated numerically. For each realization of the initial imperfection, the buckling load was found by transformation of the two-point non-linear boundary-value problem to an initial-value problem via the Qiria-Davidenko method. When the linear elastic stiffness modulus was taken such that the associated linear structure had a buckling mode with one half-wave, in the context of a special class of auto-correlation functions of the initial imperfections, Fraser's single-mode solution (1965) was in excellent agreement with the results obtained by the Monte Carlo method. By contrast, when the modulus was such that the associated linear structure had a buckling mode with more than one half-wave, and/or the initial imperfection function was not co-configurational with the buckling mode of the associated linear structure, the single-mode solution turned out to be insufficient. When the linear "spring" constant of the foundation increases as the non-linear spring constant is kept constant, the variance of the buckling loads of the non-linear structure decreases throughout the rather wide range under consideration. Still, this decrease does not conflict with the conclusion of Fraser and Budiansky (1969) that for the infinite column the buckling load is a deterministic quantity.

We must digress here and mention that the Monte Carlo method was applied by Fraser (1965) in the particular case when the buckling mode of the associated linear structure was represented by one half-wave. He also considered the buckling of structures involving exponentially correlated random imperfections (private communication to I. Elishakoff, 1978).

3.2.2 The Linear Problem

We start with the following differential equation:

$$EI\frac{d^4w}{dx_4} + P\frac{d^2(w+\bar{w})}{dx^2} + k_1 w = 0 \tag{3.1}$$

where E is the Young's modulus, I is the section moment of inertia, $\bar{w}$ is the initial imperfection function (see Figure 3.1), $w(x)$ is the additional deflection due to the axial load P, k_1 is the linear "spring" constant of the foundation, and x is the axial co-ordinate. The column is simply supported, so that the boundary conditions are

$$w = \frac{d^2w}{dx^2} = 0 \quad \text{at} \quad x = 0 \quad \text{and} \quad x = L \tag{3.2}$$

where L is the length of the column.

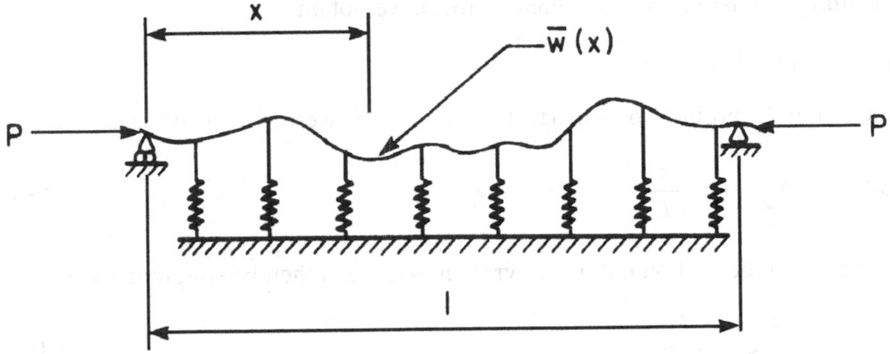

Figure 3.1 Column on non-linear elastic foundation.

If $\bar{w}(x) \equiv 0$, we obtain an associated equation for a perfect column on a linear foundation. We consider, for the buckling problem, solutions of the form

$$w = a \sin \frac{m\pi x}{L} \tag{3.3}$$

where m is the number of half-waves and a is an arbitrary constant. Substituting (3.3) into (3.1), it is determined that P must have a value of P_m given by

$$P_m = P_E\left(m^2 + \frac{k_1 L^4}{m^2 EI\pi^4}\right), \qquad P_E = \frac{\pi^2 EI}{L^2} \tag{3.4}$$

where P_E is the buckling load of the column without elastic foundation. The value of $m = m_*$, which makes P_m a minimum, determines the number of half-waves during buckling of the linear structure. Consequently, the classical buckling load of a column on a linear elastic foundation is

$$P_{cl} = P_E\left(m_*^2 + \frac{\kappa_1}{\pi^4 m_*^2}\right), \qquad \kappa_1 = \frac{k_1 L^4}{EI} \tag{3.5}$$

Unlike the column without elastic foundation, m_* does not necessarily equal unity. Assuming that the non-dimensional linear foundation coefficient is κ_1, we arrive at a situation where P_{cl} in Equation (3.5) is smaller for $m_* = 1$ than for m_*. The limiting value of κ_1, at which transition from $m_* + 1$ occurs, is found from the condition that the corresponding expression (3.5) should yield the same value for P_{cl} for $m_* + 1$, namely

$$m_*^2 + \frac{\kappa_1}{\pi^4 m_*^2} = (m_* + 1)^2 + \frac{\kappa_1}{\pi^4 (m_* + 1)^2}$$

and m_* is determined from

$$\pi^4 m_*^2 (m_* + 1)^2 = \kappa_1 \tag{3.6}$$

As κ_1 increases, so does the number of half-waves during buckling, and for $m_* \gg 1$, Equation (6) reads

$$\kappa^4 m_*^4 - \kappa_1 = 0, \qquad m_* = [m + 1] \tag{3.7}$$

where $[\kappa]$ refers to the integer part of κ.

Substituting Equation (3.7) in Equation (3.5), we obtain

$$P_{cl} = \left(\sqrt{\kappa_1}/\pi^2\right) P_E \tag{3.8}$$

Concerning the problem of imperfection sensitivity, we consider the case where

$$\bar{w}(x) = \Delta \sum_{k=1}^{\infty} \bar{\xi}_k \sin \frac{k\pi x}{L} \tag{3.9}$$

$\Delta = \sqrt{I/A}$ being the section radius of gyration; $w(x)$ can then be sought in the form

$$w(x) = \Delta \sum_{k=1}^{\infty} \xi_k \sin \frac{k\pi x}{L} \tag{3.10}$$

Substituting Equations (3.9) and (3.10) in Equation (3.1), we obtain as the relation between ξ and

$$\xi_k = \frac{P}{P_k - P} \bar{\xi}_k \tag{3.11}$$

which is meaningful, provided P is sufficiently small compared to $P_{cl} = \min P_k$.

3.2.3 Deterministic Imperfection Sensitivity: Mixed Quadratic-Cubic Foundation

The differential equation for the column's displacement $w(x)$ is now

$$EI\frac{d^4 w}{dx^4} + P\frac{d^2(w + \bar{w})}{dx^2} + k_1 w - k_2 w^2 - k_3 w^3 = 0 \tag{3.12}$$

where the original notation is adhered to and the new symbols k_2 and k_3 represent non-linear "spring" constants, assumed positive (Figure 3.1).

Equation (3.12) can be modified by introducing dimensionless independent and dependent variables in the form

$$\begin{aligned} &\eta = \frac{x}{L}, \qquad u = \frac{w}{\Delta}, \qquad \bar{u} = \frac{\bar{w}}{\Delta}, \qquad \alpha = \frac{P}{P_{cl}}, \\ &\kappa_2 = \frac{k_2 \Delta L^4}{EI}, \qquad \kappa_3 = \frac{k_3 \Delta^2 L^4}{E}, \qquad \gamma(\kappa_1) = m_*^2 \pi^2 + \frac{\kappa_1}{m_*^2 \pi^2} \end{aligned} \tag{3.13}$$

whereas P_{cl} is defined in Equation (3.5), and κ_1 is in Equation (3.5). Equations (3.12) and (3.2) then become

$$\frac{d^4 u}{d\eta^4} + \alpha\gamma(\kappa_1)\frac{d^2 u}{d\eta^2} + \kappa_1 u - \kappa_2 u^2 - \kappa_3 u^3 = \alpha\gamma(\kappa_1)\frac{d^2 \bar{u}}{d\eta^2} \tag{3.14}$$

$$u = \frac{d^2 u}{d\eta^2} = 0 \qquad \text{at} \qquad \eta = 0 \quad \eta = 1 \tag{3.15}$$

The problem is defined as follows: Given the spring coefficients of the foundation κ_1, κ_2, and κ_3 and the function $\bar{u}(\eta)$, find the buckling load $\alpha = \alpha^*$, defined from the requirement (Fraser and Budiansky, 1969)

$$\frac{d\alpha}{dF} = 0, \qquad F = F(u, \bar{u}) \tag{3.16}$$

here $F(u, \bar{u})$ is some functional of non-dimensional deflections; in the present case, we use the end shortening of the column (i.e., the distance the column ends move closer together under load)

$$d \cong \int_0^L \left[\frac{1}{2}\left(\frac{dw}{dx}\right)^2 + \left(\frac{d\bar{w}}{dx}\right)\left(\frac{dw}{dx}\right)\right]dx \tag{3.17}$$

In terms of the non-dimensional quantities, the functional could be written as

$$F(u, \bar{u}) = \frac{d}{\Delta^2} = \int_0^1 \left[\frac{1}{2}\left(\frac{du}{d\eta}\right)^2 + \left(\frac{d\bar{u}}{d\eta}\right)\left(\frac{du}{d\eta}\right)\right]d\eta \tag{3.18}$$

resorting to Boobnov-Galerkin's method, we expand $u(\eta)$ and $\bar{u}(\eta)$ in series in terms of the modes of stability loss of associated linear structure, given by Equations (3.9) and (3.10). Substituting them in Equation (3.14), multiplying the resulting equation by $\sin(m\pi\eta)$ and integrating, we arrive at the following infinite set of coupled non-linear algebraic equations for ξ_m:

$$\alpha_m \xi_m - \alpha(\xi_m + \bar{\xi}_m) - s_1\left(\frac{m_*}{m}\right)^2 J_m - \left(\frac{s_2}{8}\right)\left(\frac{m_*}{m}\right)^2 I_m = 0, \qquad m = 1, 2, \ldots \tag{3.19}$$

where

$$J_m = \sum_{p=1}^{\infty}\sum_{q=1}^{\infty} B_{pqm}\xi_p\xi_q, \qquad I_m = 8\sum_{p=1}^{\infty}\sum_{q=1}^{\infty}\sum_{r=1}^{\infty} A_{pqrm}\xi_p\xi_q\xi_r \tag{3.20}$$

$$\begin{aligned} B_{pqm} &= \int_0^1 \sin(p\pi\eta)\sin(q\pi\eta)\sin(m\pi\eta)d\eta \\ &= B(p+q, m) + B(m+q, p) + B(m+p, q) - B(p+q, -m) \end{aligned} \tag{3.21}$$

$$B(p+q, m) = \begin{cases} 0, & m = p+q \\ \dfrac{1}{4\pi}\quad \dfrac{1-(-1)^{p+q-m}}{p+q-m}, & m \neq p+q \end{cases}$$

$$\alpha_m = \frac{(m\pi)^2 + \kappa_1(m\pi)^{-2}}{(m_*\pi)^2 + \kappa_1(m_*\pi)^{-2}}, \qquad s_1 = \frac{2\kappa_2}{\kappa_1 + m_*^4\pi^4}, \qquad s_2 = \frac{2\kappa_3}{\kappa_1 + m_*^4\pi^4} \tag{3.22}$$

and δ_{ij} is the Kronecker delta.

Closed solution of the infinite set of non-linear equations in Equations (3.19) seems to be unfeasible; these equations must be truncated and solved numerically. Before doing so, let us consider the analytic solution obtainable in the single-mode case.

3.2.4 Single-Mode Solution

Retaining only the m_*th term in the series in Equations (3.9) and (3.10), Equation (3.19) reduce to a single equation, namely (asterisks are omitted):

$$\alpha(\xi_m + \bar{\xi}_m) = \xi_m \left\{ \alpha_m - s_1 \xi_m - \frac{3}{8}\left(\frac{m_*}{m}\right)^2 s_2 \xi_m^2 \right\},$$
$$F = \frac{1}{4} m^2 \pi^2 \xi_m (\xi_m + 2\bar{\xi}_m) \tag{3.23}$$

where

$$s_1 = \frac{4\kappa_2[1(-1)^m]}{3m^2\pi^3\gamma(\kappa_1)}, \qquad s_2 = \frac{2\kappa_3}{(m_*\pi)^4 + \kappa_1} = \frac{2\kappa_3}{(m_*\pi)^2\gamma(\kappa_1)} \tag{3.24}$$

When $\kappa_2 = 0$ and/or m is even, Equation (3.23) coincides formally with Equation (2.18) in Fraser's thesis (1965) for a column on a cubic foundation. It is also worth noting that according to Equation (3.23), ξ_m and $\bar{\xi}_m$ have the same sign (i.e., on further deformation of the column, the total $\bar{\xi}_m + \xi_m$ increases by the absolute value of the additional displacement); this is seen from the fact that the assumption $\bar{\xi}_m \xi_m < 0$ may imply $\alpha < 0$ for $0 < |\xi_m| < \bar{\xi}_m$ (i.e., a tensile force), which contradicts the terms of the problem. Moreover, it is seen that $\alpha_m \geq a_{m_*} = 1$.

For a perfect column ($\bar{\xi}_m = 0$), Equation (3.23) reduces to

$$\xi_m \left[\alpha - \alpha_m + s_1 \xi_m + \frac{3}{8}\left(\frac{m_*}{m^2}\right)^2 s_2 \xi^2 \right] = 0 \tag{3.25}$$

It represents two branches, the straight line $\xi_m = 0$ and the parabola

$$\alpha = -\frac{3}{8}\left(\frac{m_*}{m}\right)^2 s_2 \xi m^2 - s_1 \xi_m + \alpha_m \tag{3.26}$$

which intersects the α-axis at $\alpha = \alpha_m$, the maximum being

$$\alpha_{\max} = \alpha_m + \frac{s_1^2}{4\gamma_1} \tag{3.27}$$

Obviously, for a perfect column ($\bar{\xi}_m = 0$), there is no sign restriction on additional displacements; hereinafter we will confine ourselves to the case $\kappa_2 > 0$, $\kappa_3 > 0$.

Differentiating Equation (3.23) with respect to F and setting

$$\frac{d\alpha}{dF} = 0, \qquad \alpha = \alpha^* \tag{3.28}$$

we obtain, after lengthy algebraic manipulations, the relation between the buckling load α^* and the initial imperfection amplitude $\bar{\xi}_m$

$$\left(\alpha_m - \alpha^* + \frac{s_1^2}{3\gamma_1}\right)^3 = \frac{81}{32}\left(\frac{m_*}{m}\right)^2 s_2 \left[-(\alpha_m - \alpha^*)\frac{s_1}{3\gamma_1} - \frac{2s_1^3}{27\gamma_1^2} - \alpha^* \bar{\xi}_m \right]^2$$
$$\gamma_1 = \frac{3}{8}\left(\frac{m_*}{m}\right)^2 s_2 \tag{3.29}$$

When $\kappa_2 = 0$ and/or m is even, Equation (3.23) reduces to Equation (30) of Elishakoff (1979a) and formally coincides with Equation (2.18) of Fraser work (1965)

$$(\alpha_m - \alpha^*)^3 = \frac{81}{32}\left(\frac{m_*}{m}\right)^2 s_2 \xi_m^2 (\alpha^*)^2 \tag{3.30}$$

Note that Equations (3.30) and (3.31) are particular cases of Equation (26) of Elishakoff (1980a). The latter deals with static and dynamic buckling of a simple non-symmetric structure, namely a three-hinge, rigid-rod system, constrained laterally by a non-linear spring, the force in the latter including both quadratic and cubic terms. In other words, it is a generalization of the model originally proposed by Budiansky and Hutchinson (1964). The additional displacement corresponding to α^* is given by

$$\xi_{m(1,2)} = \frac{1}{3\gamma_1}\left[-s_1 \pm \sqrt{s_1^2 + (\alpha_m - \alpha^*)3\gamma_1}\right] \tag{3.31}$$

Note that $\xi_{m(1,2)}$ depends on $\bar{\xi}_m$ via α^*:

$$\bar{\xi}_{m(1,2)} = (\alpha_m - \alpha^*)\frac{s_1}{3\gamma_1\alpha^*} - \frac{2s_1^3}{27\gamma_1\alpha^*} \pm \frac{4}{9\alpha}\sqrt{\frac{2}{s_2}\frac{m}{m_*}}\left(\alpha_m - \alpha^* + \frac{s_1^2}{3\gamma_1}\right)^{3/2} \tag{3.32}$$

For $s_1 = 0$, Equation (3.30) coincides with Equation (41) of Elishakoff (1979a). $\bar{\xi}_{m,1}$ and $\bar{\xi}_{m,2}$ are not necessarily meaningful for any α^*; to begin with, α^* is bounded by the following value

$$\alpha^{**} = \alpha_m + \frac{s_1^2}{3\gamma_1}$$

Consider the case $\kappa_2 > 0$; then for

$$\alpha_m < \alpha^* < \alpha^{**} \tag{3.33}$$

Equation (3.30), as is readily shown by expansion of its last term in Taylor series in the vicinity of $(\alpha_m - \alpha^*)$, yields $\bar{\xi}_{m,(1,2)} > 0$, whereas $\xi_{m(1,2)} < 0$ [Equation (3.32)]. This conflicts with the earlier observation that $\xi_m\bar{\xi}_m > 0$; hence, only the branch associated with $\bar{\xi}_m < 0$ has a physical sense.

Figure 3.2 shows a typical $\alpha^* - \bar{\xi}_m$ curve for $m = m_* = 1$, $a = 1$, $\kappa_1 = \pi^4$, $\kappa_2 = 0.4\,\kappa_1$, $\kappa_3 = 0.1\,\kappa_1$. For $\kappa_2 = -0.4\kappa^4$ with other data as before, the $\alpha^* - \bar{\xi}_m$ graph is the mirror image of its counterpart for $\kappa_2 = 0.4\,\pi^4$ with respect to the α^*-axis. Dashed lines represent the meaningless branches of Equation (3.30). Note that the buckling load can exceed that of a perfect structure; for $\bar{\xi}_m = 0$, the buckling load sets in at the values

$$\xi_m = 0 \qquad \text{and} \qquad \xi_m = \frac{s_1}{2\gamma_1} \tag{3.34}$$

which in turn are associated with the maxima of α^* on the left- and right-hand branches of the $\alpha^* - \bar{\xi}_m$ curves, respectively,

$$\alpha_{\max} = \alpha_m \qquad \text{and} \qquad \alpha_{\max} = \alpha_m + \frac{s_1^2}{4\gamma_1} \tag{3.35}$$

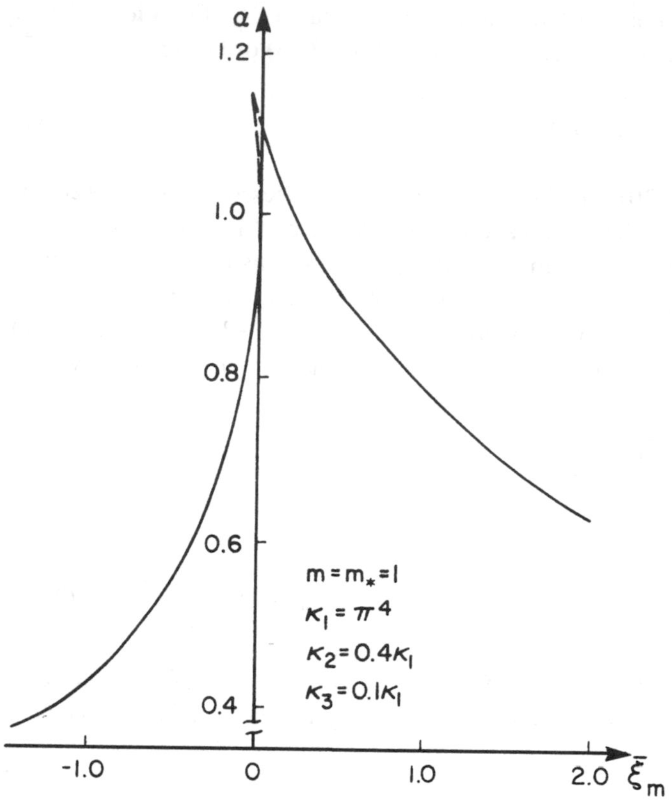

Figure 3.2 Buckling load parameter α as a function of the critical imperfection amplitude.

Note also that for the case of a cubic foundation $s_1 = 0$, both branches of $\alpha^* - \bar{\xi}_m$ curves intersect at $\alpha_{\max} = \alpha_m$.

3.2.5 Multi-Mode Solution

First, we truncate Equations (3.19) and (3.20) to some N (number of terms taken into consideration). We then apply the Qiria-Davidenko method (1951, 1953) for reducing the set of non-linear algebraic equations to the initial value problem. We denote

$$B_m(\xi_1, \xi_2, \ldots, \xi_N, \alpha) \equiv (\alpha_m - \alpha)\xi_M - \alpha\bar{\xi}_m - s_1\left(\frac{m_*}{m}\right)^2 J_m^{(N)} - \left(\frac{s_2}{8}\right)\left(\frac{m_*}{m}\right)^2 I_m^{(N)}$$
$$m = 1, 2, \ldots, N \tag{3.36}$$

here

$$I_m^{(N)} = 8\sum_{p=1}^{N}\sum_{q=1}^{N}\sum_{r=1}^{N} A_{pqrm}\xi_p\xi_q\xi_r, \qquad J_m^{(N)} = \sum_{p=1}^{N}\sum_{q=1}^{N} B_{pgm}\xi_p\xi_q \tag{3.37}$$

With F as a new independent variables, and α and ξ_m as its functions, we differentiate

Equation (3.36) with respect to F and obtain

$$\frac{\partial B_m}{\partial \alpha}\frac{d\alpha}{dF} + \sum_{k=1}^{N} \frac{\partial B_m}{\partial \xi_k}\frac{d\xi_k}{dF} = 0 \tag{3.38}$$

Equation (3.38), in conjunction with the equation

$$\sum_{k=1}^{N} \frac{\partial F}{\partial \xi_p}\frac{d\xi_p}{dF} = 1 \tag{3.39}$$

can be considered as a set of ordinary linear differential equations in α and $\xi_k(k = 1, 2, \ldots, N)$. Initial conditions are obtainable from the unstressed state of the column, in the absence of load

$$\alpha = 0, \qquad \xi_1 = \xi_2 = \cdots = \xi_N = 0 \qquad \text{at} \qquad F = 0 \tag{3.40}$$

Equations (3.38) and (3.39) take the following final form:

$$-(\xi_m + \bar{\xi}_m)\frac{d\alpha}{dF} + \sum_{p=1}^{N}\left[(\alpha_m - \alpha)\delta_{m,p} - 2\sum_{i=1}^{N} e_{imp}\xi_i - 3\sum_{i=1}^{N}\sum_{j=1}^{N} C_{ijmp}\xi_i\xi_j\right]\frac{\partial \xi_p}{\partial F} = 0 \tag{3.41}$$

$$\sum_{p=1}^{N}(\xi_p + \bar{\xi}_p)\frac{p^2\pi^2}{2}\frac{d\xi_p}{dF} - 1 = 0 \tag{3.42}$$

where

$$e_{imp} = s_1\left(\frac{m_*}{m}\right)^2 B_{imp} \qquad c_{ijmp} = s_2\left(\frac{m_*}{m}\right)^2 A_{ijmp} \tag{3.43}$$

The set of Equations (3.41) and (3.42) subject to (3.40) yields the entire $\alpha - F$ curve, the buckling load being defined as the maximum attainable load on the portion of the curve originating at zero load. Note that the final equations used to find the $\alpha - F$ curves, Equations (3.41) and (3.42), differ from the comparable Equations (33) and (34) of Elishakoff (1979a), in that they contain an additional term $-2\sum_{i=1}^{N} e_{imp}\xi_i$.

3.2.6 Buckling Under Random Imperfections – Monte Carlo Method

Assume now that the initial imperfection function is a Gaussian random function of the position η with given mean functions

$$\bar{U}(\eta) = E[\bar{u}(\eta)] \tag{3.44}$$

and auto-covariance function

$$K_{\bar{u}}(\eta_1, \eta_2) = E\{[u(\eta_1) - \bar{U}(\eta_1)][u(\eta_2) - \bar{U}(\eta_2)]\} \tag{3.45}$$

where η_1 and η_2 are two (generally distinct) points on the column axis ("observation" points), and $E(\cdots)$ denotes mathematical expectation. In these circumstances, the dimensionless buckling load α becomes a random variable. The problem consists in determining the reliability of the structure, defined as the probability of the α^* exceeding some prescribed level α':

$$R(\alpha') = \text{Prob}(\alpha^* > \alpha') \tag{3.46}$$

Purely analytic solution of this problem seems to be unfeasible. With the advent of high-speed digital computers, statistical simulation, the so-called Monte Carlo method (Shreider, 1964; Hammersley and Handscomb, 1964; Rubinstein, 1981) can be used as the logical remedy. Using this method, one usually starts with a simulation of the stochastic variables involved. Then, the numerical solution of the boundary-value problem has to be carried out repeatedly for every realization of those simulated stochastic variables. Finally, statistical analysis is performed on the data generated by the Monte Carlo simulation by calculating the relative frequencies.

In view of Equation (3.9), the mean imperfection function can be written as

$$\bar{U}(\eta) = \sum_{m=1}^{N} E(\bar{\xi}_m) \sin(m\pi\eta) \tag{3.47}$$

$$E(\bar{\xi}_m) = 2\int_0^1 \bar{U}(\eta)\sin(m\pi\eta)\,d\eta \tag{3.48}$$

Keeping Equations (3.47) and (3.48) in mind, the auto-covariance function is then written as

$$K_{\bar{u}}(\eta_1, \eta_2) = \sum_{m=1}^{N}\sum_{n=1}^{N} \sigma_{mn} \sin(m\pi\eta_1)\sin(n\pi\eta_2) \tag{3.49}$$

where

$$\sigma_{mn} = E\{[\bar{\xi}_m - \langle\bar{\xi}_m\rangle][\bar{\xi}_n - \langle\bar{\xi}_n\rangle]\} \tag{3.50}$$

is the covariance of $\bar{\xi}_m$ and $\bar{\xi}_n$, given in terms of $K_{\bar{u}}(\eta_1, \eta_2)$ as

$$\sigma_{mn} = 4\int_0^1\int_0^1 K_{\bar{u}}(\eta_1, \eta_2)\sin(m\pi\eta_1)\sin(n\pi\eta_2)\,d\eta_1\,d\eta_2 \tag{3.51}$$

A simulation procedure for the random initial imperfections was outlined in Elishakoff (1979b) and applied to various problems (Elishakoff, 1978b, 1979a). Here, we give the final result: The random initial imperfections $\bar{u}(\eta)$ with the specified mean function $\bar{U}(\eta)$ and auto-correlation function $K_{\bar{u}}(\eta_1, \eta_2)$, can be simulated by

$$\bar{u}(\eta) = \sum_{m=1}^{N} \bar{\xi}_m \sin(m\pi\eta) \tag{3.52}$$

here

$$\bar{\xi}_m = E(\bar{\xi}_m) + \sum_{k=1}^{N} c_{mk}\,d_k \tag{3.53}$$

$d_1, d_2, \ldots, d_N$ being independent normal variables with zero mean value and unit variances, and c_{mk} being elements of the lower triangular matrix C. The matrix $[\Sigma] = \{\sigma_{mn}\}_{N\times N}$ is positive definite and uniquely decomposable in the form $[\Sigma] = [C][C^T]$. With $[\Sigma]$ known, we obtain $[C]$ by Cholesky's procedure for factoring a positive definite matrix; the well-known algorithm is not reproduced here (for further details see Elishakoff, 1983b).

Thus, with different realizations of the independent normal variables $d_1, d_2, \ldots, d_N$, we obtain corresponding realizations of the dependent random variables $\bar{\xi}_1, \bar{\xi}_2, \ldots, \bar{\xi}_N$ – the Fourier coefficients of the random initial imperfection function. For each such realization, the initial value problem, Equations (3.40)–(3.43) must be solved in order to find the realization of α^*. The empirical function $R^*(\alpha')$ is then obtainable as

$$R^*(\alpha') = \frac{1}{M} u_M(\alpha') \tag{3.54}$$

where $u_M(\alpha')$ is the number of α^* values exceeding α', and M is the ensemble size (number of trials).

With a single-term approximation, a closed expression is obtainable for $R(\alpha')$. In this case, $\bar{\xi}_m$ is a normally distributed random variable with mean value $\langle \bar{\xi}_m \rangle$ as per Equation (3.48) and variance σ_{mn} as per Equation (3.50). It can be seen from Equation (3.33) (for $\alpha' < \alpha_{\mathrm{m}}$) that

$$R(\alpha') = \mathrm{Prob}(\alpha^* > \alpha') = \mathrm{Prob}(\bar{\xi}_{m,1} < \bar{\xi}_m < \bar{\xi}_{m,2}) \tag{3.55}$$

where $\bar{\xi}_{m,1}$ and $\bar{\xi}_{m,2}$ are defined as

$$\bar{\xi}_{m(1,2)} = -(\alpha_m - \alpha')\frac{s_1}{3\gamma_1 \alpha'} - \frac{2s_1^3}{27\gamma_1^2 \alpha'} \pm \frac{4}{9\alpha'}\sqrt{\frac{2}{s_2}\frac{m}{m_*}}\left(\alpha_m - \alpha' + \frac{s_1^2}{3\gamma_1}\right)^{3/2} \tag{3.56}$$

Finally, we have

$$R(\alpha') = \mathrm{erf}\left(\frac{\bar{\xi}_{m,2} - \langle \bar{\xi}_m \rangle}{\sqrt{\sigma_{mn}}}\right) - \mathrm{erf}\left(\frac{\bar{\xi}_{m,1} - \langle \bar{\xi}_m \rangle}{\sqrt{\sigma_{mn}}}\right) \tag{3.57}$$

Note that for a column on a cubic foundation and $\langle \xi_m \rangle = 0$, Equation (3.57) reduces to Equation (41) of the paper by Elishakoff (1979a):

$$R(\alpha') = \mathrm{erf}\left(\frac{\bar{\xi}'_m}{\sqrt{\sigma_{mn}}}\right), \quad \bar{\xi}'_m = \left(\frac{4}{9}\right)\sqrt{\frac{2}{s_2}}(1 - \alpha')^{3/2}\alpha' \tag{3.58}$$

and is formally identical to Equation (17) of Fraser (1965).

For $\alpha' \geq \alpha_m$, $k_2 > 0$, we have

$$R(\alpha') = \mathrm{erf}\left(\frac{\bar{\xi}_{m,2} - \langle \bar{\xi}_m \rangle}{\sqrt{\sigma_{mn}}}\right) + \mathrm{erf}\left(\frac{\langle \bar{\xi}_m \rangle}{\sqrt{\sigma_{mn}}}\right) \tag{3.59}$$

Consequently, there is some non-zero probability of $\alpha^* > \alpha_m$ (i.e., of the buckling load exceeding that of perfect structure)

$$R(\alpha_m) = \mathrm{erf}\left(\frac{\bar{\xi}_{m,2} - \langle \bar{\xi}_m \rangle}{\sqrt{\sigma_{mn}}}\right) + \mathrm{erf}\left(\frac{\langle \bar{\xi}_m \rangle}{\sqrt{\sigma_{mn}}}\right) \tag{3.60}$$

where

$$\bar{\xi}_{m,2} = -\frac{2s_1^3}{27\gamma_1^2 \alpha_m} + \frac{4s_1^3}{27\alpha_m}\frac{m}{m_*}\sqrt{\frac{2}{3s_2\gamma_1^3}} \tag{3.61}$$

Now the question arises regarding how to choose the imperfection auto-covariance function $K_{\bar{u}}(\eta_1, \eta_2)$ so that it is non-homogeneous, subject to boundary conditions

$$\bar{u}(0) = \bar{u}(1) = 0 \tag{3.62}$$

Following Krenk's suggestion (private communication, 1979), we visualize that the finite structure is obtained by cutting an infinitely long one to a given length and fixing the ends afterward. The initial deviations $\nu(\eta)$ from a middle line before cutting to unit length and fixing the ends can be assumed to be a normal field with mean function $E[\nu(\eta)]$ and the autocorrelation function

$$K_\nu(\eta_1, \eta_2) = E\{[\nu(\eta_1) - E(\nu(\eta_1))][\nu(\eta_2) - E(\nu(\eta_2))]\} \tag{3.63}$$

The structure is now cut, and the line between the end points is used as the axis. Assumption of small angles yields

$$\bar{u}(\eta) = \nu(\eta) - \nu(0) - \eta[\nu(1) - \nu(0)] \tag{3.64}$$

and Equation (3.62) is satisfied.

The covariances σ_{mn} in Equation (3.51) take the form

$$\begin{aligned}\sigma_{mn} &= \tilde{\sigma}_{mn} + \frac{4}{mn\pi^2}K_\nu(0)[(-1)^m - 1][(-1)^n - 1]\\ &+ \frac{4}{mn\pi^2}[K_\nu(0) - K_\nu(1)][(-1)^m + (-1)^n]\\ &- \frac{4}{\pi}[1 + (-1)^{m+n}]\left[\frac{1}{m}(K_\nu, \sin(n\pi\eta)) + \frac{1}{n}(K_\nu, \sin(m\pi\eta))\right]\end{aligned} \tag{3.65}$$

where $(K_\nu, \sin(m\pi n))$ denotes the inner product and $\tilde{\sigma}_{mn}$ is determined by a relation quite analogous to Equation (3.51) with $K_{\bar{u}}(\eta_1, \eta_2)$ replaced by $K_\nu(\eta_1, \eta_2)$.

3.2.7 Numerical Examples

The numerical examples were worked out using the auto-covariance function

$$K_{\bar{u}}(\eta_1, \eta_2) = \frac{A \sin B(\eta_1 - \eta_2)}{\eta_1 - \eta_2} \tag{3.66}$$

with the product AB as the variance of the initial imperfections, and with the mean function chosen as zero. The Fourier coefficients σ_{mn} were approximated from Equation (3.51) by numerical quadrature, and the elements of matrix $[C]$ were obtained by the Cholesky decomposition procedure. Equations (3.41), (3.42), and (3.43) were integrated for numerous "trial" values of the random Fourier coefficients $\bar{\xi}_1, \bar{\xi}_2, \ldots, \bar{\xi}_N$, using Hamming's predictor-corrector method with varying step size (Elishakoff, 1979a). The sign of the quantity of $d\alpha/dF$, the $(N+1)$ component of the unknown vector ξ,

$$\xi^T = (\xi_1, \xi_2, \ldots, \xi_N, \alpha)$$

was checked throughout the integration process to yield the interval containing the buckling load, which was in turn identified as the maximal value of α. Continuous nondimensional end-shortening curves are shown in Figure 3.3 for different $\bar{\xi}_m$

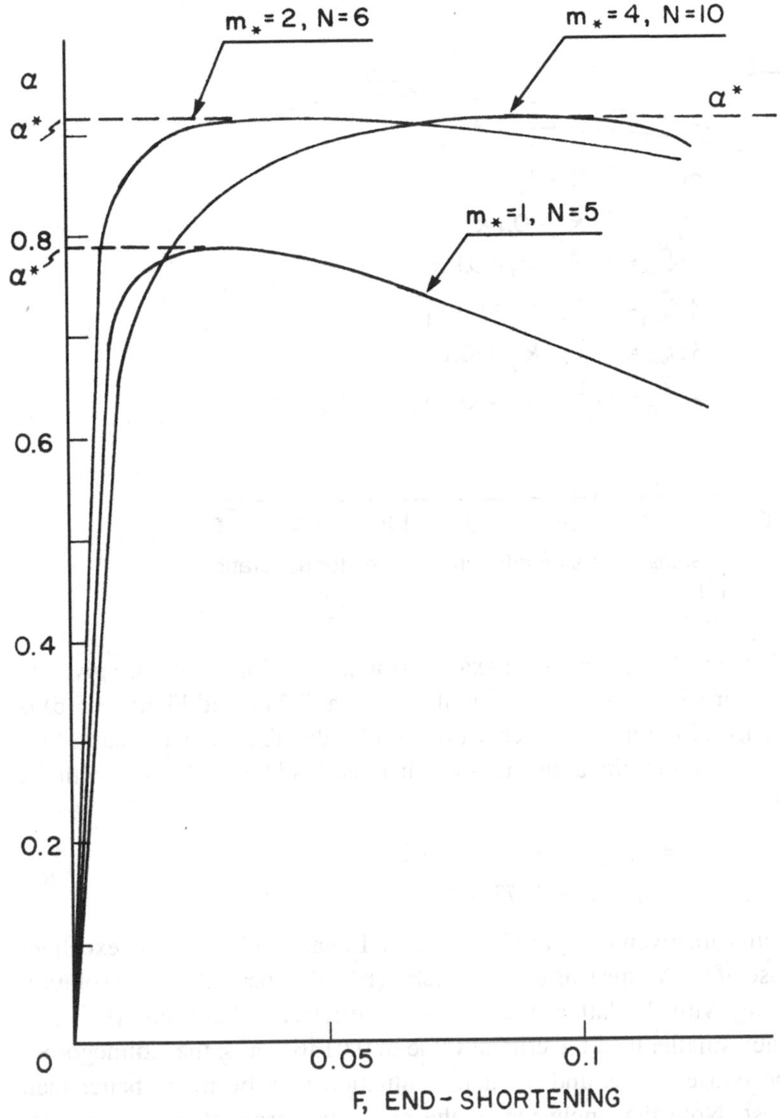

Figure 3.3 Typical load end-shortening curves (purely cubic foundation).

values with $A = 0.01/2\pi$, $B = 1$; for $m_* = 1$, the single-mode approximation suffices for the buckling load approximation (Elishakoff, 1979a). The attendant load-end shortening curves are portrayed in Figure 3.4. With the buckling loads known for a sufficient number of trials, the empirical reliability function was determined by Equation (3.55). The number of realizations used was taken as 10,000 so that, according to the Kolmogorov-Smirnov test of goodness of fit (Massey, 1951), the critical values of the maximum absolute difference between the theoretical $R(\alpha')$ and the empirical $R^*(\alpha')$ reliability function (obtained by the Monte Carlo method) is $1.36/\sqrt{10,000} = 0.0136$ at a level of significance of 0.05. To be able to check the Monte Carlo method against the exact solution, a single-term Boobnov-Galerkin

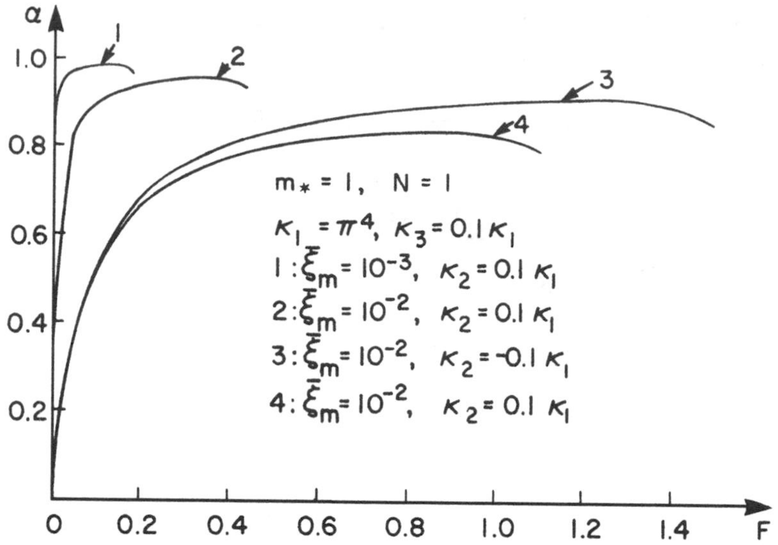

Figure 3.4 Load end-shortening curves for different $\bar{\xi}$ values for quadratic-cubic foundation ($\kappa_1 = \pi^4$).

approximation was used. Results from the exact formula Equation (3.58) are given in Figure 3.5, where Figure 3.5(a) is the solution of Equation (3.57), and Figure 3.5(b) is the probability density of the initial imperfection amplitude. The shaded area in Figure 3.5(b) represents the reliability at the non-dimensional load level $\alpha' = 0.8$ with the following parameters:

$$\begin{aligned} &m = m^* = 1, \qquad \kappa_1 = \pi_4, \qquad \kappa_2 = \kappa_3 = 0.1\kappa_1, \\ &\alpha_{\max} = 1.012008, \qquad \sigma_{11} = 0.231477 \times 10^{-3} \end{aligned} \tag{3.67}$$

The calculation results are given in Figure 3.6 based on Equation (3.58) are in excellent agreement with those of the Monte Carlo method (shown by the stars), the exact solution practically coinciding with the latter. The maximum difference between $R(\alpha')$ and $R^*(\alpha')$ is 0.005, much smaller than the critical value of 0.0136. Thus, the Kolmogorov-Smirnov test is too conservative, and the actual situation may be much better than predicted by the test. Note that inclusion of the quadratic terms skews the single-mode imperfection-sensitivity curve [Figure 3.5(a) by comparison with Figure 2 of Elishakoff, 1979a], so as to produce more sensitivity for positive values of $\bar{\xi}_m$ and less for negative.

Figure 3.7 shows the influence of the coefficient B on $R^*(\alpha')$, which is seen to decrease as B increases (for constant $A = 0.03$). For example, at load level 0.95 the reliability equals 1.0 for $B = 1$ (i.e., almost none of the columns buckles), 0.975 for $B = 2$ (i.e., about 2.5% of columns buckle), and 0.425 for $B = 3$ (i.e., about 57.5% of the columns buckle).

Figure 3.8 portrays the influence of the coefficient A on $R^*(\alpha')$, which decreases with the increase of A. This is also understandable because a larger A means a larger variance of the initial imperfections. As is seen in Figure 3.8, the reliability function furnishes a basis for design of the stochastically imperfect structures. The criterion is for such a design the reliability should be greater than some required reliability R. The

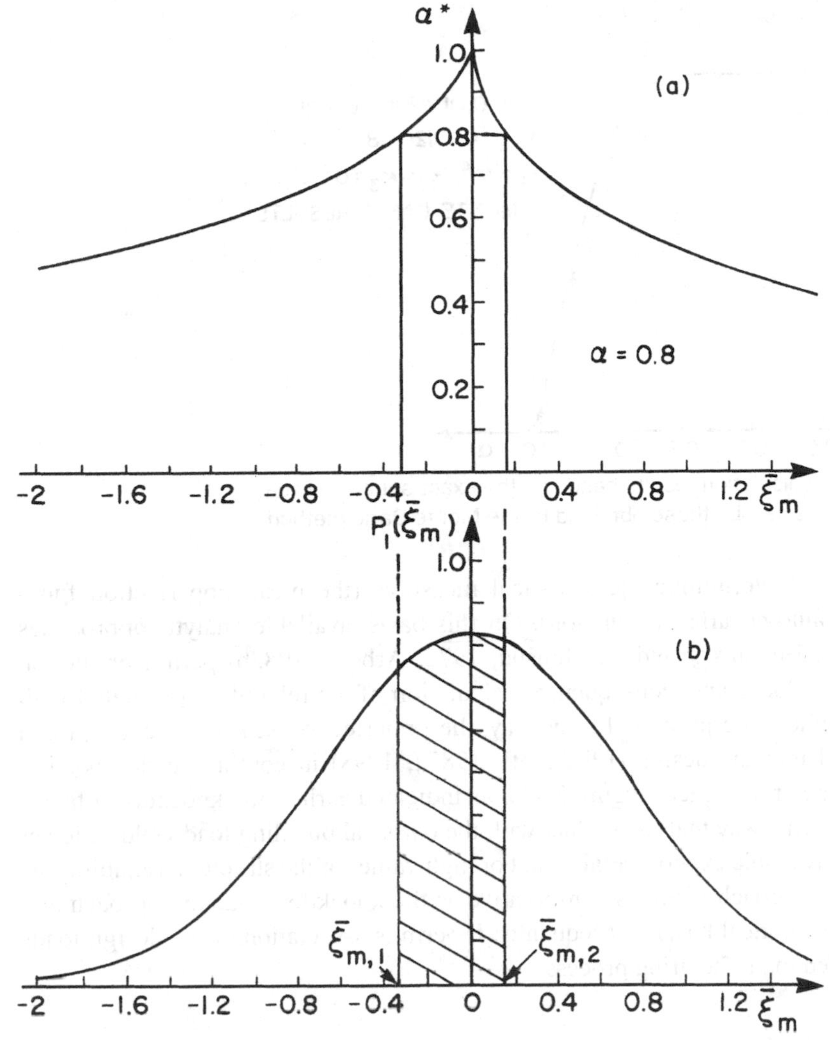

Figure 3.5 Single-term Boobnov-Galerkin solution: (a) buckling load parameter as function of critical imperfection amplitude; (b) probability density of critical imperfection amplitude, with the shaded area equal to the reliability of the structure at non-dimensional load level $\alpha = 0.8$.

maximum non-dimensional load α_{design}, which satisfies the equation

$$R(\alpha_{\text{design}}) = R \tag{3.68}$$

is then the design strength for the ensemble of the columns. If, for example, $R = 0.9$, then Figure 3.8 provides us with the design load levels 0.888 for $A = 0.03$, 0.904 for $A = 0.02$, and 0.914 for $A = 0.01$. Figure 3.9 shows a histogram of non-dimensional buckling loads, while Figure 3.10 portrays the sample variance.

The method shown in this section can be extended to other structures, and is deemed to provide a sound theoretical basis for engineering design (Elishakoff, 1983a, 1983b, 1999). The first and most difficult step is to compile extensive experimental information on imperfections, classified according to the manufacturing process,

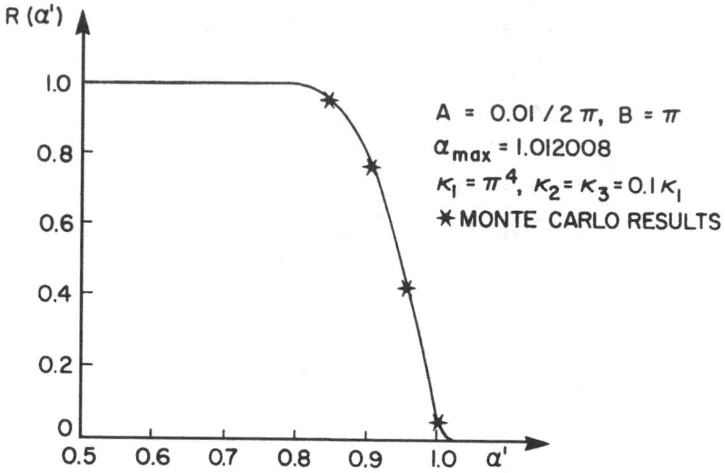

Figure 3.6 The calculation results based on the exact solution [Equation (3.57)] coincide with those obtained by the Monte Carlo method.

with a view to determining the statistical measures (the mean imperfection function and the auto-covariance function). On this basis, available analytic approaches (Koiter, 1945; Budiansky and Hutchinson, 1972; Arbocz, 1982b) permit prediction of the buckling loads and, consequently, calculation of the reliability associated with a given manufacturing process. In this way, the imperfection-sensitivity concept can be introduced into the design (Elishakoff, 1983a, 1998) in contrast to the existing knockdown-factor approach (Figure 3.11); as indicated earlier, the knockdown factor is chosen in such a way that its product with the classical buckling load yields a lower bound on the available experimental data. For high values of the structural reliability R, the reliability approach is not as conservative as the knockdown-factor approach and has a sound theoretical basis; consequently, it permits association of the design loads with a specified manufacturing process.

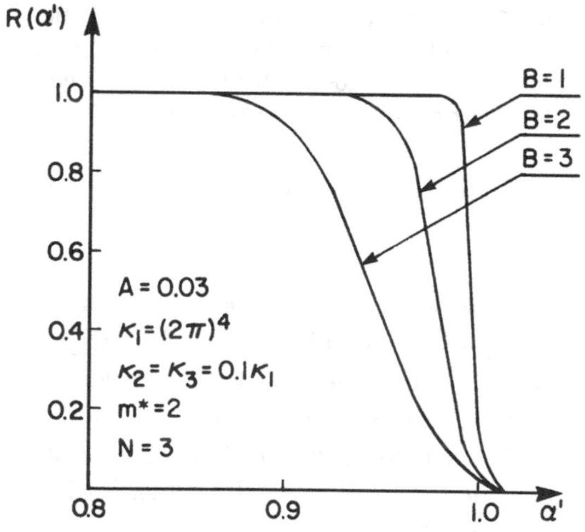

Figure 3.7 The reliability function $R(\alpha')$ versus the non-dimensional actual load α'.

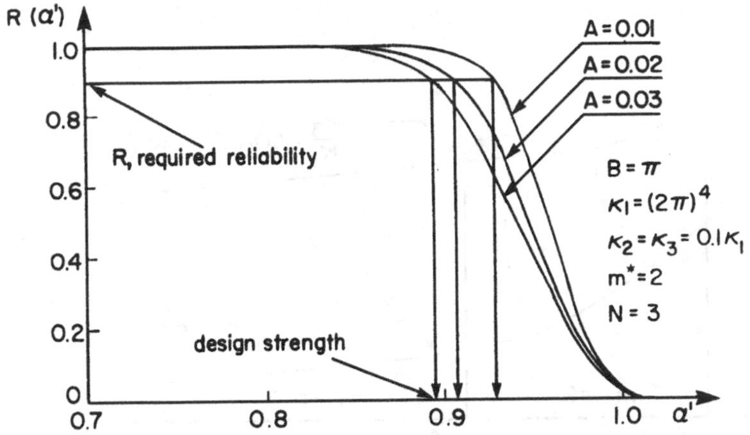

Figure 3.8 Calculation of the non-dimensional design loads associated with specific required reliability, of the structure (after Elishakoff, *Acta Mechanica*, reprinted with permission, Springer Verlag, 1985).

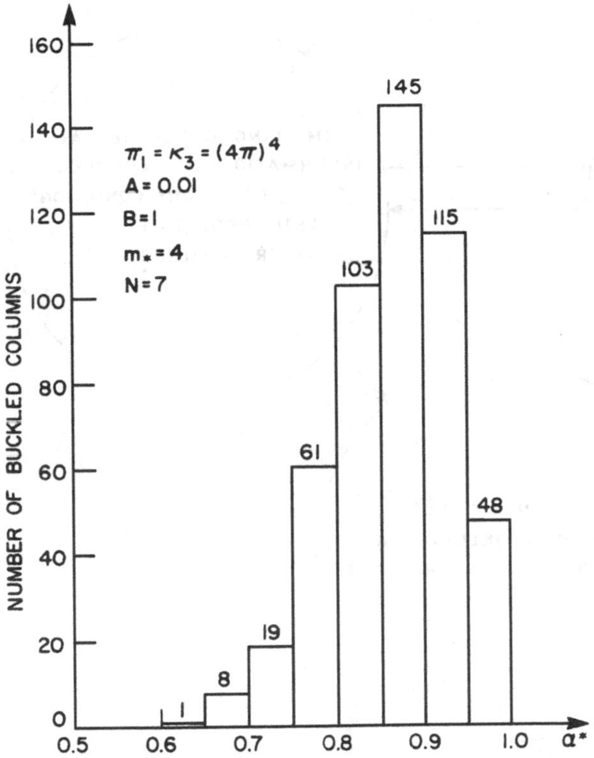

Figure 3.9 Histogram of non-dimensional buckling loads (after Elishakoff, *Journal of Applied Mechanics*, 1979a).

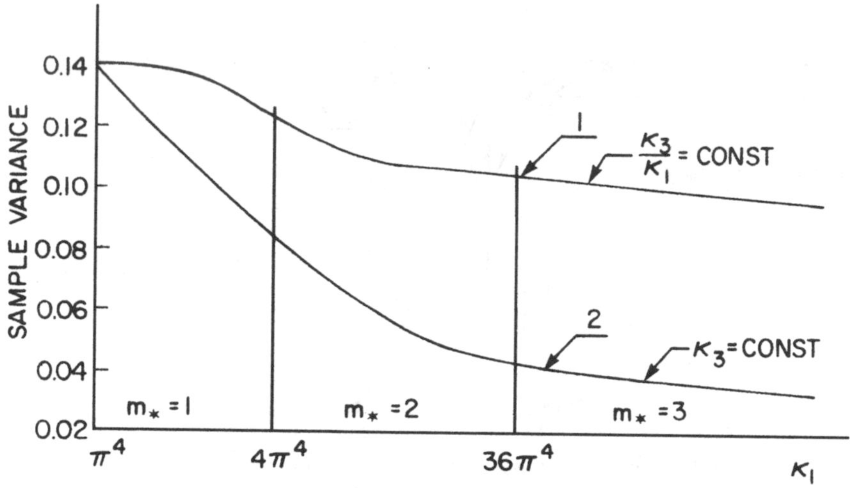

Figure 3.10 Sample variance of the non-dimensional buckling load versus κ_1 (after Elishakoff, *Journal of Applied Mechanics*, 1979a).

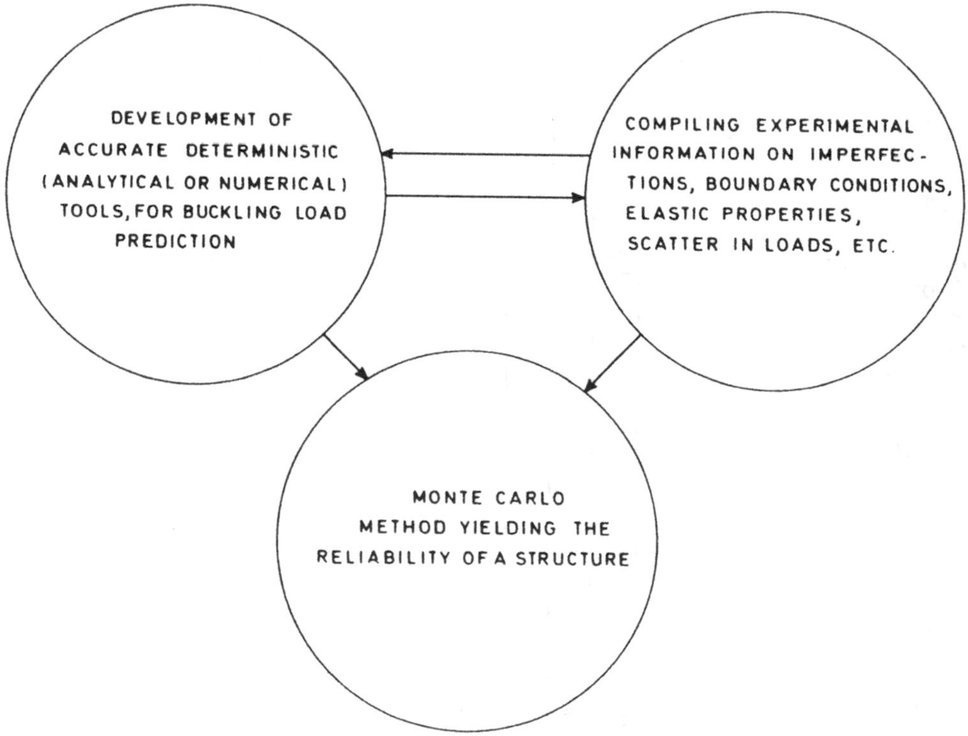

Figure 3.11 How to introduce the imperfection-sensitivity concept into design (after Elishakoff, 1983a).

3.3 Non-Linear Buckling of a Structure with Random Imperfection and Random Axial Compression by a Conditional Simulation Technique

In Section 3.2 as well as in almost all other studies, the uncertainty associated with the external load was not investigated. However, as discussed in Li et al. (1995), in real situations, the external load is always subject to some amount of variability. Two exceptions, which treated some simple cases including the uncertainty in loads in addition to that of the initial imperfection, were offered by Roorda (1980) and Elishakoff (1983a). In this section, we postulate a probabilistic model for the external load and assume it to be a random variable. Due to the complexity of the deterministic procedure involved in the evaluation of the buckling load, we must again resort to a solution by a simulation technique. Here, an improved simulation technique is introduced to reduce significantly the computational effort in deriving the reliability of the structure.

3.3.1 Deterministic Procedure

The differential equation for the deflection of the imperfect column on a non-linear cubic foundation is (Elishakoff, 1979a),

$$EI\frac{d^4w}{dx^4} + P\frac{d^2w}{dx^2} + k_1w - k_3w^3 = -P\frac{d^2\bar{w}}{dx^2} \tag{3.69}$$

For the simply supported column, the boundary conditions read:

$$w = \frac{d^2w}{dx^2} = 0 \quad \text{at} \quad x = 0 \quad \text{and} \quad x = L \tag{3.70}$$

where $\bar{w}(x)$ is the initial imperfection, $w(x)$ is the additional deflection due to the axial load P, k_1 and k_3 are the "spring" constants of the foundation, and EI is the flexural rigidity of the column.

The classical buckling load of the problem can be obtained by letting $\bar{w}(x) = 0$ and $k_3 = 0$ and substituting the expression of the classical buckling mode $w(x) = \sin(m\pi x/L)$ into Equation (3.69). Thus, one obtains

$$P(m) = \frac{EI}{L^2}\left(\pi^2m^2 + \frac{k_1L^4}{EI\pi^2m^2}\right) \tag{3.71}$$

We denote by m_* the integer m, which corresponds to the smallest value of P in (3.71), that is,

$$P(m_*) = \frac{EI}{L^2}\left(\pi^2m_*^2 + \frac{k_1L^4}{EI\pi^2m_*^2}\right) \tag{3.72}$$

where m_* takes the value of the integer nearest to $(k_1/EI)^{1/4}L/\pi$.

By introducing the non-dimensional quantities,

$$u = \frac{w}{\Delta}, \quad \bar{u} = \frac{\bar{w}}{\Delta}, \quad \eta = \frac{x}{L}, \quad \alpha = \frac{P}{P_{cl}}$$
$$\bar{k}_1 = \frac{k_1l^4}{EI}, \quad \bar{k}_3 = \frac{k_3l^4\Delta^2}{EI}, \quad \gamma = \pi^2m_* + \frac{\bar{k}_1}{\pi^2m_*} \tag{3.73}$$

where Δ is the radius of gyration of the cross section, Equations (3.69) and (3.70) can be transformed into their non-dimensional forms,

$$\frac{d^4u}{d\eta^4} + \alpha\gamma\frac{d^2u}{d\eta^2} + \bar{k}_1 u - \bar{k}_3 u^3 = -\alpha\gamma\frac{d^2\bar{u}}{d\eta^2} \tag{3.74}$$

and

$$u = \frac{d^2u}{d\eta^2} = 0 \quad \text{at} \quad \eta = 0 \quad \text{and} \quad \eta = 1 \tag{3.75}$$

We expand u and $\bar{u}$ in classical buckling modes,

$$u = \sum_{m=1}^{\infty} \xi_m \sin(m\pi\eta) \tag{3.76}$$

$$\bar{u} = \sum_{m=1}^{\infty} \bar{\xi}_m \sin(m\pi\eta) \tag{3.77}$$

Substituting Equations (3.75) and (3.76) into Equation (3.73) and employing Boobnov-Galerkin's method yields the following set of coupled nonlinear algebraic equations for ξ_m:

$$\alpha_m\xi_m - \alpha(\xi_m + \bar{\xi}_m) - \frac{s}{8}\frac{m_*^2}{m^2} I_m = 0 \tag{3.78}$$

in which

$$s = \frac{2\bar{k}_3}{\bar{k}_1 + m_*^4\xi^4}, \qquad \alpha_m = \frac{\pi^2 m^2 + \bar{k}_1/(\pi^2 m^2)}{\pi^2 m_*^2 + \bar{k}_1/(\pi^2 m_*^2)}$$

$$I_m = \sum_{p=1}^{\infty}\sum_{q=1}^{\infty}\sum_{r=1}^{\infty} \xi_p\xi_q\xi_r[\delta_{p+q,r+m} - \delta_{|p-q|,r+m} - \delta_{p+q,|r-m|} + \delta_{|p-q|,|r-m|} + \delta_{p,q}\delta_{r,m}] \tag{3.79}$$

Here m_* is the buckling wave number, and $\delta_{p,q}$ is the Kronecker delta.

An approximate solution to Equation (3.67) can be obtained by properly truncating these equations and retaining the terms that have the most important contribution to the buckling load. Fraser and Budiansky (1969) as well as Elishakoff, Cai, and Starnes (1994a, 1994b) have pointed out that the most significant contribution to the solution comes from the m_*th term and its neighboring terms. This is especially true when k_3 is much smaller than k_1, and the system possesses a "weak" nonlinearity. Hence, retaining a few terms on either side of the m_*th term, one can solve Equation (3.67) step by step for incrementally assumed ξ_m, using a Newton-Raphson type of iteration procedures. The buckling load P^* is determined as the maximum load the system can carry. Because the column falls into the category of the imperfection-sensitive structures, the unavoidable presence of initial imperfection lowers the buckling load, thus reducing the load-bearing capacity of the structure. This means when initial imperfection is present, the non-dimensional parameter $\alpha(= P^*/P_{cl})$ is below unity.

3.3.2 Formulation of Basic Random Variables

As in the previous section, the initial imperfection expression Equation (3.66) needs to be truncated,

$$\bar{u}(\eta) = \sum_{m=1}^{N} \bar{X}_m \sin(m\pi\eta) \tag{3.80}$$

where N is the number of retained terms; ξ_m is a possible value the random variable $\bar{X}_m$ can assume.

Generally speaking, if the mean initial imperfection $\bar{U}(\eta) = E[\bar{u}(\eta)]$ is known through measurements, the mean values $\mu_m = E[\bar{X}_m]$ can be calculated as

$$\mu_m = 2\int_0^1 \bar{U}(\eta)\sin(m\pi\eta)d\eta \tag{3.81}$$

and the auto-covariance function of the imperfections is

$$C_{\bar{u}}(\eta_1, \eta_2) = \sum_{m=1}^{K}\sum_{n=1}^{K} E[(\bar{X}_m - \mu_m)(\bar{X}_n - \mu_n)]\sin(m\pi\eta)\sin(n\pi\eta) \tag{3.82}$$

The elements v_{mn} of the variance-covariance matrix $[V]$ are determined by

$$\begin{aligned} v_{mn} &= E[(\bar{X}_m - \mu_m)(\bar{X}_n - \mu_n)] \\ &= 4\int_0^1\int_0^1 C_{\bar{u}}(\eta_1, \eta_2)\sin(m\pi\eta_1)\sin(n\pi\eta_2)\,d\eta_1\,d\eta_2 \end{aligned} \tag{3.83}$$

In this case, the initial imperfection coefficients $\{\xi\}$ can be simulated in terms of the following formula (Elishakoff, 1983b):

$$\{\bar{X}\} = [C]\{Z\} + \{\mu\} \tag{3.84}$$

where $[C][C]^T = [V]$, and $\{Z\}$ is an independent standard normal distribution vector, $\{\mu\}^{\mathrm{T}} = \{\mu_1, \mu_2, \ldots, \mu_N\}$.

When the initial imperfection mean function $\bar{U}(\eta)$ is not easily available, Elishakoff, Cai, and Starnes (1994a, 1994b) proposed the use of a truncated normal distribution for each random variable, that is,

$$f_{\bar{X}_m}(\xi_m) = \begin{cases} C_m \exp\left(-\dfrac{\xi_m^2}{b_m^2}\right), & |\xi_m| \le A_m \\ 0, & |\xi_m| > A_m \end{cases} \tag{3.85}$$

where $f_{\bar{X}_m}(\xi_m)$ is the probability density function of ξ_m, A_m is the maximum value that can be taken by the random variable $\bar{X}_m$, b_m is a parameter, and the normalization constant C_m is given by

$$C_m = \left[2b_m \operatorname{erf}\left(\frac{A_m}{b_m}\right)\right]^{-1} \tag{3.86}$$

where erf(·) is the error function, defined as

$$\operatorname{erf}(x) = \frac{1}{\sqrt{2\pi}} \int_0^x \exp\left(-\frac{t^2}{2}\right) dt \tag{3.87}$$

The advantage of this probability density function lies in its flexibility. As is easily verified, when the value of parameter A_m increases while b_m remains constant, $\bar{\xi}_m$ approaches the (untruncated) normal distribution. As another extreme, when $b^2 \gg A^2$, $\bar{\xi}_m$ tends to be nearly uniformly distributed.

In real situations, the external loadings are also subject to a degree of uncertainty. Consequently, the structure may be subjected to overloads. The effect of the actual load variation in addition to the randomness of the initial imperfection was previously considered by Roorda (1980) and Elishakoff (1983b) for model structures that admitted the closed-form analytical solution. When multi-mode analysis is needed, as is generally the case with realistic structures, such exact analysis is unfeasible, and simulation techniques should be utilized. In this investigation, we assume that the external load P satisfies Type I extreme-value distribution, the probabilistic distribution function of which is as follows (Thoft-Christensen and Baker, 1982):

$$F_P(p) = \exp\{-\exp[-a(p-c)]\}, \qquad a > 0 \tag{3.88}$$

where parameters a and c are related to the mean value μ_P and the standard deviation σ_P in the following manner:

$$\mu_P = c + \frac{0.5772}{a}, \qquad \sigma_P = \frac{\pi}{\sqrt{6}a} \tag{3.89}$$

3.3.3 Probabilistic Analysis

Because of the existence of random initial imperfection, the buckling load P^* is a random variable. For each sample of the random initial imperfection $\{\bar{\xi}\}$, the procedure outlined in Section 3.2.1 can be employed to determine the buckling load $P^*(\bar{\xi})$. On the other hand, the applied load P is generally also a random variable. Thus, the criterion of the satisfactory performance of the system is expressed as

$$P^*(\bar{X}) > P \tag{3.90}$$

Hence, the probability of failure is the probability of violation of the inequality in Equation (20)

$$P_f = \text{Prob}[P > P^*(\bar{X})] = 1 - \text{Prob}[P < P^*(\bar{X})] \tag{3.91}$$

Due to the complexity of the deterministic procedure for the evaluation of the buckling load, a simulation technique can be successfully applied for the determination of the probability of failure for the bar resting on the non-linear foundation. However, in view of the small value of the desired probability of failure for real structures in engineering practice (P_f is normally in the order of 10^{-4}, 10^{-5}, or less), the number of simulations N required by the Monte Carlo method is extremely large: N is proportional to the value of $(1 - P_f)/P_f\epsilon^2$, where ϵ is the desired accuracy of the simulation result (Hammersley and Handscomb, 1964). For instance, for a structure with the probability of failure $P_f = 2 \times 10^{-4}$, over 3 million simulation trials are needed in order to ensure a 10% accuracy for the estimated result from simulations. Therefore, the use of direct Monte Carlo simulation is unfeasible in this case, and appropriate modifications are needed.

There exist several variation reduction techniques that are aimed at improving the computational efficiency of the Monte Carlo method (Rubinstein, 1981), the conditional expectation technique (Ayyub and Haldar, 1985) being one of them. Later on, Li and Gan (1992) made further modification and supplemented the conditional expectation technique; the simulation of basic variables is combined with the analytical evaluation of probability (or sometimes numerical integration), and statistical treatment of the computational results is carried out with the incorporation of the batch means technique. In this way, both the efficiency and accuracy of the computation are greatly increased such that the CPU time required for the solution of the problem is substantially reduced. Here we will apply this technique to the present problem.

As we have assumed that the external load P is a random variable with probability distribution function $F_P(p)$, Equation (3.91) can be transformed as follows, using the concept of conditional probability:

$$\{P_f \mid \bar{X}_1 = \bar{\xi}_1, \bar{X}_2 = \bar{\xi}_2, \ldots, \bar{X}_N = \bar{\xi}_N\} = 1 - \text{Prob}[P < P^*(\bar{\xi})] = 1 - F_P[P^*(\bar{\xi})] \tag{3.92}$$

where $F_P(\cdot)$ is the probability distribution of the external load P, evaluated at the value of function P^* when $\bar{X}_1 = \bar{\xi}_1, \bar{X}_2 = \bar{\xi}_2, \ldots, \bar{X}_N = \bar{\xi}_N$. The unconditional probability of failure becomes, therefore,

$$\begin{aligned} P_f &= \int_{-\infty}^{+\infty} \cdots \int_{-\infty}^{+\infty} \{P_f \mid \bar{X}_1 = \bar{\xi}_1, \bar{X}_2 = \bar{\xi}_2, \ldots, \bar{X}_N = \bar{\xi}_N\} \\ &\qquad \times f_{\bar{N}}(\bar{\xi}_1, \bar{\xi}_2, \ldots, \bar{\xi}_N)\, d\bar{\xi}_1 d\bar{\xi}_2 \ldots d\bar{\xi}_N \\ &= \int_{-\infty}^{+\infty} \cdots \int_{-\infty}^{+\infty} \{1 - F_P[P^*(\bar{\xi})]\} f_{\bar{X}}(\bar{\xi}_1, \bar{\xi}_2, \ldots, \bar{\xi}_N)\, d\bar{\xi}_1 d\bar{\xi}_2 \ldots d\bar{\xi}_N \end{aligned} \tag{3.93}$$

It is remarkable that the right-hand side of this equation represents the mathematical expectation of the expression in the parentheses with respect to the initial imperfection components $\bar{X}_1, \bar{X}_2, \ldots, \bar{X}_K$:

$$P_f = 1 - E_{\bar{X}}\{F_P[P^*(\bar{X})]\} \tag{3.94}$$

An unbiased estimator of the mathematical expectation in (3.94) is

$$\tilde{P}_f = \frac{1}{N_*} \sum_{j=1}^{N_*} \tilde{P}_{f_j} \tag{3.95}$$

where

$$\tilde{P}_{f_j} = 1 - F_P[P^*(\bar{\xi}_1, \bar{\xi}_2, \ldots, \bar{\xi}_N)]_j \tag{3.96}$$

For each sequence of the simulated random initial imperfection coefficients $\{\bar{\xi}_1, \bar{\xi}_2, \ldots, \bar{\xi}_N\}$, use of the deterministic procedure described in Section 3.2.1 results in a buckling load P^*, and substitution of this buckling load P^* into Equation (3.96) yields an estimated value of the probability of failure through an analytical calculation.

If N_* sequences of random variables $\{\bar{\xi}_1, \bar{\xi}_2, \ldots, \bar{\xi}_N\}$ are generated in terms of their respective probabilistic distributions by computer, repeating the same procedure described previously, one can obtain N_* approximate values of probability of failure $\tilde{P}_{f_j}(j = 1, 2, \ldots, N)$. The mean of these N approximate values, as indicated by Equation (3.95), constitutes another, but a better, estimate of the probability of failure P_f.

The sample deviation of the preceding simulations reads

$$s^2 = \frac{1}{N_* - 1} \sum_{j=1}^{N_*} \left(\tilde{P}_{f_j} - \bar{P}_f\right)^2 \tag{3.97}$$

One of the advantages of the present simulation technique is that here only the initial imperfection needs to be simulated, whereas by the direct Monte Carlo method one has to simulate not only the initial imperfection but the external load as well. In addition, the analytical probabilistic evaluation is performed so that the computational efficiency is further improved.

However, it should be pointed out that $\bar{P}_f$, defined by Equation (3.95), represents an approximate value of the probability of failure. So, the following question naturally arises: How far is this approximate value different from its real value P_f, or, in other words, what is the precision of this approximation? To answer this question, one has to use the theory of statistics, in particular, the interval estimation.

If $\tilde{P}_{f_1}, \tilde{P}_{f_2}, \ldots, \tilde{P}_{f_{N_*}}$ are random variables satisfying the same normal distribution and are independent of each other, then $(\bar{P}_f - P_f)/\sqrt{s^2/N_*}$ tends to $t(N_* - 1)$ as N_* increases, where $t(N - 1)$ denotes the $N - 1$ order student distribution. The confidence interval with $100(1 - \alpha)\%$ level of confidence is

$$\left(\bar{P}_f - t_{\alpha/2}(N_* - 1)\sqrt{\frac{s^2}{N_*}},\ \bar{P}_f + t_{\alpha/2}(N_* - 1)\sqrt{\frac{s^2}{N_*}}\right) \tag{3.98}$$

When $N_* > 40$, $t_{\alpha/2}(N_* - 1)$ can be replaced by $z_{\alpha/2}$, the percentile of the standard normal distribution $N(0, 1)$.

However, it should be pointed out that the confidence interval expressed by Equation (3.98) is only approximate. This is because, in most situations, we do not have sufficient evidence to state that random variables $\tilde{P}_{f_1}, \tilde{P}_{f_2}, \ldots, \tilde{P}_{f_{N_*}}$ belong to the same Gaussian distribution. Besides, $\tilde{P}_{f_1}, \tilde{P}_{f_2}, \ldots, \tilde{P}_{f_N}$ are the data of simulation tests originated from the same set of seeds, so, strictly speaking, these data are not independent. In order to use the classical theory of statistics, Law and Kelton (1982) proposed several methods for transforming the simulation results into independent random variables of the same distribution. Here we adopt the batch means technique. In the actual implementation of this technique, our procedure is sightly different. Suppose the total number of simulation tests is N_*. We can carry out the simulation tests in M batches, each batch performing L_* tests ($N_* = M \times L_*$) and obtaining L_* estimated values of the probability of failure. For the sake of clarity, we will use P to denote the estimated probability of failure in the following simulation data.

Data from the first batch is denoted by $P_1, P_2, \ldots, P_{L_*}$, and data from the second batch is denoted by $P_{L_*+1}, P_{L_*+2}, \ldots, P_{2L_*}$. In a perfect analogy, data from the Mth batch is indicated as $P_{(M-1)L_*}, P_{(M-1)L_*+1}, \ldots, P_{ML_*}$. The average of the data from each batch reads

$$\bar{P}_j = \frac{1}{L_*}\sum_{j=1}^{L_*} P_{(j-1)L_*+i} \qquad (j = 1, 2, \ldots, M) \tag{3.99}$$

We utilize

$$\check{P} = \frac{1}{M}\sum_{j=1}^{M} \bar{P}_j \tag{3.100}$$

as the final estimator of the probability of failure.

According to the central limit theorem, if L_*, the number of tests in each batch, is sufficiently large, $\bar{P}_j (j = 1, 2, \ldots, M)$ approximately follows the same Gaussian distribution. In addition, every $\bar{P}_j$ is a mean value of a batch of test data. Therefore, $\bar{P}_j (j = 1, 2, \ldots, M)$ can be viewed as "almost" independent with each other. In reality, we use a different seed for a different batch of simulation tests, thus further reducing the correlation among these data $\bar{P}_j (j = 1, 2, \ldots, M)$. The sample deviation of the preceding simulation data is

$$s_M^2 = \frac{1}{M-1}\sum_{j=1}^{M} (\bar{P}_j - \check{P})^2 \tag{3.101}$$

Thus, the confidence interval defined by Equation (3.98) can be modified as

$$\left(\check{P} - t_{\alpha/2}(M-1)\sqrt{\frac{s_M^2}{M}}, \check{P} + t_{\alpha/2}(M-1)\sqrt{\frac{s_M^2}{M}}\right) \tag{3.102}$$

When using this confidence interval, we propose that the correlatedness of the data $\bar{P}_j (j = 1, 2, \ldots, M)$ should be assessed. This can be done by utilizing the following estimator of the correlation coefficient:

$$\hat{\rho} = \frac{\sum_{j=1}^{M-1} (\bar{P}_j - \check{P})(\bar{P}_{j+1} - \check{P})}{\sum_{j=1}^{M} (\bar{P}_j - \check{P})^2} \tag{3.103}$$

When the value of $\hat{\rho}$ in expression (3.103) is large, the confidence interval expressed by Equation (3.101) is not accurate. The method of reducing the value of $\hat{\rho}$ is to appropriately add some more tests into each batch and to increase the number of batches.

If we define the ratio of the semi-length of the confidence interval to the estimated value $\check{P}$ of the probability of failure as the relative precision

$$\epsilon_r = \frac{1}{\check{P}} t_{\alpha/2}(M-1)\sqrt{\frac{s_M^2}{M}} \tag{3.104}$$

then increasing L_*, the number of tests in each batch, and M, the number of batches, makes ϵ_r attain a required accuracy.

In the course of simulation tests, the number of tests in each batch L_* and the number of batches M should be properly chosen. Usually, L_* can not be too small, otherwise the data $\bar{P}_j$ might not be independent, or it may deviate from the Gaussian distribution. Similarly, M should not be too small otherwise there might be a remarkable correlation among data $\bar{P}_j (j = 1, 2, \ldots, M)$. Besides, the choice of M and L_* is also dependent on the accuracy required and the actual value of the probability of failure of the structure. Generally speaking, the higher the required accuracy, and the smaller the failure probability itself, the more simulations are needed to perform. It is worth noticing that the adoption of the batch means technique also saves the computer storage. If the batch means technique is not employed, the total number of required storage elements in computer is approximately equal to the total number of simulation tests performed. By using the batch means technique, the total number of storage elements is reduced to the number of batches.

3.3.4 Numerical Example and Discussion

As a numerical example, we consider the case where $\bar{k}_1 = 16\pi^4$ and $\bar{k}_3 = 0.1\,\bar{k}_1$ such that $m_* = 2$. Numerical calculations show that here only four terms in the deterministic procedure are needed to obtain a sufficiently accurate solution of the buckling load P^*. For the initial imperfection simulation, we assume each $\bar{\xi}_m$ $(m = 1, 2, 3, 4)$ to have the same probability distribution density $f_{\bar{X}}(\bar{\xi}_m)$ [Equation (3.94)] with $A_1 = A_2 = A_3 = A_4 = A$ and $b_1 = b_2 = b_3 = b_4 = b$; A is fixed at 0.5, and b is fixed at 0.1. Furthermore, we assume that the applied load has a Type I extreme-value distribution with the mean value $\mu_P = 0.5P_{cl}$ and standard deviation $\sigma_P = 0.05P_{cl}$. For each sample of the simulated initial imperfection, the deterministic procedure is carried out to get the corresponding buckling load P^* and then an analytical probability evaluation is performed with respect to the external load P so that an estimated value of the probability of failure $\tilde{P}_{f_j}$ is obtained. In the actual computation, we carried out the simulations by batches. The total number of simulations was 2000; here L_* (the number of tests in each batch) and M (the number of batches) were taken to be 50 and 40, respectively. A 99% confidence interval for the probability of failure is obtained through (3.102) as follows:

$$[0.237 \times 10^{-4} - 0.108 \times 10^{-5}, 0.237 \times 10^{-4} + 0.108 \times 10^{-5}]$$

The relative precision of the approximate value $\check{P}_f = 0.239 \times 10^{-4}$ is 5%. The whole process only consumed 4 minutes of CPU time (the numerical calculation was conducted on a DEC 5000/200 workstation). By contrast, one has to do over 10 million Monte Carlo simulation tests to reach a similar precision.

In engineering practice, designers usually require a very low probability of failure. The latter should be determined with high precision. Therefore, the level of confidence for the estimated probability of failure should be consistent with the value of the probability of failure itself to result in a safe design. In our case, with 4000 simulations $(L_* = 80, M = 50)$, a 99.999% confidence interval for the probability of failure is obtained as

$$[0.237 \times 10^{-4} - 0.209 \times 10^{-5}, 0.237 \times 10^{-4} + 0.209 \times 10^{-5}]$$

The computational process requires about 8 minutes of CPU time on a DEC 5000/200 workstation.

In conclusion, it is worth mentioning that the confidence interval for the probability of failure obtained here can be directly applied to the structural design. For instance, suppose that the design code requires

$$P_f \leq [P_f] \tag{3.105}$$

where $[P_f]$ is an codified value representing the maximum tolerable probability of failure for the structure. Because the actual value of the probability of failure for the structure in consideration is generally unknown, use a more conservative expression than Equation (3.105), such as

$$\check{P} + t_{\alpha/2}\sqrt{s_M^2/M} \leq [P_f]$$

For further details, consult the study by Li et al. (1995).

In some studies there is a claim that the randomness in the applied load makes for reduced reliability. To elucidate this question, we consider a simple model structure, capable of exact solution. Consider a cantilever of Figure 3.12 – a rigid link of length a, pinned to a rigid foundation and supported by a linear extensional spring of stiffness x, which is capable of resisting both tension and compression and which retains its horizontality as the system deflects (Elishakoff, 1983a). The initial imperfection is

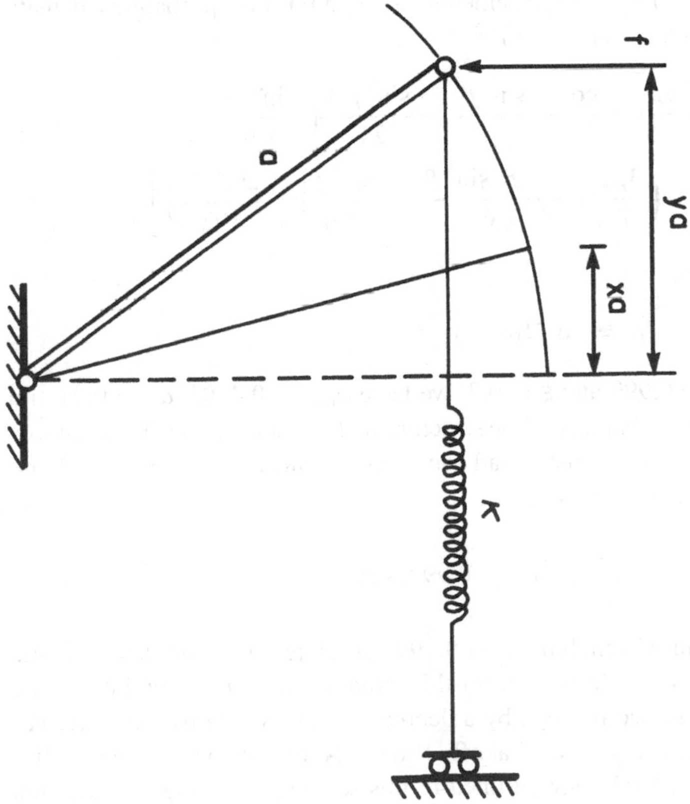

Figure 3.12 The propped cantilever.

modeled by the deflection κa from the vertical position; the total displacement is denoted as ya. Equilibrium dictates

$$\lambda y = (y - x)\sqrt{1 - y^2}$$
$$\lambda = f/_{cl}, \qquad f_{cl} = \kappa a$$

For the non-dimensional buckling load λ_s – the maximum load the structure can sustain – we put $d\lambda/dy = 0, \lambda = \lambda_s$. Hence, $y = x^{1/3}$, which yields an exact expression

$$\lambda_s = (1 - x^{2/3})^{3/2}$$

for the limit load, due to Thompson and Hunt (1973). Let the applied load be a random variable Λ, uniformly distributed over the interval (λ_1, λ_2), that is the probability density $f_\Lambda(\lambda)$ equals $(\lambda_2 - \lambda_1)^{-1}$ in the interval (λ_1, λ_2) and vanishes elsewhere. The initial imperfections are random; due to this fact, Λ_s is a random variable too, with probability distribution function $F_{\Lambda_s}(\lambda_s)$. With reasonable assumption of independence of Λ_s and Λ due to independence of the applied loads of the manufacturing process, we find [by analogy with Equation (7.4) of Elishakoff, 1983b]

$$R = \int_0^\infty [1 - F_{\Lambda_s}(z)] f_\Lambda(z)\, dz$$

The calculations yield, for initial displacements having a uniform probability density in the interval $(0, \xi)$ with $\alpha^* = (1 - \xi^{2/3})^{3/2}$

$$R = \frac{\alpha^*}{\alpha_{\max}} - \frac{3}{\alpha_{\max}\xi}\left[\frac{3\beta_2}{48} + \frac{\cos\beta_2 \sin^5\beta_2}{6} - \frac{\sin^2\beta_2}{24} + \frac{\sin^4\beta_2}{192} - \left(\frac{3\beta_1}{48} + \frac{\cos\beta_1 \sin^5\beta_1}{6} - \frac{\sin^2\beta_1}{24} + \frac{\sin^4\beta_1}{192}\right)\right]$$

where

$$\beta_1 = \cos^{-1}(\alpha^*)^{1/3}, \qquad \beta_2 = \cos^{-1}(\alpha_{\max})^{1/3}$$

For example, for $r = 0.999$ and $\xi = 0.3$, we have $\alpha_{\max} = 0.4103$, $\alpha^* = 0.41$ and $R = 0.9999995$; that is, the reliability of the structure with random applied load exceeds that of the structure under deterministic load considered. However, if $\lambda_1 = 0.4103$ and $\lambda_2 = 0.4144$, the reliability becomes

$$R = \frac{1}{(\lambda_2 - \lambda_1)} \frac{1}{\xi} \int_{\lambda_1}^{\lambda_2} (1 - z^{2/3})^{3/2}\, dz = 0.9928439$$

that is, less than the required reliability $r = 0.999$ associated with the deterministic applied load. The lesson to be learned from this example is that uncertainty in the applied load is not always accompanied by a decrease in the reliability estimate. For further details, consult the study by Elishakoff (1983a). Note that some recent studies do not follow a general methodology outlined in this section for taking into account combined randomness in initial imperfection and applied load.

3.4 Reliability of Axially Compressed Cylindrical Shells with Random Axisymmetric Imperfections

This discussion of the reliability of axially compressed cylindrical shells with random axisymmetric imperfections is based on Elishakoff and Arbocz (1982).

3.4.1 Preliminary Considerations

It has been established for some time that the theories of imperfection sensitivity of structures should be combined with the statistical analysis of the initial imperfections. The first work in this direction for the most controversial structure – the circular cylindrical shell under axial compression – was presented by Amazigo (1969). He treated infinitely long cylindrical shells with spatially homogeneous, ergodic random axisymmetric imperfections by means of a modified truncated hierarchy method. The conclusion derived was that the buckling stress is a deterministic quantity, depending only on the spectral density of the random axisymmetric imperfections. Moreover, for small values of the standard deviations of the initial imperfections, this deterministic buckling stress depended only on the value of the initial imperfection power spectral density at the spatial frequency of the classical axisymmetric buckling mode, this dependence being

$$\lambda = 1 - \left[\frac{9\pi c^2}{2\sqrt{2}}\bar{S}_{w_0}(1)\right]^{2/7} \delta^{4/7}, \qquad \lambda = \frac{P_{BIF}}{P_c} \tag{3.106}$$

where P_{BIF} is the buckling load at which the governing non-linear equations admit an asymmetric solution infinitesimally adjacent to the prebuckling axisymmetric state, P_c is the classical buckling load of the perfect structure, λ is the non-dimensional deterministic buckling load (and, therefore, also the mean buckling load), $c = [3(1-\nu^2)]^{1/2}$, ν is Poisson's ratio, $\bar{S}_{w_0}(\omega)$ is the normalized initial axisymmetric imperfection power spectral density so that $\int_{-\infty}^{\infty}\bar{S}_{w_0}(\omega)d\omega = 1$, ω is the non-dimensional spatial frequency so that $\omega = 1$ corresponds to the wave number associated with the axisymmetric buckling mode, and δ is the standard deviation of the initial imperfections. Later on, Amazigo and Budiansky (1972) modified Equation (3.106) to

$$\lambda = 1 - \left[\frac{9\pi c^2}{2\sqrt{2}}\bar{S}_{w_0}(1)\right]^{2/7} \delta^{4/7}\lambda^{4/7} \tag{3.107}$$

In the works of Amazigo (1969) and Amazigo and Budiansky (1972) as well as Tennyson et al. (1971), the normalized auto-correlation function $\bar{R}_{w_0}(x_1, x_2) = R_{w_0}(x_1, x_2)\delta^{-2}$ was assumed to be of exponential-cosine type

$$\bar{R}_{w_0}(x_1, x_2) = e^{-\beta|\eta|}\cos(\gamma\eta) \tag{3.108}$$

where η is the difference between the non-dimensional axial coordinates of the points of observation, and β and γ are some positive constants supposed to depend on the manufacturing process. The normalized spectral density associated with this auto-correlation function is

$$\bar{S}_{w_0}(\omega) = \frac{\beta}{\pi}\frac{\omega^2 + \omega^2 + \gamma^2}{\omega^4 + 2(\beta^2-\gamma^2)\omega^2 + (\beta^2+\gamma^2)^2} \tag{3.109}$$

so that $\bar{S}_{w_0}(1)$ entering into Equations (3.106) and (3.107) is

$$\bar{S}_{w_0}(1) = \frac{\beta}{\pi} \frac{1 + \beta^2 + \gamma^2}{1 + 2(\beta^2 - \gamma^2) + (\beta^2 + \gamma^2)^2} \tag{3.110}$$

A design criterion based on formula (3.107) was developed by Tennyson et al. (1971). Two approaches were suggested: to substitute the experimentally determined values of δ and $\bar{S}_{w_0}(1)$ into Equation (3.107) or to assume, like Amazigo (1969) and Tennyson et al. (1971), the exponential-cosine auto-correlation function (3.108) with parameters β and γ chosen so that the spectral density (3.109) will have a peak at the spatial frequency, which corresponds to the axisymmetric classical buckling mode of the perfect shell. Experimental results (Arbocz, 1974), however, did not yield good agreement with the proposed random axisymmetric imperfection model because the measured discrete power spectral density of initial imperfections did not have a peak at the spatial frequency which corresponds to the axisymmetric classical buckling mode of the perfect shell. A different approach was used by Roorda and Hansen (1972). They assumed that the shape of the initial imperfection was specified in the form of the axisymmetric buckling mode of the cylindrical shell and considered its magnitude as a random variable with given probability distribution. Then the following relationship derived by Koiter (1945) using his general non-linear theory of elastic stability

$$2(1 - \lambda)^2 - 3_c|\mu|\lambda = 0 \tag{3.111}$$

between the non-dimensional buckling load λ and the magnitude of non-dimensional imperfection μ was used as a transfer function to calculate the probability characteristics of buckling load in terms of the probability characteristics of the imperfection magnitude. Finally, the reliability function

$$R(\alpha) = \text{Prob}(\alpha < \Lambda \leq 1) \tag{3.112}$$

where $(\alpha < \Lambda \leq 1)$ stands for the random event that the random buckling load Λ will exceed given non-dimensional load α, was calculated.

In a subsequent paper, Roorda (1972a) proposed to consider all kinds of imperfections in a real shell of given length, radius, thickness, and boundary conditions to be equivalent to a hypothetical axisymmetric imperfection in a shell of infinite length with the same radius and thickness. This equivalent imperfection was treated as a random normal variable with its mean and variance approximated as linear function of the R/t ratio. The obtained formulas were compared with results of some 360 experiments on axially compressed cylindrical shells with different length, radius-to-thickness ratios, boundary conditions, materials, and manufacturing processes, which were reported by Hart-Smith (1970). Roorda also compared his results with the "lower bound" curve proposed by Weingarten, Morgan, and Seide (1965) for a large number of test results from different sources.

Makarov (1970) was apparently the first who performed systematic statistical analysis of the initial imperfections. He used series of 50 cylindrical shells made of steel sheets of electrical-grade pressboard. The imperfection function was represented in Fourier series, and the coefficients were treated as random variables. The analysis showed that the assumption of homogeneity of the initial imperfection in the circumferential direction was satisfactory and that the normality of their Fourier coefficients did not conflict with the experimental data. Subsequently, Makarov (1969, 1970) performed a theoretical analysis of the buckling of stochastically imperfect shells with the experimental data (see Makarov, 1969) serving as an input for the description of imperfections. The theoretical mean value of the non-dimensional buckling load was 0.31, whereas the experiments yielded 0.23.

General, non-axisymmetric random imperfections were treated by Fersht (1974) and Hansen (1977). Fersht generalized the approach by Amazigo (1969). It turned out that, for the non-axisymmetric imperfections, closed formulas of the type (1) or (2) are not obtainable and rather cumbersome numerical analysis has to be performed to yield the mean buckling load. Hansen (1977) generalized his previous deterministic results (Hansen, 1975). The main conclusion of that paper was that the imperfection parameters associated with the non-axisymmetric modes appear in three and only three distinct summations and that the system behavior is governed by the value of these summations and not by the individual imperfection parameters. It was assumed that the modal imperfection amplitudes are jointly normal random variables with zero mean. Also the strong assumption was made that these amplitudes are statistically independent and distributed identically. The Monte Carlo method was then applied: For each sample problem, the buckling load was determined (see Hansen, 1975), and the mean buckling loads as well as their confidence levels were calculated. It has been demonstrated that the non-axisymmetric imperfections play a very important role in the determination of the buckling load statistics. Another important conclusion was that the large dispersion occurs for small values of R/t and that this dispersion decreases as R/t increases. The same conclusion was accounted for by Roorda (1972a, 1972b) by postulating that the mean and the variance of the imperfection were functions decaying with increasing values of R/t.

In this chapter, contrary to other earlier works, the probabilistic properties (the autocorrelation functions or spectral densities) are not assumed, instead the mean vector and the variance-covariance matrix of the Fourier coefficients are calculated from the experimental measurements of the shell profiles. Then the Monte Carlo method is employed. Thus, at first, a large number of shells is "created." That is, the Fourier coefficients of their initial imperfection representations are simulated numerically by a special procedure. Next for each shell a deterministic analysis of buckling load evaluation is carried out (implying that the usual deterministic approach is a particular case of the probabilistic one). After the buckling loads of an ensemble of shells are available, one then proceeds by studying their probabilistic behavior. In particular, one determines the reliability function representing the relative number of shells with buckling loads exceeding the specified load. Finally, the design load for the shells produced by a given manufacturing process is obtained as that load for which the reliability function has the desired value close to unity. This section follows the study of Elishakoff and Arbocz (1982).

3.4.2 Probabilistic Properties of Initial Imperfections

We characterize the random initial imperfections W_0 by functions that are the basis of the study of random processes. The mean function $\tilde{W}_0(x)$ of a function W_0 is the expected value of the random variable $W_0(x)$

$$\tilde{W}_0(x) = E\{W_0(x)\} = \int_{-\infty}^{\infty} w_0 f(w_0; x)\, dw_0 \tag{3.113}$$

where $f(w_0; x)$ is the first-order probability density function of $W_0(x)$. In general, $\tilde{W}_0(x)$ is a function of the axial coordinate, and $E\{\ldots\}$ denotes the mathematical expectation. The auto-covariance function $R_{w_0}(x_1, x_2)$ of a random function $W_0(x)$ is the covariance of the random variables $W_0(x_1)$ and $W_0(x_2)$:

$$\begin{aligned} R_{w_0}(x_1, x_2) &= E\{[W_0(x_1) - \tilde{W}_0(x_1)][W_0(x_2) - \tilde{W}_0(x_2)]\} \\ &= \int_{-x}^{x}\int_{-x}^{x} [w_{01} - \tilde{W}_0(x_1)][w_{02} - \tilde{W}_0(x)] f(w_{01}, w_{02}; x_1, x_2)\, dw_{01}\, dw_{02} \end{aligned} \tag{3.114}$$

Here $f(w_{01}, w_{02}; x_1, x_2)$ is the second-order probability density of the random function $W_0(x)$. Assume now that the $\{\varphi_1(x)\}$ represents the complete set of orthogonal functions in $[0, L]$, where L is the shell length. Then $W_0(x)$ can be expanded in a series in terms of the $\varphi_i(x)$s:

$$W_0(x) = \sum_i A_i \varphi_i(x) \tag{3.115}$$

where A_i is a random variable for every fixed i. The mean function then becomes

$$\tilde{W}_0(x) = \sum_i E(A_i)\varphi_i(x), \qquad E(A_i) \equiv \tilde{A}_i \tag{3.116}$$

Here the $\tilde{A}_i$s are the means of the A_is and are readily found as

$$\tilde{A}_i = \frac{1}{\mu_i^2}\int_0^L \tilde{W}_0(x)\varphi_i(x)\, dx, \qquad \mu_i^2 = \int_0^L \varphi_i^2(x)\, dx \tag{3.117}$$

and they form the mean vector $\{\tilde{A}\}$. The auto-covariance function becomes

$$R_{w_0}(x_1, x_2) = \sum_i \sum_i \sigma_{ij}\varphi_i(x_i)\varphi_j(x_2) \tag{3.118}$$

where

$$\sigma_{ij} = E\{[A_i - \tilde{a}_i][A_j - \tilde{A}_j]\} \tag{3.119}$$

The σ_{ij}s are obtained as

$$\sigma_{ij} = \frac{1}{\mu_i^2 \mu_i^2}\int_0^L \int_0^L R_{w_0}(x_1, x_2)\varphi_i(x_1)\varphi_j(x_2)\, dx_1\, dx_2 \tag{3.120}$$

and they form the variance-covariance matrix $[\Sigma] = [\sigma_{ij}]$. Equations (3.113)–(3.120) imply that the knowledge of the mean and the auto-covariance function, on the one hand, and the mean vector $\{\tilde{A}\}$ and the variance-covariance matrix $[\Sigma]$, on the other hand, are equivalent.

To be able to treat also the limiting case of the infinite shell, we consider initially the complex Fourier series, so that, instead of representation (3.115), we use

$$W_0(x) = \sum_{m=-\infty}^{\infty} A_m e^{im\pi x/L} \tag{3.121}$$

For the mean function, we get

$$\tilde{W}_0(x) = \sum_{m=-\infty}^{\infty} E(A_m)e^{im\pi x/L}$$
$$\tilde{A}_m = \frac{1}{2L}\int_{-L}^{L} \tilde{W}_0(s)e^{im\pi s/L}\,ds \tag{3.122}$$

The auto-covariance function becomes in terms of the elements σ_{mn} of the variance-covariance matrix

$$R_{w_0}(x_1, x_2) = \sum_{m=-\infty}^{\infty}\sum_{n=-\infty}^{\infty} \sigma_{mn}e^{i\pi(mx_1-nx_2)/L}$$
$$\sigma_{mn} = \frac{1}{4L^2}\int_{-L}^{L}\int_{-L}^{L} R_{w_0}(s_1, s_2)e^{i\pi(-ms_1+ns_2)/L}\,ds_1\,ds_2 \tag{3.123}$$

Substitution of the σ_{mn} into the expression of the auto-covariance function yields

$$R_{w_0}(x_1, x_2) = \sum_{m=-\infty}^{\infty}\sum_{n=-\infty}^{\infty}\left[\frac{1}{4L^2}\int_{-L}^{L}\int_{-L}^{L} R_{w_0}(s_1, s_2)e^{-im\pi s_1/L}e^{\epsilon\pi s_2/L}\,ds_1\,ds_2\right]$$
$$\times e^{im\pi x_1/L}e^{-n\pi x_2/L} \tag{3.124}$$

Now, defining the spatial frequencies by

$$\omega_m = \frac{m\pi}{L}, \qquad \pi_n = \frac{n\pi}{L} \tag{3.125}$$

and the difference between the successive frequencies by

$$\Delta\omega_m = \frac{\pi}{L}, \qquad \Delta\omega_n = \frac{\pi}{L} \tag{3.126}$$

Equation (3.125) becomes

$$R_{w_0}(x_1, x_2) = \sum_{m=-\infty}^{\infty}\sum_{n=-\infty}^{\infty}\left[\frac{1}{4\pi^2}\int_{-L}^{L}\int_{-L}^{L} R_{w_0}(s_1, s_2)e^{-i\omega_m s_1}e^{i\omega_n s_2}\,ds_1\,ds_2\right]$$
$$\times e^{i\omega_m x_1}e^{-\omega_n x_2}\,\Delta\omega_m\Delta\omega_n \tag{3.127}$$

If we now define

$$S_{w_{0,L}}(\omega_m, \omega_n) = \frac{1}{4\pi^2}\int_{-L}^{L}\int_{-L}^{L} R_{w_0}(s_1, s_2)e^{-i\omega_m s_1}e^{i\omega_n s_2}\,ds_1\,ds_2 \tag{3.128}$$

then Equation (3.127) can be rewritten as follows:

$$R_{w_0}(x_1, x_2) = \sum_{m=-\infty}^{\infty}\sum_{n=-\infty}^{\infty} S_{w_{0,L}}(\omega_m, \omega_n)e^{i\omega_m x_1}e^{-\omega_n x_2}\,\Delta\omega_m\Delta\omega_n \tag{3.129}$$

Under very general conditions the limit of a sum of the form (3.129) as $\Delta\omega_m \to 0$, $\Delta\omega_n \to 0$ is the integral

$$R_{w_0}(x_1, x_2) = \int_{-\infty}^{\infty}\int_{-\infty}^{\infty} S_{w_0}(\omega_1, \omega_2)e^{i\omega_1 x_1}e^{-\omega_2 x_2}\,d\omega_1\,d\omega_2 \tag{3.130}$$

Therefore, since $L \to \infty$ implies $\Delta\omega_m \to 0$, we have

$$\lim_{L\to\infty} S_{w_{0,L}}(\omega_m, \omega_n) = S_{w_0}(\omega_1, \omega_2)$$

where

$$S_{w_0}(\omega_1, \omega_2) = \frac{1}{4\pi^2} \int_{-\infty}^{\infty} \int_{-\infty}^{\infty} R_{w_0}(s_1, s_2) e^{-i\omega_1 s_1} e^{i\omega_2 s_2} \, ds_1 \, ds_2 \tag{3.131}$$

so that $S_{w_0}(\omega_1, \omega_2)$ and $R_{w_0}(s_1, s_2)$ constitute the double Fourier transform pair, $S_{w_0}(\omega_1, \omega_2)$ is referred to as the generalized power spectral density of the initial imperfections of the infinite shell, and, as was shown earlier, it can be deduced from the elements of the variance-covariance matrix associated with the finite shell.

Consider now the particular case when the initial imperfections of the finite shell form a stationary random function, meaning that the mean imperfection is constant and the auto-covariance depends only on $s_2 - s_1$:

$$\tilde{W}_0(x) = \text{constant}, \qquad R_{w_0}(x_1, x_1 + \xi) = R_{w_0}(\xi) \tag{3.132}$$

Substituting into Equation (3.31), $R_{w_0}(s_2 - s_1)$ instead of $R_{w_2}(s_1, s_2)$, making the change of coordinates

$$s_1 - s_2 = \xi, \qquad s_2 = \eta$$

and bearing in mind that

$$\int_{-\infty}^{\infty} e^{i(\omega_2 - \omega_1)\eta} \, d\eta = 2\pi \, \delta(\omega_2 - \omega_1)$$

yields

$$S_{w_0}(\omega_1, \omega_2) = S_{w_0}(\omega_1)\delta(\omega_2 - \omega_1)$$

where $S_{w_0}(\omega_1)$ is the power spectral density of the weakly homogeneous initial imperfections (homogeneous in the wide sense). Then with $\omega_1 \to \omega$, one obtains

$$S_{w_0}(\omega) = \frac{1}{2\pi} \int_{-\infty}^{\infty} R_{w_0}(\xi) \, e^{-i\omega\xi} \, d\xi \tag{3.133}$$

with Equation (3.130) transforming to

$$R_{w_0}(\xi) = \int_{-\infty}^{\infty} S_{w_0}(\omega) \, e^{i\omega\xi} \, d\omega \tag{3.134}$$

Equations (3.133) and (3.134) constitute the Wiener-Khintchine relationship for the weakly homogeneous random functions. The concept of the power spectral density was used in the studies by Amazigo (1969) and Amazigo and Budiansky (1972) to represent the initial imperfections.

Strictly speaking, the initial imperfections of the finite shell cannot be homogeneous. Tennyson et al. (1971) used measurements performed on the finite shell in order to develop the design criterion based on formula (3.107), which was derived from the infinite shell with weakly homogeneous imperfections. That is, the assumption was made that the initial imperfections are homogenous in the interval $[0, L]$, that is, conditions (3.132) are valid when x_1, x_2 are in this interval. Consider this case in more detail. Assume, after Tennyson et al. (1971), that the mean imperfection function is identically zero. Then the initial imperfections (as those of the finite shell) are characterized by the variance-covariance matrix, Equation (3.119). Assume now in Equation (3.120) that

$R_{w_0}(s_1, s_2) = R_{w_0}(s_2 - s_1)$ and introduce a new coordinate system $z_1 = (s_2 - s_1)/\sqrt{2}$, $z_2 = (s_1 + s_2)/\sqrt{2}$ to finally arrive at

$$\sigma_{mn} = \frac{(-1)^{n-m}}{\pi(n-m)} \int_0^1 R_{w_0}(2\alpha L)(\sin 2m\pi\alpha - \sin 2n\pi\alpha)\, d\alpha, \qquad \text{if } m \neq n \tag{3.135}$$

and

$$\sigma_{mn} = 2 \int_0^1 R_{w_0}(2\alpha L)(1-\alpha)\cos 2m\pi\alpha\, d\alpha \qquad \text{if } m = n$$

where

$$\alpha = \frac{z_1}{\sqrt{2}L} = \frac{s_2 - s_1}{2L} = \frac{\gamma}{2L} \tag{3.136}$$

We will show that for $L \to \infty$, the diagonal terms σ_{mm} of the variance-covariance matrix reduce to Tennyson's discrete power spectral density. Indeed, substituting (3.136) into the expression for σ_{mn} results in

$$\sigma_{mm} = \frac{1}{L} \int_0^{2L} R_{w_0}(\gamma)\left(1 - \frac{\gamma}{2L}\right) \cos m\omega_0\gamma\, d\gamma$$

where the dimensional spatial frequency $\tilde{\omega}_0 = \pi/L$ was introduced. Then

$$\frac{\sigma_{mm}}{\tilde{\omega}_0} = \frac{1}{\pi} \int_0^{2L} R_{w_0}(\gamma)\left(1 - \frac{\gamma}{2L}\right) \cos m\tilde{\omega}_{0_\gamma}\, d\gamma$$

but now, with $L \to \infty$, $\tilde{\omega} \to 0$ and

$$\lim_{\tilde{\omega}_0 \to 0} m\tilde{\omega}_0 = \tilde{\omega}$$

hence,

$$\lim_{\tilde{\omega}_0 \to 0} \frac{\sigma_{mm}}{\tilde{\omega}_0} = \frac{1}{\pi} \int_0^{\infty} R_{w_0}(\gamma) \cos \tilde{\omega}\gamma\, d\gamma = S_{w_0}(\tilde{\omega}) \tag{3.137}$$

which is the desired result. This expression can be related also to the non-dimensional discrete power spectral density discussed by Arbocz and Williams (1976). Indeed $S_{w_0}(\tilde{\omega})$ is $p(\tilde{\omega})$. In order to non-dimensionalize $S_{w_0}(\tilde{\omega})$, we introduce the dimensionless spatial frequency ω defined as

$$\omega = \frac{m\tilde{\omega}_0}{\sqrt{(2c/Rt)}} = \frac{m}{i_{cl}}, \qquad i_{cl} = \frac{L}{\pi}\sqrt{\left(\frac{2c}{Rt}\right)}$$

where i_{cl} is the number of half-waves in the classical axisymmetric buckling mode for isotropic shells. Then the non-dimensional power spectral density $\bar{S}_{w_0}(\omega) = S_{w_0}(\tilde{\omega})/\sqrt{(Rt/2c)}$ becomes

$$\bar{S}_{w_0}(\omega) = \frac{1}{\sqrt{(Rt/2c)}} \lim_{\tilde{\omega} \to 0} \frac{\sigma_{mm}}{\tilde{\omega}_0} = \lim_{\tilde{\omega} \to 0} \sigma_{mm} i_{cl} \tag{3.138}$$

which coincides with formula (A12) in the paper by Arbocz and Williams (1976).

3.4.3 Simulation of Random Imperfections with Given Probabilistic Properties

Assume that the mean function $\tilde{W}_0(x)$ and the auto-covariance function $R_{w_0}(x_1, x_2)$ are given. We then calculate analytically or numerically the vector $\{\tilde{A}\}$ and the variance-covariance matrix $[\Sigma]$. Having these quantities, we then proceed to the simulation of the initial imperfections. Our aim is to "create" the desired number M of initial imperfection profiles having given probabilistic properties. The method of simulation, suitable for computer realization for a random normal function was developed by Elishakoff (1979a). Here we are only giving a short outline of the method. We assume that the mean vector $\{\tilde{A}\}$ is identically zero. This entails no loss of generality because, if vector $\{A\}$ is simulated as is shown later, the vector $\{A + \tilde{A}\}$ has the mean $\{\tilde{A}\}$ and the same variance-covariance matrix $[\Sigma]$. The latter matrix is positive semidefinite and can be uniquely decomposed in the form

$$[\Sigma] = [C][C]^T \tag{3.139}$$

where T means transpose and $[C]$ is a lower triangular matrix. $[C]$ is found by the Cholesky's algorithm (Elishakoff, 1983b). Now, we can form vector $\{B\}$, its elements being normally distributed, statistically independent with zero means and unit variances. Then the vector of the Fourier coefficient of the initial imperfections is simulated as follows:

$$\{A\} = [C]\{B\} + \{\tilde{A}\} \tag{3.140}$$

Having realizations of vector $\{B\}$, we obtain the same number of realizations of $\{A\}$. The main feature of this simulation technique is that it is applicable for homogeneous as well as for non-homogeneous random functions with given mean and auto-covariance functions. This procedure was applied to different static and dynamic buckling problems, involving random imperfection sensitivity (see Elishakoff, 1978b, 1979a, 1980b).

Using the auto-covariance function (3.108), we are able to compare the results of the Monte Carlo method with the expressions (3.106) or (3.107) for the mean buckling load. To begin with, we assume, after Amazigo (1969), that the initial imperfection auto-covariance function is of the exponential cosine type,

$$R_{w_0}(x_1, x_2) = \delta^2 e^{-A(|x_2 - x_1|L)} \cos B\frac{(x_2 - x_1)}{L} \tag{3.141}$$

In order to get Amazigo's representation in the form of Equation (3.141), the constants A and B must have the following values:

$$A = \beta i_{cl}\pi, \qquad B = \gamma i_{cl}\pi \tag{3.142}$$

where

$$i_{cl} = \frac{L}{\pi}\frac{\sqrt{2c}}{RT} \qquad \text{and} \qquad c = \sqrt{3(1-\nu^2)} \tag{3.143}$$

The elements of the variance-covariance matrix are calculated via Equation (3.120), with $\varphi_i = \cos(i\pi x/L)$ (i.e., the half-wave cosine representation is used for the initial imperfections). As is shown (Elishakoff, 1979b), simplification during the calculation of σ_{ij} is possible due to the weak homogeneity of W_0. Namely, the σ_{ij}s are given by

$$\sigma_{ij} = \frac{1}{\mu_i^2 \mu_j^2} \int_0^L R_{w_0}(\xi) M_{ij}(\xi) d\xi \tag{3.144}$$

where

$$M_{ij}(\xi) = P_{ij}(\xi) + P_{ji}(\xi), \qquad \mu_i^2 = \int_0^L \varphi_i^2(x)\, dx$$

$$P_{ij}(\xi) = \int_{\xi/2}^{L-\xi/2} \varphi_i\left(\eta - \frac{1}{2}\xi\right) \varphi_i\left(\eta + \frac{1}{2}\xi\right) d\eta$$

(see Equations (18)–(20) of Elishakoff, 1979b). Upon substitution and carrying out of the integrals, these equations yield the following formulas:

$$\bar{\sigma}_{ii} = \sigma_{ii} i_{cl} = \frac{4\delta^2}{\pi} \frac{\Omega_i}{\omega_i^4 + 2\omega_i^2(\beta^2 - \gamma^2) + (\beta^2 + \gamma^2)^2}$$

$$\sigma_{ij} = \sigma_{ij} i_{cl} = \begin{cases} \dfrac{8\delta_2}{i_{cl}} \dfrac{(\omega_i^2 \bar{A}_i - \omega_j^2 \bar{A}_j)}{\omega_i^2 - \omega_j^2}, & \text{if } i \neq j \text{ and } i + j = \text{even} \\ 0, & \text{if } i + j = \text{odd} \end{cases} \tag{3.145}$$

where

$$\Omega_i = \beta \bar{U}_i - \frac{1}{2\pi i_{cl}} \left(\frac{\bar{P}_i \bar{Q}_i}{\bar{R}_i} + \frac{\bar{T}_i \bar{R}_i}{\bar{Q}_i} + 2\bar{V}_i \right) [1 - (-1)^i e^{-\beta i_{cl} \pi} \cos \gamma i_{cl} \pi]$$

$$- \frac{\beta}{\pi} i_{cl} \left[\frac{(\omega_i + \gamma)\bar{Q}_i}{\bar{R}_i} - \frac{(\omega_i - \gamma)\bar{R}_i}{\bar{Q}_I} + 2\gamma \right] (-1)^i e^{-\beta i_{cl} \pi} \sin \gamma i_{cl} \pi$$

$$\bar{A}_i = \frac{1}{\pi^2} \frac{1}{\omega_i^4 + 2\omega_i^2(\beta^2 - 15^2) + (\beta^2 + \gamma^2)^2} \{ \bar{V}_i [(-1)^i e^{-\beta i_{cl} \pi} \cos \gamma i_{cl} \pi - 1]$$

$$+ -2\beta\gamma(-1)^i e^{\beta i_{cl} \pi} \sin \gamma i_{cl} \pi \}$$

$$\bar{P}_i = \beta^2 - (\omega_i + \gamma)^2, \qquad \bar{Q}_i = \beta^2 + (\omega_i - \gamma)^2$$

$$\bar{R}_i = \beta^2 + (\omega_i + \gamma)^2, \qquad \bar{T}_i = \beta^2 - (\omega_i - \gamma)^2$$

$$\bar{U}_i = \omega_i^2 + \beta^2 + \gamma^2, \qquad \bar{V}_i = \omega_i^2 + \beta^2 - \gamma^2$$

$$\omega_i = i / i_{cl} \tag{3.146}$$

In order to compare the results of the Monte Carlo method with those obtained by the application of formulas (3.106) and (3.107), the following shell was used: $R = L = 101.6$ mm, $t = 0.944931$ mm and $\nu = 0.3$, resulting in $i_{cl} = 6$. This shell, with the initial imperfection auto-covariance function used by Amazigo [see Equation (3.108) with $\beta = 0.2$ and $\gamma = 1.0$], will be referred to as the Amazigo shell. Figure 3.13 shows the normalized variance of the initial imperfections back-calculated with the aid of the σ_{ij}s given in Equation (3.109). Twelve terms have been retained in the summation given by Equation (3.118). The uniformity is disturbed in the vicinity of the edges due to Gibb's phenomenon. The auto-covariance function possessed by 100 simulated Amazigo shells is given in Figure 3.14(a), and the elements of the variance-covariance

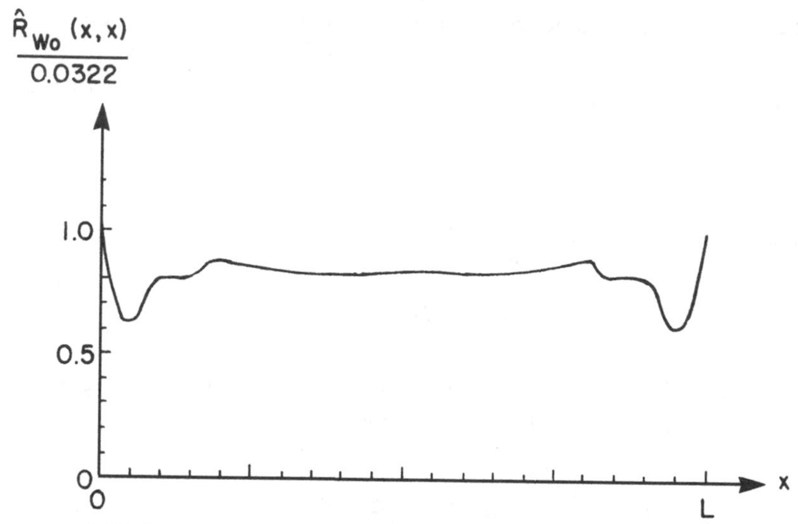

Figure 3.13 Recalculated variance of Amazigo's shell with exponential-cosine auto-correlation function (after Elishakoff and Arbocz, 1982; Copyright © Elsevier Science Ltd., reprinted with permission).

matrix are shown in Figure 3.14(b). Since $i_{cl} = 6$, it is seen that, on the diagonal at $i = j = i_{cl}$, the elements of the variance-covariance matrix reach a maximum. This maximum corresponds to the well-known peak in the spectral density [Equation (3.109)] used by Amazigo in his analysis of the infinite shell.

Notice that, for sufficiently long shells, the number of half-waves associated with the classical buckling mode $i_{cl} \gg 1$. Then ω_i tends to the continuous spatial frequency ω, and in the expression for $_i$ [see Equation (3.146)] the second and the third terms become vanishingly small in comparison with the leading term $\beta \bar{U}_i$. Thus, as $L \to \infty$, σ_{ii} in Equation (3.145) reduces to the spectral density $\bar{S}_{w_0}(\omega)$ used by Amazigo for the infinite shell [see Equation (3.109)]. The extra factor of 4 is due to the fact that the half-wave cosine representation was employed for $\varphi_i(x)$ in deriving Equation (3.145) instead of the complex representation used by Amazigo. Further, as can be seen from Figure 3.14(b), some off-diagonal terms $\sigma_{ij}(i \neq j)$ are different from zero. However, it can easily be shown using the expression given in Equation (3.146) that as $L \to \infty$ the off-diagonal terms approach zero.

When comparing the results of the Monte Carlo method for the mean buckling load with the prediction based on Equations (3.106) and (3.107), the critical value of the maximum absolute difference between the unknown theoretical and the obtained simulated distributions of the buckling loads is $1.36/\sqrt{100} = 0.136$ according to the Kolmogorov-Smirnov test of goodness of fit (see Massey, 1951) at a level of significance of 0.05. The variance of the simulated initial imperfections was fixed at $\delta^2 = 0.005$. The histogram of the buckling loads and the reliability function are shown in Figures 3.15(a) and 3.15(b), respectively. The non-dimensional buckling loads λ were distributed between 0.376 and 0.886 so that the design buckling load at the required reliability, 0.98 say, equals 0.37. Obviously, the design buckling load associated with the high level of required reliability is a more powerful design criterion than the mean buckling load.

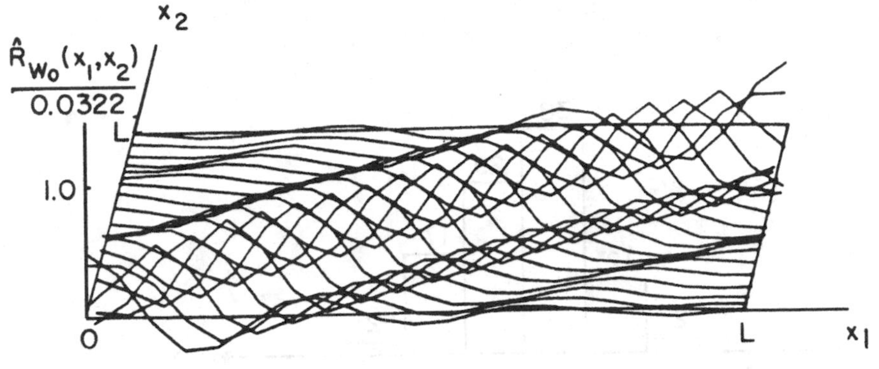

a. The back-calculated autocovariance function

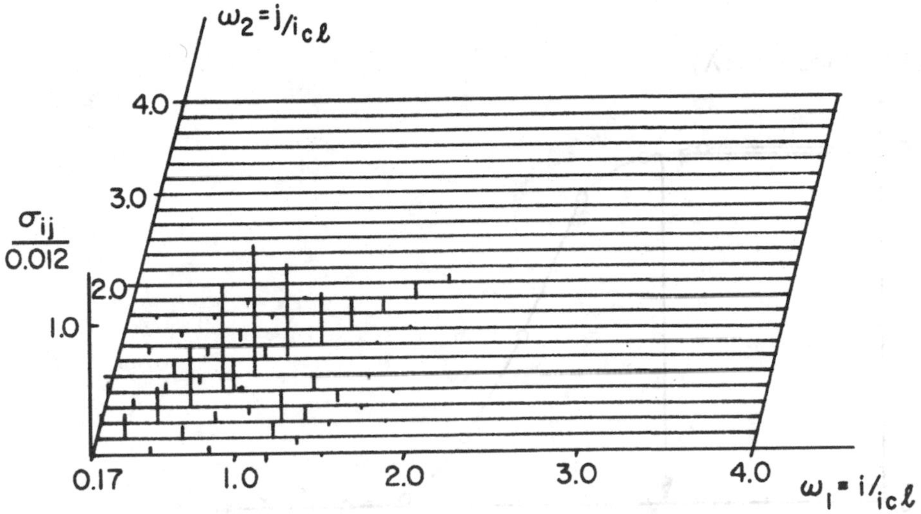

b. The elements of the variance-covariance matrix

Figure 3.14 Statistical properties of the simulated Amazigo shells (after Elishakoff and Arbocz, 1982; Copyright © Elsevier Science Ltd., reprinted with permission).

The mean buckling load possessed by the "created" shells is 0.608. Formula (3.106) by Amazigo (1969) predicts 0.468 and formula (3.107) by Amazigo and Budiansky (1972) yields 0.602. As is seen, the mean buckling load given by Equation (3.107) is much more reliable than the one predicted by Equation (3.106). This is in agreement with the conjecture of Amazigo and Budiansky (1972).

It should be remarked here that in the present work no use was made of the ergodicity assumption adopted by Amazigo and Budiansky (1972), but ensemble averaging was employed to find the characteristics of the buckling load, which turns out to be a random variable. Also, for the Monte Carlo simulation, a shell of finite length was used, not an infinite one as in the papers of Amazigo (1969) and Amazigo and Budiansky (1972).

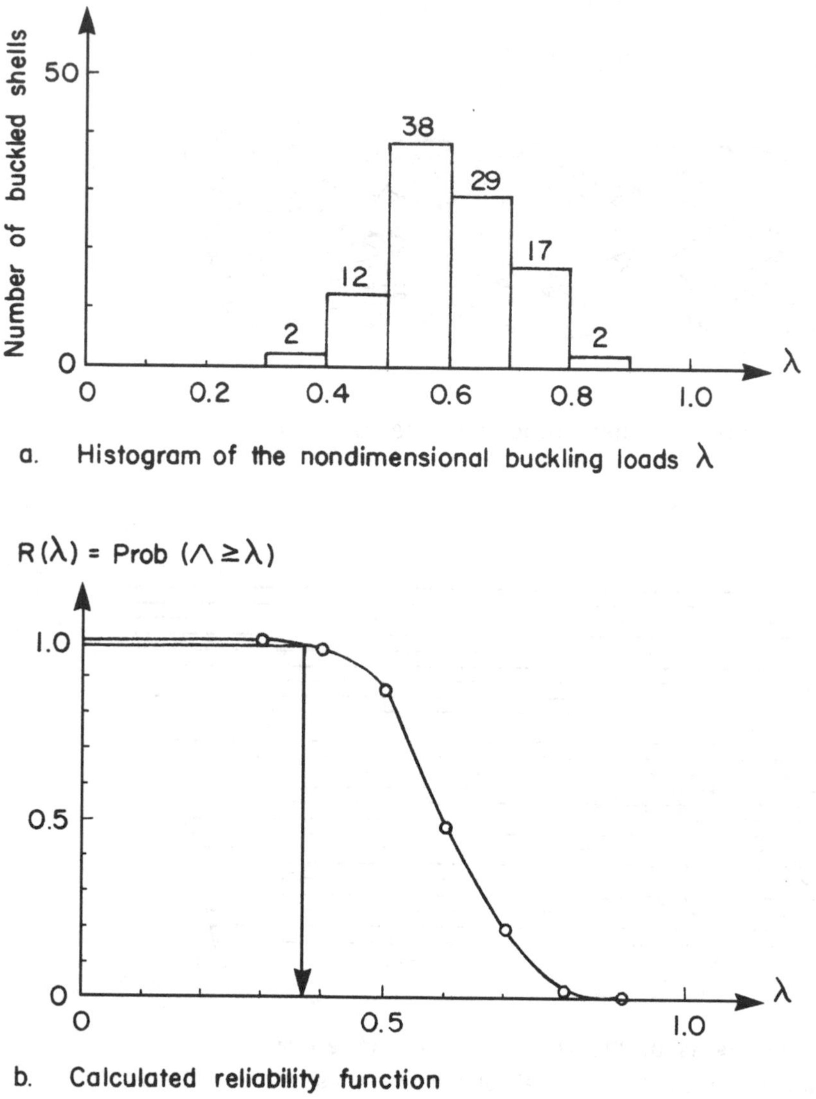

Figure 3.15 The reliability function associated with the simulated Amazigo shells (after Elishakoff and Arbocz, 1982; Copyright © Elsevier Science Ltd., reprinted with permission).

3.4.4 Simulation of Random Initial Imperfections from Measured Data

The techniques for measuring the initial imperfections of shells are well established (see, for example, Arbocz and Babcock, 1968) and the results are collected in the papers of Singer, Abramovich, and Yaffe (1978) and Arbocz and Abramovich (1979). The techniques employed measure the deviation of the shell outer surface relative to an imaginary cylindrical reference surface (the "perfect shell" associated with the imperfect shell under consideration). Consider, for example, cylindrical shells of $L = 176.02$ mm, $R = 101.6$ mm, and $t = 0.1160$ mm, manufactured by electroplating from pure cooper and tested in a controlled end-displacement type compression-testing machine. We can

Table 3.1. First nine Fourier coefficients $a_i^{(m)}$ of the initial imperfections expanded in the series

$$w_0^{(m)}(x) = tW_0^{(m)}(x) = t\sum_{i=0}^{8} a_i^{(m)} \cos\frac{i\pi x}{L}$$

Shell number, i	A-7	A-8	A-9	A-10	A-12	A-13	A-14
0	0.0176	0.0343	0.0226	0.0108	0.0023	0.0018	0.0029
1	0.0669	0.6534	0.0832	−0.0231	0.0158	0.0242	0.0662
2	−0.0164	0.1033	−0.0437	−0.0265	−0.0164	0.0095	0.0041
3	−0.0176	−0.0696	−0.0079	0.0054	0.0274	0.0006	0.0237
4	−0.0403	−0.1997	−0.0519	−0.0232	0.0092	0.0048	0.0013
5	−0.0031	−0.1637	0.0015	−0.0055	−0.0194	−0.0021	−0.0438
6	−0.0313	−0.0787	−0.0347	−0.0187	0.0062	−0.0047	−0.0349
7	−0.0050	−0.0092	−0.0080	−0.0106	0.0115	0.0060	0.0042
8	−0.0326	−0.0821	−0.0370	−0.0158	−0.0116	0.0038	−0.0041

For the group of A-shells ($R = 101.6$ mm, $t = 0.1160$ mm, $L = 176.02$ mm, $E = 1.0441 \times 10^5$ N/mm^2, $\nu = 0.3$, $i_{cl} = 30$). Note: Here $w_0^{(0)}(x)$ is positive outward.

visualize that a suitable stock of such shells, referred to as the A-shells (Arbocz and Abramovich, 1979), is available. Due to the very nature of the manufacturing process, each realization of the shell will have a different initial shape that cannot be predicted in advance. The imperfections represent deviations of the initial shape from the perfect circular cylinder amounting to a fraction of the wall thickness. They can be picked up and recorded by the special experimental setup developed at Caltech (Arbocz and Babcock, 1968). The scanning device, moving in both the axial and the circumferential directions, yields a complete surface map of the shells. Any two shells produced by the same manufacturing process may have totally different imperfection profiles, as can be seen from the three-dimensional plot of initial imperfections, shown in Figure 3.16. Measured imperfection surfaces are represented by different Fourier series. The integrals involved in the determination of the Fourier coefficients are carried out numerically using the trapezoidal rule. The axisymmetric Fourier coefficients $a_i^{(m)}$ of the different A-shells are listed in Table 3.1. Now we are looking at the $a_i^{(m)}$s as realizations of the random variable A_i in Equation (116). Then the sample mean is estimated as

$$A_i^{(e)} = \frac{1}{N}\sum_{m=1}^{N} a_i^{(m)} \tag{3.147}$$

where N is the number of sample shells.

The elements σ_{ij} of the variance-covariance matrix are estimated as

$$\sigma_{ij}^{(e)} = \frac{1}{N-1}\sum_{m=1}^{N}[a_i^{(m)} - \tilde{A}_i^{(e)}][a_j^{(m)} - \tilde{A}_j^{(m)}] \tag{3.148}$$

which is an unbiased estimate (see Table 3.2).

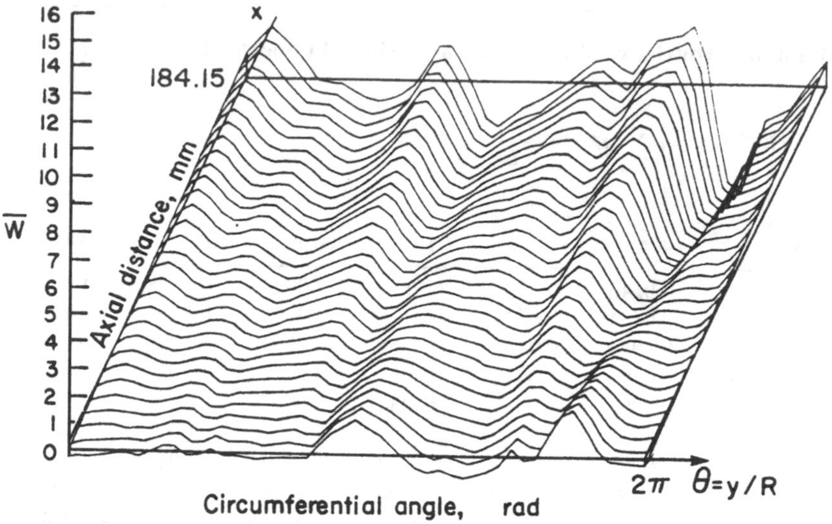

(a) Shell A-12 (Arbocz and Babcock, 1968)

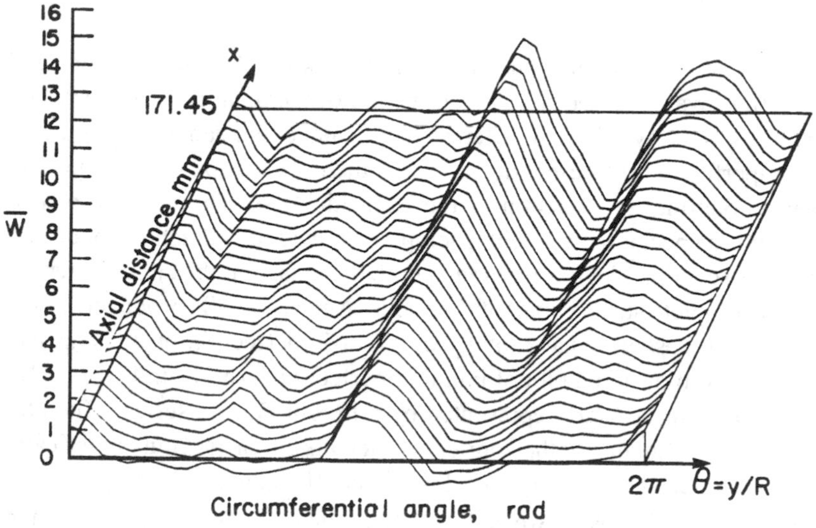

(b) Shell A-13 (Arbocz and Abramovich, 1979)

Figure 3.16 Three-dimensional plots of measured initial imperfections (after Elishakoff and Arbocz, 1982; Copyright © Elsevier Science Ltd., reprinted with permission).

The estimates of the mean initial imperfection function and of the auto-covariance functions become, respectively [see Equations (3.116) and (3.118)]

$$\begin{aligned} \tilde{W}_0^{(e)}(x) &= \sum_i \tilde{A}_i^{(e)} \varphi_i(x) \\ R_{w_0}^{(e)}(x_1, x_2) &= \sum_i \sum_j \sigma_{ij}^{(e)} \varphi_i(x_i) \varphi_j(x_2) \end{aligned} \tag{3.149}$$

Table 3.2. Elements of the variance-covariance matrix $[\Sigma^{(e)}]$ for the group of A-shells

	i/j							
	0	**1**	**2**	**3**	**4**	**5**	**6**	**7**
0	0.015	0.230	0.030	−0.035	−0.083	−0.046	−0.029	−0.008
1	0.230	5.530	1.031	−0.658	−1.644	−1.338	−0.556	−0.079
2	0.030	1.031	0.232	−0.113	−0.277	−0.270	−0.092	−0.005
3	−0.035	−0.658	−0.113	0.096	0.217	0.140	0.069	0.016
4	−0.083	−1.644	−0.277	0.217	0.533	0.375	0.180	0.039
5	−0.046	−1.338	−0.270	0.140	0.375	0.353	0.130	0.013
6	−0.029	−0.556	−0.092	0.069	0.180	0.130	0.074	0.016
7	−0.008	−0.079	−0.005	0.016	0.039	0.013	0.016	0.007

Each entry should be multiplied by a factor 10^{-2}.

Since $[\Sigma] = [\sigma_{ij}^{(e)}]$ is a positive–semi-definite matrix, therefore, according to Sylvester's theorem (Chetaev, 1961), all principal minor determinants associated with matrix $[\Sigma]$ are non-negative. The same property must be possessed by the estimate $[\Sigma^{(e)}]$. This property is used in order to "correct" initial values of $\sigma_{ij}^{(e)}$. If, for example, the rth-order principal minor determinant is non-negative

$$\begin{vmatrix} \sigma_{11}^{(e)} & \cdots & \sigma_{1r}^{(e)} \\ \vdots & & \\ \sigma_{rl}^{(e)} & \cdots & \sigma_{rr}^{(e)} \end{vmatrix} \geq 0$$

$$\begin{vmatrix} \sigma_{11}^{(e)} & \cdots & \sigma_{1r}^{(e)} & \Sigma_{1,4+1}^{(e)} \\ \vdots & & & \\ \sigma_{r1}^{(e)} & \cdots & \sigma_{rr}^{(e)} & \sigma_{r,r+1}^{(e)} \\ \sigma_{r+1,1}^{(e)} & \cdots & \sigma_{r+1,r}^{(e)} & \sigma_{r+1,r+1}^{(e)} \end{vmatrix} \leq 0$$

but the $(r+1)$th-order principal minor determinant is negative

$$\begin{vmatrix} \sigma_{11}^{(e)} & \cdots & \sigma_{1r}^{(e)} & \Sigma_{1,4+1}^{(e)} + x \\ \vdots & & & \\ \sigma_{r1}^{(e)} & \cdots & \sigma_{rr}^{(e)} & \sigma_{r,r+1}^{(e)} + x \\ \sigma_{r+1,1}^{(e)} + x & \cdots & \sigma_{r+1,r}^{(e)} + x & \sigma_{r+1,r+1}^{(e)} + x \end{vmatrix} \leq 0$$

This "correction" can be used if $\max_j \sigma_{j,r+1}^{(e)} + x$ lies within the confidence interval for the element matrix $\sigma_{j,r+1}^{(e)}$, and, moreover, if $x \ll \max_j \sigma_{j,r+1}^{(e)}$.

Having the estimates of $\tilde{A}_i^{(e)}$ and $\sigma_{ij}^{(e)}$, we are proceeding to the simulation of the initial imperfections, as described in the previous section. Instead of $\{\tilde{A}\}$ and $[\Sigma]$ in Equations (3.139) and (3.140), we use $\{\tilde{A}^{(e)}\}$ and $[\Sigma^{(e)}]$, respectively. As a result we obtain the desired number $M > N$ initial imperfections profiles, which are statistically "equivalent" to the initial sample N shells.

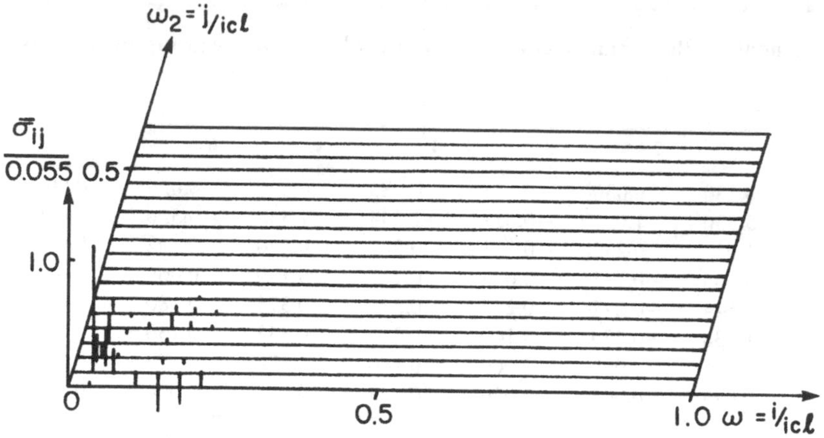

Figure 3.17 Elements of the variance-covariance matrix $[\Sigma^e]$ of the group of A shells (after Elishakoff and Arbocz, 1982; Copyright © Elsevier Science Ltd., reprinted with permission).

To check the simulation procedure, the auto-covariance function of the simulated sample of shells, defined as

$$R_{w_0}^{(s)}(x_1, x_2) = \sum_m \sum_n \sigma_{ij}^{(s)} \varphi_i(x_1)\varphi_j(x_2) \tag{3.150}$$

with

$$\sigma_{ij}^{(s)} = \frac{1}{N-1} \sum_{m=1}^{M} [a_i^{(s)(m)} + \tilde{A}_i^{(s)}][a_j^{(s)(m)} - \tilde{A}_j^{(s)}] \tag{3.151}$$

must be compared with the auto-covariance function of the initial sample $R_{w_0}^{(e)}$. The variance-covariance matrix for the group of A-shells is given in the paper by Elishakoff and Arbocz (1982) and its elements are displayed in Figure 3.17. The auto-covariance function estimated from the measured data is displayed in Figure 3.18. The

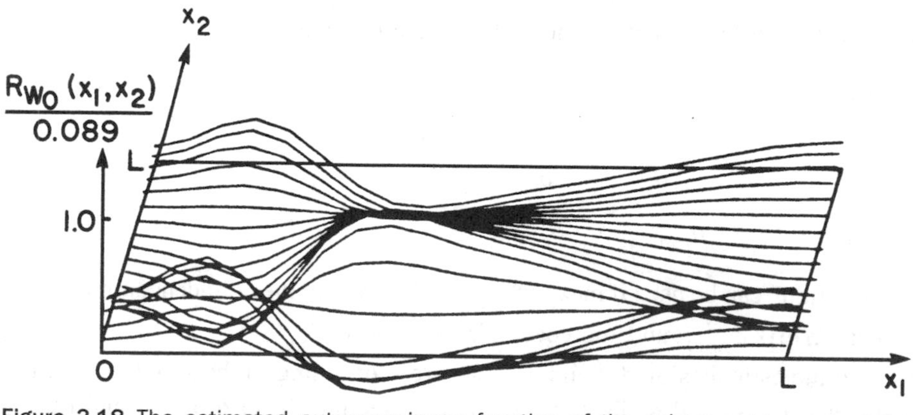

Figure 3.18 The estimated auto-covariance function of the axisymmetric part of the measured initial imperfections of the group of A shells (after Elishakoff and Arbocz, 1982; Copyright © Elsevier Science Ltd., reprinted with permission).

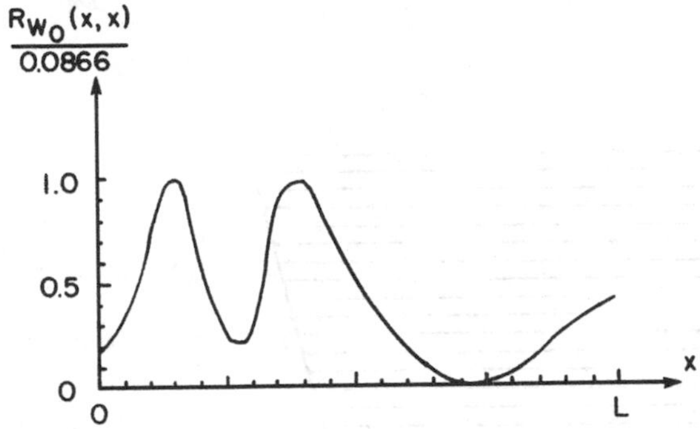

Figure 3.19 Estimated variance of the measured initial imperfections as a function of the axial coordinate for the group of *A* shells (after Elishakoff and Arbocz, 1982; Copyright © Elsevier Science Ltd., reprinted with permission).

auto-covariance function of the simulated sample turns out to be indistinguishable from that of the measured initial sample. The estimated variance of the measured initial imperfection versus the axial coordinate is shown in Figure 3.19.

A group of B-shells (Arbocz and Abramovich, 1979) was also considered. These shells were manufactured by putting pieces of thick-walled, seamless, brass tubes onto a mandrel and then machining them to the desired wall thickness. The average dimensions of this group of shells are $L = 134.37$ mm, $R = 101.6$ mm, $t = 0.2007$ mm, and $\nu = 0.3$. The auto-covariance function estimated from the measured data is displayed in Figure 3.20. Also, in this case, the auto-covariance function of the simulated sample showed excellent agreement with that of the measured initial sample. The elements of the variance-covariance matrix are displayed in Figure 3.21. The estimated variance of the measured initial data versus the axial coordinate is depicted in Figure 3.22. As shown in Figures 3.19 and 3.22, the variances associated with A- and B-shells are non-uniform. This implies that their initial imperfections form non-homogeneous random fields. Note that $i_{cl} = 30$ for the group of A-shells and $i_{il} = 17$ for the group of B-shells.

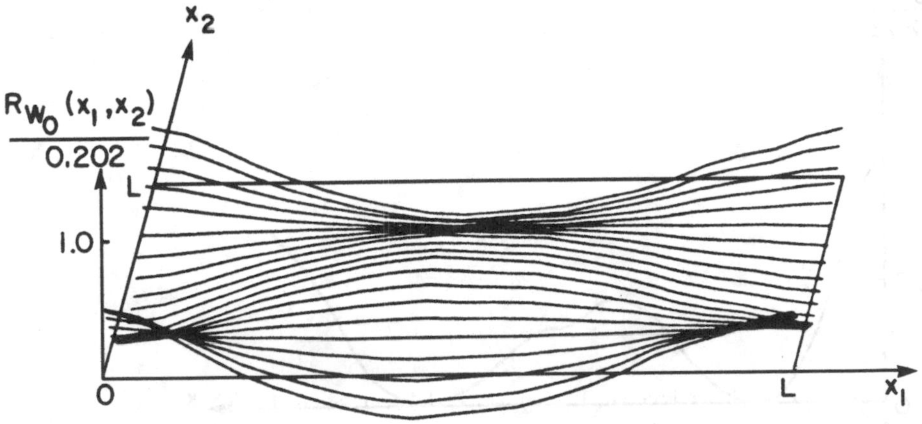

Figure 3.20 The estimated auto-covariance function of the axisymmetric part of the measure initial imperfections for the group of *B*-shells (after Elishakoff and Arbocz, 1982; Copyright © Elsevier Science Ltd., reprinted with permission).

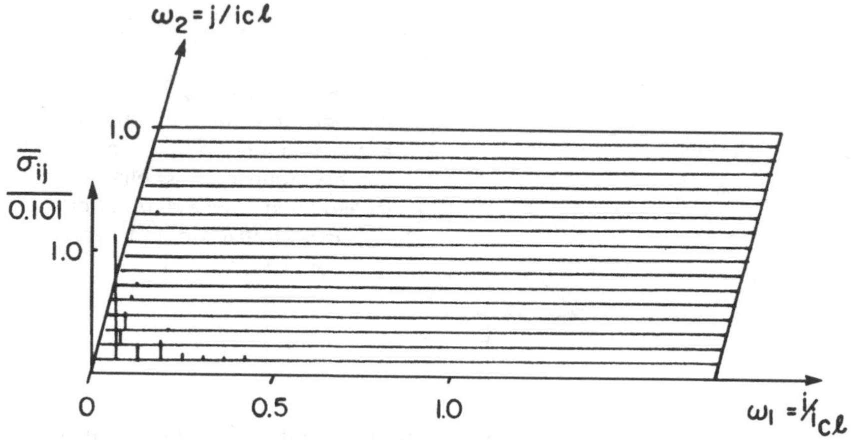

Figure 3.21 Elements of the variance-covariance matrix $[\Sigma^e]$ of the group of B-shells (after Elishakoff and Arbocz, 1982; Copyright © Elsevier Science Ltd., reprinted with permission).

As can be seen from Figures 3.17 and 3.21, the Fourier coefficients of the axisymmetric part of the initial imperfections "peak" near $\omega = 0$ and become vanishingly small near $\omega = 1$. That means that the variance-covariance matrices of the shells investigated are dominated by the lower order modes and not by the classical axisymmetric buckling mode. Next we proceed to the second step of the Monte Carlo method: evaluation of the buckling load for each "created" shell.

3.4.5 Computation of the Buckling Loads

The buckling loads are calculated by using Koiter's special theory (Koiter, 1963). We are considering a cylindrical shell with an axisymmetric initial imperfection of the

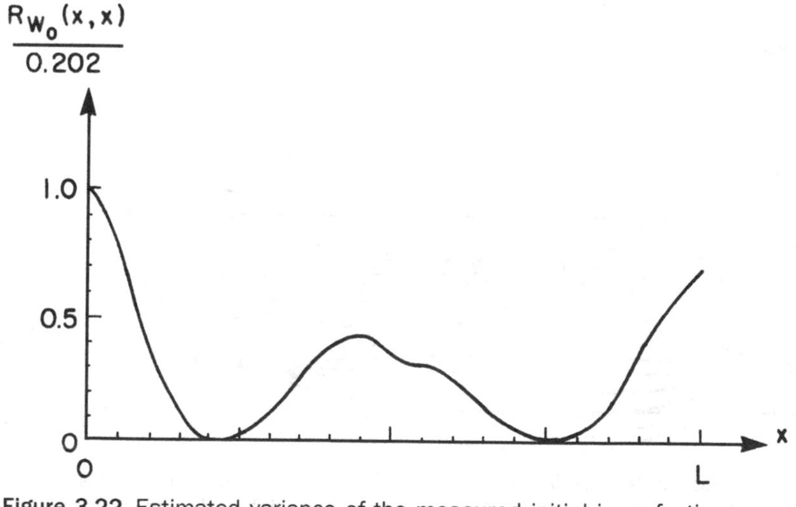

Figure 3.22 Estimated variance of the measured initial imperfections as a function of the axial coordinate for the group of B-shells (after Elishakoff and Arbocz, 1982; Copyright © Elsevier Science Ltd., reprinted with permission).

form

$$w_0(x) = t\bar{\xi}_i \cos\frac{i\pi x}{L} \tag{3.152}$$

where $\bar{\xi}_i$ denotes the magnitude of the imperfection as a fractional value of the shell thickness t and i is an integer denoting the number of half-waves in the axial direction. Using Koiter's special non-linear theory, one can derive a relationship between the non-dimensional axial load level λ (at which the resulting fundamental equilibrium state bifurcates into an asymmetric deformation pattern) and the imperfection amplitude $\bar{\xi}_i$. If one assumes the buckling mode

$$w(x, y) = tC_{ki} \sin\frac{k\pi x}{L} \cos\frac{ly}{R} \tag{3.153}$$

where k and l are integers denoting the number of half-waves and the number of full waves in the axial and in the circumferential direction, respectively, then the following non-linear transfer function between λ and $\bar{\xi}_i$ is obtained:

$$(\lambda_{ci} - \lambda)^2(\lambda_{c_{ki}} - \lambda) + (\lambda_{ci} + \lambda)\frac{c}{2}\frac{\beta_l^2}{\alpha_k^2}\left[\lambda + \lambda_{c_i}\frac{8\alpha_k^4}{(\alpha_k^2 + \beta_l^2)^2}\right]\bar{\xi}_i\delta_{ij}$$
$$+ 8c^2\alpha_k^2\beta_l^4\left[\frac{1}{(9\alpha_k^2 + \beta_l^2)^2} + \frac{1}{(\alpha_k^2 + \beta_l^2)^2}\right]\lambda_{ci}^2\bar{\xi}_i^2 = 0 \tag{3.154}$$

where

$$\lambda_{c_i} = \frac{1}{2}\left(\alpha_i^2 + \frac{1}{\alpha_i^2}\right), \qquad \lambda_{c_{ki}} = \frac{1}{2}\left[\frac{(\alpha_k^2 + \beta_k^2)^2}{\alpha_k^2} + \frac{\alpha_k^2}{(\alpha_k^2 + \beta_l^2)^2}\right]$$
$$\alpha_i^2 = i^2\frac{Rt}{2c}\left(\frac{\pi}{L}\right)^2, \qquad \alpha_k^2 = k^2\frac{Rt}{2c}\left(\frac{\pi}{L}\right)^2, \qquad \beta_l^2 = l^2\frac{Rt}{2c}\left(\frac{1}{R}\right)^2 \tag{3.155}$$

and δ_{ij} is the Kronecker delta with $j = 2k$.

It can be shown that Equation (3.154) can be reduced to Equation (5.2) of Koiter (1963). From Equation (3.154) for an imperfection-sensitive structure, $\bar{\xi}_i$ must be negative and i must be an even integer. The use of Equation (3.154) to calculate the critical buckling load for a given shell then proceeds as follows. Initially with $i = 2$, $k = 1$ (since $i = 2k$) and for different values of l, Equation (3.154) is solved repeatedly, and the lowest bifurcation buckling load λ is determined. Next the process is repeated with $i = 4, 6, \ldots,$ until all the available imperfection harmonics have been tested. The absolute minimum bifurcation buckling load is then identified as the critical buckling load for the shell under consideration.

Once the critical buckling loads for the large number of simulated shells ($M = 100$) are available, one can then proceed to calculate the histogram of the buckling loads and to determine the corresponding reliability function. Figures 3.23 through 3.26 display the histograms of the buckling loads and the calculated reliability functions for the group of A- and B-shells, respectively. A comparison of Figures 3.24 and 3.26 shows that

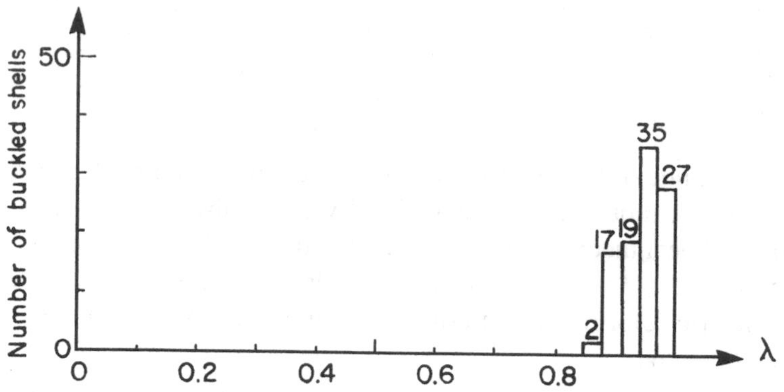

Figure 3.23 Histogram of the non-dimensional buckling loads λ for the group of *A*-shells (after Elishakoff and Arbocz, 1982; Copyright © Elsevier Science Ltd., reprinted with permission).

the reliability of the *B*-shells is less than that of the *A*-shells, meaning that machining thin-walled shells out of thick-walled, seamless, brass tubes is a "rougher" procedure than electroplating.

Note that the mean buckling load for the simulated sample of *A*-shells is 0.946, whereas that for the *B*-shells is 0.724. These loads are considerably higher than the experimentally observed mean buckling loads for the corresponding initial samples (0.643 for the *A*-shells and 0.592 for the *B*-shells). The reason is that, as was pointed out by Arbocz and Babcock (1976), for accurate buckling load predictions, the asymmetric imperfections must also be taken into account. This work could be viewed as the first step toward a more general analysis. The investigation of the effect of general, non-axisymmetric imperfections will be discussed in Section 3.5.

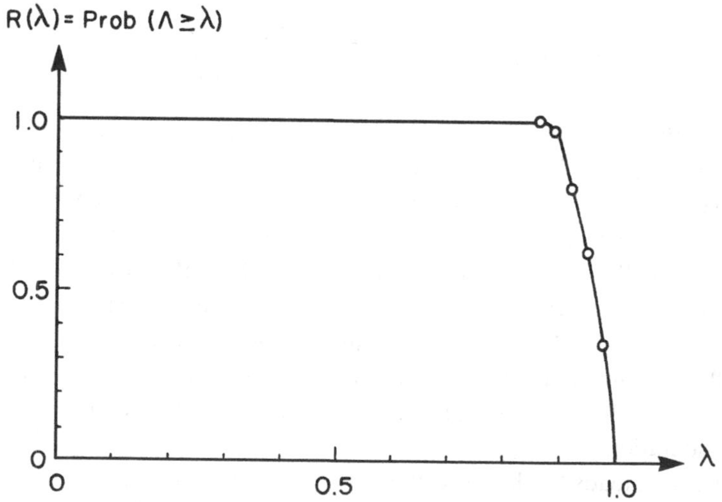

Figure 3.24 Calculated reliability function versus the non-dimensional buckling load λ for the group of *A*-shells (after Elishakoff and Arbocz, 1982; Copyright © Elsevier Science Ltd., reprinted with permission).

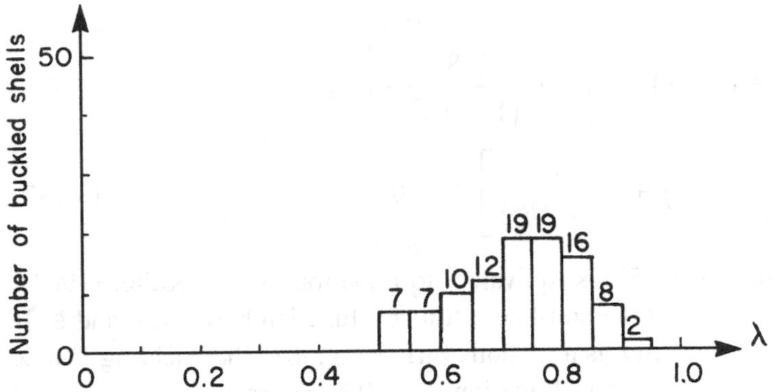

Figure 3.25 Histogram of the non-dimensional buckling loads λ for the group of *B*-shells (after Elishakoff and Arbocz, 1982; Copyright © Elsevier Science Ltd., reprinted with permission).

3.4.6 Comparison of the Monte Carlo Method with the Benchmark Solution

At this point, a natural question arises: What is the accuracy of the Monte Carlo method? In order to provide a check of this method when applied to shell structures, let us consider a case amenable of an exact solution. If one assumes that the initial imperfections are of the form

$$w_0(x) = t\bar{\xi}\cos i_{cl}\frac{\pi x}{L} \tag{3.156}$$

where $\bar{\xi}$ is a random variable $\bar{X}$ with given probability density $f_{\bar{x}}(\bar{\xi})$, then the shape of the axisymmetric imperfection coincides with the axisymmetric buckling mode of the perfect shell. Hence, $\lambda_{c_i} = 1$, $\alpha_k^2 = \frac{1}{4}$ (since $i = 2k$), and Equation (3.144)

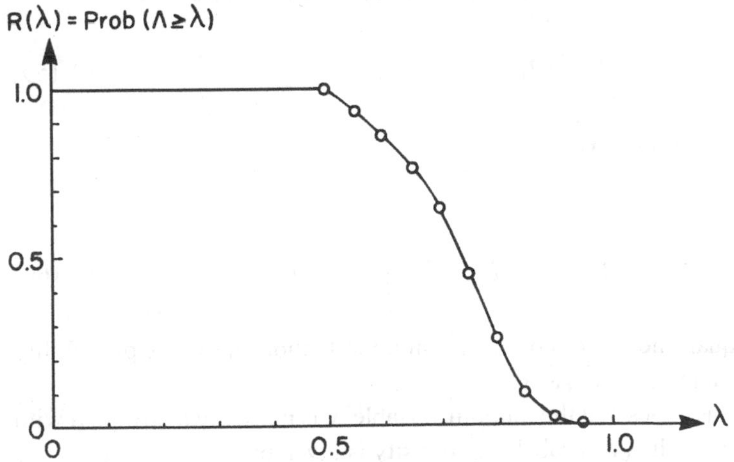

Figure 3.26 Reliability function versus the non-dimensional load for the group of *B*-shells (after Elishakoff and Arbocz, 1982; Copyright © Elsevier Science Ltd., reprinted with permission).

reduces to

$$(1-\lambda)^2(\lambda_{c_{kl}}-\lambda)+(1-\lambda)2c\beta_l^2\left[\lambda+\frac{8}{\left(1+4\beta_l^2\right)^2}\right]\bar{\xi}$$
$$+32c^2\beta_l^4\left[\frac{1}{\left(9+4\beta_l^2\right)^2}+\frac{1}{\left(1+4\beta_l^2\right)}\right]\bar{\xi}^2=0 \qquad (3.157)$$

One can show that Equation (3.157) is equivalent to Equation (5.4) of Koiter (1963). Equation (3.157) then represents the non-linear transfer function between λ and ξ. If the amplitude of the imperfection $\bar{\xi}$ is a random variable $\bar{X}$, then the buckling load λ will also be a random variable Λ. Thus, Equation (3.157) becomes

$$(1-\lambda)^2(\lambda_{c_{kl}}-\Lambda)-(1-\Lambda)2c\beta_l^2\left[\Lambda+\frac{8}{\left(1+4\beta_l^2\right)^2}\right]|\bar{X}|$$
$$+32c^2\beta_l^4\left[\frac{1}{\left(9+4\beta_l^2\right)^2}+\frac{1}{\left(1+4\beta_l^2\right)}\right]\bar{X}^2=0 \qquad (3.158)$$

The minus sign has been introduced (as discussed earlier) in order to obtain an imperfection-sensitive structure. The reliability is then defined as the probability that the buckling load Λ will be greater or equal to some specified value α. It follows from Equation (3.158) that this is equivalent to the probability that the absolute value of the imperfection $\bar{X}$ be less than or equal to some value $\bar{\xi}^*$, where $\bar{\xi}^*$ is the smallest root of the transfer function (3.158) for the specified value of $\Lambda=\alpha$.

It should be noted here that Equation (3.158) contains as a free parameter l, the number of full waves in the circumferential direction. Thus, finding the smallest root $\bar{\xi}^*$ for a given value of $\Lambda=\alpha$ involves repeated solution of Equation (3.158) for different values of l.

If we now introduce the distribution function $F_{\bar{x}}(\bar{\xi})$, then by definition

$$\text{Prob}\{\bar{X}\le\bar{\xi}^*\}=F_{\bar{x}}(\bar{\xi}^*)=\int_{-\infty}^{\bar{\xi}^*}f_{\bar{x}}(\bar{\xi})\,d\bar{\xi} \qquad (3.159)$$

and the reliability can be written as

$$R(\alpha)=\text{Prob}\{\Lambda\ge\alpha\}$$
$$=\text{Prob}\{|\bar{X}|\le\bar{\xi}^*\}=\int_{-\bar{\xi}^*}^{\bar{\xi}^*}f_{\bar{x}}(\bar{\xi})d\bar{\xi}=F_{\bar{x}}(\bar{\xi}^*)-F_{\bar{x}}(-\bar{\xi}^*) \qquad (3.160)$$

Thus, the reliability equals the integral of the initial imperfection amplitude probability density $f_{\bar{x}}(\bar{\xi})$ over the interval $(-\bar{\xi}^*,\bar{\xi}^*)$.

If one now further assumes that the random variable $\bar{X}$ is normally distributed with mean $m_{\bar{x}}$ and variance $\sigma_{\bar{x}}^2$, then its probability density is given by

$$f_{\bar{x}}(\bar{\xi})=\frac{1}{\sigma_{\bar{x}}\sqrt{2\pi}}e^{[-(\bar{\xi}-m_{\bar{x}})/2\sigma_{\bar{x}}]^2} \qquad (3.161)$$

and we get from Equation (3.161) the following closed form solution for the reliability:

$$R(\alpha) = \frac{1}{2}\left\{\operatorname{erf}\left[\frac{\bar{\xi}^* - m_{\bar{x}}}{\sigma_{\bar{x}}\sqrt{2}}\right] - \operatorname{erf}\left[\frac{-(\bar{\xi}^* - m_{\bar{x}})}{\sigma_{\bar{x}}\sqrt{2}}\right]\right\} = \operatorname{erf}\left(\frac{\bar{\xi}^* - m_{\bar{x}}}{\sigma_{\bar{x}}\sqrt{2}}\right) \quad (3.162)$$

where erf (x) is an error function defined as

$$\operatorname{erf}(x) = \frac{2}{\sqrt{2}}\int_0^x e^{-t^2}\, dt \quad (3.163)$$

The reliability calculation for this benchmark case is illustrated in Figure 3.27,

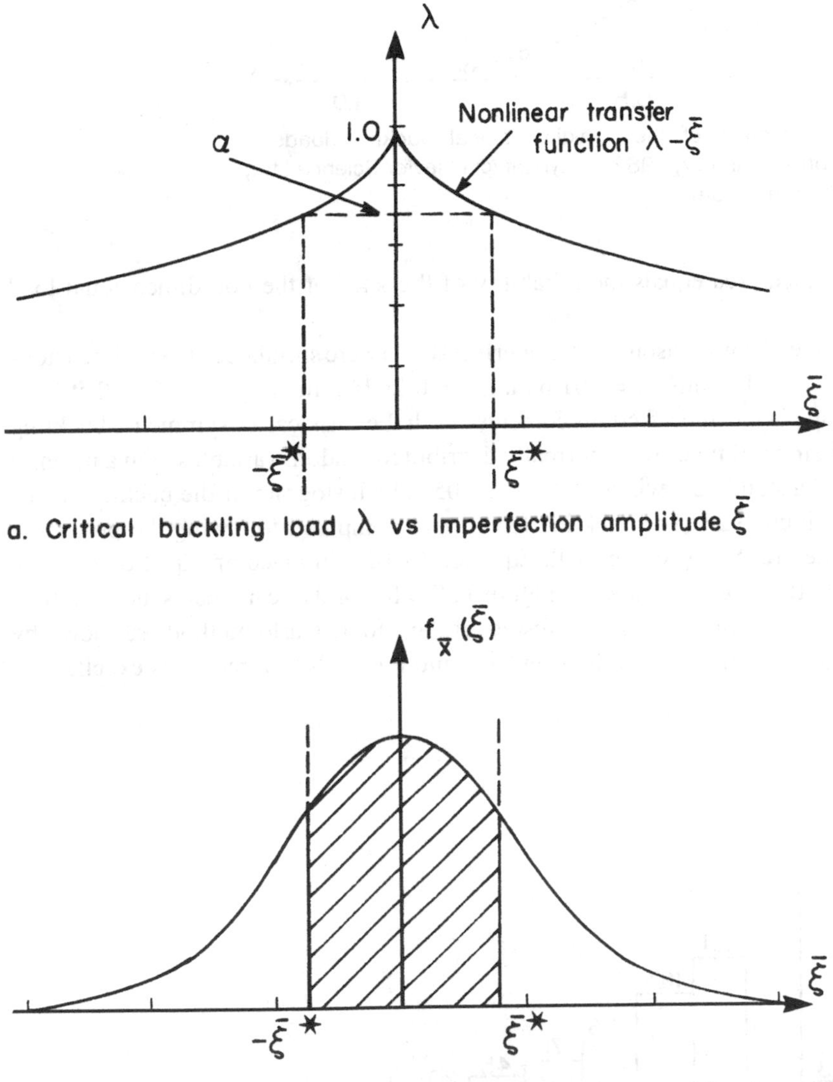

Figure 3.27 Illustration of the reliability calculation for the benchmark case (after Elishakoff and Arbocz, 1982; Copyright © Elsevier Science Ltd., reprinted with permission).

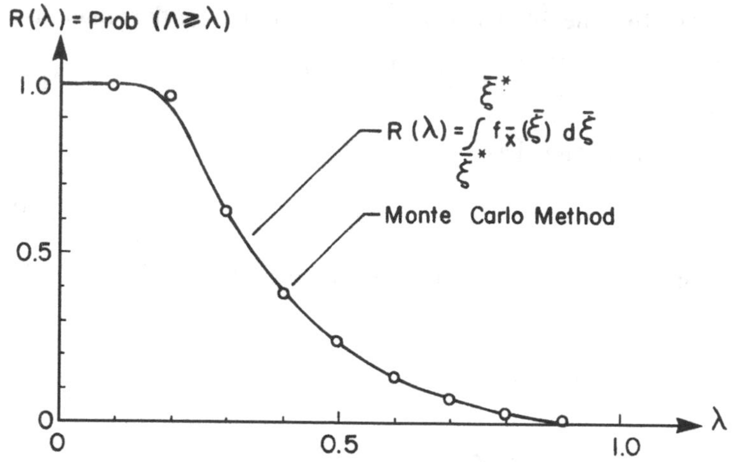

Figure 3.28 Histogram of the non-dimensional buckling loads λ (after Elishakoff and Arbocz, 1982; Copyright © Elsevier Science Ltd., reprinted with permission).

where the shaded area equals the reliability of the shell at the non-dimensional load level α.

For the sake of comparison, 1089 imperfect shells were simulated. Their dimensions were set at $L = 141.0$ mm, $R = 101.6$ mm, $t = 0.2634$ mm, and $\nu = 0.3$ so that $i_{cl} = 16.0$. The shape of the imperfections coincided with the classical axisymmetric buckling mode, and their amplitudes were normally distributed random variables with a mean of $m_{\bar{x}} = 0.1$ and a standard deviation of $\sigma_{\bar{x}} = 0.05$. The histogram of the buckling loads is shown in Figure 3.28. The buckling loads were computed following the procedure outlined in the preceding section with Equation (3.157) in place of Equation (3.165). The reliability functions are shown in Figure 3.29. The solid line indicates the analytical solution from Equation (3.157). The results of the Monte Carlo method are shown by circles. The agreement between the simulated and the analytical results is excellent.

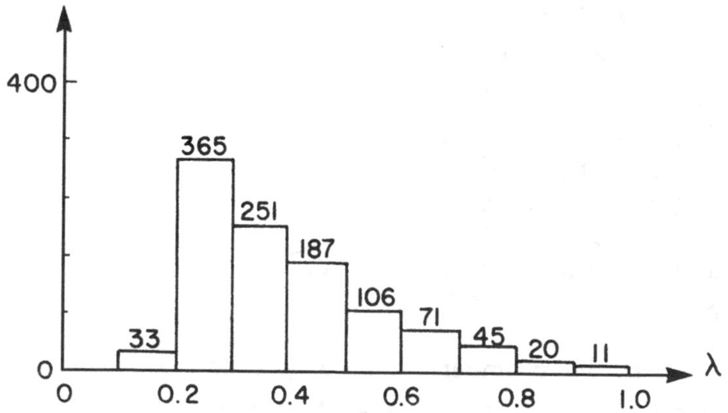

Figure 3.29 Comparison of analytical reliability function with results of the Monte Carlo method (after Elishakoff and Arbocz, 1982; Copyright © Elsevier Science Ltd., reprinted with permission).

This section has demonstrated that the Monte Carlo method can be used successfully for the investigation of the stochastic imperfection sensitivity of axially compressed cylindrical shells with axisymmetric initial imperfections. We believe that the reliability function – the final product of such an analysis – represents a more powerful design criterion than the ones based on deterministic or mean buckling load formulas.

Computation of the variance-covariance matrices of the Fourier coefficients as ensemble averages of the experimentally determined values have shown that the measured initial imperfections of finite shells are non-homogeneous, and hence non-ergodic. Thus the ergodicity assumption used in many previous investigations is not appropriate for realistic structures of finite length.

As has been seen, the results of the existing initial imperfection data banks can be incorporated directly into the Monte Carlo method. Thus, the results presented in this section further reinforce the need for compiling extensive experimental information on imperfections classified according to the manufacturing processes.

Using the technique developed in the earlier papers (Elishakoff, 1978b, 1979b, 1980b; Elishakoff and Arbocz, 1982) and in this section (as well as in the following section where the discussion has been extended to include general non-symmetric imperfections), one can then calculate via Monte Carlo method the reliability functions associated with different manufacturing processes. Thus, through this approach, the imperfection-sensitivity concept may finally be introduced into the design procedure.

3.5 Reliability of Axially Compressed Cylindrical Shells with Nonsymmetric Imperfections

This discussion of the reliability of axially compressed cylindrical shells with non-symmetric imperfections is based on Elishakoff and Arbocz (1985).

3.5.1 Probabilistic Properties and Simulation of the Initial Imperfections for a Finite Shell

Let us represent the initial imperfection functions as the following series:

$$W_n(\xi,\theta) = \sum_{i=0}^{N_1} A_i \cos(i\pi xi) + \sum_{k=l}^{N_2} \sum_{i=l}^{N_3} [C_{kl}\sin(k\pi\xi)\cos(l\theta) + D_{kl}\sin(k\pi\xi)\sin(l\theta)] \tag{3.164}$$

where

$$W_n(\xi,\theta) = \frac{W_n(\xi,\theta)}{t}, \qquad \xi = \frac{x}{L}, \qquad \theta = \frac{y}{R}, \qquad 0 \le x \le L, \qquad 0 \le \theta \le 2\pi$$

$W_n(\xi,\theta)$ and $W_n(\xi,\theta)$ are dimensional and non-dimensional initial imperfections: t, L, and R are the thickness, the length, and the radius of the shell, respectively; x is the axial and y is the circumferential coordinate. Notice that in Equation (3.165) the first summation represents the axisymmetric part of the initial imperfection profile, whereas the second, a double summation, is associated with its non-symmetric part. The axisymmetric part is expressed in the half-range cosine series, whereas the non-symmetric part is represented by the half-range sine series so that the series (3.165) sums up to the measured imperfection profile in the range $0 \leq x \leq L, 0 \leq \theta \leq 2\pi$.

The mathematical expectation of $W_n(\xi, \theta)$ is given by

$$E[w_0(\xi,\theta)] = \sum_{i=0}^{N_1} E(A_i)\cos(i\pi\xi) + \sum_{k=l}^{N_2}\sum_{l=1}^{N_3}[E(C_{kl})\sin(k\pi\xi)\cos(l\theta) + E(D_{kl})\sin(k\pi\xi)\sin(l\theta)] \tag{3.165}$$

where $E(\cdot)$ denotes a mathematical expectation.

The auto-covariance function becomes

$$\begin{aligned} C_{W_0}(\xi_1,\theta_1;\xi_2,\theta_2) &= E\{[W_0(\xi_1,\theta_1) - E(W_0(\xi_1,\theta_1))][W_0(\xi_2,\theta_2) - E(W_0(\xi_2,\theta_2))]\} \\ &= E\left\{\left[\sum_{i=0}^{N_1}(A_i - E(A_i))\cos(i\pi\xi_1)\right.\right. \\ &\quad + \sum_{k=1}^{N_2}\sum_{l=1}^{N_3}(C_{kl} - E(C_{kl}))\sin(k\pi\xi_1)\cos(l\theta_1) \\ &\quad \left. + \sum_{k=1}^{N_2}\sum_{l=1}^{N_3}(D_kl - E(D_kl))\sin(k\pi\xi_1)\sin(l\theta_1)\right] \\ &\quad \times\left[\sum_{j=0}^{N_1}(A_j - E(A_j))\cos(j\pi\xi_2)\right. \\ &\quad + \sum_{m=1}^{N_2}\sum_{n=1}^{N_3}(C_{mn} - E(C_{mn}))\sin(m\pi\xi_2)\cos(n\theta_2) \\ &\quad \left.\left. + \sum_{m=1}^{N_2}\sum_{n=1}^{N_3}(D_{mn} - E(D_{mn}))\sin(m\pi\xi_2)\sin(n\theta_2)\right]\right\} \end{aligned} \tag{3.166}$$

For the sake of simplicity, we rewrite Equation (3.164) in an alternative way, replacing the double summation in Equation (3.164) by a single summation

$$W_0(\xi,\theta) = \sum_{i=1}^{N_1} A_i\cos(i\pi\xi) + \sum_{r=1}^{N}[C_r\sin(k_r\pi\xi)\cos(l_r\theta) + D_r\sin(k_r\pi\xi)\sin(l_r\theta)] \tag{3.167}$$

where the quantities indexed with r are chosen to ensure the equivalence of the two series given by Equations (3.164) and (3.167) and $N = N_2 \times N_3$. The auto-covariance function can be written as

$$\begin{aligned} C_{W_0}(\xi_1,\theta_1;\xi_2,\theta_2) &= \sum_{i=0}^{N_1}\sum_{j=0}^{N_1} K_{A_iA_j}\cos(i\pi\xi_1)\cos(j\pi\xi_2) \\ &\quad + \sum_{i=0}^{N_1}\sum_{x=1}^{N} K_{A_iC_s}\cos(i\pi\xi_1)\sin(k_s\pi\xi_2)\cos(l_s\theta_2) \\ &\quad + \sum_{i=0}^{N_1}\sum_{s=0}^{N} K_{A_{iD_s}}\cos(i\pi\xi_1)\sin(k_s\pi\xi_2)\sin(l_s\theta_2) \end{aligned}$$

$$+\sum_{r=1}^{N}\sum_{j=0}^{N_1} K_{C_r A_j} \sin(k_r \pi x u_1)\cos(l_r\theta_1)\cos(j\pi\xi_2)$$
$$+\sum_{r=1}^{N}\sum_{j=0}^{N_1} K_{D_r A_j} \sin(k_r \pi \xi_1)\sin(l_r\theta_1)\cos(j\pi\xi_2)$$
$$+\sum_{r=1}^{N}\sum_{s=1}^{N} K_{C_r C_s} \sin(k_r \pi \xi_1)\cos(l_r\theta_1)\sin(k_s\pi\xi_2)\cos(l_s\theta_2)$$
$$+\sum_{r=1}^{N}\sum_{s=1}^{N} K_{C_r D_s} \sin(k_r \pi \xi_1)\cos(l_r\theta_1)\sin(k_s\pi\xi_2)\sin(l_s\theta_2)$$
$$+\sum_{r=1}^{N}\sum_{s=1}^{N} K_{D_r C_s} \sin(k_r \pi \xi_1)\sin(l_r\theta_1)\sin(k_s\pi\xi_2)\cos(l_s\theta_2)$$
$$+\sum_{r=1}^{N}\sum_{s=1}^{N} K_{D_r D_s} \sin(k_r \pi \xi_1)\sin(l_r\theta_1)\sin(k_s\pi\xi_2)\sin(l_s\theta_2) \tag{3.168}$$

where the variance-covariance matrices $K_{A_iA_j}, \ldots,$ etc. are defined as follows:

$$\begin{aligned}
K_{A_iA_j} &= E[((A_i - E(A_i))(A_j - E(A_j))],\\
K_{A_jC_s} &= E[(A_j - E(A_i))(C_s - E(C_s))]\\
K_{A_iD_s} &= E[A_1 - E(A_i))(D_s - E(D_s))],\\
K_{C_rC_s} &= E(C_r - E(C_r))(C_s - E(C_s))]\\
K_{C_rD_s} &= E[(C_r - E(C_r))(D_s - E(D_s))],\\
K_{D_rD_s} &= E[(D_r - E(D_r)))(D_s - E(D_s))]
\end{aligned} \tag{3.169}$$

If the auto-covariance function $C_{W_0}(\xi_1, \theta_1; \xi_2, \theta_2)$ is known, then these quantities can be obtained as follows:

$$\begin{aligned}
K_{A_iA_j} &= \frac{1}{\pi^2}\int_0^{2\pi}\int_0^1\int_0^{2\pi}\int_0^1 C_{W_0}(\xi_1,\theta_1;\xi_2,\theta_2)\\
&\quad\times\cos(i\pi\xi_1)\cos(j\pi\xi_2)\,d\xi_1\,d\theta_1\,d\xi_2\,d\theta_2\\
K_{A_iCmn} &= \frac{2}{\pi^2}\int_0^{2\pi}\int_0^1\int_0^{2\pi}\int_0^1 C_{W_0}(\xi_1,\theta;\xi_2,\theta_2)\cos(i\pi\xi_1)\\
&\quad\times\sin(m\pi\xi_2)\cos(n\theta_2)\,d\xi_1\,d\theta_1\,d\xi_2\,d\theta_2\\
K_{A_iDmn} &= \frac{2}{\pi^2}\int_0^{2\pi}\int_0^1\int_0^{2\pi}\int_0^1 C_{W_0}(\xi_1,\theta_1;\xi_2,\theta_2)\cos(i\pi\xi_1)\\
&\quad\times\sin(m\pi\xi_2)\sin(n\theta_2)\,d\xi_1\,d\theta_1\,d\xi_2\,d\theta_2\\
K_{A_jC_{kl}} &= \frac{2}{\pi^2}\int_0^{2\pi}\int_0^1\int_0^{2\pi}\int_0^1 C_{W_0}(\xi_1,\theta_1;\xi_2,\theta_2)\cos(j\pi\xi_2)\\
&\quad\times\sin(k\pi\xi_1)\cos(l\theta_1)\,d\xi_1\,d\theta_1\,d\xi_2\,d\theta_2
\end{aligned}$$

$$
\begin{aligned}
K_{A_jC_{kl}} &= \frac{2}{\pi^2}\int_0^{2\pi}\int_0^1\int_0^{2\pi}\int_0^1 C_{W_0}(\xi_1,\theta_1;\xi_2,\theta_2)\cos(j\pi\xi_2)\\
&\quad\times\sin(k\pi\xi_1)\sin(l\theta_1)\,d\xi_1\,d\theta_1\,d\xi_2\,d\theta_2\\
K_{C_{kl}C_{mn}} &= \frac{4}{\pi^2}\int_0^{2\pi}\int_0^1\int_0^{2\pi}\int_0^1 C_{W_0}(\xi_1,\theta_1;\xi_2,\theta_2)\sin(k\pi\xi_1)\cos(l\theta_1)\\
&\quad\times\sin(m\pi\xi_2)\cos(n\theta_2)\,d\xi_1\,d\theta_1\,d\xi_2\,d\theta_2\\
K_{D_{kl}C_{mn}} &= \frac{4}{\pi^2}\int_0^{2\pi}\int_0^1\int_0^{2\pi}\int_0^1 C_{W_0}(\xi_1,\theta_1;\xi_2\theta_2)\sin(k\pi\xi_1)\sin(l\theta_1)\\
&\quad\times\sin(m\pi\xi_2)\cos(n\theta_2)\,d\xi_1 d\theta_1 d\xi_2 d\theta_2\\
K_{D_{kl}D_{mn}} &= \frac{4}{\pi^2}\int_0^{2\pi}\int_0^1\int_0^{2\pi}\int_0^1 C_{W_0}(\xi_1,\theta_1;\xi_2,\theta_2)\sin(k\pi\xi_1)\sin(l\theta_1)\\
&\quad\times\sin(m\pi\xi_2)\sin(n\theta_2)\,d\xi_1\,d\theta_1\,d\xi_2\,d\theta_2
\end{aligned}
\tag{3.170}
$$

Notice that if $W_0(\xi,\theta)$ is constant and

$$K_{A_iC_s} = K_{A_iD_s} = K_{C_rA_j} = K_{D_rA_j} = K_{C_rD_s} = K_{D_rC_s} = 0 \tag{3.171}$$

for any combination of indices and, moreover,

$$
\begin{aligned}
K_{C_rC_s} &= K_{D_rD_s}\\
K_{C_rC_s} &= K_{C_rC_s}\delta_{l_rl_s}
\end{aligned}
\tag{3.172}
$$

where $\delta_{l_rl_s}$ is a Kronecker delta, then the initial imperfection is a weakly homogeneous function in the circumferential direction. Under the conditions (3.171) and (3.172), the auto-covariance function takes the form

$$
\begin{aligned}
C_{W_0}(\xi_1,\theta_1;\xi_2,\theta_2) &= \sum_{i=0}^{N_1}\sum_{j=0}^{N_1} K_{A_iA_j}\cos(i\pi\xi_1)\cos(j\pi\xi_2)\\
&\quad+\sum_{r=1}^{N}\sum_{s=1}^{N} K_{C_rC_s}\sin(k_r\pi\xi_1)\sin(k_s\pi\xi_2)\cos[l_r(\theta_2-\theta_1)]
\end{aligned}
\tag{3.173}
$$

(i.e., it depends on $\theta_2-\theta_1$ rather than on θ_1 and θ_2 separately). It can be shown that Equations (3.172) and (3.173) represent not only the sufficient conditions but also the necessary ones. One could argue that for the closed, nominally circular, seamless cylindrical shell the probabilistic properties would not be affected by a shift of the origin of the coordinate axes in the circumferential direction. Interestingly, in the experiments performed by Makarov (1971), this weak homogeneity was preserved even for the series of shells with pronounced seams. Due to the frequent use of this property, we will first show how to simulate the initial imperfections possessing weak homogeneity in the circumferential direction.

To simulate the large number of initial imperfection profiles needed for the Monte Carlo method, first the mean values and the variance-covariance matrices of the measured initial imperfections must be determined. This involves the evaluation of the following ensemble averages for a sample of experimentally measured initial imperfections:

$$\tilde{A}_i^{(e)} = \frac{1}{M}\sum_{m=1}^{M} A_i^{(m)}; \qquad \tilde{C}_r^{(e)} = \frac{1}{M}\sum_{m=1}^{M} C_r^{(m)}$$

$$K_{A_iA_j}^{(e)} = \frac{1}{M-1}\sum_{m=1}^{M}[A_i^{(m)} - \tilde{A}_i^{(e)}][A_j^{(m)} - \tilde{A}_j^{(e)}] \tag{3.174}$$

$$K_{C_rC_s}^{(e)} = \frac{1}{M-1}\sum_{m=1}^{M}[C_r^{(m)} - \tilde{C}_r^{(e)}][C_s^{(m)} - \tilde{C}_s^{(e)}]$$

where M is the number of sample shells, and m is the serial number of the shells. The variance-covariance matrices are positive–semi-definite and can be uniquely decomposed in the form

$$\left[K_{A_iA_j}^{(e)}\right] = [G][G]^T, \qquad \left[K_{C_rC_s}^{(e)}\right] = [G'][G']^T \tag{3.175}$$

where T means transpose and $[G]$, $[G']$ are lower triangular matrices found by the Cholesky decomposition algorithm. Now we form the random vectors $[B]$ and $[B']$, the elements of which are normally distributed and statistically independent with zero means and unit variance. Then the vectors of the Fourier coefficients of the initial imperfections are simulated as follows:

$$[A] = [G]\{B\} + \{\tilde{A}^{(e)}\}, \qquad \{C\} = [G']\{B'\} + \{\tilde{C}^{(e)}\} \tag{3.176}$$

Having the desired large number of realizations of the vectors $\{B\}$ and $\{B'\}$, one obtains the same number of realizations of $\{A\}$ and $\{C\}$. The main feature of this simulation technique (Elishakoff, 1979b) is that it is applicable for homogeneous as well as non-homogeneous random functions with given mean and auto-covariance functions.

Equation (3.176) represents the simulated vectors $\{A\}$ and $\{C\}$ for the random imperfections, weakly homogeneous in the circumferential direction. For the imperfections that form a general non-homogeneous random field, the refined simulation procedure (Elishakoff, 1988) has to be utilized. The essence of this refinement is the replacement of the multiple summations in Equations (3.165) and (3.167) by a single "string" and the dealing with the resultant mixed series.

3.5.2 Multi-Mode Deterministic Analysis for Each Realization of Random Initial Imperfections

The buckling load for each created shell is then calculated by the so-called multi-mode analysis (Arbocz and Babcock, 1974), which allows the incorporation of imperfection shapes in the form of the double Fourier series given in Equation (3.164). By definition, the value of the loading parameter λ corresponding to the limit point of the prebuckling states is the theoretical buckling load. The number of modes of deformation included in the analysis is limited by practical considerations, like the available core size and the time required for obtaining the solution. Thus, since the shell buckling load will be determined by solving the governing equations for a particular set of modes, an attempt

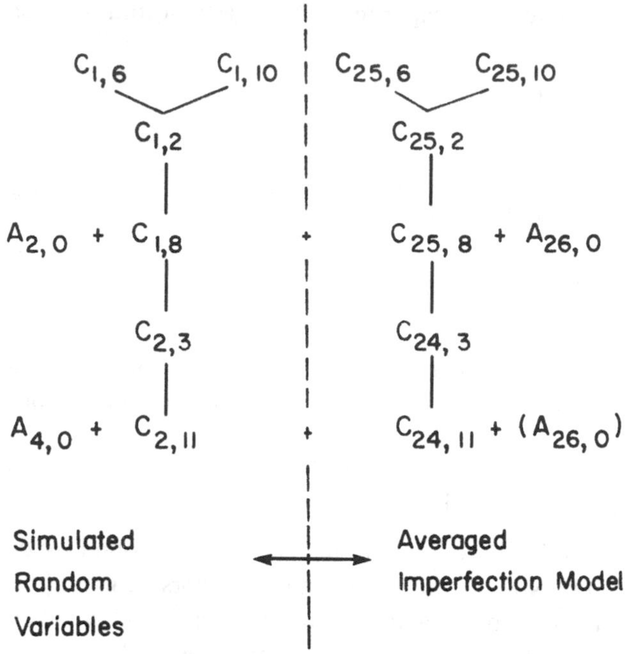

Figure 3.30 Modified 15-mode imperfection model used for the analysis by the Monte Carlo method (after Elishakoff and Arbocz, *Journal of Applied Mechanics*, 1985).

at optimizing the selection of these modes must be made. That is, it is necessary to locate those modes that dominate the prebuckling and buckling behavior of the shell. Previous investigations by Arbocz and Babcock (1974, 1976) have shown that to yield a noticeable decrease from the buckling load of the perfect shell, the initial imperfection harmonics used must include at least one mode with a significant initial amplitude and an associated eigenvalue that is close to the buckling load of the perfect shell. Furthermore, if the modes are so selected that the non-linear coupling conditions are satisfied, then the resulting buckling load of the shell generally will be lower than the buckling loads obtained with each mode considered separately.

Based on these considerations and the results of Arbocz and Babcock (1974), we choose the imperfection model shown in Figure 3.30 for the Monte Carlo simulation. In this model $A_{2.0}$ stands for a half-wave cosine axisymmetric Fourier coefficient, with two half-waves in the axial direction and no waves in the circumferential direction. On the other hand, $C_{1.10}$ stands for an asymmetric Fourier coefficient with a single half-wave in the axial direction and 10 full waves in the circumferential direction.

As pointed out by Arbocz and Babcock (1976), the chosen imperfection model requires imperfection amplitudes at wave numbers that were not measured. This is due to the fact that in the early experimental work the mesh-spacing used was not sufficiently close to resolve all the harmonic amplitudes of interest. Therefore, the Donnell-Imbert imperfection model was fitted over the wave numbers actually measured, and then the amplitudes of the harmonics of interest were obtained by extrapolation. It should be stressed here that the averaged (in the axial direction short wavelength)

Table 3.3. First 15 Fourier coefficients $a_i^{(m)}$ of the initial imperfections expanded as

$$w_0^{(m)} = tW_0^{(m)}(x) = t\sum_{i=0}^{14} a_i^{(m)} \cos\frac{i\pi x}{L}$$

Shell number, *i*	*B*-1	*B*-2	*B*-3	*B*-4
0	0.0010	0.0028	0.0111	0.0080
1	−0.0333	−0.1889	−0.6231	0.1096
2	−0.0108	−0.0272	−0.0899	−0.0176
3	−0.0190	−0.0276	−0.0803	0.0407
4	0.0226	−0.0078	−0.0255	−0.0092
5	−0.0025	−0.0097	−0.0230	0.0132
6	0.0018	−0.0049	−0.0223	−0.0059
7	−0.0062	−0.0080	−0.0189	0.0120
8	0.0076	−0.0074	0.0052	−0.0125
9	−0.0050	−0.0023	0.0053	0.0254
10	−0.0013	−0.0041	−0.0133	−0.0265
11	0.0015	−0.0007	−0.0033	0.0200
12	−0.0062	−0.0009	−0.0085	0.0176
13	0.0084	−0.0031	0.0092	0.0103
14	0.0020	−0.0027	−0.0136	−0.0114

For the group of B-shells ($R = 101.6$ mm, $t = 0.2007$ mm, $L = 134.30$ mm, $E = 1.065 \times 10^5$ N/mm^2, $\nu = 0.3$, $i_{cl} = 17$). Note: Here $w_0^{(m)}(x)$ is positive outward.

modes of the imperfection model shown in Figure 3.30 must be included in the analysis to satisfy the non-linear coupling conditions. Their initial amplitudes are actually insignificant.

For the purpose of the Monte Carlo method, the MIUTAM code (see Arbocz and Babcock, 1976) was incorporated into a new program, which then one by one automatically starts the calculations for the simulated imperfections and at the end lists all the buckling loads obtained.

3.5.3 Numerical Results and Discussion

The procedure described in the previous sections was applied to the group of shells, referred by Arbocz and Abramovich (1979) as B-shells. These shells were originally cut from thick-walled seamless brass tubing; the pieces were mounted on a mandrel and the outer surface was machined to the desired dimensions. The geometric and material properties of the B-shells are summarized in Table 3.3, whereas Table 3.4 lists elements of the variance-covariance matrix.

As is seen from Figure 3.30, the simulation procedure was applied only to the eight lower order modes, namely $A_{2,0}$, $A_{4,0}$, $C_{1,2}$, $C_{1,6}$, $C_{1,8}$, $C_{1,10}$, $C_{2,3}$, and $C_{2,11}$. The remaining higher order modes, namely $A_{26,0}$, $C_{24,3}$, $C_{24,11}$, $C_{25,2}$, and $C_{25,10}$ were

Table 3.4. Elements of the variance-covariance matrix $[\Sigma^{(e)}]$ or the group of *B*-shells.

	i/*j*										
$[\Sigma^{(m)}]$	**0**	**1**	**2**	**3**	**4**	**5**	**6**	**7**	**8**	**9**	**10**
0	0.002	−0.079	−0.013	−0.006	−0.008	−0.002	−0.004	−0.001	−0.0005	0.003	−0.004
1	−0.079	10.059	1.095	1.489	0.398	0.454	0.279	0.364	−0.160	0.145	−0.070
2	−0.013	1.095	0.132	0.145	0.057	0.044	0.036	0.033	−0.013	0.003	0.005
3	−0.006	1.489	0.145	0.247	0.035	0.074	0.033	0.063	−0.034	0.041	0.028
4	−0.008	0.398	0.057	0.035	0.040	0.012	0.018	0.006	0.006	−0.012	0.012
5	−0.002	0.454	0.044	0.074	0.012	0.023	0.010	0.019	0.009	0.012	0.008
6	−0.004	0.279	0.036	0.033	0.018	0.010	0.010	0.007	0.002	−0.002	0.004
7	−0.001	0.364	0.033	0.063	0.006	0.019	0.007	0.016	−0.009	0.012	0.009
8	−0.0005	−0.160	−0.013	−0.034	0.006	−0.009	−0.002	−0.009	0.009	−0.010	0.007
9	0.003	0.145	0.003	0.041	−0.012	0.012	−0.002	0.012	−0.010	0.017	0.014
10	−0.004	−0.070	0.005	−0.028	0.012	−0.008	0.004	−0.009	0.007	−0.014	0.013

Each entry should be multiplied by a factor 10^{-2}.

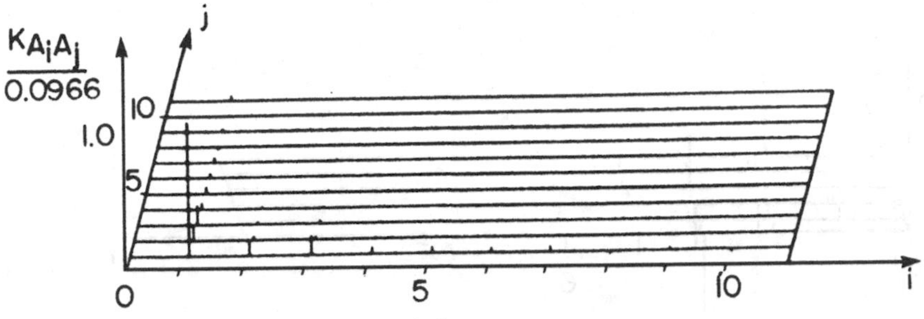

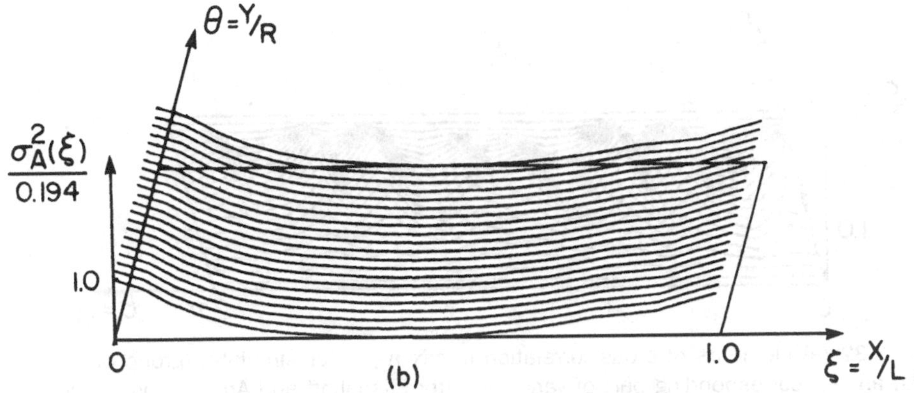

Figure 3.31 (a) Elements of cross-correlation matrix $K_{A_iA_j}$ for simulated group of 500 *B*-shells, and (b) corresponding part of variance (after Elishakoff and Arbocz, *Journal of Applied Mechanics*, 1985).

obtained by extrapolation from the corresponding Donnell-Imbert imperfection model (Arbocz and Babcock, 1976; Imbert, 1971).

Figure 3.31 shows the elements of the variance-covariance matrices $K_{A_iA_j}$ and the corresponding part of the variance, denoted by $\sigma_A^2(\xi)$

$$\sigma_A^2(\xi) = \sum_{i=0}^{N_1}\sum_{j=0} K_{A_iA_j}\delta_{ij}\cos(\pi i\xi)\cos(\pi j\xi) \tag{3.177}$$

Figures 3.32 and 3.33 show the elements of the variance-covariance matrices $K_{C_rC_s}$ and $K_{D_rD_s}$, respectively, and the associated parts of the variance, denoted by $\sigma_C^2(\xi,\theta)$ and $\sigma_D^2(\xi,\theta)$:

$$\sigma_C^2(\xi,\theta) = \sum_{r=1}^{N}\sum_{s=1}^{N} K_{C_rC_s}\sin(k_r\pi\xi)\sin(k_s\pi\xi)\cos(l_r\theta)\cos(l_s\theta) \tag{3.178}$$

$$\sigma_D^2(\xi,\theta) = \sum_{r=1}^{N}\sum_{s=1}^{N} K_{D_rD_s}\sin(k_r\pi\xi)\sin(k_s\pi\xi)\sin(l_r\theta)\sin(l_s\theta) \tag{3.179}$$

Figure 3.34 portrays the probabilistic characteristics of the 500 simulated shells. Figure 3.35(a) shows the mean imperfection function, whereas Fig. 3.35(b) displays the

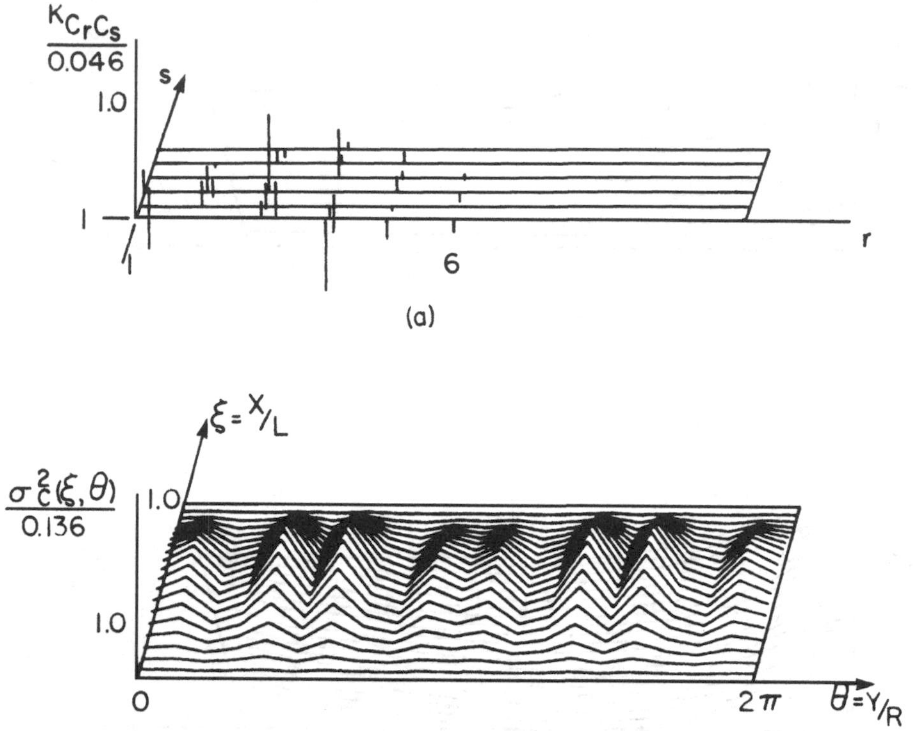

Figure 3.32 (a) Elements of cross-correlation matrix $K_{C_rC_s}$ for simulated group of *B*-shells, and (b) corresponding part of variance (after Elishakoff and Arbocz, *Journal of Applied Mechanics*, 1985).

variance. As is seen from Figure 3.35 neither the mean function nor the variance are constant in the circumferential direction, implying that the random imperfections do not consitute a circumferentially homogeneous field. This conclusion can also be deduced from examination of values $K_{C_rC_s}$ and $K_{D_rD_s}$, which reveals that not only do the corresponding elements of these matrices not coincide but a ratio between them also may well exceed 10. Thus, the homogeneity assumption adopted in the work of Amazigo (1974) and other investigators turns out to be unjustifiable even for seamless shells.

To calculate the reliability functions shown in Figure 3.36, the following dimensions corresponding to shell B-1 (Arbocz and Abramovich, 1979) were used: length of 196.85 mm, radius of 101.60 mm, and thickness of 0.205 mm. In addition, for the buckling load calculations of shells with axisymmetric imperfections, the one-sided transfer-function shown in Figure 3.37 was chosen. This was done due to the fact, pointed out by Babcock and Sechler (1962), that for a finite length shell the buckling load is sensitive only to those axisymmetric imperfections that point inward at the mid-plane of the shell (at $x = L/2$).

In Figure 3.36, curve 1 represents the reliability function for the case of purely axisymmetric imperfections (with an estimated mean buckling load of 0.935), whereas curve 2 shows the reliability function for the 15-modes non-symmetric imperfection mode (with an estimated mean buckling load of 0.739). As can be seen, the reliability estimate depends strongly on the number of terms taken into account (i.e., it is sensitive

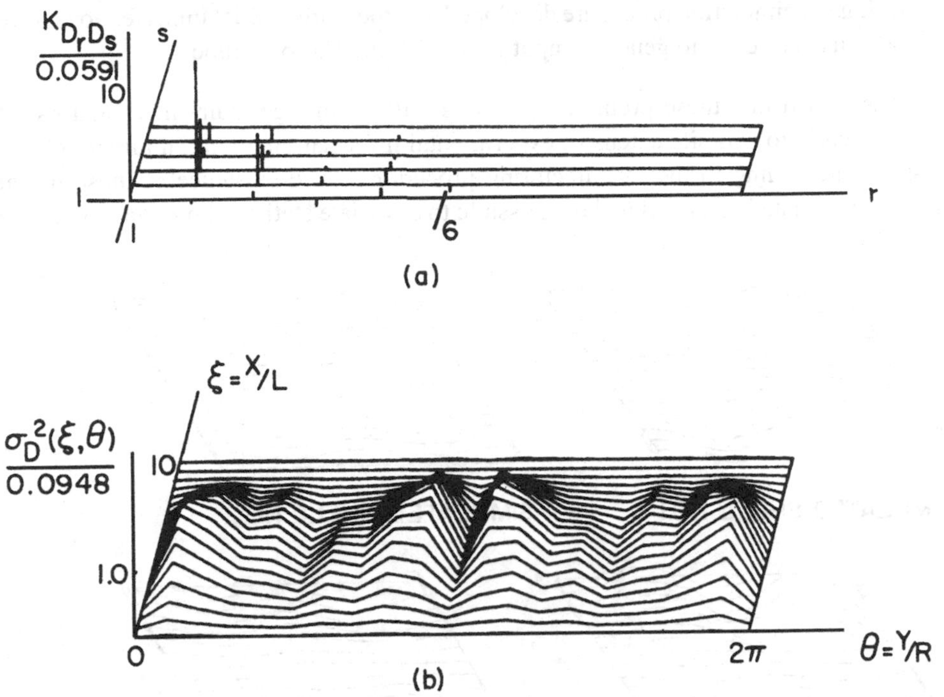

Figure 3.33 (a) Elements of cross-correlation matrix $K_{D_r D_s}$ for simulated group of 500 *B*-shells, and (b) corresponding part of variance (after Elishakoff and Arbocz, *Journal of Applied Mechanics*, 1985).

to the adequacy of the underlying deterministic model). As a check on the correctness of the simulated results the 15-modes imperfection model shown in Figure 3.30 was used to calculate the collapse loads of the original four *B*-shells involved, using the experimentally measured initial imperfections in place of the simulated random variables. These computations yielded for the shells *B*-1, *B*-2, *B*-3, and *B*-4, respectively, the following collapse loads 0.751, 0.746, 0.740, and 0.781 with a mean of 0.756. Closeness to the simulated results is remarkable. It should also be mentioned here that the theoretical collapse load of $\rho_s = 0.66$, reported in Arbocz and Babcock (1976) for the shell *B*-1, was entirely based on the Donnell-Imbert imperfection model. Considering the results shown in Figure 3.36 further, one sees that the inclusion of some asymmetric imperfection components has reduced the estimate of the mean buckling load considerably, though it is still higher than the experimental mean buckling load for the group of *B*-shells of 0.592 (Arbocz and Abramovich, 1979). However, further refinements in the non-symmetric random imperfection model may lead to lower simulated buckling loads.

In conclusion, one may summarize the results obtained in this section so far as follows:

1. It has been demonstrated that the Monte Carlo method can be used successfully to obtain reliability functions for shells with axisymmetric as well as non-symmetric imperfections.
2. It has been found that for shells of finite length, non-homogeneous probabilistic characteristic must be used (thus ergodicity assumption is not applicable).

3. Using the simulation procedure developed, the measured initial imperfections have been used directly to generate input for the Monte Carlo method.

It is hoped that these preliminary results will encourage many investigators all over the world to compile extensive experimental information on initial imperfections classified according to the manufacturing procedures. The existence of these initial imperfection data banks will make it possible to associate statistical measures with the

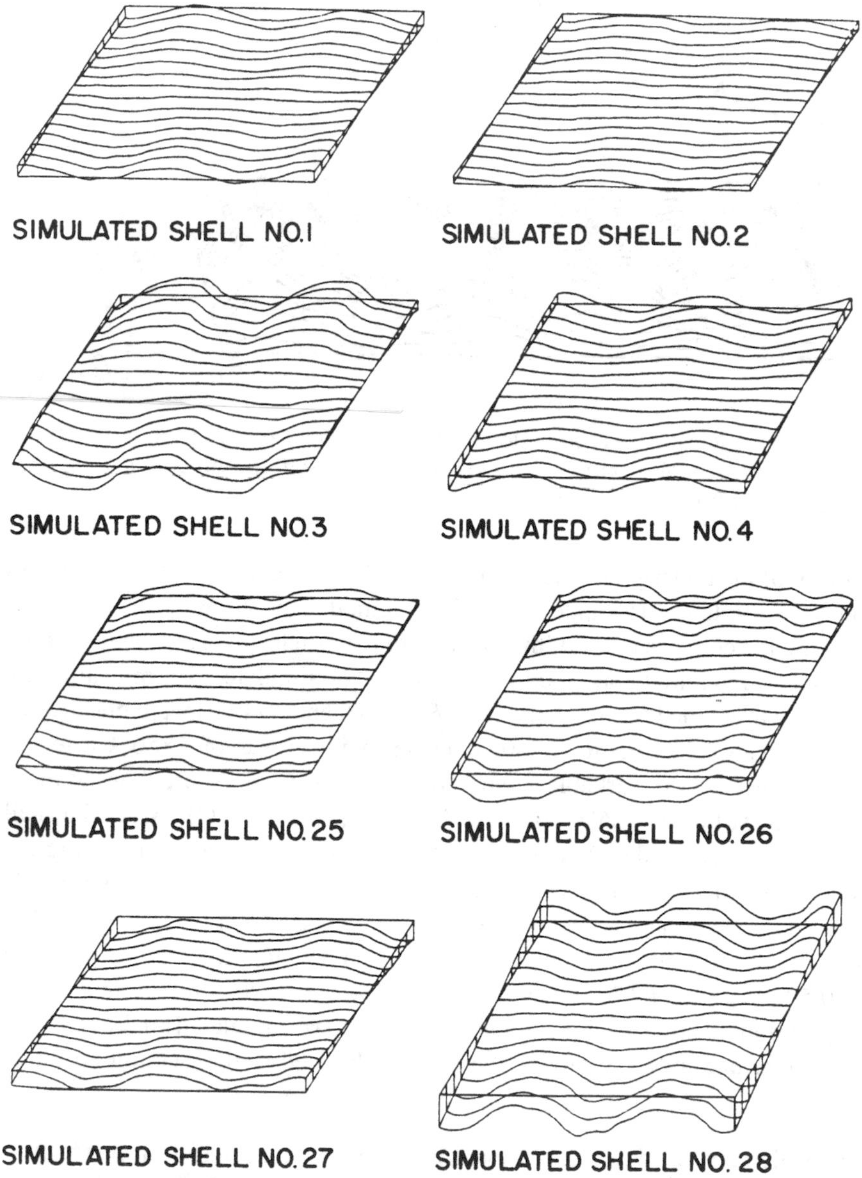

Figure 3.34 Three-dimensional plots of initial imperfection profiles of selected simulated shells (after Elishakoff, 1988; Copyright © Elsevier Science Ltd., reprinted with permission).

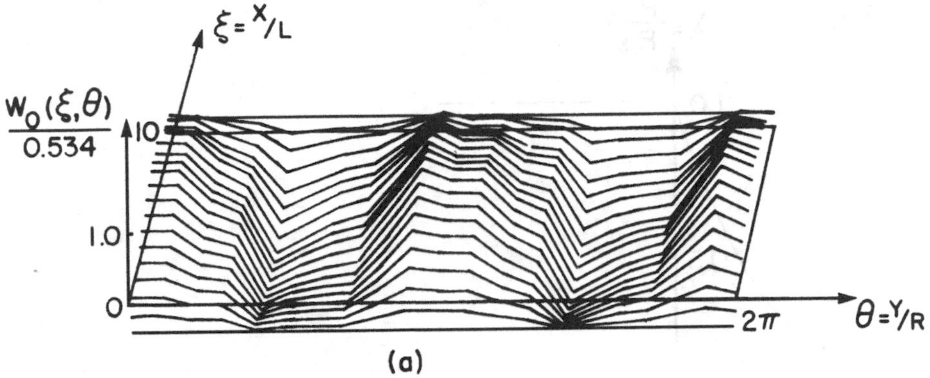

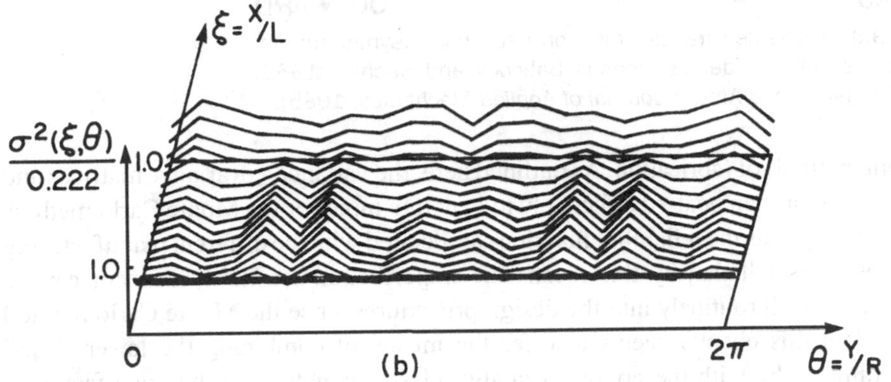

Figure 3.35 Probabilistic characteristics of simulated group of 500 *B*-shells; (a) mean function, (b) variance (after Elishakoff and Arbocz, *Journal of Applied Mechanics*, 1985).

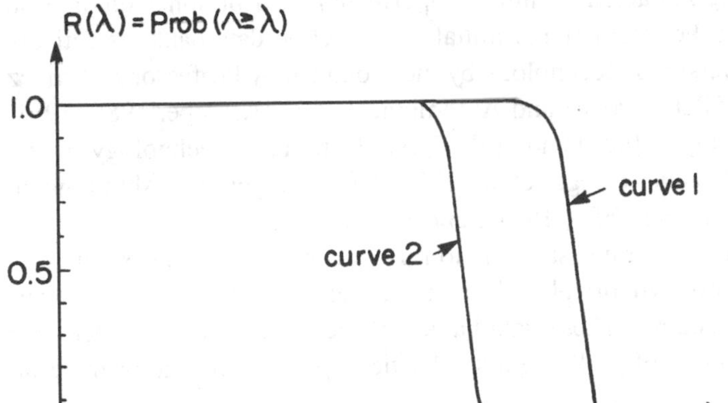

Figure 3.36 Reliability functions for simulated group of 500 *B*-shells (after Elishakoff and Arbocz, *Journal of Applied Mechanics*, 1985).

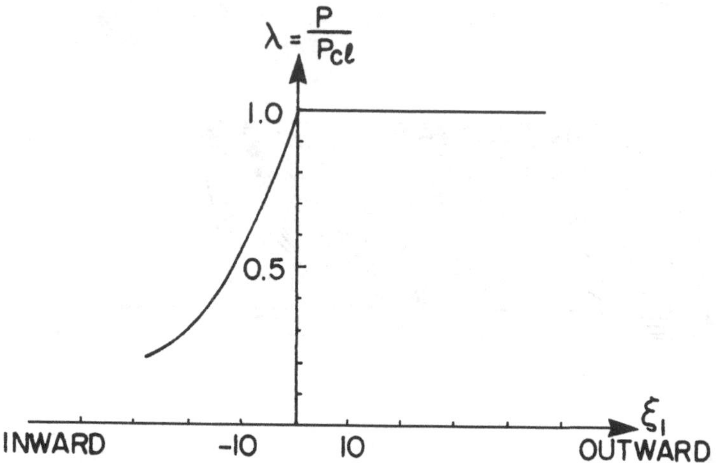

Figure 3.37 One-sided transfer function used for axisymmetric imperfections only (for details, consult Babcock and Sechler, 1962) (after Elishakoff and Arbocz, *Journal of Applied Mechanics*, 1985).

different methods of fabrication. As outlined here, the variance-covariance matrices and the mean vectors can be used effectively to generate input for the Monte Carlo method, which in turn yields the reliability functions associated with the different manufacturing processes. It is felt that by this means the imperfection-sensitivity concept can be finally introduced routinely into the design procedures since the Monte Carlo method described in this chapter seems to offer the means of combining the lower bound design philosophy with the notion of quality class depending on the manufacturing process. Thus, shells manufactured by a process, which produces inherently a less damaging initial imperfection distribution, will not be penalized because of the lower experimental results obtained from shells produced by another process, which may generate a more damaging characteristic initial imperfection distribution. Compilation and extensive analysis of the international initial imperfection data banks is actively pursued at the Delft University of Technology by the group led by Professor J. Arbocz (Arbocz, 1981, 1982a, 1982b; Arbocz and Abramovich, 1979; Klompé, 1986, 1988, 1989; Klompe and Den Reyer, 1989) and at the Israel Institute of Technology by the group led by Professor J. Singer (Singer et al., 1978; Yaffe, Singer, and Abramovich, 1981; Abramovich, Singer, and Yaffe, 1981), and by other investigators. These data banks provide unique material for investigators to further contribute to the explanation of the perplexing behavior of cylindrical shells and to develop novel design methods of imperfection-sensitive structures. These data banks can be directly incorporated also for the finite element analysis of shells with stochastic imperfections and/or material properties.

CHAPTER FOUR

Stochastic Buckling of Structures: Analytical and Numerical Non–Monte Carlo Techniques

Analytical procedures should be consistent with the complexity and importance of the structure. Judgement and experience may be guides as reliable as analytical procedures based on simplifying assumptions.

E. E. Sechler

It is remarkable that a science which began with the consideration of games of chance should have become the most important object of human knowledge.

P. Laplace

Probabilistic investigations are essential for any real progress in the design of imperfection-sensitive structures.

G. Augusti and A. Baratta

In this chapter, we continue our discussion on stochastic buckling of structures. However, instead of resorting to the Monte Carlo method – the only universal technique to deal with highly non-linear stochastic boundary value problems – we resort to approximate analytical techniques. These techniques, although lacking versatility, are still useful (a) to get additional insights into the problem, (b) to collaborate results obtained via the Monte Carlo method, and (c) to devise simplified engineering methods for analysis and design of engineering structures.

4.1 Asymptotic Analysis of Reliability of Structures in Buckling Context

The stochasticity of initial imperfections of structures has been given considerable attention. As first postulated by Bolotin (1958), the buckling load Λ_c of a structure can be expressed as a function of a number of stochastic parameters $X_i (i = 1, \ldots, p)$ representing the initial imperfections:

$$\Lambda_c = \phi(X_1, \ldots, X_p) \tag{4.1}$$

Since X_i are random variables, the critical buckling load turns out to be a random variable. Hence, uppercase notations are used for both X_i and Λ_c. The evaluation of the probability density of Λ_c through the use of Equation (4.1) entails two major difficulties: (a) the probability densities of all initial imperfections $X_i (i = 1, \ldots, p)$ are difficult to obtain; (b) the function ϕ usually is a complicated non-linear function, obtainable only in the form of the sophisticated numerical code.

In order to tackle these difficulties, extensive research has been conducted (see, for example, the bibliography of Roorda, 1972b, and Amazigo, 1976). The *initial imperfection data bank* has been developed by measuring the initial imperfections of shells (Arbocz and Abramovich, 1979; and Abramovich et al., 1981). Elishakoff and Arbocz (1982) investigated the effect of random axisymmetric imperfections on the buckling of circular cylindrical shells under axial compression. The stochastic properties of the shells were evaluated based on the measured data, and the reliability function of the shells was computed by means of the Monte Carlo method. Later on, on the basis of an assumption that the initial imperfections are represented by normally distributed random variables, the first-order second-moment method was employed, in addition to the Monte Carlo method, to greatly reduce computational costs (see, for example, Elishakoff et al., 1987; Arbocz and Hol, 1991). To sum up, the assumption of Gaussianity of imperfections has been successfully used to resolve the first difficulty. In the case of imperfections that do not follow normal distribution, the use should be made of methods pertinent to non-Gaussian random variables and functions (Grigoriu, 1995).

A series of *asymptotic* theories on initial imperfections has been developed to tackle the second difficulty. Roorda and Hansen (1972) extended Koiter's theory (1945) to a single mode normally distributed initial imperfection. The *critical initial imperfection* vector, which achieves the steepest decline of critical load under the constraint of a constant norm, was explicitly obtained (Ikeda and Murota, 1990a, 1990b; Murota and Ikeda, 1991). Based on results of the latter work, with an initial imperfection vector being assumed to be randomly distributed under the constraint of a constant norm, explicit forms of the probability density function of critical loads were obtained (Ikeda and Murota, 1991b; Murota and Ikeda, 1992). As the logical sequel to this, a theoretical method of deriving reliability of structures with normally distributed initial imperfections was presented by Ikeda and Murota (1993).

This method, which combines the simplicity of both the normal distribution and the asymptotic theory, appears to be a promising way to overcome the aforementioned two difficulties, associated with Bolotin's postulate in the form of Equation (4.1). The method by Ikeda and Murota (1993) is applied here to the results of the stochastic studies on realistic structural models. including the buckling of a stochastically imperfect column on a non-linear elastic foundation (Elishakoff, 1979a) and buckling of the axially compressed cylindrical shells with random axisymmetric imperfections (Elishakoff and Arbocz, 1982). The initial imperfections on columns were assumed to be normally distributed. The column problem is re-visited here to ensure the validity of the present method, whereas the shell problem is studied to assess applicability of the asymptotic method to realistic structures. This section closely follows the study by Ikeda, Murota, and Elishakoff (1995).

4.1.1 Overview of the Work by Ikeda and Murota

We will start the formulation of the present theory by summarizing the study by Ikeda and Murota (1993). A critical point may appear in a different type depending on the specific structural and loading condition; three types of simple critical points, namely, limit, asymmetric bifurcation, and unstable-symmetric bifurcation points are discussed.

We consider a system of non-linear equilibrium equations

$$\boldsymbol{H}(\lambda, u, v) = 0 \tag{4.2}$$

where λ denotes a loading parameter, u indicates an N-dimensional nodal displacement (or position) vector, and ν is a p-dimensional imperfection pattern vector. We assume $\boldsymbol{H}$ to be sufficiently smooth. For a fixed ν, a set of solutions (λ, u) of the preceding system of equations makes up equilibrium paths. Let $(\lambda_c, u_c) = (\lambda_c(\nu), u_c(\nu))$ denote the first critical point on the main path of engineering interest $[(\cdot)_c$ refers to the critical point], governing the critical load (defined as the buckling load) of the structure with an imperfection pattern vector ν. In particular, $(\lambda_c^0, u_c^0, \nu^0)$ denotes the critical point of the perfect system (a quantity with superscript 0 refers to the perfect system).

We write

$$\nu = \nu^0 + \epsilon \boldsymbol{X}, \qquad \lambda_c = \lambda_c^0 + \tilde{\lambda}_c \tag{4.3}$$

where $\epsilon(>0)$ denotes the magnitude of the initial imperfection, $\boldsymbol{X} = (X_1, \ldots, X_p)^T$ is a vector indicating the pattern of initial imperfections, and $\tilde{\lambda}_c$ means the increment of the critical load.

We are interested in the stochastic behavior of the critical load Λ_c for random imperfections ν. To be more specific, we consider the case where the imperfection pattern vector $\boldsymbol{X}$ represents a multi-variate normally distributed vector $N(0, W^{-1})$ with zero mean and variance-covariance matrix characterized by a positive definite matrix W^{-1}. The analyses are asymptotic in the sense that they are valid only when ϵ is small.

As has been made clear by Koiter (1945), the asymptotic behavior for the increment (increase or decrease) $\tilde{\lambda}_c$ of the critical load λ_c of the imperfect system is known to be expressed as

$$\tilde{\lambda}_c = \lambda_c - \lambda_c^0 \sim C_0 a^\rho \epsilon^\rho \tag{4.4}$$

when ϵ is small. Here C_0 is positive constant, and ρ varies with the type of points as follows:

$$\begin{cases} \rho = 1, & \text{at limit point} \\ \rho = \frac{1}{2}, & \text{at asymmetric bifurcation point} \\ \rho = \frac{2}{3}, & \text{at unstable-symmetric bifurcation point} \end{cases} \tag{4.5}$$

The variable a, which we call the "effective initial imperfection," depends on the imperfection pattern vector X through

$$a = \boldsymbol{c}^T \boldsymbol{X} = \sum_{i=1}^{p} c_i X_i \tag{4.6}$$

with some constant vector $\boldsymbol{c} = (c_1, \ldots, c_p)^T$. The variable a of Equation (4.6), which is a sum of normally distributed variables $c_i d_i (i = 1, \ldots, p)$, is itself normally distributed $N(0, \tilde{\sigma}^2)$ with zero mean and variance $\tilde{\sigma}^2 = \sum_{i=1}^{p} \sum_{j=1}^{p} c_i c_j \operatorname{cov}(X_i, X_j)$. The probability density function $f_a(a)$ of a is given as

$$f_a(a) = \frac{1}{\sqrt{2\pi}\tilde{\sigma}} \exp\left(-\frac{a^2}{2\tilde{\sigma}^2}\right), \qquad -\infty < a < \infty \tag{4.7}$$

With the use of (4.7), the probability density function of the critical load Λ_c is evaluated as follows:

$$f_{\Lambda_c}(\lambda_c) = \begin{cases} f_a(a)\dfrac{da}{d\lambda_c} = g_1(\lambda_c), & -\infty < \lambda_c < \infty, \quad \text{at limit point} \\ 2 f_a(a)\dfrac{da}{d\lambda_c} = 2 g_{1/2}(\lambda_c), & -\infty < \lambda_c < \lambda_c^0, \quad \text{at asymmetric bifurcation point} \\ 2 f_a(a)\dfrac{da}{d\lambda_c} = 2 g_{2/3}(\lambda_c), & -\infty < \lambda_c < \lambda_c^0, \quad \text{at unstable-symmetric bifurcation point} \end{cases} \tag{4.8}$$

where

$$g_\rho(\lambda_c) = \frac{\left|\lambda_c - \lambda_c^0\right|^{1/\rho - 1}}{\sqrt{2\pi}(C_0 \sigma^\rho)^{1/\rho}} \exp\left(\frac{-1}{2}\left|\frac{\lambda_c - \lambda_c^0}{C_0 \sigma^\rho}\right|^{2/\rho}\right) \tag{4.9}$$

and $\sigma = \tilde{\sigma}\epsilon$. The mean $E[\Lambda_c]$ and the variance $Var[\Lambda_c]$ of Λ_c are expressed respectively as

$$E[\Lambda_c] = \lambda_c^0 + E[\zeta] C_0 \sigma^\rho, \quad Var[\Lambda_c] = Var[\zeta](C_0 \sigma^\rho)^2, \tag{4.10}$$

where $\zeta = \tilde{\lambda}_c / (C_0 \sigma^\rho)$ is now a normalized critical load increment. Its mean $E[\zeta]$ and variance $Var[\zeta]$ for various kinds of simple critical points are listed in Table 4.1.

Table 4.1. $E[\zeta]$ and Var$[\zeta]$ for various kinds of simple critical points [$\Gamma(x)$ is the gamma function]

Type of critical points	$E[\zeta]$	Var$[\zeta]$
Limit point	0	1^2
Asymmetric bifurcation point	$\dfrac{-2^{3/4}}{\sqrt{2\pi}}\Gamma\left(\dfrac{3}{4}\right) = -0.822$	$(0.349)^2$
Symmetric bifurcation point	$\dfrac{-2^{5/6}}{\sqrt{2\pi}}\Gamma\left(\dfrac{5}{6}\right) = -0.802$	$(0.432)^2$

Integration of (4.8) leads to the reliability function:

$$R(\lambda) = \begin{cases} 1 - \Phi\left(\dfrac{\lambda - \lambda_c^0}{C_0\sigma}\right), & -\infty < \lambda < \infty, \quad \text{at limit point} \\ 1 - 2\Phi\left(-\left(\dfrac{\lambda - \lambda_c^0}{C_0\sigma^{1/2}}\right)^2\right), & -\infty, \lambda < \lambda_c^0, \quad \text{at asymmetric bifurcation point} \\ 1 - 2\Phi\left(-\left|\dfrac{\lambda - \lambda_c^0}{C_0\sigma^{2/3}}\right|^{3/2}\right), & -\infty, \lambda < \lambda_c^0, \quad \text{at unstable-symmetric bifurcation point} \end{cases} \tag{4.11}$$

Here

$$\Phi(\zeta) = \int_{-\infty}^{\zeta} \frac{1}{\sqrt{2\pi}} \exp\left(-\frac{\zeta^2}{2}\right) d\zeta \tag{4.12}$$

denotes the cumulative probability distribution function of standard normal variable.

It is to be emphasized here that by the present method a mere calculation of the sample mean $E[\Lambda_c]$ and variance $\mathrm{Var}[\Lambda_c]$ of the critical loads will yield the parameters λ_c^0 and $C_0\sigma^\rho$ in (4.9), and, in turn, the probability density function in (4.8) and the reliability function in (4.11). The present method thus is quite simple. As we have seen in (4.8) and (4.11), the forms of probability density function and reliability function vary with the type of critical points. The present method, which is based on a firm theoretical background, can implement such a variation in a systematic way.

4.1.2 Finite Column on a Non-Linear Foundation

In this section we deal again with a model structure of the column on a non-linear foundation (Ikeda, Morota, and Elishakoff, 1995). This model was extensively studied in Chapter 3 in the context of the Monte Carlo method. In this section, we will utilize this model to illustrate the theory of Ikeda and Murota (1993). This section closely follows the recent article by Ikeda et al. (1995). It is instructive to consider the non-dimensional form of the equation

$$\frac{d^4u}{d\eta^4} + \lambda\gamma(\kappa_1)\frac{d^2u}{d\eta^2} + \kappa_1 u - \kappa_3 u^3 = -\lambda\gamma(\kappa_1)\frac{d^2\tilde{u}}{d\eta^2} \tag{4.13}$$

for the deflection of an imperfect column on a non-linear "softening" foundation with simply supported boundary conditions

$$u = \frac{d^2u}{d\eta^2} = 0, \quad \text{at} \quad \eta = 0 \quad \text{and} \quad \eta = 1 \tag{4.14}$$

where

$$u = \frac{w}{\Delta}, \quad \tilde{u} = \frac{\tilde{w}}{\Delta}, \quad \eta = \frac{x}{l}, \quad \lambda = \frac{P}{P_{cl}}$$
$$\gamma(\kappa_1) = n_*^2\pi^2 + \frac{\kappa_1}{(n_*\pi)^2}, \quad \kappa_1 = \frac{k_1 l^4}{EI}, \quad \kappa_3 = \frac{k_3 l^4 \Delta^2}{EI} \tag{4.15}$$

$\tilde{w}(x)$ is the initial imperfection function; $w(x)$ is the additional deflection due to the axial load P; λ is the non-dimensional buckling load normalized with respect to the (classical) buckling load P_{cl} of a column on a linear elastic foundation; κ_1 and κ_3 are, respectively, the linear and non-linear spring coefficients of the foundation; x is the axial coordinate; l is the length of the column; E is the Young's modulus; I is the cross-section moment of inertia; Δ is the cross-section radius of gyration; and n_* denotes the number of the half-waves for the buckling mode of the linear structure.

Here, a procedure to obtain the stochastic properties of the column is introduced because a summary of Elishakoff (1979a) will be recapitulated. The normalized initial imperfection $\tilde{u}(\eta)$ is assumed to be a Gaussian random function of the position η with given mean function $\bar{U}(\eta)$ and auto-correlation function $K_{\tilde{u}}(\eta_1, \eta_2)$, η_1, and η_2 being two (generally distinct) points on the column axis. To employ the Boobnov-Galerkin-type solution procedure, we resolve the displacement and initial imperfections, respectively, into finite Fourier series

$$u(\eta) = \sum_{\kappa=1}^{N} \xi_\kappa \sin(\kappa\pi\eta), \qquad \tilde{u}(\eta) = \sum_{\kappa=1}^{N} \tilde{\xi}_\kappa \sin(\kappa\pi\eta) \tag{4.16}$$

compatible with the boundary condition (4.14), where N denotes the number of terms to be implemented in the formulation. We take $(\tilde{\xi}_1, \ldots, \tilde{\xi}_N)$ as the vector of imperfections, [i.e., $\nu = (\tilde{\xi}_1, \ldots, \tilde{\xi}_N)^T$ (with $p = N$) in the notation of Section 4.1.1].

The mean values $X_k = E[\tilde{\xi}_k](k = 1, \ldots, N)$ are obtainable as

$$X_k = 2\int_0^1 \bar{U}(\eta)\sin(k\pi\eta)\,d\eta \tag{4.17}$$

The auto-correlation function of the imperfections takes the form

$$\begin{aligned} K_{\tilde{u}}(\eta_1, \eta_2) &= E\{[u(\eta_1) - \bar{U}(\eta_1)][u(\eta_2) - \bar{U}(\eta_2)]\} \\ &= \sum_{k=1}^{N}\sum_{r=1}^{N} E[(\tilde{\xi}_k - \bar{X}_k)(\tilde{\xi}_r - \bar{X}_r)]\sin(k\pi\eta_1)\sin(r\pi\eta_2) \end{aligned} \tag{4.18}$$

Hence the elements $(W^{-1})_{kr}$ of the variance-covariance matrix W^{-1} of $(\tilde{\xi}_1, \ldots, \tilde{\xi}_N)$ are determined by

$$(W^{-1})_{kr} = 4\int_0^1\int_0^1 K_{\tilde{u}}(\eta_1, \eta_2)\sin(k\pi\eta_1)\sin(r\pi\eta 2)\,d\eta_1 d\eta_2 \tag{4.19}$$

The probabilistic characteristics of imperfections $\tilde{u}(\eta)$ are specified such that

$$\nu^0 = \bar{\boldsymbol{X}} = (X_1, \ldots, X_N)^T = 0 \tag{4.20}$$

$$K_{\tilde{u}}(\eta_1, \eta_2) = A(\eta_1 - \eta_2)^{-1}\sin B(\eta_1 - \eta_2) \tag{4.21}$$

with positive constants A and B (due to Boyce, 1961). The substitution of (4.21) into (4.19) yields the variance-covariance matrix W^{-1}.

In Elishakoff (1979a), the differential equations (4.13) of the column was discretized with the use of (4.16) and was numerically solved to arrive at the non-dimensional buckling load λ_c, which was governed by the simple unstable-symmetric

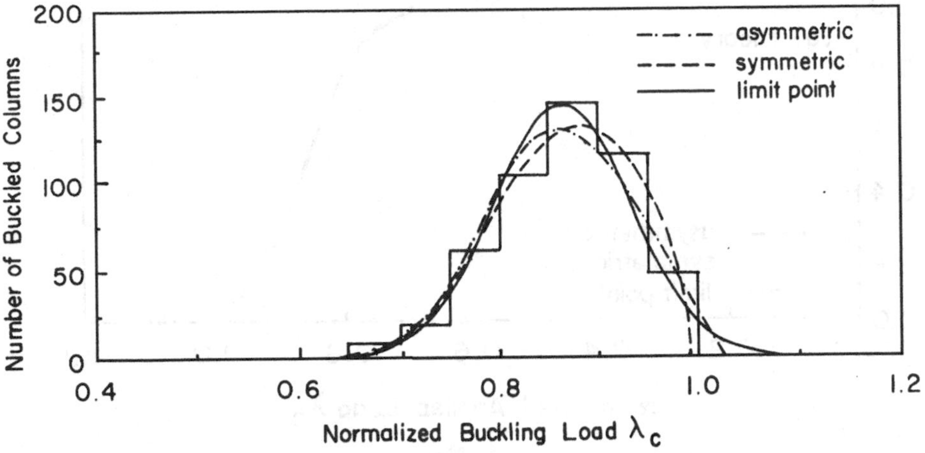

Figure 4.1 Comparison of theoretical probability density function f_{λ_c} (λ_c) numerical histogram of the critical load λ_c for the finite column on a non-linear elastic foundation. Histogram after Elishakoff, *Journal of Applied Mechanics*, 1979a ($\kappa_1 = \kappa_3 = (4\pi)^4$, $A = 0.01$, $B = 1$, $m^* = 4$, $N = 7$) (after Ikeda, Murota, and Elishakoff, 1995; Copyright © Elsevier Science Ltd., reprinted with permission).

bifurcation point. Figure 4.1 shows the histogram of λ_c produced by the Monte Carlo method for an ensemble of 1000 columns with the Gaussian imperfections with the probabilistic characteristics of (4.17) and (4.18). The non-dimensional buckling load λ_c^0 for the perfect system satisfies

$$\lambda_c^0 = 1 \tag{4.22}$$

4.1.3 Application of Ikeda-Murota Theory

The sample mean $E[\Lambda_c]$ and variance $Var[\lambda_c]$ of buckling load Λ_c are calculated respectively, as $E[\Lambda_c] = 0.86$ and $Var[\lambda_c] = 0.069$ based on the histogram of Λ_c in Figure 4.1. The substitution of these values into (4.10) results in the value of λ_c^0 and $C_0\sigma^\rho$ listed in Table 4.2 for various kinds of critical points. For the unstable simple-symmetric bifurcation point, which governs the critical load in this case, the value of $\lambda_c^0 = 1.03$ computed in this manner is close to its theoretical value, which is equal to the unity by (4.22).

Table 4.2. Parameters for the finite column computed for various kinds of simple critical points

Type of critical points	λ_c^0	$C_0\sigma^\rho$	Difference
Limit point	0.86	0.069	16.9
Asymmetric bifurcation point	0.99	0.16	22.8
Symmetric bifurcation point	1.03	0.20	6.5

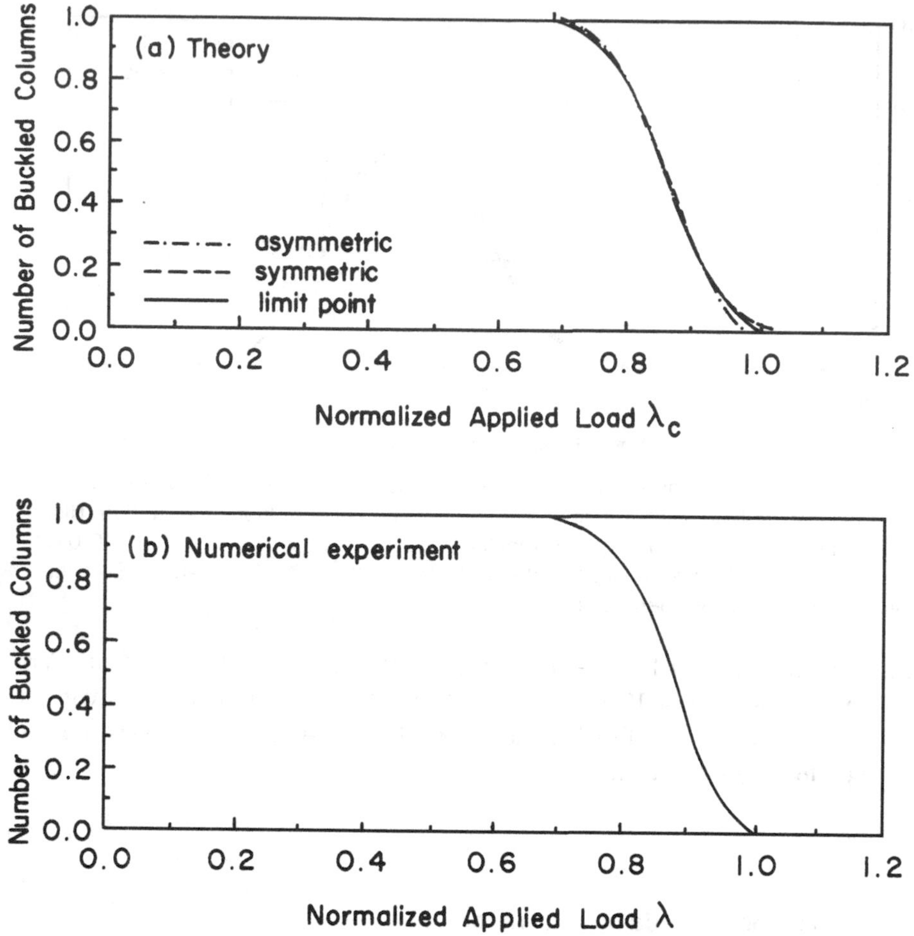

Figure 4.2 Comparison of theoretical and numerical reliability functions $R(\lambda)$ for the finite column on a non-linear elastic foundation: (a) theory (unstable-symmetric bifurcation point) and (b) Monte Carlo method (Elishakoff, *Journal of Applied Mechanics*, 1979a) ($\kappa_1 = \kappa_3 = (4\pi)^4$, $A = 0.01$, $B = 1$, $m^* = 4$, $N = 7$) (after Ikeda, Murota, and Elishakoff, 1995; Copyright © Elsevier Science Ltd., reprinted with permission).

With the use of the values of λ_c^0 and $C_0\sigma^\rho$ in Table 4.2, we computed the curves of the probability density function for various kinds of critical points in Figure 4.1, and those of the reliability function in Figure 4.2(a); Figure 4.2(b) shows the empirical reliability curve computed by Elishakoff (1979a). The curves for the unstable simple-symmetric bifurcation point serve as theoretical ones, whereas those for the other points are used here for comparison. According to χ^2 test of goodness of fit (Kendall and Stuart, 1973), the critical value of the difference between the observed and the theoretical probability density function at a level of significance of 0.05 is 11.1 or less. (Strictly speaking, the true critical value should be slightly smaller than 11.1 because our procedure estimates the location and scale parameters of the distribution. But 11.1 is employed here to approximate it.) The value of difference for the unstable simple-symmetric bifurcation point is equal to 6.5 and, hence, is significantly smaller than the critical value 11.1. The reliability curve by the present method in Figure 4.2(a) is very close to the empirical curve in Figure 4.2(b). These may suffice to show the validity of the present method, which achieves the accuracy, while retaining the desired simplicity.

4.1.4 Axially Compressed Cylindrical Shells

We consider two groups of cylindrical shells, called A-shells and B-shells after Ikeda et al. (1995). A group of A-shells of $L = 176.02$ mm, $R = 101.6$ mm, and $t = 0.1160$ mm were manufactured by electroplating from pure copper (where L is the shell length, R is its radius, and t is its wall thickness) (Arbocz and Abramovich, 1979). A group of B-shells of $L = 134.37$ mm, $R = 101.6$ mm, and $t = 0.2007$ mm were manufactured by putting pieces of thick-walled, seamless, brass tubes onto a mandrel and then machining them to the desired wall thickness (Babcock and Sechler, 1962). Because of the very nature of the manufacturing process, each realization of the shell will have a different initial shape. The imperfections represent deviations of the initial shape from the perfect circular cylinder. They were recorded by the special experimental setup developed at Caltech (Arbocz and Babcock, 1968) and later sent to the data bank of a complete surface map of the shells (Singer et al., 1978; Arbocz and Abramovich, 1979).

This subsection offers a procedure to arrive at the reliability of the shells (Elishakoff and Arbocz, 1982). We consider a cylindrical shell with an axisymmetric initial imperfection of the form

$$v(x) = t\xi_i \cos\frac{i\pi x}{L} \tag{4.23}$$

where i is an integer denoting the number of half-waves in the (axial) x-direction. We further assume the following buckling mode:

$$w(x, y) = tC_k l \sin\frac{k\pi x}{L}\cos\frac{ly}{R} \tag{4.24}$$

where k and l are integers denoting the number of half-waves in the axial and the circumferential directions, respectively. With the use of Koiter's special theory (Koiter, 1963), one can derive the following relationship between the non-dimensional axial load λ (at which the resulting fundamental equilibrium state bifurcates into an asymmetric pattern) and the imperfection magnitude ξ_i

$$(\lambda_i - \lambda)^2(\lambda_{kl} - \lambda) + (\lambda_i - \lambda)\frac{c\beta_l^2}{2\alpha_k^2}\left[\lambda + \lambda_i\frac{8\alpha_k^4}{(\alpha_k^2 + \beta_l^2)}\right]\xi_i\delta_{i,2k}$$
$$+ 8c^2\alpha_k^2\beta_l^4\left[\frac{1}{(9\alpha_k^2 + \beta_l^2)^2} + \frac{1}{(\alpha_k^2 + \beta_l^2)^2}\right]\lambda_i^2\xi_i^2 = 0 \tag{4.25}$$

where

$$\lambda_i = \frac{1}{2}\left(\alpha_i^2 + \frac{1}{\alpha_i^2}\right), \qquad \lambda_{kl} = \frac{1}{2}\left[\frac{(\alpha_k^2 + \beta_l^2)^2}{\alpha_k^2} + \frac{\alpha_k^2}{(\alpha_k^2 + \beta_l^2)^2}\right],$$
$$\alpha_i^2 = i^2\frac{Rt}{2c}\left(\frac{\pi}{L}\right)^2, \qquad \alpha_k^2 = k^2\frac{Rt}{2c}\left(\frac{\pi}{L}\right)^2, \qquad \beta_l^2 = l^2\frac{Rt}{2c}\left(\frac{1}{R}\right)^2 \tag{4.26}$$

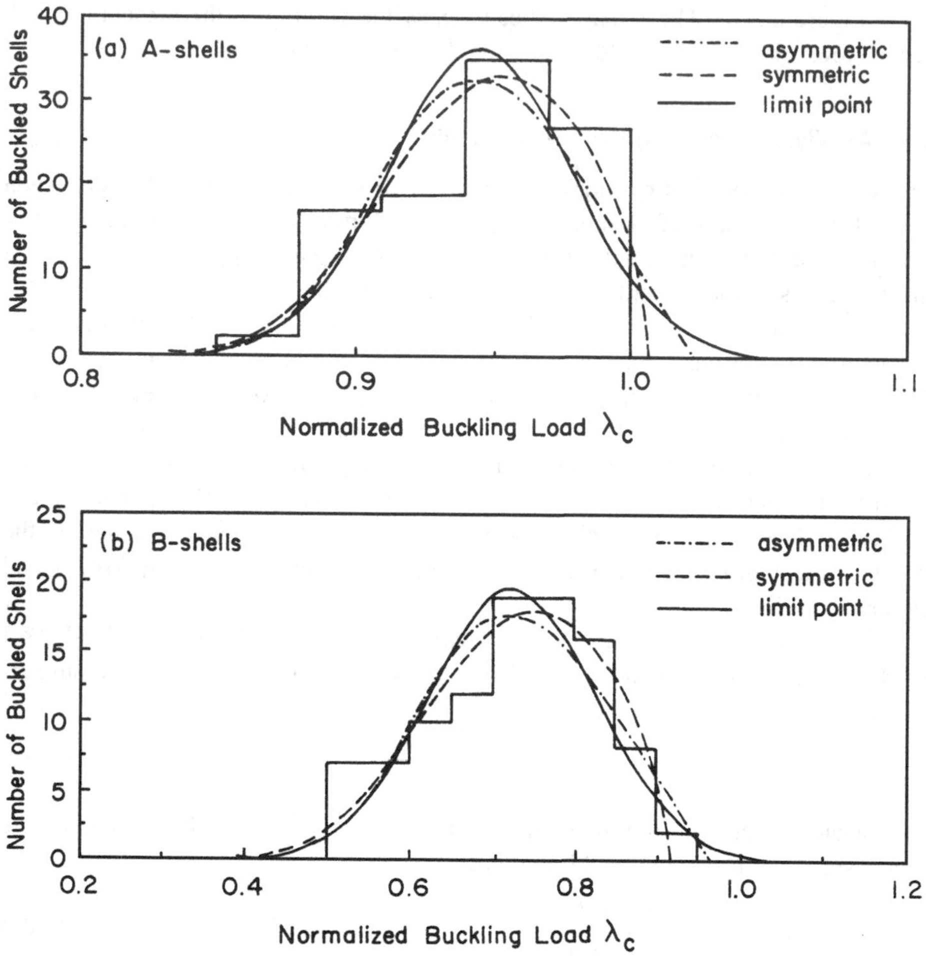

Figure 4.3 Comparison of theoretical probability density function f_{λ_c} (λ_c) and numerical histogram (Elishakoff and Arbocz, *International Journal of Solids and Structures*, 1982): (a) *A*-shells and (b) *B*-shells (after Ikeda, Murota, and Elishakoff, 1995; Copyright © Elsevier Science Ltd., reprinted with permission).

c is a constant, and $\delta_{i,2k}$ is the Kronecker delta, having a value of unity or zero depending on whether $i = 2k$.

In Equation (4.25) for an imperfection-sensitive structure, $\tilde{\xi}_i$ must be negative, and i must be an even integer. It also implies that the critical load for this shell is governed by the simple asymmetric bifurcation point. The buckling loads were obtained by solving (4.25) for different values of k and $l(i = 2k)$ until all the available imperfection harmonics have been tested. The absolute minimum bifurcation buckling load is then identified as the critical buckling load for the shell under consideration.

The critical buckling loads for the groups of A- and B-shells, respectively, were computed for an ensemble of 100 shells, and their histograms and the reliability curves were computed as shown in Figures 4.3 and 4.4.

Table 4.3 lists the value of λ_c^0 and $C_0\sigma^\rho$ for the A- and B-shells computed for various kinds of critical points. The values of $\lambda_c^0 = 1.02$ and 0.97 for the asymmetric

Table 4.3a. The parameters for *A*-shells computed for various kinds of simple critical points

Type of critical points	λ_c^0	$C_0\sigma^\rho$	Difference[a]
Limit point	0.95	0.033	7.3
Asymmetric bifurcation point	1.02	0.095	5.5
Symmetric bifurcation point	1.01	0.077	5.0

[a] Critical value at a significance level of 0.05 is 6.0 or less.

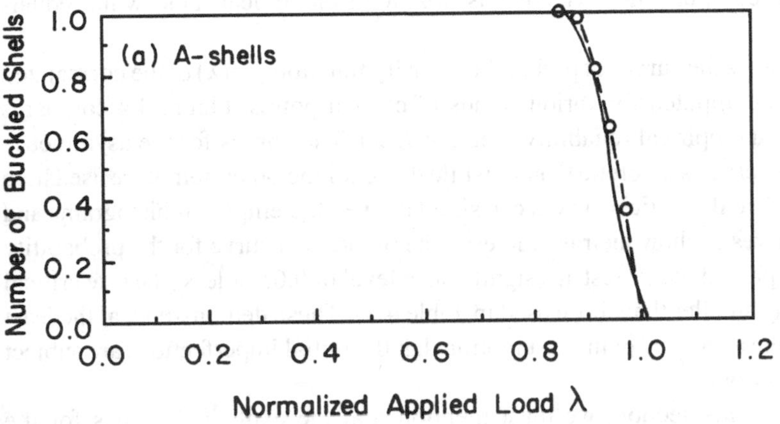

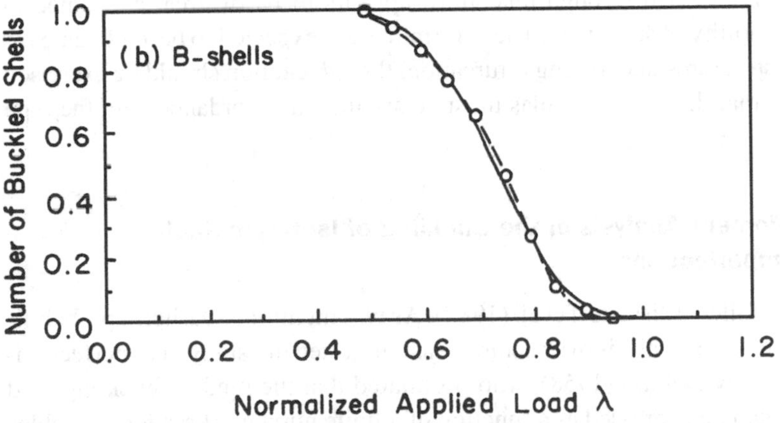

Figure 4.4 Comparison of theoretical and numerical reliability functions $R(\lambda)$: (a) *A*-shells and (b) *B*-shells. Solid line: theory (asymmetric bifurcation point); dashed line: Monte Carlo method – Elishakoff and Arbocz, 1985 (after Ikeda, Murota, and Elishakoff, 1995; Copyright © Elsevier Science Ltd., reprinted with permission).

Table 4.3b. The parameters for *B*-shells computed for various kinds of simple critical points

Type of critical points	λ_c^0	$C_0\sigma^p$	Difference[a]
Limit point	0.73	0.102	5.7
Asymmetric bifurcation point	0.97	0.29	5.5
Symmetric bifurcation point	0.91	0.24	1.6

[a] Critical value at a significance level of 0.05 is 11.1 or less.

bifurcation point, computed in this manner is close to its theoretical value, which equals the unity.

Figure 4.3 shows the curves of probability density function $f_{\Lambda_c}(\lambda)$ of the normalized buckling load λ_c computed for various kinds of critical points. Figure 4.4 compares the theoretical and empirical reliability function $R(\lambda)$. The curves for the asymmetric bifurcation point serve as theoretical ones, while those for the other points are used here for comparison. The theoretical curves can simulate well the empirical histograms and the reliability curves to show their usefulness. The theoretical curve for the probability density function passed the χ^2 test at a significance level of 0.05 or less (the true critical values are slightly smaller than those used in Table 4.3). These demonstrate, at the least for these shells, the adequacy of the assumption that the initial imperfections are subject to normal distribution.

In concluding this section, we must mention that the explicit formulas for the probability density function (4.7) and (4.8) of the critical buckling loads, the validity of which has been assessed through this study, appears to be of great assistance in evaluating the reliability of structures. These formulas are expected to be applicable for various kinds of structures undergoing bifurcation, though caution should be exercised based on the fact that different formulas must be applied in accordance with the type of critical points.

4.2 Second-Moment Analysis of the Buckling of Isotropic Shells with Random Imperfections

In this section, we follow Elishakoff et al. (1987). As was mentioned earlier, a probabilistic approach to buckling analysis of structures was suggested in a study of imperfection-sensitive structures by Bolotin (1958), who postulated that the random buckling load Λ_c of a structure can be expressed as a function of a finite number of random variables X_i representing the initial imperfection Fourier coefficients, Equation (4.1).

Bolotin applied this method to a cylindrical panel under a uniform compressive load along its curved edges, with the initial imperfections being represented by a single normally distributed amplitude parameter. A single-term Boobnov-Galerkin approximation yielded Equation (4.1). Makarov (1969) used a simplified expression of Equation (4.1) to derive the statistical properties of the buckling loads. However, as Amazigo (1976) mentioned, "it is . . . a non-trivial problem to obtain a relation of type Eq. (4.1) and to perform the above analysis for $p > 2$, say. It is this difficulty that limits the effectiveness of this method." Indeed, because there is no formula of Equation (4.1) in the literature for multi-parametric imperfections, the direct, analytical evaluation of the buckling loads seems to be a formidable task.

On the other hand, procedures (see Babcock, 1983; Almroth and Brogan, 1978; Arbocz and Babcock, 1974, 1978) have been developed, dealing with the determination of the buckling loads numerically. We limit ourselves to mentioning the multi-mode analysis (MIUTAM) by Arbocz and Babcock (1974), the well-known general-purpose code STAGS (Almorth and Brogan, 1978; Arbocz and Babcock, 1978), and the Dutch multi-purpose finite-element package DIANA (Borst et al., 1983). In addition, the recent results of extensive initial imperfection surveys have been directly incorporated into the probabilistic analysis of shells with random imperfections (Elishakoff and Arbocz, 1982, 1985) without resorting to the number of restrictive assumptions used in the literature on the probabilistic buckling of imperfect structures.

The question that arises in this context follows: Is it possible to develop a simple but rational method of checking the reliability of the shells, using some statistical measures of the imperfections involved, to provide an estimate of the structural reliability without recourse to the Monte Carlo method? Of course, even with the relation of type of Equation (4.1) available, it would still be an enormous task to use it for the reliability calculations in view of the cumbersome integration in a multi-dimensional space. Alternatively, the order zero second-moment approach (Rzhanitsin, 1949; Cornell, 1969) has been known for many years to those engaged in the probabilistic analysis of structures. This method, requiring only the knowledge of mean values and elements of a variance-covariance matrix of the basic variables (imperfection Fourier coefficients), will be adopted in this section.

The cornerstone of the method is the deterministic state equation

$$Z = Z(X_1, X_2, \ldots, X_p) \tag{4.27}$$

where the nature of the so-called performance function $Z(\ldots)$ depends on the type of the structure and the limit state considered. According to the definition, the equation

$$Z = 0 \tag{4.28}$$

determines the failure boundary; $Z < 0$ implies failure, and $Z > 0$ indicates non-failure (successful performance). The use of the second-moment method then requires linearization of the function Z at the mean point and knowledge of the distribution of the random vector $\boldsymbol{X}$. Calculations are relatively simple if $\boldsymbol{X}$ is normally distributed. If $\boldsymbol{X}$ is not normally distributed, an appropriate normal distribution must be substituted instead of the actual one.

Because of the randomness of initial imperfection parameters, the buckling load in Equation (4.1) is a random variable, denoted by Λ_c. In the present case, we are interested in knowing the reliability of the structure at any given load λ, that is

$$R(\lambda) = \text{Prob}(\Lambda_c > \lambda) \tag{4.29}$$

A function Z can be defined as

$$Z(\lambda) = \Lambda_c - \lambda = \psi(X_1, X_2, \ldots, X_p) - \lambda \tag{4.30}$$

where λ is the applied deterministic load. At first glance, it appears that, due to the absence of straightforward deterministic relation connecting Λ_c and the X_i, the first-order second-moment analysis is unfeasible.

Indeed, it is impossible to perform such an analysis analytically. However, it can be done numerically, as has been performed for a different problem, as reported by Kazadeniz, van Manen, and Vrouwenvelder (1982). To combine the numerical codes with the mean value first-order second-moment method, we need to know the lower order probabilistic characteristics of Z. In the first approximation, the mean value will be determined as follows:

$$\begin{aligned} E(Z) &= E(\Lambda_p) - \lambda \\ &= E[\psi(X_1), E(X_2), \ldots, E(X_p)] - \lambda \\ &\cong \psi[E(X_1), E(X_2), \ldots, E(X_p)] - \lambda \end{aligned} \tag{4.31}$$

This corresponds to the use of the Laplace approximation of the moments of the non-linear functions. The value of

$$\psi[E(X_1), E(X_2), \ldots, E(X_p)] \tag{4.32}$$

should be calculated numerically by any of the codes reported by Arbocz and Babcock (1974, 1978) and Borst et al. (1983). It corresponds to the deterministic buckling load of the structure possessing mean imperfection amplitudes.

The variance of Z is given by

$$Var(Z) = Var(\Lambda_s) \cong \sum_{j=1}^{p}\sum_{k=1}^{p} \left(\frac{\partial\psi}{\partial\xi_j}\right)_{\xi_j=E(X_j)} \left(\frac{\partial\psi}{\partial\xi_k}\right)_{\xi_k=E(X_k)} v_{jk} \tag{4.33}$$

where $v_{jk} = \text{cov}\,(X_j, X_k)$ is a covariance.

Calculation of the derivatives of $\partial\psi/\partial\xi_j$ are performed numerically by using the following numerical differentiation formula:

$$\frac{\partial\psi}{\partial\xi_j} \cong \frac{\psi(\xi_1, \xi_2, \ldots, \xi_{j-1}, \xi_j + \Delta\xi_j, \xi_{j+1}, \ldots, \xi_p) - \psi(\xi_1, \xi_2, \ldots, \xi)}{\Delta\xi_j} \tag{4.34}$$

at values of $\xi_j = E(\bar{X}_j)$. To find these derivatives it is necessary to carry out p calculations of the buckling problem.

Having estimated $E(Z)$ and $Var(Z)$ we obtain the estimate for the probability of failure:

$$P_f(\lambda) = \text{Prob}(Z < 0) = \Phi(-\beta) \tag{4.35}$$

at the load level λ. In Equation (4.35), Φ is the standard normal probability distribution function and

$$\beta = E(Z)/\sqrt{Var(Z)}$$

is the reliability index. Accordingly, reliability will be estimated as

$$P(\lambda) = \text{Prob}(Z < 0) = 1 - \Phi(-\beta) = \Phi(\beta) \tag{4.36}$$

Table 4.4. Geometrical and material data on the group of *B*-shells (Arbocz and Babcock, 1974)

Shell[a]	R, mm	t, mm	L, mm	L_{HA}, mm[b]	P_{exp}, mm
B-1	101.60	0.2050	196.85	171.45	11326.0
B-2	101.60	0.1852	144.78	121.92	7178.5[c]
B-3	101.60	0.1491	140.97	121.92	—
B-4	101.60	0.2634	140.97	121.92	16661.2

[a] For all shells $E = 1.065 \times 10^5$ N/mm² and $\nu = 0.3$.
[b] L_{HA} = length used for harmonic analysis.
[c] Shell had a visible stable prebuckle at a spot where the surface had been scratched by a sharp object.

For the actual calculations, we used the data associated with the so-called B-shells (Arbocz and Babcock, 1974). The geometrical and material properties of the B-shells are provided in Table 4.4.

Since the measurements included only 41 points in the axial direction and 49 points in the circumferential direction, it was not feasible to compute those Fourier coefficients the order of which exceeds the "cut-off" values of $k = 20$ and $l = 24$. For these coefficients, the Donnell-Imbert imperfection model (Imbert, 1971) was used:

$$X_{r,s} = X/k^r l^s \tag{4.37}$$

where k is the number of axial half-waves and l is the number of circumferential full waves. Values of measured imperfections are given in Table 4.5. Also shown is the complete imperfection model used with an indication of which coefficients were actually

Table 4.5. Imperfection model and values of the Fourier coefficients (Elishakoff and Arbocz, 1985)

Simulated random variables					Donnell-Imbert imperfection model		
$C_{1,6}$			$C_{1,10}$	$C_{25,6}$			$C_{25,10}$
		$C_{1,2}$			$C_{25,2}$		
$A_{2,0}$	+	$C_{1,8}$	+		$C_{25,8}$	+	$A_{26,0}$
		$C_{2,3}$			$C_{24,3}$		
$A_{4,0}$	+	$C_{2,11}$	+		$C_{24,11}$	+	$(A_{26,0})$

	B-1	B-2	B-3	B-4
$A_{2,0}$	−0.010809	−0.027238	−0.089906	−0.017560
$A_{4,0}$	0.022578	−0.007836	−0.089906	−0.009239
$C_{1,2}$	0.417400	0.392870	0.741280	0.222900
$C_{1,6}$	−0.077872	−0.143490	0.017483	0.077668
$C_{1,8}$	−0.263690	−0.009405	0.112470	0.101510
$C_{1,10}$	0.036568	0.043628	−0.245610	−0.008853
$C_{2,3}$	−0.101290	0.034018	−0.064766	−0.001887
$C_{2,11}$	0.009732	−0.008685	−0.028261	0.013545

Table 4.6. Sample mean vector and sample variance-covariance matrix (Elishakoff et al., 1987)

Mean vector		Elements of variance-covariance matrix (All elements of the variance-covariance matrix are multiplied by 100)							
−0.0364	1	0.1319							
−0.0050	2	0.0566	0.0402						
0.4436	3	−0.7074	−0.1916	4.686					
−0.0316	4	−0.0926	−0.0810	−0.0872	0.9670				
−0.0148	5	−0.3646	−0.3327	0.6154	0.9956	3.057			
−0.0436	6	0.4771	0.1986	−2.478	−0.6529	−1.372	1.868		
−0.0335	7	0.0384	−0.0518	−0.5978	−0.0833	0.5647	0.2623	0.3710	
−0.0034	8	0.0646	0.0272	−0.3739	0.0205	−0.1497	0.2068	0.0022	0.0369

measured and which ones were extrapolated in accordance with Equation (4.37). The parameters X, r, and s were determined by the least-square fitting, the distribution of the measured Fourier coefficients.

The mean vector and the variance-covariance matrix of the Fourier coefficients, treated as random variables, are given in Table 4.6. In order to apply the first-order second-moment method, the mean buckling load has to be calculated first. For this purpose, the multi-mode analysis is used.

Let us give a brief overview of the multi-mode analysis. We can write the Donnell-type non-linear equations for imperfect stiffened cylindrical shells in the form:

$$L_H(F) - L_Q(W) = -\frac{1}{R}w_{,xx} - \frac{1}{2}L_{NL}(W, W + 2\bar{W}) \tag{4.38}$$

$$L_Q(F) + L_D(W) = \frac{1}{R}F_{,xx} + L_{NL}(F, W + \bar{W}) \tag{4.39}$$

where W is positive inward, and the linear operators are

$$\begin{aligned} L_D(\,) &= D_{xx}(\,)_{,xxxx} + D_{xy}(\,)_{,xxyy} + D_{yy}(\,)_{,yyyy} \\ L_H(\,) &= H_{xx}(\,)_{,xxxx} + H_{xy}(\,)_{,xxyy} + H_{yy}(\,)_{,yyyy} \\ L_Q(\,) &= Q_{xx}(\,)_{,xxxx} + Q_{xy}(\,)_{,xxyy} + Q_{yy}(\,)_{,yyyy} \end{aligned} \tag{4.40}$$

and the non-linear operator is

$$L_{NL}(S, T) = S_{,xx}T_{,yy} - 2S_{,xy}T_{,xy} + S_{,yy}T_{,xx} \tag{4.41}$$

Commas in the subscripts denote repeated partial differentiation with respect to the independent variables following the comma. The stiffener properties have been "smeared out" to arrive at effective bending, stretching, and torsional stiffness. The stiffener parameters D_{xx}, H_{xx}, Q_{xx}, $D_{xy}, \ldots$, etc., are defined as in Babcock (1983). Here $\bar{W}$ is the initial radial imperfection, W is the component of displacement normal to the shell mid-surface and F is the Airy stress function.

Let us represent the $(m+1)$ approximation to a solution to Equations (4.38) and (4.39) by

$$W_{m+1} = W_m + \delta W_m, \qquad F_{m+1} = F_m + \delta F_m \tag{4.42}$$

where W_m, F_m = mth approximation to the solution, and δW_m, δF_m = correction to the mth approximation. Substitution of Equation (4.42) into Equations (4.38) and (4.39) and omission of products of the correction quantities yields a set of linear partial differential equations for determining the correction terms. If one represents the initial imperfections by

$$\bar{W} = t\sum_{i=1}^{N_1} \bar{W}_{io}\cos(i\bar{x}) + t\sum\sum_{k,l=1}^{N_2} \bar{W}_{kt}\sin(k\bar{x})\cos(l\bar{y}) \tag{4.43}$$

where $\bar{x} = \pi x/L$ and $\bar{y} = y/R$, then the linearized governing equations admit separable solutions of the form

$$\begin{Bmatrix} W_m \\ \delta W_m \end{Bmatrix} = t\begin{Bmatrix} W_v \\ 0 \end{Bmatrix} + \sum_{i=1}^{N_1}\begin{Bmatrix} W_{io} \\ \delta W_{io} \end{Bmatrix}\cos(i\bar{x}) + t\sum\sum_{k,t=1}^{N_2}\begin{Bmatrix} W_{kt} \\ \delta W_{kt} \end{Bmatrix}\sin(k\bar{x})\cos(l\bar{y})$$

$$\begin{Bmatrix} F_m \\ \Delta F_m \end{Bmatrix} = \frac{ERt^2}{c}\begin{Bmatrix} -\lambda\bar{y}^2/2 \\ 0 \end{Bmatrix} + \frac{ERt^2}{c}\sum_{i=1}^{N_1}\begin{Bmatrix} F_{io} \\ \delta F_{io} \end{Bmatrix}\cos(i\bar{x})$$

$$+\frac{ERt^2}{c}\sum\sum_{k,l=1}^{N_2}\begin{Bmatrix} F_{kt} \\ \delta F_{kt} \end{Bmatrix}\sin(k\bar{x})\cos(l\bar{y}) \tag{4.44}$$

where $W_v = (-\nu/c)(\bar{H}_{xx}/1 + \mu_1)\lambda$ and $c = \sqrt{3(1-\nu^2)}$.

The set of linear partial differential equations are satisfied approximately by Boobnov-Galerkin's procedure yielding a set of linear algebraic equations in terms of the unknown correction terms. In matrix notation,

$$\begin{aligned} [A]\{\delta F\} + [B]\{\delta W\} &= -\{E^{(1)}\} \\ [C]\{\delta F\} + [D]\{\delta W\} &= -\{E^{(2)}\} \end{aligned} \tag{4.45}$$

To obtain the buckling load for a given imperfect cylindrical shell, one begins by making an initial guess for $\{W\}$ and $\{F\}$ at a small initial load level λ. Iteration is then carried out until the correction vectors are smaller than some preselected value. The converged solutions then are used as the initial given at the next higher axial load level $\lambda + \Delta\lambda$. The entire process is repeated for increasing values of the axial load parameter λ. The non-linear analysis then will locate the limit point of the prebuckling states. By definition the value of the loading parameter λ corresponding to the limit point will be the theoretical buckling load.

It is shown (Babcock, 1983) that the solution satisfies the circumferential periodicity. It also contains details of the coefficient matrices A, B, C, and D and the error vectors $E^{(1)}$ and $E^{(2)}$.

The result of the calculation of the mean buckling load $E(\Lambda_c)$ is given in Figure 4.5; $E(\Lambda_c) = 0.746$ (i.e., 74.6% of the classical buckling load). The mean buckling load calculated via the Monte Carlo method is $E(\Lambda_c) = 0.739$; thus, the difference between mean buckling loads, derived by these two methods is only 0.007 or 0.95%. Next the sensitivity derivatives were calculated. For the increment of the Fourier coefficients in

Table 4.7. Derivatives of with respect to the Fourier coefficients (Elishakoff et al., 1987)

X_j	$A_{2,0}$	$A_{4,0}$	$C_{1,2}$	$C_{1,6}$	$C_{1,8}$	$C_{1,10}$	$C_{2,3}$	$C_{2,11}$
$\partial\varphi/\partial X_j$	0.09668	0.00340	−0.01854	−0.05687	−0.24686	−0.08183	−0.01214	−0.07173

Equation (4.34), 10% of their original values are used, so that $\Delta\xi_j = 0.10X_j$. The calculated derivatives are listed in Table 4.7. In this study the increments of end-shortening were chosen in such a way that the limit loads were found to an accuracy of 0.0001.

The results of the mathematical expectation and variance of Z are $E(Z) = 0.746 - \lambda$ and $\mathrm{Var}(Z) = 0.0175$, respectively. The reliability is calculated directly from Equation (4.36). The reliability functions calculated with the Monte Carlo method and with the first-order second-moment method are both given in Figure 4.6. These curves appear to be in excellent agreement, however, in the higher reliability region, which is important for design load derivation. The deviation is more noticeable here.

From the results of the calculations it may be concluded that:

1. The first-order second-moment method can be successfully used for determining the reliability function of axially compressed shells.
2. The number of buckling load calculations necessary for first-order second-moment method is significantly less than with the Monte Carlo method.
3. The mean buckling loads due to both methods are in excellent agreement, but still higher than the experimental value. This is caused by the simplified deterministic buckling load analysis. Because the present method does not need as many calculations as the Monte Carlo method, a more advanced and expensive method

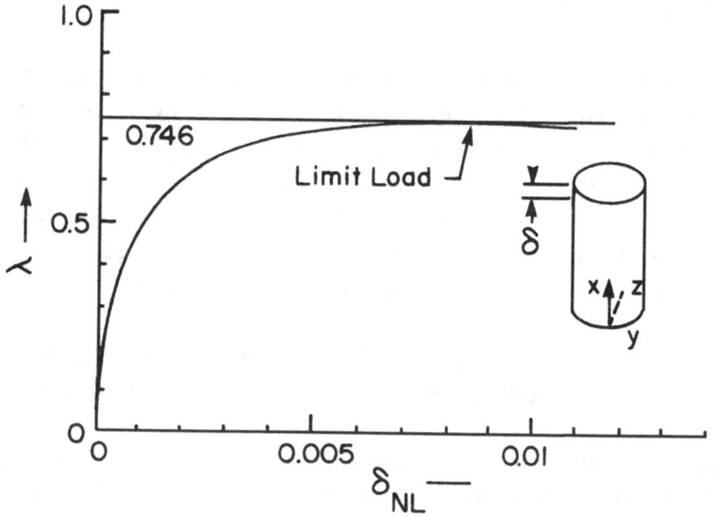

Figure 4.5 Response curve with mean imperfections in the λ-$\delta_{\underline{N}}$ plane; dimensions used for the analysis are the dimensions of shell B-1 (after Elishakoff et al., Copyright © 1987 AIAA, reprinted with permission).

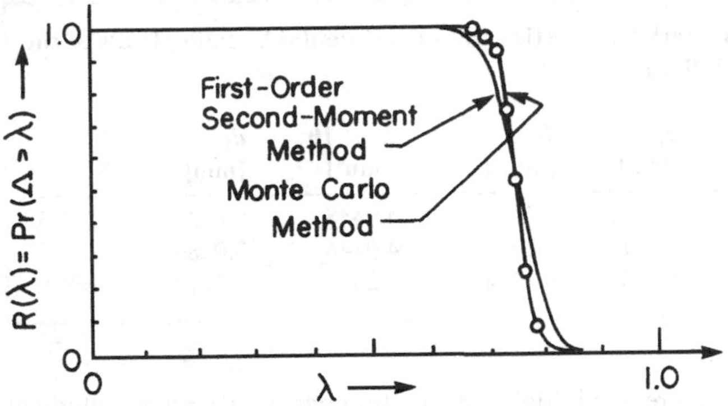

Figure 4.6 The reliability curves for a group of *B*-shells, calculated via the Monte Carlo method, and the mean value, first-order second-moment method (after Elishakoff et al., 1987; Copyright © AIAA, reprinted with permission).

(in terms of the computer time) can be used for the deterministic analysis of buckling load calculations.

The first-order second moment method was implemented by van den Nieuwendijk (1997).

4.3 Use of STAGS to Derive Reliability Functions by Arbocz and Hol

In a previous analysis, following Elishakoff et al. (1987), we utilized the multi-mode analysis for derivation of shell reliability. The highest level of computational complexity and accuracy in deterministic analysis at present is to employ a two-dimensional nonlinear shell analysis code such as STAGS. STAGS was utilized for reliability analysis by Arbocz and Hol (1990b). With the use of this type of numerical tool, one can, in principle, determine the buckling load of a complete shell structure including the effect of arbitrary initial imperfections represented by a double Fourier series. The computations by Arbocz and Hol (1990b) were carried out with a STAGS-A code (Almroth et al., 1973) modified so that, in addition to increments of load and increments of axial displacement, increments of a 'path parameter' (Riks, 1984) can also be used as a load parameter.

The number of imperfection modes included in the analysis is limited by practical considerations, like the time required for obtaining the solutions of all buckling problems needed for calculations of the derivatives $\partial\psi/\partial X_i$. Thus, because the shell buckling load is determined by solving the governing equations for a particular set of initial imperfections, an attempt to select a optimal combination of these modes must be made (i.e., there is a need to locate those imperfection modes which dominate the prebuckling and the collapse behavior of the shell). Recently, novel group-theoretical methods have been developed by Wohlever and Healey (1995) to address this question.

Examples of attempts to locate *critical imperfection modes*, defined as that combination of axisymmetric and asymmetric imperfection modes which would yield the lowest buckling load have been reported in the literature (Arbocz 1974; Arbocz and Babcock, 1976, 1980b). These studies have shown that, to yield a decrease from the

Table 4.8. Geometric and material properties and experimental buckling loads of the AS-shells (Arbocz ad Hol, 1990b)

Shells	t [mm]	A_1 [mm²]	e_1 [mm]	$I_{11} \times 10^2$ [mm⁴]	$I_{t_1} \times 10^2$ [mm⁴]	d_1 [mm]	P_{EXP} [N]
AS-2	0.1966	0.7987	0.3368	1.5038	4.9448	8.0239	14286.3
AS-3	0.2807	0.7432	0.3614	1.2033	4.0146	8.0289	22357.1
AS-4	0.2593	0.4890	0.2758	0.3474	1.2383	8.0112	17074.9

buckling load of a perfect structure, the initial imperfection harmonics must include at least one mode with a significant initial amplitude and an associated eigenvalue that is close to the critical buckling load of the perfect structure. For the calculations, the data associated with the integrally stringer stiffened aluminum alloy shells tested at California Institute of Technology; the so-called *AS*-shells (Arbocz and Abramovich, 1979) will be used. The shell properties are listed in Table 4.8. For the numerical calculations the properties of shell *AS*-2 will be used. Relying on the results of earlier investigations of the buckling behavior of the imperfect *AS*-2 shell (Arbocz and Babcock, 1978), it was decided to employ the following initial imperfection model for the collapse load calculations:

$$w(x,\theta) = hX_1 \cos\frac{2\pi x}{L} - \sin\frac{\pi x}{L}(X_2 \cos 2\theta + X_3 \cos 9\theta + X_4 \cos 10\theta + X_5 \cos 11\theta + X_6 \cos 19\theta + X_7 \cos 21\theta) \tag{4.46}$$

where $\theta = y/R$, $y =$ circumferential coordinate, $R =$ radius. Note that the shape of this imperfection model is symmetric in the axial direction about the center of the shell; hence, only half of the shell length needs to be modeled. However, the imperfection model includes modes with both even and odd number of circumferential waves. This implies that, to be able to use the symmetry conditions, half the shell perimeter must be modeled. Based on the results of the numerical convergence studies (Arbocz and Babcock, 1978), this leads to the use of the discrete model consisting of 21×131 mesh points (21 mesh points in the axial direction, and 131 mesh points in the circumferential direction). The foregoing imperfection model requires eight collapse load calculations to be able to evaluate derivatives $\partial\psi/\partial X_i$.

To apply the first-order, second moment method, the mean buckling load must be calculated first. Using the foregoing imperfection model with the mean values of the corresponding equivalent imperfection amplitudes listed in Table 4.9, the calculation result is $E(\lambda_c) = 0.87538$, whereby the mean buckling load is normalized by -223.079 N/cm, the buckling load of the perfect *AS*-2 shell computed using non-linear prebuckling condition. The derivatives $\partial\psi/\partial X_i$ are listed in Table 4.10. Increments of the path parameter were chosen in such a way that the limit loads were determined with accuracy of 0.01%. The variance-covariance matrix of initial imperfection parameters are listed in Table 4.11. Mathematical expectation and

Table 4.9. Values of the equivalent Fourier coefficients and the reduced sample mean vector (Arbocz and Hol, 1990b)

X_j	*AS-2*	*AS-3*	*AS-4*	$E(X_i)$
1	0.00455	0.01378	−0.01126	0.00236
2	0.33691	0.08298	0.54217	0.32069
3	0.08843	0.02445	0.00297	0.03862
4	0.05524	0.03148	0.00414	0.03028
5	0.05494	0.01912	0.00502	0.02636
6	0.01106	0.00689	0.00424	0.00740
7	0.00879	0.00475	0.00095	0.00483

Table 4.10. Derivatives of ψ with respect to the equivalent Fourier coefficients (seven-mode imperfection model) (Arbocz and Hol, 1990b)

	$\partial\psi/\partial X_j$	
X_j	SS-3	C-4
1	−0.6354	−0.5986
2	0.1498	0.1582
3	0.6924	0.3678
4	0.9138	0.6672
5	0.6233	1.0844
6	0.2449	0.2922
7	0.1811	0.4202

Table 4.11. The reduced sample variance-covariance matrix (all terms are multiplied by 100) (Arbocz and Hol, 1990b)

0.01604				Symmetric		
0.28485	5.29110					
−0.02165	−0.18590	0.19763				
−0.02122	0.28341	0.10789	0.0537			
−0.01354	−0.12712	0.11436	0.06313	0.06625		
−0.00226	−0.02590	0.01511	0.00866	0.00879	0.00118	
−0.00303	−0.0384	0.01683	0.01001	0.00983	0.00134	0.00154

variance of the performance function Z are $E(Z) = 0.87538 - \lambda$, $Var(Z) = 0.00486$ Note that, at the reliability level of 0.98, one obtains a theoretical estimate of the design load $\lambda_{\text{design}} = 0.73$, which also represents a theoretical estimate of the *knockdown factor*.

These results are associated with so-called SS-3 boundary conditions stipulating that at the boundaries

$$N_x = v = w = M_x = 0 \tag{4.47}$$

Another set of boundary conditions, designated in the literature as C-4 boundary conditions were also studied. In this case, the boundary conditions read

$$u = v = w = \partial w/\partial x = 0 \tag{4.48}$$

The derivatives $\partial\psi/\partial X_i$ are listed in Table 4.10. In this case, the calculated mean buckling load equals $E(\Lambda_c) = 0.96298$, a value normalized by -315.323 N/cm, the buckling load of the perfect *AS*-2 shell computed using non-linear prebuckling and the same C-4 boundary conditions. The computation of the mathematical expectation and the variance of Z yields: $E(Z) = 0.96298 - \lambda$, $Var(Z) = 0.00400$, respectively. Notice that in this case for a reliability of 0.98 one obtains a knockdown factor of $\lambda_{\text{design}} = 0.84$, where now λ_{design} is normalized by -315.323 N/cm, the buckling load of the perfect *AS*-2 shell using the C-4 boundary conditions.

Comparing the buckling loads predicted for a reliability of 0.98 of $N_{\text{ss}-3} = -162.848$ N/cm and $N_{c-4} = -264.871$ N/cm based on the seven-modes imperfection model of Equation (4.46) with the experimental buckling load $N_{\text{exp}} = -223.793$ N/cm, one notices that the calculated results seem to support the suggestion made by Arbocz (1982b) that the experimental boundary conditions of the test setup used to buckle the *AS*-shells at the California Institute of Technology imposed a special sort of elastic boundary conditions. For all shells $E = 6.895 \times 10^4$ N/mm^2, $\nu = 0.3$, $R = 101.60$ mm, $L = 139.70$ mm, NR $\times$ NC $= 21 \times 49$, 80 stringers.

4.4 Reliability of Composite Shells by STAGS

The methodology developed by Elishakoff et al. (1987) has been applied to cylindrical shells made of composite materials by Arbocz and Hol (1989). They have investigated the glass-epoxy layered (30°, 0°, −30°) composite shell, previously studied by Booton (1976) in the deterministic setting. Values of the Fourier coefficients and their mean values are listed in Table 4.12. Reliability of composite shells was derived using STAGS computer code. The following initial imperfection model was utilized for the collapse load calculations:

$$\bar{w}(x,\theta) = h \sin\frac{\pi x}{L}\left[\sum_{i=5}^{10} C_{1,i}\cos i\theta + \sum_{j=5}^{10} D_{ij}\sin j\theta\right] \tag{4.49}$$

This imperfection model requires 13 calculations of the collapse load to evaluate the derivatives $\partial\psi/\partial X_j$. Note that the required number of the collapse load calculations

Table 4.12. Values of the Fourier coefficients and the sample mean vector (Arbocz and Hol, 1989)

	X_j	*B*-1	*B*-2	*B*-3	*B*-4	Mean vector
$C_{1,5}$	1	−0.081	0.115	0.024	0.166	0.0560
$C_{1,6}$	2	−0.078	−0.143	0.017	0.078	−0.0315
$C_{1,7}$	3	0.005	0.258	−0.062	−0.014	0.0538
$C_{1,8}$	4	−0.264	−0.009	0.112	0.102	−0.0148
$C_{1,9}$	5	−0.116	0.025	−0.051	−0.002	−0.0350
$C_{1,10}$	6	0.037	0.044	−0.246	−0.009	−0.0435
$D_{1,5}$	7	0.115	0.177	0.100	0.025	0.1043
$D_{1,6}$	8	−0.390	0.058	0.128	−0.248	−0.1130
$D_{1,7}$	9	−0.185	−0.086	0.213	0.060	0.0005
$D_{1,8}$	10	−0.029	0.087	0.185	−0.050	0.0483
$D_{1,9}$	11	0.063	0.018	0.102	0.064	0.0618
$D_{1,10}$	12	−0.013	0.039	−0.051	−0.051	−0.0140

can be reduced to 7 if one replaces the foregoing imperfection model with

$$\bar{w}(x,\theta) = h \sin\frac{\pi x}{L} \sum_{i=5}^{10} \xi_{1,i} \cos(i\theta - \varphi_{1,i}) \tag{4.50}$$

where $\xi_{kl} = (C_{kl} + D_{kl})^{1/2}$ and $\varphi_{kl} = \tan^{-1}(D_{kl}/C_{kl})$. The coefficients ξ_{kl} are listed in Table 4.13. The mean vector and the variance-covariance matrix of coefficients ξ_{kl}, treated as random variables, are given in Tables 4.13 and 4.14, respectively. For calculation of the mean buckling load, initially the reduced six-mode imperfection in Equation (4.50) is utilized, with $\varphi_{1,i}$ set to zero. The result of calculation is $E(\Lambda_c) = 0.5942$, whereby the mean buckling load is normalized by −730.939 N/cm, the buckling load of the perfect "Booton-shell" computed using linear prebuckling analysis. The derivatives $\partial\psi/\partial\xi_i$ are listed in Table 4.15. The results of the calculation of the mathematical expectation and variance of the performance function Z are $E(Z) = 0.5942 - \lambda$, $\text{Var}(Z) = 0.00926$. The specified reliability of 0.98 corresponds to the non-dimensional design load, or theoretical knockdown factor of 0.39, a value that appears to be too low for the published experimental results for composite shells. Analysis of Arbocz and Hol (1989) shows that the imperfection model in Equation (4.50) results

Table 4.13. Values of the equivalent Fourier coefficients and the reduced sample mean vector (Arbocz and Hol, 1989)

	X_j	*B*-1	*B*-2	*B*-3	*B*-4	Mean vector
$\xi_{1,5}$	1	0.1407	0.2111	0.1028	0.1679	0.1556
$\xi_{1,6}$	2	0.3977	0.1543	0.1291	0.2600	0.2353
$\xi_{1,7}$	3	0.1851	0.2720	0.2218	0.0616	0.1851
$\xi_{1,8}$	4	0.2656	0.0875	0.2163	0.1136	0.1707
$\xi_{1,9}$	5	0.1320	0.0308	0.1140	0.0640	0.0852
$\xi_{1,10}$	6	0.0392	0.0588	0.2479	0.0518	0.0994

in a strongly localized collapse pattern. This, however, disagrees with the observed global collapse phenomenon, reported in experimental investigations.

Arbocz and Hol (1989) have repeated the calculations of the reliability function $R(\lambda)$ using the 12-mode imperfection model in Equation (4.49). Elements of variance-covariance matrix read:

$$
\begin{array}{lll}
v_{11} = 1.1791 & v_{21} = 0.3428 & v_{22} = 0.9646 \\
v_{31} = 0.6020 & v_{32} = -1.0158 & v_{33} = 1.9691 \\
v_{41} = 1.4424 & v_{42} = 0.9960 & v_{43} = -0.1996 \\
v_{44} = 3.0618 & v_{51} = 0.6406 & v_{52} = 0.0117 \\
v_{53} = 0.5528 & v_{54} = 0.7609 & v_{55} = 0.3939 \\
v_{61} = 0.1470 & v_{62} = -0.6514 & v_{63} = 1.2005 \\
v_{64} = -1.3733 & v_{65} = 0.1082 & v_{66} = 1.8778 \\
v_{71} = -0.1921 & v_{72} = -0.5832 & v_{73} = 0.5992 \\
v_{74} = -0.4017 & v_{75} = 0.0210 & v_{76} = 0.1786 \\
v_{77} = 0.3902 & v_{81} = 0.8492 & v_{82} = -0.3093 \\
v_{83} = 0.8634 & v_{84} = 2.8270 & v_{85} = 0.7949 \\
v_{86} = -2.0265 & v_{87} = 0.6379 & v_{88} = 0.6759 \\
v_{91} = 0.6685 & v_{92} = 1.1697 & v_{93} = -1.1862 \\
v_{94} = 2.6540 & v_{95} = 0.2879 & v_{96} = -2.1160 \\
v_{97} = -0.4635 & v_{98} = 2.6591 & v_{99} = 3.0196 \\
v_{10,1} = -0.0771 & v_{10,2} = -0.1618 & v_{10,3} = -0.0081 \\
v_{10,4} = 0.8447 & v_{10,5} = 0.0920 & v_{10,6} = -1.1303 \\
v_{10,7} = 0.3065 & v_{10,8} = 2.4748 & v_{10,9} = 1.1397 \\
v_{10,10} = 1.1941 & v_{11,1} = -0.1264 & v_{11,2} = 0.2340 \\
v_{11,3} = -0.4582 & v_{11,4} = 0.1600 & v_{11,5} = -0.1096 \\
v_{11,6} = -0.3934 & v_{11,7} = -0.1173 & v_{11,8} = 0.0523 \\
v_{11,9} = 0.4080 & v_{11,10} = 0.1164 & v_{11,11} = 0.1180 \\
v_{12,1} = -0.0179 & v_{12,2} = -0.3611 & v_{12,3} = 0.4738 \\
v_{12,4} = -0.2140 & v_{12,5} = 0.0667 & v_{12,6} = 0.2295 \\
v_{12,7} = 0.2290 & v_{12,8} = 0.3228 & v_{12,9} = -0.3528 \\
v_{12,10} = 0.1096 & v_{12,11} = -0.1028 & v_{12,12} = 0.1489
\end{array}
\tag{4.51}
$$

The calculations of the mean buckling load yields, in this case, $E(\Lambda_c) = 0.8854$, whereby again -730.939 N/cm is used for non-linearization purposes. The results of the computations of the derivatives $\partial\psi/\partial\xi_i$, are listed in Table 4.16. The mathematical expectation and the variance of Z are, in this case, $E(Z) = 0.8854 - \lambda$, $\mathrm{Var}(Z) = 0.00014$, respectively. In this case for the targeted reliability of 0.98, the use of the knockdown factor $\lambda_{\text{design}} = 0.79$ is implied; it is a value that is considerably higher than the value obtained for the reduced six-mode imperfection model. Utilizing the initial imperfection pattern given in Equation (4.49), or, alternatively, inclusion of

Table 4.14. The reduced sample variance-covariance matrix (all terms are multiplied by 100) (Arbocz and Hol, 1989)

$$\begin{bmatrix} 0.2078 & & & & & \\ -0.0338 & 1.4941 & & & & \\ 0.0455 & -0.4663 & 0.8048 & & & \\ -0.3047 & 0.5302 & 0.0497 & 0.7090 & & \\ -0.1833 & 0.2807 & -0.0351 & 0.3830 & 0.2143 & \\ -0.3258 & -0.7811 & 0.0260 & 0.2385 & 0.1561 & 0.9868 \end{bmatrix}$$

the phase $\varphi_{1,i}$ in the imperfection model [Equation (4.50)] results in a global deformation pattern and in a less localized collapse mode. The results agree better with existing experimental evidence. Comparison of the results obtained through the use of 6- or 12-mode imperfection models clearly illustrates that one should exercise extreme caution when choosing the deterministic model for predicting the buckling loads for performing the stochastic analysis.

An analogous conclusion was also arrived at in another study by Arbocz and Hol (1989). On the other hand, the success of the deterministic buckling load analysis very heavily depends on the appropriate choice of the non-linear model, which in turn requires considerable knowledge by the analyst of the expected physical behavior of imperfect shell structures.

We digress, in words of Arbocz and Hol (1989) that "... only a shell design specialist who is aware of the latest theoretical developments and who is familiar with the theories upon which the non-linear structural analysis codes he uses are based, can achieve the accurate modeling of the collapse behavior of complex structures that guarantees a successful application" of stochastic theories of buckling as developed by Elishakoff (1978b, 1979b, 1980b), Elishakoff and Arbocz (1982, 1985), Elishakoff et al. (1987), and other investigators. According to Arbocz and Hol (1989),

> ... one can not repeat this warning often enough. The danger of incorrect predictions lies in the use of sophisticated computational tools by persons of inadequate theoretical background. For a successful implementation of the proposed improved shell design procedure the companies involved in the production of shell structures must be prepared to do the initial investment in carrying out complete imperfection curves on a small sample of shells that are representative of their production line. With the modern measuring and data acquisition systems complete surface maps of very large shells can be carried out, at a negligibly small fraction of their production cost.

Table 4.15. Derivatives of ψ with respect to the equivalent Fourier coefficients (6-mode imperfection model) (Arbocz and Hol, 1989)

Serial number, j	X_j	$\partial\psi/\partial X_j$	Serial number, j	X_j	$\partial\psi/\partial X_j$
1	$\xi_{1,5}$	−0.1558	4	$\xi_{1,8}$	−0.5342
2	$\xi_{1,6}$	−0.3492	5	$\xi_{1,9}$	−0.4527
3	$\xi_{1,7}$	−0.5001	6	$\xi_{1,10}$	−0.2997

Table 4.16. Derivatives of ψ with respect to the Fourier coefficients (12-mode imperfections model) (Arbocz and Hol, 1990b)

Serial number, j	X_j	$\partial\psi/\partial X_j$	Serial number, j	X_j	$\partial\psi/\partial X_j$
1	$C_{1,5}$	0.2719	7	$D_{1,5}$	0.0251
2	$C_{1,6}$	0.0689	8	$D_{1,6}$	0.4428
3	$C_{1,7}$	−0.5224	9	$D_{1,7}$	0.2000
4	$C_{1,8}$	−0.3292	10	$D_{1,8}$	−0.4526
5	$C_{1,9}$	0.3150	11	$D_{1,9}$	−0.03892
6	$C_{1,10}$	0.3854	12	$D_{1,10}$	0.1856

4.5 Buckling Mode Localization in a Probabilistic Setting

In this and the following section, we will combine the probabilistic treatment with the finite element method, although in different contexts (Xie, 1995). Here we will consider mode localization in a randomly disordered multi-span continuous beam with deterministic elastic modulus. In Section 4.6, in contrast, we will deal with the situation where the geometric disorder is not present, but the elastic modulus constitutes a random field. In both cases, the discretization through the finite element method will be conducted. One must stress, however, that the disorder (or irregularity) that can occur in geometry of configuration and material properties of the structure is generally of a uncertain structure. In Chapter 1, we dealt with the disorder in the geometry of configuration, in the deterministic setting. Here uncertain disorder is modeled as having a random nature. Buckling mode localization in probabilistic setting was apparently first treated by Ariaratnam and Xie (1996). This section follows the study by Xie (1995).

An N-span continuous beam under compressive load P as shown in Figure 4.7(a) is considered. The length of the ith span is L_i, and the flexural stiffness is $k_i = EI_i/L_i$. The nominal span length and flexural stiffness are L and k, respectively. The torsional spring connecting span $i-1$ and span i has spring stiffness T_i, which reflects the

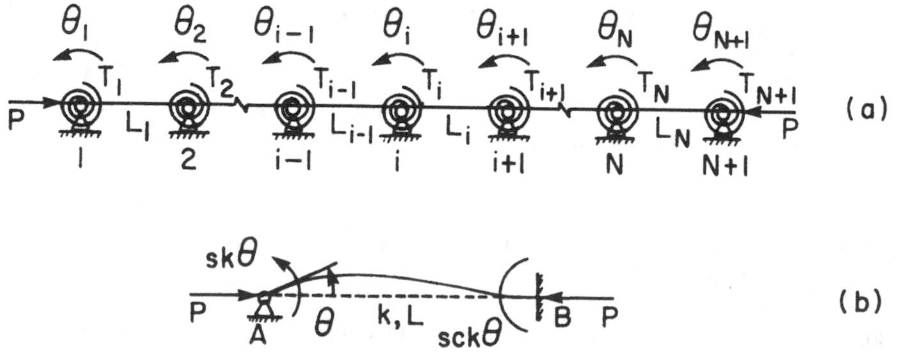

Figure 4.7 (a) Multi-span continuous beam under compressive axial load, (b) stability functions (s) and (c) (after Xie, 1995; Copyright © 1995 AIAA, reprinted with permission).

coupling between these two spans. When T_i is small, span $i-1$ and span i are strongly coupled; when T_i is large, the two spans are weakly coupled. As an extreme case, when T_i approaches infinity, each span is individually clamped: there is no coupling between the adjacent spans in this case.

Consider a beam of length L and flexural stiffness $k = EI/L$ as shown in Figure 4.7(b). It is known (Horne and Merchant, 1965) that in order to have rotation θ at support A and no rotation at support B, a moment $M_A = sk\theta$ at support A and a moment $M_B = sck\theta$ at support B are required, where s and c are the *stability functions* defined as

$$s = \frac{(1-2\alpha\cot 2\alpha)\alpha}{\tan\alpha - \alpha}, \qquad c = \frac{2\alpha - \sin 2\alpha}{\sin 2\alpha - 2\alpha\cos 2\alpha}, \qquad \alpha = \frac{\pi}{2}\sqrt{\rho} \tag{4.52}$$

For the N-span continuous beam, the rotation at support i is denoted by θ_i. Employing the preceding result, the equilibrium condition at support i requires

$$s_{i-1}c_{i-1}k_{i-1}\theta_{i-1} + (s_{i-1}k_{i-1} + s_ik_i + T_i)\theta_i + s_ic_ik_i\theta_{i+1} = 0 \tag{4.53}$$

where

$$s_i = \frac{(1-2\alpha_i\cot 2\alpha_i)\alpha_i}{\tan\alpha_i - \alpha_i}, \qquad c_i = \frac{2\alpha_i - \sin 2\alpha_i}{\sin 2\alpha_i - 2\alpha_i\cos 2\alpha_i},$$

$$\alpha_i = \frac{\pi}{2}\sqrt{\rho_i}, \qquad \rho_i = \frac{P}{P_{E,i}} = \frac{P}{\hat{\rho}_i}, \qquad \hat{\rho}_i = \frac{P_{E,i}}{P_E} \tag{4.54}$$

$$P_{E,i} = EI_i(\pi/L_i)^2$$

From Equation (4.52), θ_{i+1} may be expressed as

$$\theta_{i+1} = -\frac{s_{i-1}\hat{k}_{i-1} + s_i\hat{k}_i + t_i}{s_ic_i\hat{k}_i}\theta_i - \frac{s_{i-1}c_{i-1}\hat{k}_{i-1}}{s_ic_i\hat{k}_i}\theta_{i-1} \qquad (i = 2, 3, \ldots, N) \tag{4.55}$$

where $\bar{k}_i = k_i/k$, $t_i = T_i/k$, or, in the matrix form,

$$\boldsymbol{x}_i = \boldsymbol{T}_i\boldsymbol{x}_{i-1}, \qquad \boldsymbol{x}_i = \begin{Bmatrix}\theta_{i+1}\\ \theta_i\end{Bmatrix}$$

$$\boldsymbol{T}_i = \begin{bmatrix} -\dfrac{s_{i-1}\hat{k}_{i-1} + s_i\hat{k}_i + t_i}{s_ic_i\hat{k}_i}\theta_i & -\dfrac{s_{i-1}c_{i-1}\hat{k}_{i-1}}{s_ic_i\hat{k}_i} \\ 1 & 0 \end{bmatrix} \tag{4.56}$$

in which $\boldsymbol{T}_i$ is the transfer matrix. The state vector $\boldsymbol{x}_n = \{\theta_{n+1}, \theta_n\}^T$ can be related to the initial state vector $\boldsymbol{x}_1 = \{\theta_2, \theta_1\}^T$ by a product of transfer matrices $\boldsymbol{x}_n = \boldsymbol{T}_n\boldsymbol{T}_{n-1}\cdots\boldsymbol{T}_2\boldsymbol{x}_1$, where the superscript T denotes transpose.

It is assumed that the disorders of spans are random, are statistically independent of disorders in other spans, and have a common probability distribution. Then the transfer matrices $\boldsymbol{T}_i$ will also be independent and identically distributed. The rate of growth of the state vector $\boldsymbol{x}_n$ or θ_{n+1} is governed by the behavior of the product of random matrices $\boldsymbol{T}_n\boldsymbol{T}_{n-1}\cdots\boldsymbol{T}_2$. The asymptotic properties of such a product have been studied by many researchers. In this paper, Furstenberg's theorem (Furstenberg, 1963) on the limiting behavior of products of random matrices will be utilized. Furstenberg's theorem may be stated as follows.

Furstenberg's Theorem: Let $T_1, T_2, \ldots, T_n$ be non-singular, independent, identically distributed 2×2 matrices, where $\boldsymbol{T}_i = \boldsymbol{T}_i(\epsilon)$ is a function of the random vector ϵ with probability density $p(\epsilon)$. If at least two of the random transfer matrices do not have common eigenvectors, and if

$$\lim_{n\to\infty} \frac{1}{n} \ln \left(\prod_{i=1}^{n} |\det \boldsymbol{T}_i| \right) = 0 \tag{4.57}$$

then there exists a constant $\delta > 0$ such that, for each $\boldsymbol{x}_0 \neq 0$,

$$\lim_{n\to\infty} \frac{1}{n} \ln \|\boldsymbol{T}_n \boldsymbol{T}_{n-1} \cdots \boldsymbol{T}_1 \boldsymbol{x}_0\| = \delta \tag{4.58}$$

with probability 1, and

$$\lim_{n\to\infty} \frac{1}{n} \ln \|\boldsymbol{T}_n \boldsymbol{T}_{n-1} \cdots \boldsymbol{T}_1\| = \delta \tag{4.59}$$

with probability unity, where $\|\boldsymbol{x}\|$ is a suitable norm of the vector $\boldsymbol{x}$ and $\|\boldsymbol{T}\|$ denotes a suitable norm of the matrix $\boldsymbol{T}$.

It can easily be seen that the transfer matrices defined in Equation (4.56) satisfy the condition (4.57). Applying Furstenberg's theorem to Equation (4.56), one obtains, with probability unity,

$$\lim_{n\to\infty} \frac{1}{n-1} \ln \|\boldsymbol{x}_n\| = \lim_{n-1} \ln \|\boldsymbol{T}_n \boldsymbol{T}_{n-1} \cdots \boldsymbol{T}_2 \boldsymbol{x}_1\| = \delta > 0 \tag{4.60}$$

implying that, for large n,

$$\|\boldsymbol{x}_n\| = e^{\delta(n-1)} \|\boldsymbol{x}_1\| \tag{4.61}$$

The positive number δ characterizes the average exponential rate of growth of the norm of the state vector $x_n = \{\theta_{n+1}, \theta_n\}^T$; Λ is called the *Lyapunov exponent*. It can be shown that Equation (4.61) implies for large n

$$|\theta_{n+1}| = e^{\delta(n-1)} |\theta_1| \tag{4.62}$$

Therefore, the Lyapunov exponent λ characterizes the exponential rate of growth of the angles of rotations. The positivity of the Lyapunov exponent δ for randomly disordered structure results in the *localization* in the buckling modes because the non-zero angles of rotation growing exponentially from each end of the large multi-span beam must match at the maximum; in other words, the buckling mode is localized with amplitudes decaying exponentially at the average rate δ on either side of some region. The Lyapunov exponent δ is, therefore, the *localization factor*.

On the other hand, letting $B_n = \boldsymbol{T}_n \boldsymbol{T}_{n-1} \cdots \boldsymbol{T}_2$, one obtains $\boldsymbol{x}_n = \boldsymbol{B}_n \boldsymbol{x}_1$. The Euclidean norm of $\boldsymbol{x}_n$ is $\|\boldsymbol{x}_n\|^2 = \boldsymbol{x}_1^T \boldsymbol{B}_n^T \boldsymbol{B}_n \boldsymbol{x}_1$. Let $\sigma_{\max}^2 = \sigma_1^2 \geq \sigma_2^2 = \sigma_{\min}^2$ be the eigenvalues of $\boldsymbol{B}_n^T \boldsymbol{B}_n$ and $\boldsymbol{e}_1$ and $\boldsymbol{e}_2$ the corresponding orthonormal eigenvectors. By Rayleigh's principle,

$$\sigma_{\max}^2 = \frac{\boldsymbol{e}_1^T \boldsymbol{B}_n^T \boldsymbol{B}_n \boldsymbol{e}_1}{\|\boldsymbol{e}_1\|^2} \geq \frac{\boldsymbol{x}_1^T \boldsymbol{B}_n^T \boldsymbol{B}_n \boldsymbol{x}_1}{\|\boldsymbol{x}_1\|^2} \geq \frac{\boldsymbol{e}_2^T \boldsymbol{B}_n^T \boldsymbol{B}_n \boldsymbol{e}_2}{\|\boldsymbol{e}_2\|^2} = \sigma_{\min}^2 \tag{4.63}$$

or

$$\sigma_{\max} \|\boldsymbol{x}_1\| \geq \|\boldsymbol{x}_n\| \geq \sigma_{\min} \|\boldsymbol{x}_1\| \tag{4.64}$$

The upper and lower bounds are reached when $\boldsymbol{x}_1$ takes the values $\boldsymbol{e}_1$ and $\boldsymbol{e}_2$, respectively. Hence, taking the natural logarithm, dividing by $n-1$, and taking the limit as $n \to \infty$ results in

$$\lim_{n\to\infty} \frac{1}{n-1} \ln \sigma_{\max} \geq \lim_{n\to\infty} \frac{1}{n-1} \ln \|\boldsymbol{x}_n\| \geq \lim_{n\to\infty} \frac{1}{n-1} \ln \sigma_{\min} \tag{4.65}$$

The larger Lyapunov exponent is then given by

$$\delta_1 = \delta_{\max} = \lim_{n\to\infty} \frac{1}{n-1} \ln \sigma_{\max} \tag{4.66}$$

From the well-known multiplicative ergodic theorem of Oceledec (1968), $\|\boldsymbol{x}_n\|$ grows exponentially at the average rate $\delta_{\max}$ for any $\boldsymbol{x}_1$ except when $\boldsymbol{x}_1 = \boldsymbol{e}_2$, $\|\boldsymbol{x}_n\|$ grows at the rate

$$\delta_2 = \delta_{\min} = \lim_{n\to\infty} \frac{1}{n-1} \ln \sigma_{\min} \tag{4.67}$$

It may be noted that $\delta_2 = -\delta_1$ because the determinant of matrix $\boldsymbol{B}_n^T \boldsymbol{B}_n$ is unity.

If a monocoupled structure is perfectly periodic, $\boldsymbol{T}_2 = \boldsymbol{T}_3 = \cdots = \boldsymbol{T}_n = \boldsymbol{T}$. Let $\Lambda_1, \Lambda_2 (|\Lambda_1| \geq |\Lambda_2|)$ be the eigenvalues of $\boldsymbol{T}$ and $\boldsymbol{E}_1, \boldsymbol{E}_2$ be the corresponding orthonormal eigenvectors. It can be shown that $\sigma_{\max}^2 = |\Lambda_1|^{2n-2}$, $\sigma_{\min}^2 = |\Lambda_2|^{2n-2}$ are the eigenvalues of $\boldsymbol{B}_n^T \boldsymbol{B}_n$. Hence, from Equations (4.66) and (4.67), for any $\boldsymbol{x}_1$ not parallel to $\boldsymbol{E}_2$, $\|\boldsymbol{x}_n\|$ grows exponentially at the rate $\delta_1 = \ln |\Lambda_1|$, whereas when x_1 is parallel to $\boldsymbol{E}_2$, $\|\boldsymbol{x}_n\|$ grows exponentially at the rate $\delta_2 = \ln |\Lambda_2|$.

If the multi-span continuous beam shown in Figure 4.7 is perfectly periodic (i.e., $L_1 = L_2 = \cdots = L_N = L$, $k_1 = k_2 = \cdots = k_N = k$, $t_1 = t_2 = \cdots = t_{N+1} = t$), the transfer matrix given by Equation (4.56) is then

$$\boldsymbol{T} = \begin{bmatrix} 2\gamma & 1 \\ 1 & 0 \end{bmatrix}, \qquad \gamma = -\frac{1}{c}\left[1 + \frac{t}{2S}\right] \tag{4.68}$$

If the absolute value of the trace of the matrix T satisfies the inequality, $|\mathrm{tr}(\boldsymbol{T})| < 2$ (i.e., $|\gamma| < 1$), the eigenvalues of $\boldsymbol{T}$ are complex and the Lyapunov exponents are $\delta_{1,2} = 0$. For the values of axial compressive load P corresponding to $|\gamma| < 1$, buckling can take place. These ranges are known as the *pass bands*.

If $|\mathrm{tr}(\boldsymbol{T})| < 2$ (i.e., $|\gamma| < 1$), the eigenvalues of $\boldsymbol{T}$ are real, and $\|\boldsymbol{x}_n\|$ grows exponentially at the rate

$$\delta = \begin{cases} \mathrm{sgn}(\gamma) \ln |\gamma + \sqrt{\gamma^2 - 1}|, & \text{if } \theta_2/\theta_1 \neq \gamma - \mathrm{sgn}(\gamma)\sqrt{\gamma^2 - 1} \\ \mathrm{sgn}(\gamma) \ln |\gamma + \sqrt{\gamma^2 - 1}|, & \text{if } \theta_2/\theta_1 = \gamma - \mathrm{sgn}(\gamma)\sqrt{\gamma^2 - 1} \end{cases} \tag{4.69}$$

For the values of the axial load P corresponding to $|\gamma| > 1$, the larger Lyapunov exponent is positive; therefore, buckling cannot take place. These regions are known as the *stop bands*.

When the structures are randomly disordered, the transfer matrices $\boldsymbol{T}_2, \boldsymbol{T}_3, \ldots, \boldsymbol{T}_n$ are random matrices. The Lyapunov exponents or the localization factors γ can be determined using the following algorithm. An arbitrary non-zero unit vector $\hat{\boldsymbol{x}}_1$ is chosen first. The state vector $\boldsymbol{x}_i$ is determined iteratively. At the ith iteration,

$$\boldsymbol{x}_i = \boldsymbol{T}_i \hat{\boldsymbol{x}}_{i-1} \tag{4.70}$$

$\boldsymbol{x}_i$ is then normalized to give $\boldsymbol{x}_i = \|\boldsymbol{x}_i\| \hat{\boldsymbol{x}}_i$. It is easy to show that

$$\|\boldsymbol{T}_n \boldsymbol{T}_{n-1} \cdots \boldsymbol{T}_2 \hat{\boldsymbol{x}}_1\| = \prod_{i=2}^{n} \|\boldsymbol{x}_i\| \tag{4.71}$$

From Equation (4.59), the Lyapunov exponent or the localization factor is obtained as

$$\delta = \lim_{n\to\infty} \frac{1}{n-1} \sum_{i=2}^{n} \ln \|\boldsymbol{x}_i\| \tag{4.72}$$

For perfectly periodic multi-span continuous beams, the Lyapunov exponents or localization factors δ are calculated by Equation (4.69) and are plotted in Figure 4.8 for different values of the non-dimensional torsional spring stiffness t. The pass bands are determined by $|\text{tr}(\boldsymbol{T})| \leq 2$. The values of ρ corresponding to the upper ends of the pass bands are given by $1/s = 0$ and $c = -1$ for odd-numbered pass bands or $c = 1$ for

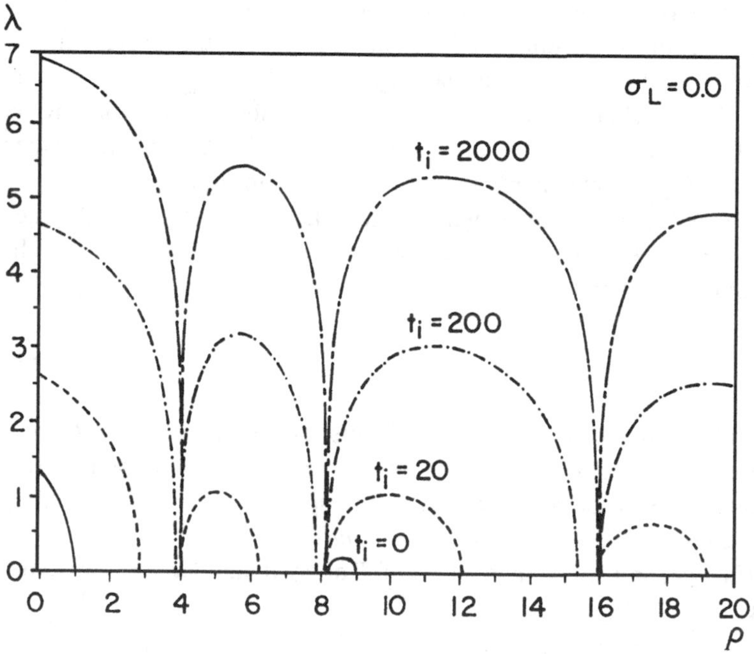

Figure 4.8 Localization factors for the periodic structure, $\sigma_L = 0.0$ (after Xie, 1995; Copyright © 1995 AIAA, reprinted with permission).

even-numbered pass bands. The values of ρ corresponding to the lower ends of the pass bands are given by the roots of $|[1 + t/(2s)]/c| = 1$, in which $1/s \neq 0$. It is seen that the larger the value of the non-dimensional torsional spring stiffness t or the weaker the coupling between the adjacent spans is, the smaller the width of the pass bands is. As an extreme case, when t approaches infinity, the width of the pass bands becomes zero; the pass bands are pass points. The pass bands corresponding to different values of t have the same upper-end value P or ρ because the corresponding buckling modes have zero slope at all the supports and the torsional springs do not affect the buckling of the multi-span continuous beam.

As an example of disordered periodic structures, the lengths of the spans are assumed to be uniformly distributed random numbers with a mean value L and a standard deviation ρ_L, whereas other parameters are constants. Equation (4.72) is employed to determine the Lyapunov exponents or the localization factors numerically. In performing the simulation, for the ith iteration, a standard uniformly distributed random number R_i is generated, the length of the ith span is calculated as $L_i = L(1 + \sigma_L R_i)$, and the entries of the transfer matrix are calculated by Equation (4.72). Iteration is carried out for a large number of transfer matrices (e.g., $n = 10^5$). Numerical results are plotted in Figure 4.9 for $\sigma_L = 0.01$ and in Figure 4.10 for $\sigma_L = 0.1$ and different values of t_i. It can be seen that when $t_i = 0$ (i.e., when the adjacent spans are strongly coupled), the localization factors are very small, and localization in the buckling modes is weak. However, if t_i is large or if the adjacent spans are weakly coupled, the localization factors are large, especially when $\rho\tau$ is close to the ends of pass bands, which means

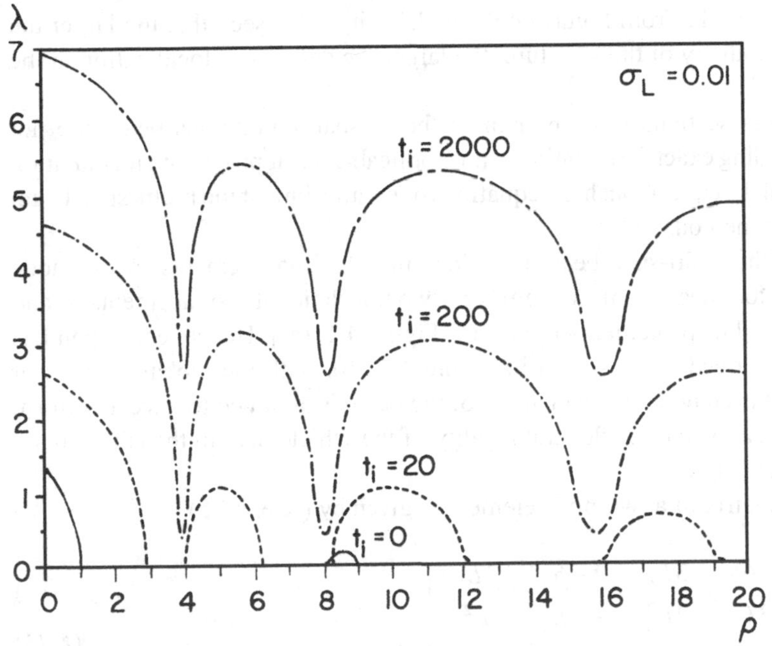

Figure 4.9 Localization factors for the disordered structure, $\sigma_L = 0.01$ (after Xie, 1995; Copyright © 1995 AIAA, reprinted with permission).

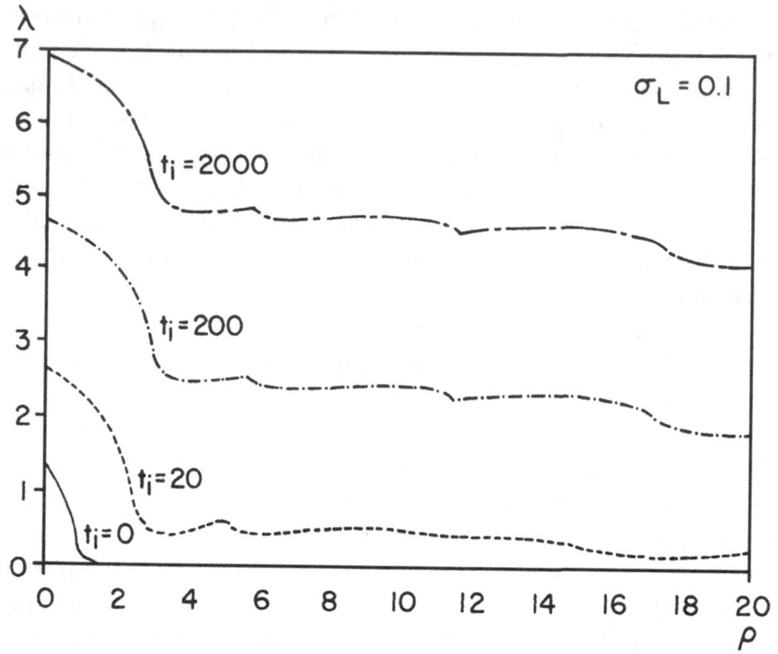

Figure 4.10 Localization factors for the disordered structure, $\sigma_L = 0.1$ (after Xie, 1995; Copyright © 1995 AIAA, reprinted with permission).

that localization in the corresponding buckling modes is strong. At the middle of the pass band, the localization factors are relatively small, and localization in the buckling modes is relatively weak. From Figures 4.9 and 4.10, it is also seen that the larger the disorder in the periodicity of the structure, the larger the degrees of localization in the buckling modes.

When the cross sections in each span of the N-span continuous beam are not uniform, the preceding exact formulation is not applicable; an approximate formulation has to be employed. Let us establish the equations of equilibrium of the multi-span beam by a finite element method.

Each span of the multi-span beam is divided into M finite elements. A two-node beam element has four degrees of freedom [i.e., two translational displacements ν_1^e and ν_3^e and two rotational displacements ν_2^e and ν_4^e; Figure 4.11(a)]. Hence, each span has $2M-1$ degrees of freedom as shown in Figure 4.11(b), with the global degrees of freedom marked at each node. Each element of the beam is assumed to have a uniform cross section; the length and the flexural rigidity of the jth element in the ith span are L_{ij} and EI_{ij}, respectively.

The stiffness matrix of a two-node element is given by (Xie, 1995)

$$K_{ij}^e = \frac{2EI_{ij}}{L_{ij}^3}\begin{bmatrix} 6 & 3L_{ij} & -6 & 3L_{ij} \\ 3L_{ij} & 2L_{ij}^2 & -3L_{ij} & L_{ij}^2 \\ -6 & -3L_{ij} & 6 & -3L_{ij} \\ 3L_{ij} & L_{ij}^2 & -3L_{ij} & 2L_{ij}^2 \end{bmatrix} \tag{4.73}$$

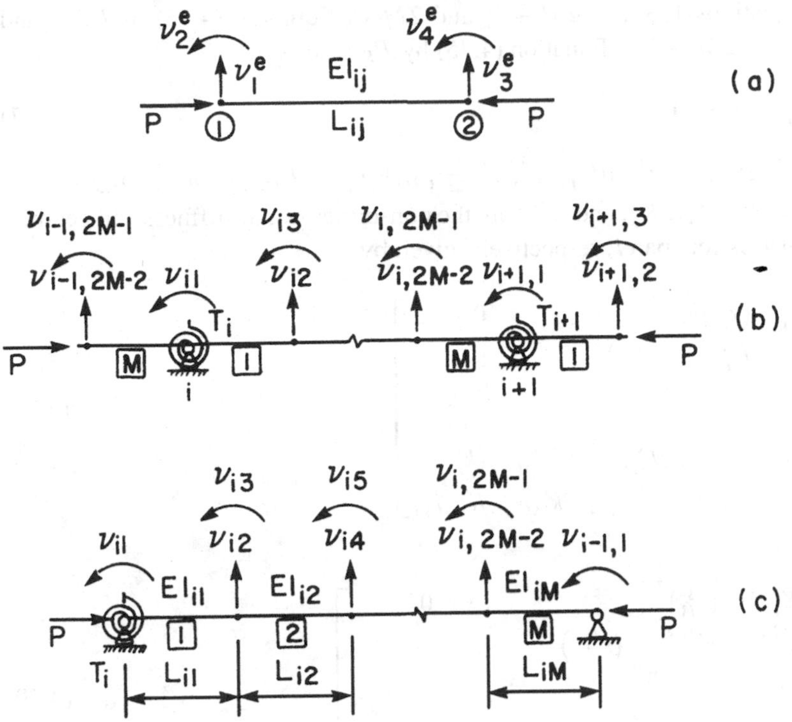

Figure 4.11 (a) Two-node beam elements, (b) global coordinates of a multi-span beam, (c) M-element span (after Xie, 1995; Copyright © 1995 AIAA, reprinted with permission).

whereas the geometric stiffness matrix under axial compressive load P is

$$K_{ij}^{G,e} = \frac{P}{30L_{ij}} \begin{bmatrix} 36 & 3L_{ij} & -36 & 3L_{ij} \\ 3L_{ij} & 4L_{ij}^2 & -3L_{ij} & -L_{ij}^2 \\ -36 & -3L_{ij} & 36 & -3L_{ij} \\ 3L_{ij} & -L_{ij}^2 & -3L_{ij} & 4L_{ij}^2 \end{bmatrix} \tag{4.74}$$

where the superscript e denotes element.

For the span i as shown in Figure 4.11(c), let the displacement vector be $\nu_i^s = \{\nu_{i1}^T, \nu_{i2}^T, \ldots, \nu_{iM}^T, \nu_{i,M+1}^T\}^T$, where $\nu_{i1}^T = \{\nu_{i1}\}$, $\nu_{ij}^T + \{\nu_{i,2j-2}\nu_{i,2j-1}\}$, $j = 2, 3, \ldots, M$, and $\nu_{i,M+1}^T = \{\nu_{i+1,1}\}$. By assembling the element stiffness matrix (4.73) and the element geometric stiffness matrix (4.74), one obtains the stiffness matrix $\tilde{K}_i^s$ and the geometric stiffness matrix $\tilde{K}_i^{G,s}$ for span i, in which the superscript s stands for span. The equations of equilibrium are then given by

$$\left(\tilde{K}_i^2 - \tilde{K}_i^{G,s}\right)\nu_i^s = 0 \tag{4.75}$$

Employing the following notation

$$u_{i1} = \nu_{i1}, \qquad u_{i,2j-2} = \nu_{i,2j-2}/l, \qquad u_{i,2j-1} = \nu_{i,2j-1}$$
$$j = 2, 3, \ldots, M, \qquad u_{i+1,1} = \nu_{i+1,1} \tag{4.76}$$

and dividing equations $1, 3, \ldots, 2M-1$, and $2M$ of Equation (4.75) by $l^2 P_E$, and equations $2, 4, \ldots, 2M-2$ of Equation (4.75) by P_E results in

$$\left(K_I^s - \nu K_i^{G,s}\right) u_i^s = 0 \tag{4.77}$$

where $u_i^s = \{u_{i1}^T, u_{i2}^T, \ldots, u_{im}^T, u_{i,M+1}^T\}^T$, $u_{i1}^T = \{u_{i1}\}$, $u_{ij}^T = \{u_{i,2j-2}, u_{i,2j-1}\}$, $j = 2, 3, \ldots, M$, $u_{i,M+1}^T = \{u_{i+1,1}\}$, K_i^s, and $K_i^{G,s}$ are the non-dimensional stiffness and geometric stiffness matrices for span i, respectively, given by

$$K_i^s = \begin{bmatrix} K_{i1}^s & (k_{i1}^s)^T & & \cdots & 0 \\ k_{i1}^s & K_{i2}^s & (k_{i2}^s)^T & & \vdots \\ & \ddots & \ddots & \ddots & \\ \vdots & & k_{i,M-1}^s & K_{iM}^s & (k_{iM}^s)^T \\ 0 & \cdots & & k_{iM}^s & K_{i,M+1} \end{bmatrix} \tag{4.78}$$

$$K_i^{G,s} = \begin{bmatrix} K_{i1}^{G,s} & (k_{i1}^s)^T & & \cdots & 0 \\ k_{i1}^{G,s} & K_{i2}^{G,s} & (k_{i2}^{G,s})^T & & \vdots \\ & \ddots & \ddots & \ddots & \\ \vdots & & k_{i,M-1}^{G,s} & K_{iM}^{G,s} & (k_{iM}^{G,s})^T \\ 0 & \cdots & & k_{iM}^{G,s} & K_{i,M+1}^{G,s} \end{bmatrix} \tag{4.79}$$

For the elements of K_i^s and $K_i^{G,s}$ one should consult the study of Xie (1995). Equation (4.77) may also be written as

$$A_i^s u_i^s = 0 \tag{4.80}$$

where $A_i^s = K_i^s - \nu K_i^{G,s}$.

For the span considered, $u_i^B = \{u_{i1}, u_{i+1,1}\}^T$ are boundary degrees of freedom, and $u_i^I = \{u_{i2}, u_{i3}, \ldots, U_{i,2M-1}\}^T$ are interior degrees of freedom. Using the method of condensation, the interior degrees of freedom may be expressed in terms of the boundary degrees of freedom. Equation (4.80) can be rearranged as

$$\begin{bmatrix} A_i^{BB} & A_i^{BI} \\ A_i^{IB} & A_i^{II} \end{bmatrix} \begin{Bmatrix} u_i^B \\ u_i^I \end{Bmatrix} = 0 \tag{4.81}$$

or

$$A_i^{BB} u_i^B + A_i^{BI} u_i^I = 0 \tag{4.82}$$

$$A_i^{IB} u_i^B + A_i^{II} u_i^I = 0 \tag{4.83}$$

From Equation (4.83) u_i^I may be expressed in terms of u_i^B, for non-singular A_i^{II}, $u_i^I =$

$-(A_i^{II})^{-1} A_i^{IB} u_i^B$. Substituting u_i^I into Equation (4.82) yields

$$A_i^B u_i^B = 0 \tag{4.84}$$

where A_i^B is a 2×2 symmetric matrix given by

$$A_i^B = A_i^{BB} - A_i^{BI}\left(A_i^{II}\right)^{-1} A_i^{IB} = \begin{bmatrix} \alpha_{i1} & \beta_i \\ \beta_i & \alpha_{i2} \end{bmatrix} \tag{4.85}$$

To compare with the exact results obtained for uniform beams in the following numerical examples the M elements in each span are taken to be identical; for example, $\kappa\kappa_{i1} = \kappa_{i2} = \cdots = \kappa_{iM} = \kappa_i$, $l_{i1} = l_{i2} = \cdots = l_{iM} = l_i$, and $r_{i1} = r_{i2} = \cdots = r_{iM} = r_i$. The elements of matrix A_i^B are

$$\begin{aligned} &\alpha_{i1} = \alpha_{i2} + t_i/M, \qquad \alpha_{i2} = [2(\kappa_i - 2\nu_i) + N_\alpha/D]l_i^2 \\ &\beta_i = N_\beta l_i^2/D \end{aligned} \tag{4.86}$$

where $\nu_i = \nu r_i$, and for a two-element span (i.e., $M = 2$),

$$\begin{aligned} N_\alpha &= 2\left(3\nu_i^3 - \kappa_i\nu_i^2 + 4\kappa_i^2\nu_i - \kappa_i^3\right) \\ N_\beta &= \kappa_i\left(13\nu_i^2 - 4\kappa_i\nu_i + \kappa_i^2\right) \\ D &= 2(6\nu_i - \kappa_i)(2\nu_i - \kappa_i) \end{aligned} \tag{4.87}$$

whereas for a four-element span (i.e., $M = 4$),

$$\begin{aligned} N_\alpha &= 3\big(11475\nu_i^7 - 24330\kappa_i\nu_i^6 + 32890\kappa_i^2\nu_i^5 - 27248\kappa_i^3\nu_i^4 \\ &\quad + 11123\kappa_i^4\nu_i^3 - 2066\kappa_i^5\nu_i^2 + 160\kappa_i^6\nu_i - 4\kappa_i\big) \\ N_\beta &= -\big(10125\nu_i^7 - 33300\kappa_i\nu_i^6 + 37590\kappa_i^2\nu_i^5 - 21818\kappa_i^3\nu_i^4 \\ &\quad + 6301\kappa_i^4\nu_i^3 - 960\kappa_i^5\nu_i^2 + 64\kappa_i^6\nu_i - 2\kappa_i\big) \\ D &= 8(6\nu_i - \kappa_i)\left(2025\nu_i^5 - 3645\kappa_i\nu_i^4 + 2195\kappa_i^2\nu_i^3 - 516\kappa_i^3\nu_i^2 + 42\left(\kappa_i^4\nu_i - \kappa_i^5\right)\right) \end{aligned} \tag{4.88}$$

For the N-span beam as shown in Figure 4.7, assembling Equations (4.85) gives the equations of equilibrium

$$\begin{bmatrix} \alpha_1 & \beta_1 & & \cdots & 0 \\ \beta_1 & \alpha_2 & \beta_2 & & \vdots \\ & \ddots & \ddots & \ddots & \\ \vdots & & \beta_{N-1} & \alpha_N & \beta_N \\ 0 & \cdots & & \beta_N & \alpha_{N+1} \end{bmatrix} \begin{Bmatrix} \theta_1 \\ \theta_2 \\ \vdots \\ \theta_N \\ \theta_{N+1} \end{Bmatrix} = 0 \tag{4.89}$$

where $\alpha_1 = \alpha_{11}$, $\alpha_i = \alpha_{i-1,2} + \alpha_{i1}$, $i = 2, 3, \ldots, N$, $\alpha_{N+1} = \alpha_{N2} + t_{N+1}/M$, and $\theta_i = u_{i1}, i = 1, 2, \ldots, N + 1$.

Equation (4.89) may be written in the form of finite difference equations. From Equation (4.90), one writes

$$\beta_{i-1}\theta_{i-1} + \alpha_i\theta_i + \beta_i\theta_{i+1} = 0 \tag{4.90}$$

$$\boldsymbol{x}_i = \boldsymbol{T}_i x_{i-1}, \qquad x_i = \begin{Bmatrix} \theta_{i+1} \\ \theta_i \end{Bmatrix}$$

$$\boldsymbol{T}_i = \begin{bmatrix} -\dfrac{\alpha_i}{\beta_i} & -\dfrac{\beta_{i-1}}{\boldsymbol{B}_i} \\ 1 & 0 \end{bmatrix} \tag{4.91}$$

Equations (4.53) and (4.91) are of identical form; the only differences are the entries of the transfer matrices T_i. Therefore, the method employed earlier may be used to determine the localization factors for the buckling analysis. Numerical results are evaluated and critically discussed by Xie (1995). It turns out that if only two finite elements are taken for each span, the buckling loads in the first-pass bands are reasonably accurate. When more finite elements are taken from each span, the buckling loads in the higher pass bands turn out to be more accurate. Xie (1995) shows that when four elements are taken for each span, the buckling loads in the first two pass bands are quite accurate; the numerical results obtained from a finite element formulation turn out to agree quite well with those obtained from an exact formulation. To sum up, the method of finite elements is suitable for studying the localization phenomenon in buckling modes of multi-span beams having non-uniform cross sections. Xie (1995) also utilized Green's function formulation, to determine the localization factor for one-dimensional disordered systems, whose governing equations form a tridiagonal system. The drawback of this method is that, when the number of finite elements is increased, the size of the matrices involved becomes very large and, hence, may create a computational problem. If a multi-span beam is randomly disordered, the localization factor turns out to be always positive, with attendant localization in the buckling modes. When the adjacent spans are strongly coupled, the localization factors are small, and localization in the buckling modes is weak. If the adjacent spans are weakly coupled, the localization factors are large, especially when ρ is close to the ends of the pass bands, with attendant strong localization in corresponding buckling modes. Generalization of the study by Xie (1995) to rib-stiffened plates with randomly misplaced stiffeners, by the Kantorovich method, was performed by Xie and Elishakoff (2000).

4.6 Finite Element Method for Buckling of Structures with Stochastic Elastic Modulus

In previous sections, we concentrated on the initial geometric imperfections, thickness variations, or misplacements in the stiffeners. A natural question arises: how to treat the variation of the elastic moduli? As we saw in Section 2.5, the dissimilarity in elastic moduli may cause the additional imperfection sensitivity. In general, elastic modulus is not a constant but rather a function of axial and circumferential parameters x and y. Also, it can be considered to be an uncertain function. If there is a sufficient data available on elastic moduli, they can be treated as random fields of the coordinates. At present, a large body of literature treats the elastic modulus as a random field, in the context of the finite element method for stochastic problems (FEMSP; i.e. structures with random elastic moduli). At present there exist three monographs, namely by Nakagiri and Hisada (1985), Ghanem and Spanos (1991), and Kleiber and Hien (1993). Much of the analyses on the FEMSP is concerned with the second-moment analysis of response (i.e., computing the mean and the variance of the displacement, strain, or stress). Such analyses involve employment of the perturbation method (Cambou, 1975), Neumann expansion (Yamazaki and Shinozuka, 1988), Karhunen-Loeve expansion by Ghanem and Spanos (1991), and Monte Carlo simulation by Shinozuka and his associates. For further references, consult with the aforementioned monographs and the review papers by Vanmarke et al. (1986), Benaroya and Rehak (1988), Brenner (1991), Ghanem and Spanos (1997), Matthies and Bucher (1999), Der Kiureghian and Zhang (1999), and Elishakoff and Ren (1999).

Here we will concentrate on the orthogonal series expansion of random fields. This method is chosen because, in our analysis of random initial imperfections, we have extensively used the orthogonal series expansion in terms of the eigenfunctions of the appropriate linear operators (Elishakoff, 1979b, 1980b, 1983b). We will concentrate in this section primarily on the Karhunen-Loeve expansion representation, extensively used in the FEMSP context by Ghanem and Spanos (1991). The Karhunen-Loeve expansion was derived apparently independently by a number of investigators (Karhunen, 1947; Loeve, 1948; Kac and Siegert, 1947).

The Karhunen-Loeve expansion involves a set of orthogonal deterministic functions, the eigenfunctions of the auto-covariance function of the random field, and uncorrelated random variables. The Karhunen-Loeve expansion can be implemented only if the eigenvalues and eigenfunctions of the auto-covariance function are known. The determination of these eigenvalues and eigenfunctions involves the solution of an integral equation. For illustration purposes, let us study a one-dimensional problem. Consider a continuous random function $E(x)$, with mean $M[E(x)]$ and the auto-covariance function $C_E(x_1, x_2)$; the operator of mathematical expectation is denoted as $M[\cdot]$, in contrast to the previous use for this purpose of notation $E[\cdot]$; the new notation is used so that the elastic modulus is not confused with the notation of the mathematical expectation. We expand the function $E(x)$ in a series

$$E(x) = M[E(x)] + \sum_{i=0}^{\infty} c_i \nu_i h_i(x) \tag{4.92}$$

where c_i = constant coefficients; ν_i = random variables with zero means; and $h_i(x)$ = any complete set of orthogonal deterministic functions. The auto-covariance function is expressed as

$$\begin{aligned} C_E(x_1, x_2) &= M\langle\{E(x_1) - M[E(x_1)]\}\{E(x_2) - M[E(x_2)]\}\rangle \\ &= \sum_{i=1}^{\infty}\sum_{j=1}^{\infty} c_{ic_j} M\langle \nu_i \nu_j\rangle h_i(x_1) h_j(x_2) \end{aligned} \tag{4.93}$$

Multiplying both sides of Equation (4.93) by $h_m(x_1)$, integrating over domain Ω of variation of x, and using the orthogonality property of $h_i(x)$

$$\int_\Omega h_i(x_1)h_m(x_1)dx = \begin{cases} 0, & for\ i \neq m \\ 1, & for\ i = m \end{cases} \tag{4.94}$$

yields

$$\int_\Omega C_E(x_1, x_2)h_m(x_1)dx_1 = \sum_{j=1}^{\infty} c_m c_j \boldsymbol{M}\langle \nu_m \nu_j \rangle h_j(x_2) \tag{4.95}$$

Multiplying both sides of Equation (4.95) by $h_n(x_2)$ and integrating over domain Ω with respect to x_2 results in

$$\int_\Omega \int_\Omega C_E(x_1, x_2)h_m(x_1)h_n(x_2)dx_1dx_2 = c_m c_n M\langle \nu_m \nu_n \rangle \tag{4.96}$$

Hereinafter we will follow Zhang and Ellingwood (1994). Equation (4.92) becomes a Karhunen-Loeve expansion when the eigenvalue and eigenfunction of $C_E(x_1, x_2)$ are chosen as the coefficient c_i^2 and function $h_i(x)$, respectively. Let c_i^2 and $h_i(x)$ be the ith eigenvalue and eigenfunction of the auto-covariance function $C_E(x_1, x_2)$. Then, by definition,

$$\int_\Omega C_E(x_1, x_2)h_i(x_2)dx_2 = c_i^2 h_i(x_1) \tag{4.97}$$

Rearranging Equation (4.96) gives

$$\int_\Omega h_n(x_2)dx_2 \int_\Omega C_E(x_1, x_2)h_m(x_1)\,dx_1 = c_m c_n M\langle \nu_m \nu_n \rangle \tag{4.98}$$

Using Equation (4.97) and the orthogonality of $h_i(x)$ yields

$$\begin{aligned} c_m c_n M\langle \nu_m \nu_n \rangle &= \int_\Omega h_n(x_2) c_m^2 h_m(x_2)\,dx_2 \\ c_m c_n M\langle \nu_m \nu_n \rangle &= c_m^2 \delta_{mn} \end{aligned} \tag{4.99}$$

where δ_{mn} = Kronecker delta. Thus the random variables ν_m in the Karhunen-Loeve expansion are uncorrelated:

$$M\langle \nu_m \nu_n \rangle = \delta_{mn} \tag{4.100}$$

Expansion (4.92) in the applied mechanics context was used apparently for the first time by Elishakoff (1979b), who utilized trigonometric functions

$$h_i(x) = \sin \frac{i\pi x}{l} \tag{4.101}$$

for beams simply supported at both ends. The following functions, representing the modes of free vibrations of a beam in vacuo, were used for beams clamped at both ends,

$$\begin{aligned} h_i(x) &= \sin \lambda_i \xi - \sinh \lambda_i \xi + A_i(\cos \lambda_i \xi - \cosh \lambda_i \xi) \\ A_i &= (\cos \lambda_i - \cosh \lambda_i)(\sin \lambda_i + \sinh \lambda_i)^{-1} \end{aligned} \tag{4.102}$$

where eigenvalues are the roots of the transcendental equation

$$1 - (\cos \lambda_i)(\cosh \lambda_i) = 0 \tag{4.103}$$

and have the approximate values

$$\begin{aligned} &\lambda_1 = 4.730027, \qquad \lambda_2 = 7.853185, \qquad \lambda_3 = 10.995588, \\ &\lambda_4 = 14.137146, \qquad \lambda_5 = 17.278740, \qquad \lambda_6 = 20.4203327 \\ &\lambda_i \approx \left(i + \frac{1}{2}\right)\pi, \qquad i \geq 4 \end{aligned} \tag{4.104}$$

Other representations of the random field as a sum of continuous deterministic functions with random coefficients include method of shape functions and method of optimal linear estimation. Zhang and Ellingwood (1994) used Legendre polynomials.

To elucidate the relationship between other orthogonal expansions and the Karhunen-Loeve expansion, the correlated random variables v_i in Equation (4.92) first are transformed to uncorrelated random variables. We define a random vector V as follows:

$$\begin{aligned} V^T &= (V_1, V_2, \ldots, V_i, \ldots) \\ &= (c_1 v_1, c_2 v_2, \ldots, c_i v_i, \ldots) \end{aligned} \tag{4.105}$$

The mean vector and the variance-covariance matrix of V read

$$\begin{aligned} &E(V_i) = 0 \\ &C(V_i, V_j) = \int_\Omega \int_\Omega C_E(x_1, x_2) h_i(x_1) h_j(x_2)\, dx_1 dx_2 \end{aligned} \tag{4.106}$$

We want to transform the correlated random vector V into a vector of standard uncorrelated random variables U. We first determine the eigenvectors of the variance-covariance matrix $\boldsymbol{C} = [\boldsymbol{C}(V_i, V_j)]$:

$$CA = A\Lambda \tag{4.107}$$

where Λ diagonal matrix of eigenvalues λ_i; the matrix A has columns consisting of the corresponding eigenvectors. The correlated random vector V then can be transformed into the vector U by

$$V = (A^T)^{-1}\Lambda^{1/2} U \tag{4.108}$$

Substituting Equation (4.108) into Equation (4.92) results in

$$E(x) = M[E(x)] + \sum_{i=1}^{\infty} \lambda_i^{1/2} U_i \left[\sum_{j=1}^{\infty} a_j^{(i)} h_j(x)\right] \tag{4.109}$$

where $\{a_j^{(i)}, j = 1, 2, \ldots\}$ is the ith column of $(A^T)^{-1}$. Because U = vector of standard independent random variables, Equation (4.109) represents a Karhunen-Loeve expansion. The following part of the ith term in the expansion, after truncating the series by N terms

$$E_i(x) = \sum_{j=1}^{N} a_j^{(i)} h_j(x) \tag{4.110}$$

represents the ith eigenfunction of $C_E(x_1, x_2)$.

On the other hand, the nth eigenvalue and eigenfunction that are employed in the Karhunen-Loeve expansion also can be obtained by solving the following integral equation (Ghanem and Spanos, 1991).

$$\lambda_n f_n(x) = \int_{\Omega} C_E(x_1, x_2) f_n(x_2)\, dx_2 \tag{4.111}$$

The numerical solution for each eigenfunction can be symbolically written as

$$f_i(x) = \sum_{j=1}^{N} d_j^{(i)} h_j(x) \tag{4.112}$$

where $d_j^{(i)}$ = constant coefficients. Requiring the truncation error to be orthogonal to each term in the expansion base results in the following eigenvalue problem (Ghanem and Spanos, 1991)

$$\boldsymbol{C}\boldsymbol{D} = \boldsymbol{D}\boldsymbol{\Lambda} \tag{4.113}$$

where Λ diagonal matrix, $\{d_j^{(i)}, j = 1, 2, \ldots, N\}$ is the ith column of $\boldsymbol{D}$ and

$$C_{ij} = \int_{\Omega}\int_{\Omega} C_E(x_1, x_2) h_i(x_1) h_j(x_2) dx_1 dx_2 \tag{4.114}$$

The matrix of eigenvectors for the positive symmetric C must be related by

$$[A^T]^{-1} = A = D \tag{4.115}$$

Thus, Equations (4.110) and (4.112) are identical to each other. This result implies that any orthogonal expansion of a random field $E(x)$ can be related to the Karhunen-Loeve expansion of that random field. Expanding a random function on an orthogonal base $h_n(x)$ is equivalent to expanding the random function using the Karhunen-Loeve expansion, in which the eigenfunctions are determined numerically by expanding them on the same orthogonal base.

4.7 Stochastic Finite Element Formulation by Zhang and Ellingwood

Let us turn now to the finite element formulation following Zhang and Ellingwood (1995b). Consider a simply supported beam-column subjected to axial compression P. The material property of interest is the flexural rigidity $F = E(x)I(x)$, which is modeled as a random field with mean m_F and covariance function $C_F(x_1, x_2)$. The potential energy for each element

$$\prod^{e} = \frac{1}{2}\int_{l^e} F\left(\frac{d^2 w^e}{dx^2}\right)^2 dx - \frac{1}{2} P \int_{l^e} \left(\frac{dw^e}{dx}\right)^2 dx \tag{4.116}$$

in which w^e = element displacement field, l^e = element length. The element displacements are interpolated from the element nodal degrees of freedom $(\boldsymbol{R}^e)$

$$w^e = (\boldsymbol{N}^e)(\boldsymbol{R}^e) \tag{4.117}$$

where N^e = the shape function matrix. From Equation (4.117), $\partial w^e/\partial x$ and $\partial^2 w^e/\partial x^2$ can be obtained, respectively,

$$\begin{aligned}\frac{\partial w^e}{\partial x} &= \frac{d}{dx}(N^e)(\boldsymbol{R}^e) = (\boldsymbol{C}^e)(\boldsymbol{R}^e)\\ \frac{\partial^2 w^e}{\partial x^2} &= \frac{d^2}{dx^2}(N^e)(\boldsymbol{R}^e) = (\boldsymbol{B}^e)(\boldsymbol{R}^e)\end{aligned} \tag{4.118}$$

The random rigidity F can be expanded using the truncated orthogonal series expansion:

$$F(x) = m_F + \sum_{m=1}^{M} V_m h_m(x) \tag{4.119}$$

where V = zero-mean random variables, which have the following covariance matrix:

$$E(V_m V_n) = \int_\Omega \int_\Omega C_F(x_1, x_2) h_m(x_1) h_n(x_2) dx_1 dx_2, \qquad m, n = 0, 1, 2, \ldots \tag{4.120}$$

Substituting Equations (4.118)–(4.119) into Equation (4.114), the potential energy can then be written as

$$\prod_P^e = \frac{1}{2}(\boldsymbol{R}^e)^T \left(\boldsymbol{K}_F^e + \sum_{m=1}^{M} V_m \left(\boldsymbol{K}_m^e\right)_F\right)(\boldsymbol{R}^e) - \frac{1}{2}P(\boldsymbol{R}^e)^T\left(\boldsymbol{K}_g^e\right)(\boldsymbol{R}^e) \tag{4.121}$$

$$(\bar{K}^e)_F = \int_{l^e} (\boldsymbol{B}^e)^T m_F (\boldsymbol{B}^e)\, dx; \qquad \left(\boldsymbol{K}_m^e\right)_F = \int_{l^e} (\boldsymbol{B}^e)^T h_m(x) (\boldsymbol{B}^e)\, dx \tag{4.122}$$

$$\left(\boldsymbol{K}_g^e\right) = \int_{l^e} (\boldsymbol{C}^e)^T (\boldsymbol{C}^e)\, dx \tag{4.123}$$

Applying the principle of stationary potential energy yields

$$(\bar{K}^e)_F + \sum_{m=1}^{M} V_m \left(K_m^e\right)_F - P\left(K_g^e\right)(R^e) = 0 \tag{4.124}$$

Equation (4.124) constitutes the stiffness matrix equation for a stochastic beam element with axial force P. The first-order and geometric element stiffness matrices $(\bar{\boldsymbol{K}}^e)_F$ and $(\boldsymbol{K}_g^e)$ in Equation (4.124) are exactly the same as in the deterministic case; the stochastic beam-column stiffness has the one additional term involving V.

The global stiffness matrix can be assembled using standard finite-element analysis:

$$\boldsymbol{K} = \sum_{e=1}^{N_e} (\boldsymbol{K}^e) \tag{4.125}$$

where N_e = total number of elements; the matrices in this summation involve the expanded element stiffness matrices expressed in global coordinates. Finally Equation (4.124), expressed in the global coordinate system, reads:

$$(\bar{\boldsymbol{K}})_F + \sum_{m=1}^{M} V_m (\boldsymbol{K}_m)_F - P(\boldsymbol{K}_g)(\boldsymbol{R}) = 0 \tag{4.126}$$

where $\boldsymbol{R}$ = global nodal displacement vector.

At bifurcation, the determinant of the stiffness matrix must vanish. Thus, the instability analysis involves finding the smallest eigenvalue of Equation (4.126). Because the stiffness matrix involves the random variables $V_m (m = 1, 2, \ldots, M)$, the eigenvalues and eigenvectors F also turn out to be random. Equation (4.126) can be evaluated by perturbation analysis or by Monte Carlo simulation to obtain the probabilistic characteristics of the buckling load P.

Let us now consider the perturbation method. The perturbation technique has been implemented in stochastic finite-element analysis by several researchers (e.g., Nakagiri and Hisada, 1985; Liu, Belytschko, and Mani, 1984). In the following, the second-order perturbation method is used to identify the bifurcation loads involving random parameters.

Equations (4.126) can be converted to the following general form:

$$[\boldsymbol{K} - \lambda(\boldsymbol{K}_g)](\boldsymbol{R}) = 0 \tag{4.127}$$

$$\boldsymbol{K} = \bar{\boldsymbol{K}} + \sum_{i=1}^{L} \alpha_i(\boldsymbol{K}_i) \tag{4.128}$$

$$\lambda = P \tag{4.129}$$

where L = the total number of random variables; $\bar{\boldsymbol{K}}$ = the stiffness matrix based on mean values of the random fields; $\boldsymbol{K}_g$ = the geometric stiffness matrix; $\boldsymbol{\alpha}$ = a set of zero-mean correlated random variables; and K_i = the set of corresponding stiffness matrices. Here, $\bar{\boldsymbol{K}}$, $\boldsymbol{K}_i$, and $\boldsymbol{K}_g$ are all deterministic.

The stiffness matrix $\boldsymbol{K}$ involves the random variables $\alpha(i = 1, 2, \ldots, L)$. Using the mean-centered second-order perturbation method, the stiffness matrix $\boldsymbol{K}$, eigenvalues λ, the eigenvectors R can be expanded in a Maclaurin series with respect to the random variables α_i:

$$\boldsymbol{K} = \boldsymbol{K}_0 + \sum_{i=1}^{L} (\boldsymbol{K}_i^I)\alpha_i + \frac{1}{2}\sum_{i=1}^{L}\sum_{j=1}^{L} (\boldsymbol{K}_{ij}^{II})\alpha_i\alpha_j + \cdots \tag{4.130}$$

$$\lambda = \lambda_0 + \sum_{i=1}^{L} (\lambda_i^I)\alpha_i + \frac{1}{2}\sum_{i=1}^{L}\sum_{j=1}^{L} (\lambda_{ij}^{II})\alpha_i\alpha_j + \cdots \tag{4.131}$$

$$\boldsymbol{R} = \boldsymbol{R}_0 + \sum_{i=1}^{L} (\boldsymbol{R}_i^I)\alpha_i + \frac{1}{2}\sum_{i=1}^{L}\sum_{j=1}^{L} (\boldsymbol{R}_{ij}^{II})\alpha_i\alpha_j + \cdots \tag{4.132}$$

where

$$\boldsymbol{K}_0 = \bar{\boldsymbol{K}} \tag{4.133}$$

$$\boldsymbol{K}_i^I = \left.\frac{\partial \boldsymbol{K}}{\partial \alpha_i}\right|_{\alpha=0} = \boldsymbol{K}_i \tag{4.134}$$

$$\boldsymbol{K}_{ij}^{II} = \left.\frac{\partial^2 \boldsymbol{K}}{\partial \alpha_i \partial \alpha_j}\right|_{\alpha=0} = 0 \tag{4.135}$$

and similarly for $\boldsymbol{R}$. Substituting Equations (4.130)–(4.132) into Equation (4.127), the random eigenvalues λ and eigenvectors $\boldsymbol{R}$ are found recursively as

Zero order:

$$[(\boldsymbol{K}_0) - \lambda_0(\boldsymbol{K}_g)](\boldsymbol{R}_0) = 0 \tag{4.136}$$

First order:

$$(\boldsymbol{R}_0)^T(\boldsymbol{K}_g)(\boldsymbol{R}_0)\lambda_i^I = (\boldsymbol{R}_0)^T(\boldsymbol{K}_i^I)(\boldsymbol{R}_0) \tag{4.137}$$

$$[\lambda_0(\boldsymbol{K}_g) - (\boldsymbol{K}_0)](\boldsymbol{R}_i^I) = [(\boldsymbol{K}_i^I) - \lambda_i^I(\boldsymbol{K}_g)](\boldsymbol{R}_0)) \tag{4.138}$$

Second order:

$$(\boldsymbol{R}_0)^T(\boldsymbol{K}_g)(\boldsymbol{R}_0)\lambda_{ij}^{II} = (\boldsymbol{R}_0)^T[(\boldsymbol{K}_i^I) - \lambda_i^I(\boldsymbol{K}_g)](\boldsymbol{R}_j^I) + (\boldsymbol{R}_0)^T[(\boldsymbol{K}_j^I) - \lambda_j^I(\boldsymbol{K}_g)](\boldsymbol{R}_i^I) \tag{4.139}$$

$$[\lambda_0(\boldsymbol{K}_g) - (\boldsymbol{K}_0)](\boldsymbol{R}_{ij}^{II}) = -\lambda_{ij}(\boldsymbol{K}_g)(\boldsymbol{R}_0) + [(\boldsymbol{K}_i^I) - \lambda_i^I(\boldsymbol{K}_g)](\boldsymbol{R}_j^I) + [(K_j^I) - \lambda_i^I(\boldsymbol{K}_g)](\boldsymbol{R}_i^I) \tag{4.140}$$

The mean and the variance of the buckling load can be obtained from Equation (4.131) using either the first-order or second-order approximation. The first-order or second-order approximation is obtained by considering the first two terms on the right-hand side of Equation (4.131), which results in

$$E(\lambda) \approx \lambda_0 \tag{4.141}$$

$$\mathrm{Var}(\lambda) \approx \sum_{i=1}^{I}\sum_{j=1}^{I} \lambda_i^I \lambda_j^I E(\alpha_i \alpha_j) \tag{4.142}$$

Similarly, the second-order approximation is obtained as

$$E(\lambda) \approx \lambda_0 + \sum_{i=1}^{L}\sum_{j=1}^{L} \lambda_{ij}^{II} E(\alpha_i \alpha_j) \tag{4.143}$$

$$\mathrm{Var}(\lambda) \approx \sum_{i=1}^{L}\sum_{j=1}^{L} \lambda_i^I \lambda_j^I (\alpha_i \alpha_j) + \frac{1}{4}\sum_{i=1}^{L}\sum_{j=1}^{L}\sum_{k=1}^{L}\sum_{l=1}^{L}[E(\alpha_i \alpha_l)E(\alpha_j \alpha_k) + E(\alpha_i \alpha_k)E(\alpha_j \alpha_l)]\lambda_{ij}^{II}\lambda_{kl}^{II} \tag{4.144}$$

In the second term in Equation (4.144), the fourth moments have been expressed in terms of second moments using a relation that is only exact when the random vector (α_i) is Gaussian.

To evaluate the accuracy and any limitations in the preceding formulations and to investigate the effects of uncertain material properties on elastic structural stability, consider buckling of a simply supported column. Monte Carlo simulation is used to validate the results obtained from the perturbation analysis by Zhang and Ellingwood (1995b).

The random fields in the following examples are assumed to be weakly homogeneous and are described in the second-moments sense by the exponential auto-correlation function:

$$C(x_1, x_2) = \sigma^2 \exp\left[\frac{-|x_1 - x_2|}{D_{EI}}\right] \tag{4.145}$$

where the parameter D_{EI} is commonly known as the correlation length and σ^2 is the variance of the random field. For D_{EI} tending to infinity, the random field reduces to a random variable; for D_{EI} tending to zero, the random field becomes an ideal white noise, under some conditions. Without loss of generality, the random field domain Ω is defined as $(-1, 1)$. The Legendre polynomials are chosen as the orthogonal functions $h(x)$ to represent the basis of the random field (Zhang and Ellingwood, 1994). Figure 4.12 illustrates the exponentially decaying auto-correlation function and its approximation recovered from evaluating the Legendre polynomial series expansion of the random field. The number of terms M in the expansion is chosen such that the integral error measure defined as

$$\epsilon_M = \frac{1}{A}\int_\Omega\int_\Omega \frac{1}{\sigma_g^2}\left|C(x_1, x_2) - \sum_{i=0}^{M}\sum_{j=0}^{M} E(V_i, V_j)h_i(x_1)h_j(x_2)\right| dx_1 dx_2 \tag{4.146}$$

is less than 0.005. In Equation (4.146), A is the area of the domain of the covariance function, and σ_g^2 is the variance of the random field. The correlation lengths 24.0 and 0.15 (see Figure 4.12) are used as the lower- and upper-limit correlation lengths in numerical calculations. Intensities at any two points in the random field can be

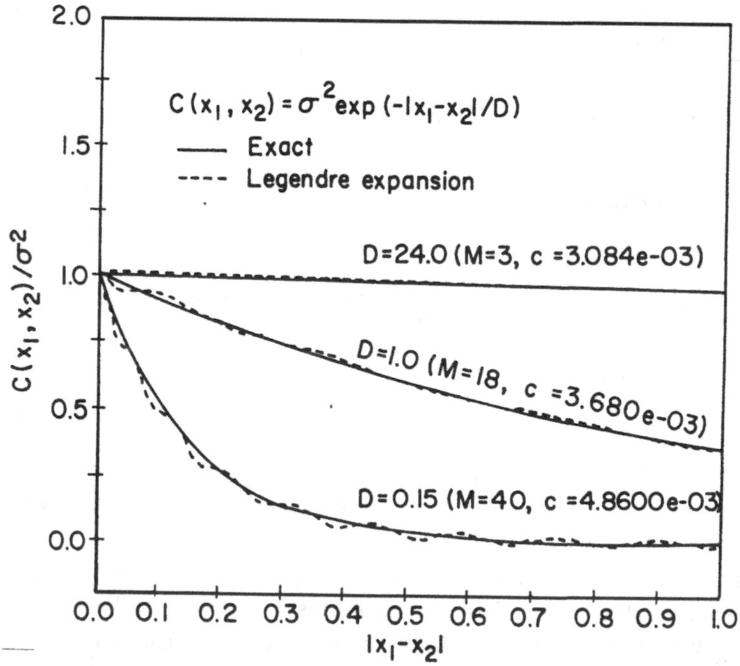

Figure 4.12 Normalized exponential auto-correlation function: exact expression versus results obtained via Legendre expansion (after Zhang and Ellingwood, 1995; Copyright © 1995 ASCE, reprinted with permission).

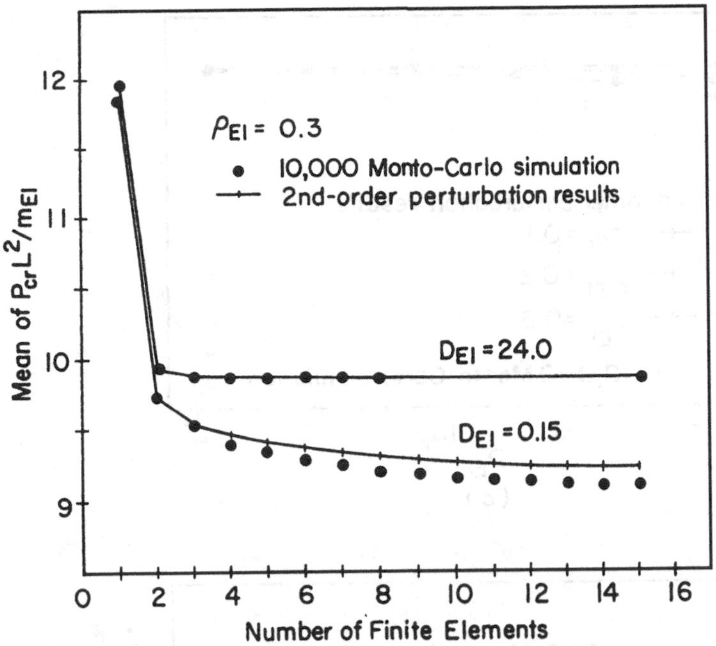

Figure 4.13 Convergence of the mean value of the buckling load of a stochastic simply supported column with number of elements (after Zhang and Ellingwood, 1995; Copyright © 1995 ASCE, reprinted with permission).

considered as almost perfectly correlated when $D_{EI} = 24$, in which case the random field can be approximated simply as a random variable. Conversely, intensities within the random field are weakly correlated, and spatial fluctuations are relatively large when $D_{EI} = 0.15$. For a constant ϵ_M, the number of terms required to represent the random field increases as γ decreases (Zhang and Ellingwood, 1994).

As an example, a simply supported column wilh stochastic flexural rigidity, subjected to the axial load P, is investigated. The flexural rigidity $EI(x)$ is assumed to be a weakly homogenous Gaussian random field with mean m_F and coefficient of variation ρ_{EI}. Figure 4.13 shows the convergence of the mean of the buckling load P_{cr} when $\rho_F = 0.3$ as the number of finite elements is increased. The results are presented in terms of the non-dimensional parameter $P_{cr}L^2/m_F$, so that this parameter equals π^2 when the column is deterministic. The convergence of P_{cr} occurs more slowly as the correlation length of the random field decreases (i.e., a finer finite-element mesh is needed to achieve convergent results when intensities of the random field at two points a given distance apart become uncorrelated). With the continuous representation of the random field, there is no need to discretize the field by a mesh of random variables, and the orthogonal series expansion of the random field is incorporated in the finite-element formulation [see Equation (4.124)]. It is apparent from Figure 4.13 that the finite-element mesh necessary to achieve satisfactory results is affected by the random field, particularly for random fields with short correlation lengths. This occurs because the shape functions used in the finite-element analysis are not exact; as γ decreases, either more finite elements or higher order shape functions are necessary to achieve the same level

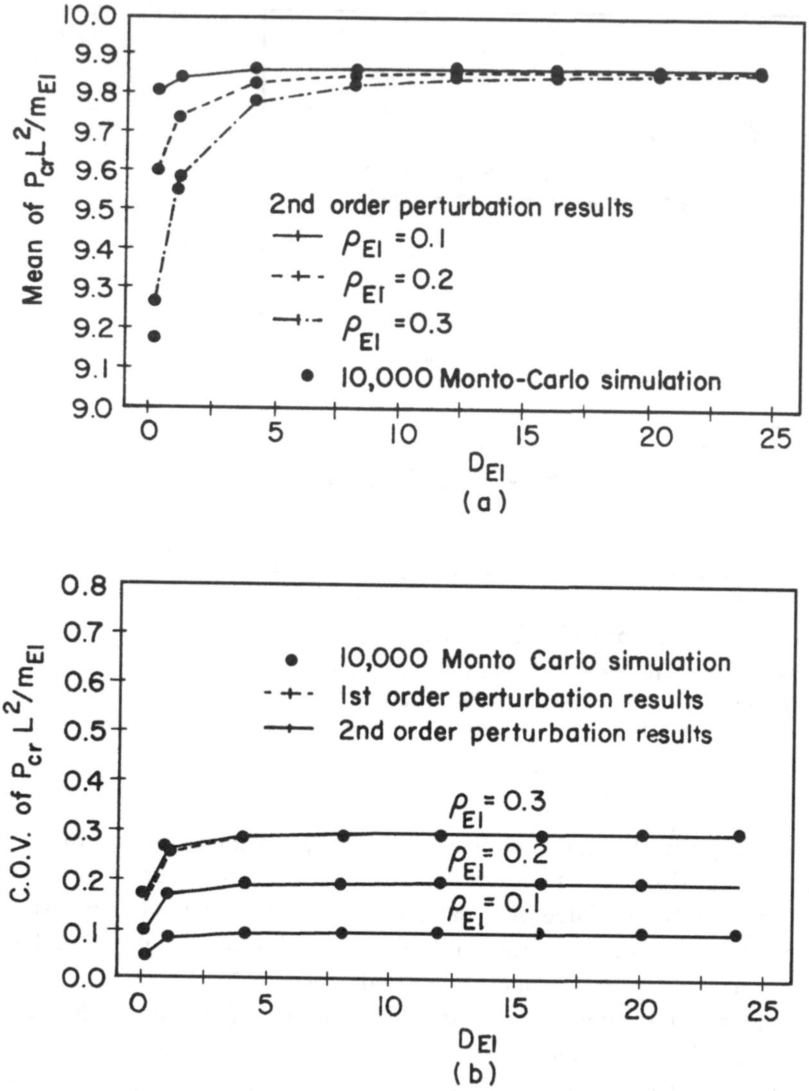

Figure 4.14 Influence of the correlation length and coefficient of variation of elastic stiffness on probabilistic characteristics of the buckling load of a simply supported column: (a) mean value of buckling load and (b) coefficient of variation of buckling load (after Zhang and Ellingwood, 1995; Copyright © 1995 ASCE, reprinted with permission).

of accuracy. The results from the second-order perturbation analysis appear to agree well with results of Monte Carlo simulation, but the accuracy of the second-order perturbation analysis decreases as the correlation length decreases. Based on these results, the column was modeled with 10 finite elements by Zhang and Ellingwood (1995b).

Figure 4.14 illustrates the effects of the correlation length D_{EI} and the coefficient of variation ρ_{EI} on the buckling load statistics. As D_{EI} decreases, the mean value and the coefficient of variation of P_{cr} decrease, indicating that higher spatial fluctuations in the rigidity of the column tend to reduce both the mean value and the variability of P_{cr}.

Conversely as the correlation length becomes large, the mean of $P_{cr}L^2/m_F$ approaches π^2, which is what one would expect if EI were treated as a random variable. As ρ_{EI} increases (i.e., the variability in the intensity of the random field increases), the mean value of P_{cr} decreases, and the coefficient of variation of P_{cr} increases. The second-order perturbation results again agree well with the Monte Carlo simulation results, but the accuracy of the perturbation analysis decreases as either D_{EI} or ρ_{EI} increases. Note that the Kahrunen-Loeve expansion method was utilized recently by Schenk, Schuëller and Arbocz (2000a, 2000b) to deal with buckling analysis of cylindrical shells with random imperfections.

We conclude this chapter on stochastic buckling by observing that according to some researchers probabilistic considerations may prove to be useful not only for design of structures with imperfections, but also for attacking, as Budiansky and Hutchinson (1979) designate it, "...the unresolved, nagging problem of the quest for a basic, general stability theorem." In their elegant overview of modern problems of buckling, these authors mention:

> But there is no doubt that at least an esthetic problem remains, and that a new, congenial definition of stability is desirable. Perhaps some statistical concepts may be fruitful. When the second-variation method fails, it appear that the minimum in φ is destroyed only by presence of obscure secret passages in function space into which no self-respecting structure would venture except by wildly improbable accident. Accordingly, an appropriately defined probability of failure should, under these circumstances, be absurdly low. But we do not have any helpful suggestions concerning this definition, which, in order to be useful in the assessment of the practical stability of structure, should permit easy evaluation of the desired probability.

For the discussion of the fundamental dilemma in the theory of elastic stability, readers are referred to articles by Koiter (1965, 1975, 1976), Budiansky (1974), Como and Grimaldi (1975, 1995), Budiansky and Hutchinson (1979), and Potier-Ferry (1981b). Interrelation between the concepts of stability and stochasticity is discussed by Bolotin (1967). Potier-Ferry (1987) stresses that

> ...difficulty no longer exists within Kelvin-Voigt viscoelasticity... or within theory of beams with moderate rotations (Ball, 1974). Since a century, the energy criterion has been extensively applied. A discrepancy between theory and experiment has never been explained by removing this criterion. Hence, those mathematical difficulties should not be used to question the validity of this stability test. On the contrary, mathematics use to comply to physical evidence and some norms or some constitutive assumptions must be rejected of the corresponding notion of stability is not equivalent to the energy criterion.

It remains to be seen if the notion of probability will prove useful for the very foundation of buckling of structures; yet we illustrated that the probabilistic concepts are useful for a more modest objective, namely, to provide the explanation both of the knockdown factors and for their numerical determination, based on the data derived from the initial imperfection data banks.

For additional aspects of stochastic imperfection sensitivity of structures, consult papers by Arbocz and Hol (1995), Stam (1996), Stam and Arbocz (1997), as well as review articles by Chryssanthopoulos (1997) and Elishakoff (1998).

CHAPTER FIVE

Anti-Optimization in Buckling of Structures

So far as the laws of mathematics refer to reality, they are not certain. And so far as they are certain, they do not refer to reality.

A. Einstein

To a person who is studying algebra, it is often more useful to solve the same problem with three or four different methods, than to solve three or four different problems. By solving problems by different methods, one can by the comparison clarify which of them is shorter and more effective.

W. W. Soyer

The subject of probability is over two hundred years old and for the whole period of its existence there has been dispute about its meaning.

D. V. Lindley

A thousand probabilities do not make one truth.

English proverb

Probability does not exist.

B. de Finetti

I see and approve better things, but follow worse.

Publius Ovidius Naso

In a traditional probabilistic analysis, the statistical parameters of uncertain quantities – initial geometric imperfections or elastic moduli – are presumed to be known, which must be inferred from on-site measurements. Because the available data of such measurements are often limited to permit the probabilistic analysis, a new discipline, called *convex modeling of uncertainty*, is applied to obtain estimates of the upper and lower bounds of the buckling loads. From a structural safety point of view, the least favorable lower buckling load should be used in design. Critical comparison of the probabilistic and convex analyses is performed.

5.1 Incorporation of Uncertainties in Elastic Moduli

Previous studies on plates and shells made of composite have been based on the assumption that the properties of the materials are characterized by predetermined elastic moduli, and no uncertainties of these moduli have been taken into account (Tennyson, 1975; Vinson and Sierakowski, 1986; Whitney, 1987). However, the composite materials are always subject to a certain amount of scatter in their measured elastic moduli (Tewary, 1978). Such uncertainties in elastic moduli of composite materials are due to the many factors that influence the actual properties of composites. For example, among other things, misalignment of fibers or imperfect bonding between the fibers and the matrix may contribute to the scattered values of the measured elastic moduli. To a large extent, the properties of composite materials are dependent on the fabrication process. But even the composite materials manufactured by the same process may demonstrate significant differences in their elastic properties. Therefore one should be aware of the potential variations in load-carrying capacity of composite structure that can arise due to the uncertainty in elastic moduli. A more realistic analysis of composite structures should be performed with the variations of the elastic moduli being taken into consideration at the same time.

The effect of scatter in material properties on buckling of structures is usually treated within the realm of the new tool in stochastic analysis of structures, namely the finite element method for stochastic structures (Ramu and Ganesan, 1993; Zhang and Ellingwood, 1994, 1995a, 1995b, 1995c, 1995d) as discussed in Chapter 4. However, as Shinozuka (1987) mentions,

> it is recognized that it is rather difficult to estimate experimentally the autocorrelation function, or in the case of weak homogeneity, the spectral density function of the stochastic variation of material properties. In view of this, the upper bound results are particularly important, since the bounds derived ... do not require knowledge of the autocorrelation function.

A somewhat analogous idea was expressed by Ariaratnam (private communication, 1992). In the stochastic context, Shinozuka and Deodatis (1988) and Deodatis and Shinozuka (1989) derived upper and lower bounds on the probabilistic characteristics of the response in terms of probabilistic characteristics of the material variability.

The present investigation represents, to the best knowledge of the authors, the first study in the literature, which develops analytical tools to incorporate the uncertainties in elastic moduli into analysis. Here, we deal with the buckling problems of laminated plates and shells by use of a *convex modeling* of uncertainty (Ben-Haim and Elishakoff, 1990; Elishakoff et al., 1994; Elishakoff, 1995). Both the upper and lower bounds of the buckling load are derived so that the designer may have a better understanding of the actual load-carrying capacities possessed by these structures.

5.1.1 Basic Equations

We start our discussion from the buckling of axially compressed composite shells. Here we use the Donnell-Mushtari-Vlasov shell theory (Timoshenko and Gere, 1961).

Strain-displacement relations are

$$\begin{aligned}
\epsilon_x &= \frac{\partial u}{\partial x}, & \kappa_x &= -\frac{\partial^2 w}{\partial x^2} \\
\epsilon_y &= \frac{\partial v}{\partial y} + \frac{w}{R}, & \kappa_y &= -\frac{\partial^2 w}{\partial y^2} \\
\gamma_{xy} &= \frac{\partial v}{\partial x} + \frac{\partial u}{\partial y}, & \kappa_{xy} &= -2\frac{\partial^2 w}{\partial x \partial y}
\end{aligned} \tag{5.1}$$

where x and y are the axial and circumferential coordinates in the shell middle surface; u and v are the shell displacement along axial and circumferential directions; w is the radial displacement, positive outward; ϵ_x, ϵ_y, and γ_{xy} are strain components; κ_x, κ_y, and κ_{xy} are middle surface curvatures of the shell; and R is the radius of the cylindrical shell. The constitutive relations for the composite laminate read

$$\begin{Bmatrix} N_x \\ N_\theta \\ N_{xy} \\ M_x \\ M_y \\ M_{xy} \end{Bmatrix} = \begin{bmatrix} A_{11} & A_{12} & A_{16} & B_{11} & B_{12} & B_{16} \\ A_{12} & A_{22} & A_{26} & B_{12} & B_{22} & B_{26} \\ A_{16} & A_{26} & A_{66} & B_{16} & B_{26} & B_{66} \\ B_{11} & B_{12} & B_{26} & D_{11} & D_{12} & D_{16} \\ B_{12} & B_{22} & B_{26} & D_{12} & D_{22} & D_{26} \\ B_{16} & B_{26} & B_{66} & D_{16} & D_{26} & D_{66} \end{bmatrix} \begin{Bmatrix} \epsilon_x \\ \epsilon_y \\ \epsilon_{xy} \\ \kappa_x \\ \kappa_y \\ \kappa_{xy} \end{Bmatrix} \tag{5.2}$$

where N_x, N_y, and N_{xy} are stress resultants; M_x, M_y, and M_{xy} are bending and twisting moments, acting on a laminate; the laminate stiffnesses A_{ij}, B_{ij}, and D_{ij} are defined as

$$(A_{ij}, B_{ij}, D_{ij}) = \int_{-h/2}^{h/2} \bar{Q}_{ij}^{(k)}(1, z, z^2)\, dz$$

where h is the total thickness of the laminate, and z is the coordinate in the direction of the laminate thickness; $\bar{Q}_{ij}$ are the transformed reduced stiffnesses and can be expressed in terms of the lamina orientation and four independent engineering material constants in principal material directions [i.e., E_1, E_2, ν_{12}, and G_{12} (Jones, 1975)].

The equations governing the buckling of the cylindrical shell under axial compression read

$$\begin{aligned}
&\frac{\partial N_x}{\partial x} + \frac{\partial N_{xy}}{\partial y} = 0 \\
&\frac{\partial N_{xy}}{\partial x} + \frac{\partial N_y}{\partial y} = 0 \\
&\frac{\partial^2 M_x}{\partial x^2} + 2\frac{\partial^2 M_{xy}}{\partial x \partial y} + \frac{\partial^2 M_y}{\partial y^2} - \frac{1}{R}N_y - N_x\frac{\partial^2 w}{\partial x^2} = 0
\end{aligned} \tag{5.3}$$

where N_x is axial loading applied at the ends of the shell.

Using Equations (5.1) and (5.2), Equation (5.3) can be written as

$$\bar{L}\bar{y} = \bar{f}$$

In Equation (5.4),

$$\bar{L} = \begin{bmatrix} L_{11} & L_{12} & L_{13} \\ L_{12} & L_{22} & L_{23} \\ L_{13} & L_{23} & L_{33} \end{bmatrix}$$

$$\bar{y} = \begin{Bmatrix} u \\ v \\ w \end{Bmatrix} \tag{5.4}$$

$$f = \begin{Bmatrix} 0 \\ 0 \\ N_x \dfrac{\partial^2 w}{\partial x^2} \end{Bmatrix}$$

where operators L_{ij} are

$$\begin{aligned}
L_{11} &= A_{11}\frac{\partial}{\partial x^2} + 2A_{16}\frac{\partial^2}{\partial x \partial y} + A_{66}\frac{\partial^2}{\partial y^2} \\
L_{12} &= A_{16}\frac{\partial^2}{\partial x^2} + (A_{12} + A_{66})\frac{\partial^2}{\partial x \partial y} + A_{26}\frac{\partial^2}{\partial y^2} \\
L_{13} &= \frac{1}{R}\left(A_{12}\frac{\partial}{\partial x} + A_{26}\frac{\partial}{\partial y}\right) - B_{11}\frac{\partial^2}{\partial x^2} - 3B_{16}\frac{\partial^3}{\partial x^2 \partial y} \\
&\quad - (B_{12} + B_{66})\frac{\partial^3}{\partial x \partial y^2} - B_{26}\frac{\partial^3}{\partial y^3} \\
L_{22} &= A_{66}\frac{\partial^2}{\partial x^2} + 2A_{26}\frac{\partial^2}{\partial x \partial y} + A_{22}\frac{\partial^2}{\partial y^2} \\
L_{23} &= \frac{1}{R}\left(A_{26}\frac{\partial}{\partial x} + A_{22}\frac{\partial}{\partial y}\right) - B_{22}\frac{\partial^3}{\partial y^3} \\
&\quad - 3B_{26}\frac{\partial^3}{\partial x \partial y^2} - (B_{12} + 2B_{66})\frac{\partial^3}{\partial x^2 \partial y} - B_{16}\frac{\partial^3}{\partial x^3} \\
L_{33} &= -\frac{2}{R}\left(B_{12}\frac{\partial^2}{\partial x^2} + 2B_{26}\frac{\partial^2}{\partial x \partial y} + B_{22}\frac{\partial^2}{\partial y^2}\right) + \frac{A_{22}}{R^2} + D_{11}\frac{\partial^4}{\partial x^4} \\
&\quad + 4D_{16}\frac{\partial^4}{\partial x^3 \partial y} + 2(D_{12} + 2D_{66})\frac{\partial^4}{\partial x^2 \partial y^2} + 4D_{26}\frac{\partial^4}{\partial x \partial y^3} + D_{22}\frac{\partial^4}{\partial y^4}
\end{aligned} \tag{5.5}$$

We consider the axially compressed cylindrical shell with simply supported boundary conditions. In this case, the boundary conditions are represented by

$$w = M_x = v = N_x = 0 \qquad \text{at} \quad x = 0, L \tag{5.6}$$

The preceding boundary conditions are satisfied by the following displacement functions:

$$\begin{Bmatrix} u \\ v \\ w \end{Bmatrix} = \sum_{m=1}^{\infty}\sum_{n=0}^{\infty} \begin{Bmatrix} U_{mn}\cos\dfrac{m\pi x}{L}\cos\dfrac{ny}{R} \\ V_{mn}\sin\dfrac{m\pi x}{L}\sin\dfrac{ny}{R} \\ W_{mn}\sin\dfrac{m\pi x}{L}\cos\dfrac{ny}{R} \end{Bmatrix} \tag{5.7}$$

Similar to Hirano (1979), here the coupling stiffnesses ($A_{16}, A_{26}, B_{16}, B_{26}, D_{16}, D_{26}$) are assumed to be zero. They are actually zero for symmetric cross-ply laminates. As for symmetric angle-ply laminates, B_{16} and B_{26} are zero, and A_{16}, A_{26}, D_{16}, and D_{26} can be neglected for laminates with many layers.

Substitution of Equations (5.1), (5.2), and (5.7) into Equation (5.3) leads to a set of homogeneous linear algebraic equations, and the existence of non-trivial solutions requires that the determinant of the coefficient matrix vanishes,

$$\det\begin{bmatrix} C_{11} & C_{12} & C_{13} \\ C_{21} & C_{22} & C_{23} \\ C_{31} & C_{32} & C_{33} - \left(\dfrac{m\pi}{L}\right)^2 N_{cl} \end{bmatrix} = 0 \tag{5.8}$$

where

$$\begin{aligned}
C_{11} &= A_{11}\left(\frac{m\pi}{L}\right)^2 + A_{66}\left(\frac{n}{R}\right)^2 \\
C_{22} &= A_{22}\left(\frac{n}{R}\right)^2 + A_{66}\left(\frac{m\pi}{L}\right)^2 \\
C_{33} &= D_{11}\left(\frac{m\pi}{L}\right)^4 + 2(D_{12}+2D_{66})\left(\frac{m\pi}{L}\right)^2\left(\frac{n}{R}\right)^2 + D_{22}\left(\frac{n}{R}\right)^4 \\
&\quad + \frac{A_{22}}{R^2} + 2\frac{B_{22}}{R}\left(\frac{n}{R}\right)^2 + 2\frac{B_{12}}{R}\left(\frac{m\pi}{L}\right)^2 \\
C_{12} &= C_{21} = (A_{12}+A_{66})\left(\frac{m\pi}{L}\right)\left(\frac{n}{R}\right) \\
C_{23} &= C_{32} = (B_{12}+2B_{66})\left(\frac{m\pi}{L}\right)^2\left(\frac{n}{R}\right) + \frac{A_{22}}{R}\left(\frac{n}{R}\right) + B_{22}\left(\frac{n}{R}\right)^3 \\
C_{13} &= C_{31} = \frac{A_{12}}{R}\left(\frac{m\pi}{L}\right) + B_{11}\left(\frac{m\pi}{L}\right)^3 + (B_{12}+2B_{66})\left(\frac{m\pi}{L}\right)\left(\frac{n}{R}\right)^2
\end{aligned} \tag{5.9}$$

Thus, we arrive at the classical buckling load,

$$N_{cl} = \left(\frac{L}{m\pi}\right)^2 \frac{C_{11}C_{22}C_{33} + 2C_{12}C_{23}C_{13} - C_{13}^2C_{22} - C_{23}^2C_{11} - C_{12}^2C_{33}}{C_{11}C_{22} - C_{12}^2} \tag{5.10}$$

To determine the critical buckling load, N_{cl}, for a cylindrical shell with given dimensions and material properties, one determines those integer values of m and n which make N_{cl} a minimum.

Let us now consider the case of the symmetrically laminated plate subjected to uni-axial loading. This case could be easily handled by putting $R \to \infty$ and $B_{ij} = 0$ in Equation (5.4). The governing equation becomes

$$D_{11}\frac{\partial^4 w}{\partial x^4} + D_{16}\frac{\partial^4 w}{\partial x^3 \partial y} + 2(D_{12} + D_{66})\frac{\partial^4 w}{\partial x^2 \partial y^2} + 4D_{26}\frac{\partial^4 w}{\partial x \partial y^3} + D_{22}\frac{\partial^4 w}{\partial y^4} = N_x \frac{\partial^2 w}{\partial x^2} \quad (5.11)$$

In the simplest case, when the behavior of the composite plate is of a special orthotropy, the coupling terms D_{16} and D_{26} vanish, and the governing equation (5.11) reduces to

$$D_{11}\frac{\partial^4 w}{\partial x^4} + 2(D_{12} + D_{66})\frac{\partial^4 w}{\partial x^2 \partial y^2} + D_{22}\frac{\partial^4 w}{\partial y^4} = N_x \frac{\partial^2 w}{\partial x^2} \quad (5.12)$$

The following displacement function satisfies the simply supported boundary conditions:

$$w(x, y) = \sum_{m=1}^{\infty}\sum_{n=1}^{\infty} A_{mn} \sin\frac{m\pi x}{a} \sin\frac{n\pi y}{b} \quad (5.13)$$

where a and b are the length and width of the plate, and m and n are integers, denoting the buckling wave numbers along x and y directions, respectively.

Substituting Equation (5.13) into Equation (5.12) yields

$$N_{cl} = -\frac{\pi^2 a^2}{m^2}\left[D_{11}\left(\frac{m}{a}\right)^4 + 2(D_{12} + 2D_{66})\left(\frac{m}{a}\right)^2\left(\frac{n}{b}\right)^2 + D_{22}\left(\frac{n}{b}\right)^4\right] \quad (5.14)$$

It is clear that the integer n should be equal to unity in order to result in the lowest value for the buckling load. However, the value of the integer m depends on the ratio a/b of side lengths as well as the material properties of the plate considered. In practice, m can be determined by a search for a smallest value of the buckling load N_{cl}.

For the general anisotropic plate, the twisting coupling stiffnesses D_{16} and D_{26} may not be neglected. In these cases, the boundary conditions for the simply supported plates become

$$w = M_x = -D_{11}\frac{\partial^2 w}{\partial x^2} - 2D_{16}\frac{\partial^2 w}{\partial x \partial y} - D_{12}\frac{\partial^2 w}{\partial y^2} = 0 \quad \text{at} \quad x = 0, a \quad (5.15)$$

$$w = M_y = -D_{12}\frac{\partial^2 w}{\partial x^2} - 2D_{26}\frac{\partial^2 w}{\partial x \partial y} - D_{22}\frac{\partial^2 w}{\partial y^2} = 0 \quad \text{at} \quad y = 0, b \quad (5.16)$$

Due to the presence of stiffness moduli D_{16} and D_{26} in the governing equation (5.11) and the boundary conditions, the closed form solution is unattainable. Whitney (1987) employs a method of weighted residuals to obtain an approximate solution of the problem. Using the displacement function as described by Equation (5.13) and taking the boundary conditions (5.11) into account, the equation of weighted residues becomes [the misprints of Whitney (1987) relative to Equations (5.17) and (5.18) have been corrected here]

$$\begin{aligned}
&\int_0^b \int_0^a \left[D_{11}\frac{\partial^4 w}{\partial x^4} + 4D_{16}\frac{\partial^4 w}{\partial x^3 \partial y} + 2(D_{12}+2D_{66})\frac{\partial^4 w}{\partial x^2 \partial y^2} + 4D_{26}\frac{\partial^4 w}{\partial x \partial y^3} \right. \\
&\quad \left. + D_{22}\frac{\partial^4 w}{\partial y^4} + N_{cl}\frac{\partial^2 w}{\partial x^2} \right] \sin\frac{m\pi x}{a}\sin\frac{n\pi y}{b} dxdy \\
&\quad - 2D_{26}\int_0^a \left[(-1)^n \left(\frac{\partial^2 w}{\partial x \partial y}\right)_{y=b} - \left(\frac{\partial^2 w}{\partial x \partial y}\right)_{y=0} \right] \frac{n\pi}{b}\sin\frac{m\pi x}{a} dx \\
&\quad - 2D_{16}\int_0^b \left[(-1)^m \left(\frac{\partial^2 w}{\partial x \partial y}\right)_{x=a} - \left(\frac{\partial^2 w}{\partial x \partial y}\right)_{x=0} \right] \frac{m\pi}{a}\sin\frac{n\pi y}{b} dy \\
&\quad = 0 \qquad \begin{cases} m = 1, 2, \ldots, M \\ n = 1, 2, \ldots, N \end{cases}
\end{aligned} \tag{5.17}$$

Further substitution of Equation (5.13) into Equation (5.17) yields the following set of homogeneous linear algebraic equations:

$$\begin{aligned}
&\pi^4 \left[D_{11}m^4 + 2(D_{12}+2D_{66})m^2n^2R^2 + D_{22}n^4R^4 - N_{cl}\frac{m^2a^2}{\pi^2} \right] A_{mn} \\
&\quad - 32mnR\pi^2 \sum_{i=1}^{M}\sum_{j=1}^{N} M_{ij}[(m^2+i^2)D_{16} + (n^2+j^2)D_{26}R^2]A_{ij} = 0
\end{aligned} \tag{5.18}$$

$$(m = 1, 2, \ldots, M; n = 1, 2, \ldots, N)$$

where

$$\begin{aligned}
M_{ij} &= \frac{ij}{(m^2-i^2)(n^2-j^2)} \qquad \begin{cases} m \pm i \text{ odd} \\ n \pm j \text{ odd} \end{cases} \\
&= 0 \qquad \begin{cases} i = m, m \pm i \text{ even} \\ j = n, n \pm j \text{ even} \end{cases} \\
R &= \frac{a}{b}
\end{aligned} \tag{5.19}$$

Again, the non-trivial solution of Equation (5. 18) depends on the condition that the determinant of the coefficient matrix must vanish, from which the uni-axial buckling load N_{cl} can be determined.

5.1.2 Extremal Buckling Load Analysis

The classical buckling load of the structure depends on the four basic elastic moduli, mentioned in Section 5.1.1. For the sake of generality, in the following analysis, a generic formula for the classical buckling load is adopted instead of relying on more concrete expressions, such as Equations (5.10) and (5.14).

Suppose that the classical buckling load N_{cl} takes the form

$$N_{cl} = F(E_1, E_2, \nu_{12}, G_{12}) \tag{5.20}$$

or, more simply,

$$N_{cl} = F(E_i), \qquad (i = 1, 2, 3, 4) \tag{5.21}$$

where $E_3 = \nu_{12}$ and $E_4 = G_{12}$. The function F in Equation (5.21) also depends on the form of structure (plate or shell), boundary conditions as well as geometric properties.

Let E_i^o be the nominal values of the elastic moduli, which might be visualized as the average values of those elastic moduli data from measurements. Then, the elastic moduli of values slightly different from those nominal values could be denoted as $E_i^o + \delta_i$, δ_i being small quantities. The classical buckling load corresponding to these elastic moduli can be written, retaining only the first order in δ_i, as follows:

$$F\left(E_i^o + \delta_i\right) = F\left(E_i^o\right) + \sum_{i=1}^{4} \frac{\partial F\left(E_i^o\right)}{\partial E_i}\delta_i \tag{5.22}$$

We introduce the following notations:

$$\boldsymbol{f}^{\boldsymbol{T}} = \left[\frac{\partial F\left(E_i^o\right)}{\partial E_1}, \frac{\partial F\left(E_i^o\right)}{\partial E_2}, \frac{\partial F\left(E_i^o\right)}{\partial E_3}, \frac{\partial F\left(E_i^o\right)}{\partial E_4}\right] \tag{5.23}$$

$$\boldsymbol{\delta}^{\boldsymbol{T}} = (\delta_1, \delta_2, \delta_3, \delta_4)$$

where T denotes transpose operation. Then Equation (5.22) can be rewritten as

$$F\left(E_i^o + \delta_i\right) = F\left(E_i^o\right) + \boldsymbol{f}^{\boldsymbol{T}}\boldsymbol{\delta} \tag{5.24}$$

The deviation $\boldsymbol{\delta}$ from the nominal elastic moduli is assumed to vary in the following *ellipsoidal* set:

$$Z(\alpha, \boldsymbol{e}) = \left\{\boldsymbol{\delta}\colon \sum_{i=1}^{4} \frac{\delta_i^2}{e_i^2} \le \alpha^2\right\} \tag{5.25}$$

where the size parameter α and the semi-axes e_i $(i = 1, 2, 3, 4)$ are based on the experimental data available and will be discussed later.

The problem is formulated as follows: given an ellipsoid of the elastic moduli [Equation (5.25)], find the extremal buckling load,

$$N_{ext} = \underset{\delta \in Z(\alpha, e)}{\text{extremum}} \left(F\left(E_i^o\right) + \boldsymbol{f}^{\boldsymbol{T}}\boldsymbol{\delta}\right) \tag{5.26}$$

In Equation (5.26), N_{ext} is the lowest or highest buckling load of the composite structure with the elastic moduli varying within the range of the ellipsoidal set Z. Because Equation (5.26) calls for finding the extremum of the linear functional $\boldsymbol{f}^{\boldsymbol{T}}\boldsymbol{\delta}$ on the convex set $Z(\alpha, \boldsymbol{e})$, the extremal values take place on the set of extreme points (Ben-Haim and Elishakoff, 1990), or the boundary, of Z, that is on the ellipsoidal shell defined as follows:

$$C(\alpha, \boldsymbol{e}) = \left\{\boldsymbol{\delta}\colon \sum_{i=1}^{4} \frac{\delta_i^2}{e_i^2} = \alpha^2\right\} \tag{5.27}$$

Thus, the extremum buckling load becomes

$$N_{ext} = \underset{\delta \in C(\alpha, e)}{\text{extremum}} \left(F\left(E_i^o\right) + \boldsymbol{f}^T \boldsymbol{\delta} \right) \tag{5.28}$$

The analysis of the extremal buckling load is quite similar mathematically to the other problems considered by Ben-Haim and Elishakoff (1990) and Elishakoff et al. (1994). Let us define ϵ as an 4×4 diagonal matrix whose ith diagonal element is $1/e_i^2$. Then the constraint given by Equation (5.27) becomes

$$\boldsymbol{\delta}^T \epsilon \boldsymbol{\delta} - \alpha^2 = 0 \tag{5.29}$$

We use the method of Lagrange multipliers, and define the Lagrangean as

$$H(\boldsymbol{\delta}) = \boldsymbol{f}^T \boldsymbol{\delta} + \lambda(\boldsymbol{\delta}^T \epsilon \boldsymbol{\delta} - \alpha^2) \tag{5.30}$$

where λ is the Lagrange multiplier. For the extremum, the derivative of the Hamiltonian must vanish,

$$\frac{\partial H}{\partial \boldsymbol{\delta}} = 0 = \boldsymbol{f}^T + 2\lambda \epsilon \boldsymbol{\delta} \tag{5.31}$$

or

$$\boldsymbol{\delta} = -\frac{1}{2\lambda} \epsilon^{-1} \boldsymbol{f}^T \tag{5.32}$$

In view of Equation (5.29), we have

$$\lambda^2 = \frac{1}{4\alpha^2} \boldsymbol{f}^T \epsilon^{-1} \boldsymbol{f} \tag{5.33}$$

and

$$\boldsymbol{\delta} = \pm \frac{\alpha}{\sqrt{\boldsymbol{f}^T \epsilon^{-1} \boldsymbol{f}}} \epsilon^{-1} \boldsymbol{f} \tag{5.34}$$

It follows from Equation (5.34) that the maximum and minimum buckling loads have the following expression:

$$\left. \begin{matrix} N_{\max} \\ N_{\min} \end{matrix} \right\} = F\left(E_i^o\right) \pm \alpha \sqrt{\boldsymbol{f}^T \epsilon^{-1} \boldsymbol{f}}$$

$$= F\left(E_i^o\right) \pm \alpha \sqrt{\sum_{i=1}^{4} \left(e_i \frac{\partial F\left(E_i^o\right)}{\partial E_i} \right)^2} \tag{5.35}$$

From Equation (5.35), the upper and lower bounds of the critical buckling load are calculated. Equation (5.35) shows explicitly that the uncertainties in elastic moduli have a direct effect on the value of the buckling load. It is remarkable that the semi-axes of the uncertainty ellipsoid, as well as the sensitivity derivatives are directly incorporated in Equation (5.35) to yield the least and most favorable buckling loads due to the uncertainty in elastic moduli.

5.1.3 Determination of Convex Set from Measured Data

Before any prediction can be made on the critical buckling load of the composite plate or shell, the values of α and e_i ($i = 1, 2, 3, 4$) should be known in advance. In fact, these values are dependent on the manufacturing process by which the composite structure has been fabricated. It is understandable that the more advanced the manufacturing process and the better the workmanship, the smaller these parameters will be in value. Besides, the evaluation of these parameters is also linked with the amount of information one has about the properties of the composite considered; in other words, with increased information on the experimental data, more precise calibration of the parameters can be performed.

In the paper by Goggin (1973), some measurements of the elastic moduli for the composite materials were described; they clearly show a scatter of the measured data for the elastic moduli. In particular, for the transverse and longitudinal Poisson's ratio, the experimental values have a large scatter. In this study, the parameter α is set equal to unity. As long as α is fixed, the other parameters could be readily determined by the evaluation of the existent experimental data. Normally, if a sufficient amount of experimental data is available, the average value of these data could be used as the nominal value E_i^0 for the corresponding elastic modulus, whereas parameters e_is could be chosen as the proper deviations from the average values of the corresponding measured data. Even if one has only limited knowledge of the uncertain elastic moduli, it is still possible to judge what value the maximum buckling load of the composite structure will be, provided the variation ranges of these elastic moduli are known.

Suppose that we deduce from the experimental data that the elastic moduli are varying in the following ranges, respectively:

$$E_i^L \leq E_i \leq E_i^U \qquad (i = 1, 2, 3, 4) \tag{5.36}$$

where E_i^U and E_i^L correspond to upper and lower bounds of the elastic moduli E_i, respectively.

We introduce the following notations:

$$\begin{aligned} E_i^o &= \frac{E_i^U + E_i^L}{2} \\ \Delta_i &= \frac{E_i^U - E_i^L}{2} \end{aligned} \tag{5.37}$$

Then Equation (5.36) can be rewritten as

$$\begin{aligned} E_i &= E^{i^0} + \delta_i \\ |\delta_i| &\leq \Delta_i \end{aligned} \tag{5.38}$$

The second of Equation (5.38) describes a "box." To use the theory developed in the last section, we need to enclose this "box" by an ellipsoid,

$$\sum_{i=1}^{4} \frac{\delta_i^2}{e_i^2} \leq 1 \tag{5.39}$$

The question arises as to how to determine the semi-axes of this ellipsoid. Naturally, such an ellipsoid should have a minimum volume. The volume of the preceding ellipsoid is given by

$$V = Ce_1e_2e_3e_4 \tag{5.40}$$

where C is a constant.

Because the corner points of the "box" [Equation (5.38)] should be on the surface of the ellipsoid, we have

$$\frac{\Delta_1^2}{e_1^2} + \frac{\Delta_2^2}{e_2^2} + \frac{\Delta_3^2}{e_3^2} + \frac{\Delta_4^2}{e_4^2} = 1 \tag{5.41}$$

We are interested in minimizing the volume V of the ellipsoid, defined by Equation (5.40), subject to the constraint of Equation (5.41). To do this, we use the Lagrange multiplier technique. The Lagrangean $\boldsymbol{L}$ reads,

$$\boldsymbol{L} = Ce_1e_2e_3e_4 + \lambda\left(\frac{\Delta_1^2}{e_1^2} + \frac{\Delta_2^2}{e_2^2} + \frac{\Delta_3^2}{e_3^2} + \frac{\Delta_4^2}{e_4^2} - 1\right) \tag{5.42}$$

Requirements

$$\frac{\partial \boldsymbol{L}}{\partial e_i} = 0 \qquad (i = 1, 2, 3, 4) \tag{5.43}$$

lead to equations

$$Ce_2e_3e_4 - \frac{2\lambda\Delta_1^2}{e_1^3} = 0 \tag{5.44}$$

$$Ce_1e_3e_4 - \frac{2\lambda\Delta_2^2}{e_2^3} = 0 \tag{5.45}$$

$$Ce_1e_2e_4 - \frac{2\lambda\Delta_3^2}{e_3^3} = 0 \tag{5.46}$$

$$Ce_1e_2e_3 - \frac{2\lambda\Delta_4^2}{e_4^3} = 0 \tag{5.47}$$

We multiply Equation (5.44) by e_1, Equation (5.45) by e_2, Equation (5.46) by e_3, Equation (5.47) by e_4, and sum up all four equations to yield

$$4V - 2\lambda\left(\frac{\Delta_1^2}{e_1^2} + \frac{\Delta_2^2}{e_2^2} + \frac{\Delta_3^2}{e_3^2} + \frac{\Delta_4^2}{e_4^2}\right) = 0 \tag{5.48}$$

Bearing in mind Equation (5.41), Equation (5.48) becomes

$$\lambda = 2V \tag{5.49}$$

Substituting Equation (5.49) into Equation (5.44) leads to

$$\frac{V}{e_1} - 4\frac{\Delta_1^2}{e_1^3}V = 0 \tag{5.50}$$

which implies that

$$e_1 = 2\Delta_1 \tag{5.51}$$

Likewise, we can determine the other semi-axes of the ellipsoid as

$$e_2 = 2\Delta_2, \qquad e_3 = 2\Delta_3, \qquad e_4 = 2\Delta_4 \tag{5.52}$$

Thus, once the size of the "box" containing the experimental data is known, the semi-axes of the ellipsoid can be determined, and the analysis of convex modeling can be carried out.

5.1.4 Numerical Examples and Discussion

We will now proceed to investigate several cases of the buckling problem of composite plates and shells with a view of gaining some insight into the effect of uncertainties in the material properties on the load-carrying capacity of these structures.

In the following analysis, the experimental data for the elastic moduli are based on the available data (Goggin, 1973). The material of the lamina is composed of carbon fibers and resin matrix, with a volume fraction of fiber being 40%. From the figures in the preceding study, the following data were deduced:

$$\begin{aligned}
E_1^U &= 100 \text{ Gpa } (14.5 \times 10^6 \text{ psi}), & E_1^L &= 90 \text{ Gpa } (13.0 \times 10^6 \text{ psi})\\
E_2^U &= 7.3 \text{ Gpa } (1.06 \times 10^6 \text{ psi}), & E_2^L &= 6.8 \text{ Gpa } (0.99 \times 10^6 \text{ psi})\\
E_3^U &= 0.28, & E_3^L &= 0.22\\
E_4^U &= 3.2 \text{ Gpa } (0.46 \times 10^6 \text{ psi}), & E_4^L &= 2.6 \text{ Gpa } (0.38 \times 10^6 \text{ psi})
\end{aligned} \tag{5.53}$$

From these data, the nominal elastic moduli E_i^0 and the semi-axes e_i can be evaluated, by using Equations (5.32), (5.46), and (5.47), as the following:

$$\begin{aligned}
E_1^o &= 13.75 \times 10^6 \text{ psi}, & e_1 &= 1.5 \times 10^6 \text{ psi}\\
E_2^o &= 1.03 \times 10^6 \text{ psi}, & e_2 &= 0.07 \times 10^6 \text{ psi}\\
E_3^o &= 0.25, & e_3 &= 0.06\\
E_4^o &= 0.42 \times 10^6 \text{ psi}, & e_4 &= 0.08 \times 10^6 \text{ psi}
\end{aligned} \tag{5.54}$$

Let us concentrate first on a specially orthotropic laminated rectangular plate subject to the simply supported boundary conditions. We investigate cross-ply laminates, in which fibers of adjacent plies are oriented at 90° to each other and parallel to the plate edges. The critical buckling load for the uni-axially compressed plate is given by Equation (5.14). Note here that the basic elastic moduli E_i are implicitly contained in the flexural stiffness D_{ij}.

Here, the width b is fixed at 10 in., while the length a varies, and the thickness of each lamina is 0.012 in. The following three cases are studied:

Case 1: A 5-ply plate, with laminate configuration (0°, 90°, 0°, 90°, 0°).

Case 2: An 11-ply plate, with odd-numbered layers oriented at 0° and even-numbered layers oriented at 90°.

Case 3: A set of 10-ply square plates ($a = b = 10$ in.), with laminate configuration as $[\theta/-\theta/\theta/-\theta/\theta]_s$, θ ranging from 0° to 90°.

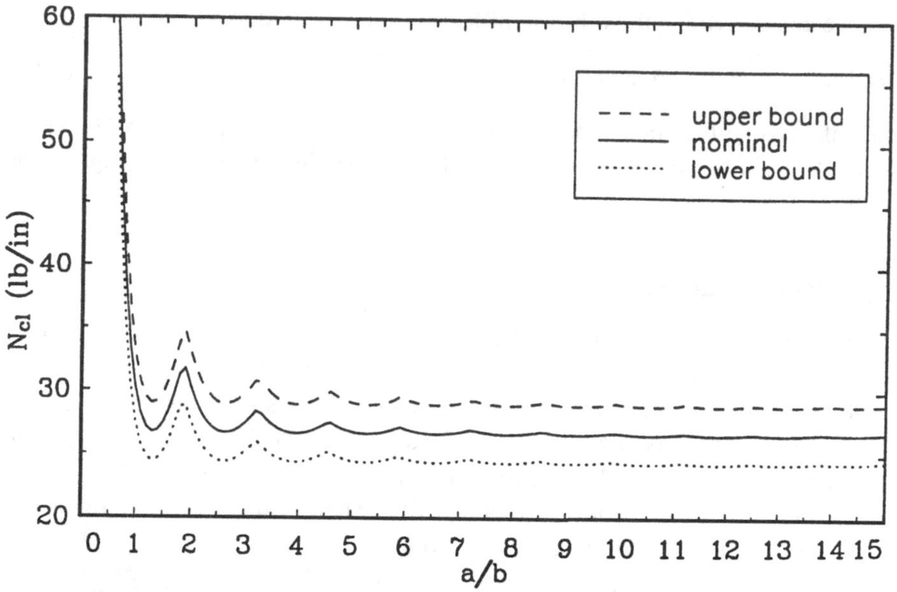

Figure 5.1 Uncertainty in buckling load for a 5-ply laminated plate (Case 1).

The analysis described in previous sections were carried out for these three cases and numerical results are shown in Figures 5.1, 5.2, and 5.3. In the first two cases, the analyses were relatively straightforward. However, the calculations in Case 3 were more involved; here the results in Figure 5.3 were calculated with $M = N = 12$ in Equation (5.18), and the differentiations as required by Equation (5.35) were performed numerically. From these figures, one can see that for different dimensions of the plate,

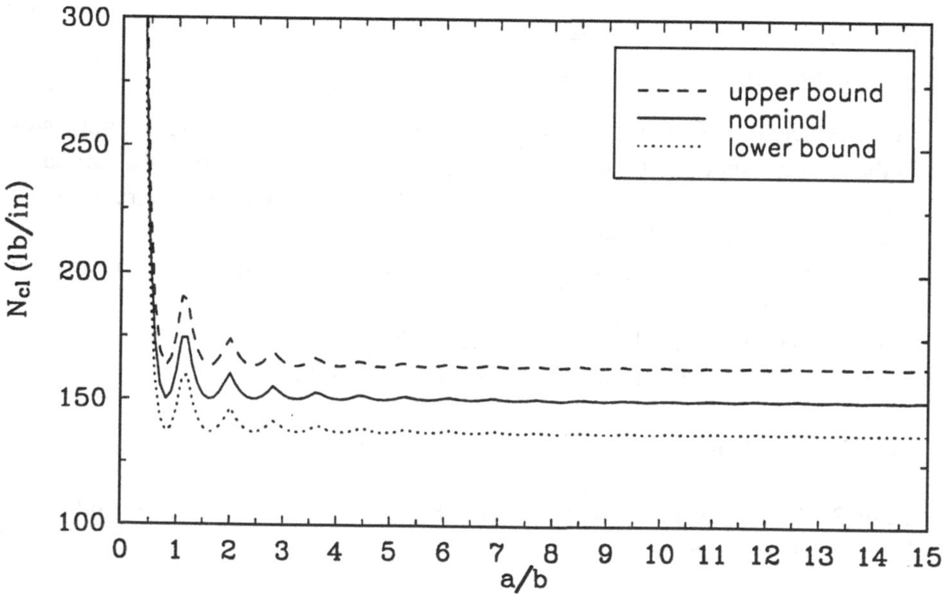

Figure 5.2 Uncertainty in buckling load for an 11-ply laminated plate (Case 2).

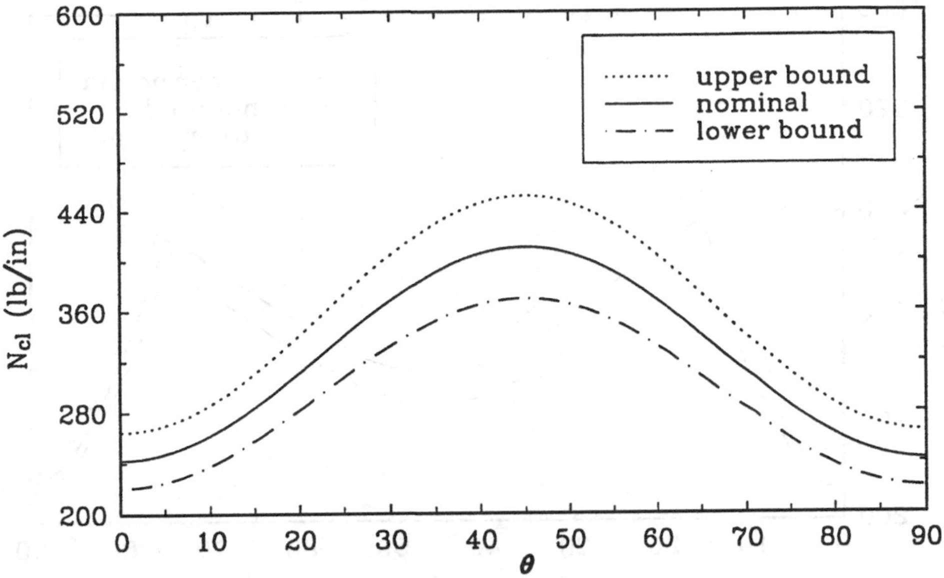

Figure 5.3 Uncertainty in buckling load for a set of 10-ply laminated plates (Case 3).

the effect of uncertainty in elastic moduli on the buckling load is different. The percentagewise variability is defined as

$$\nu = \frac{v_u - v_l}{2v_n} \times 100\% \tag{5.55}$$

where v_u, v_l, and v_n are the upper bound, the lower bound, and the nominal values, respectively. As suggested by Equation (5.53), the percentagewise variabilities are: 5.3% in E_1, 3.5% in E_2, 12% in E_3, and 10.3% in E_4. These variabilities in elastic moduli lead to up to 11% (for Case 1) and 9% (for Case 2) of variation in buckling load of the plate. However, when one of the dimensions of the plate increases, such an effect tends to stabilize; in the cases considered, the uncertainty in buckling load is about 8.5% of its nominal value. It is interesting to note that the uncertainty of buckling load induced by uncertainty in elastic moduli becomes more "stabilized" with the increase of layers. Besides, the variability of the buckling load is also dependent on the lamination configuration of the plate; Figure 5.3 indicates that the scatter in the buckling load N_{cl} is more noticeable when θ is in the vicinity of 45°.

Now we consider the buckling of the symmetric angle-ply cylindrical shell subjected to axial compression. Equation (5. 10) is used in conjunction with Equation (5.35). Here, an integer search is performed for the determination of the axial buckling wave number m. The shells investigated have a 6.0-in. radius and are composed of 0.012-in.-thick layers. The following two cases are considered:

Case 1: The 3-layer laminated shell, with ply angle being $[\theta, -\theta, \theta]$, θ ranging from 0° to 90°.

Case 2: The 5-layer laminated shell, with ply angle being $[\theta, -\theta, \theta, -\theta, \theta]$, θ ranging from 0° to 90°.

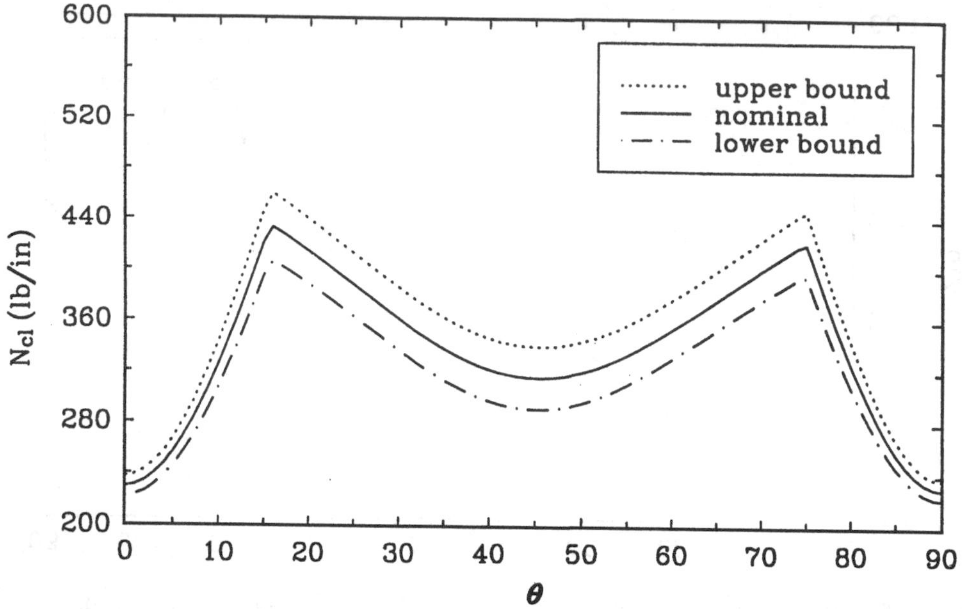

Figure 5.4 Uncertainty in buckling load for a set of 3-ply laminated shells (Case 1).

Figures 5.4, 5.5, 5.6, and 5.7 portray the variability of buckling load due to the uncertainty in elastic moduli for Cases 1 and 2. Again, the effect of uncertainty in elastic moduli on the buckling load varies with the laminate configuration and the number of layers that make up the laminated shell. The maximum variabilities in buckling load of the shell constitute 8% for Case 1 and 9% for Case 2 (Figures 5.5 and 5.7).

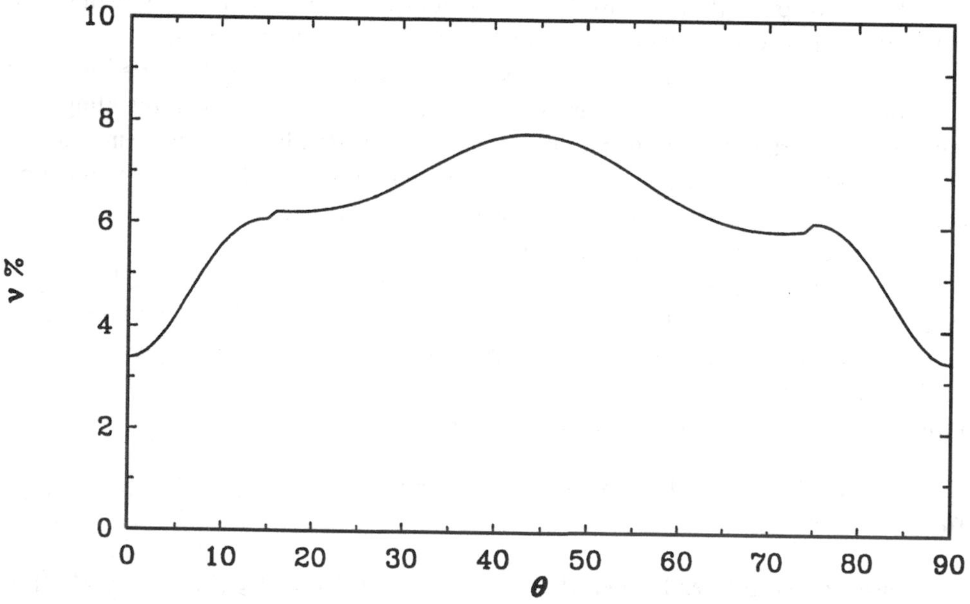

Figure 5.5 Effect of uncertainty in elastic moduli on the buckling load of a set of 3-ply laminated shells (Case 1).

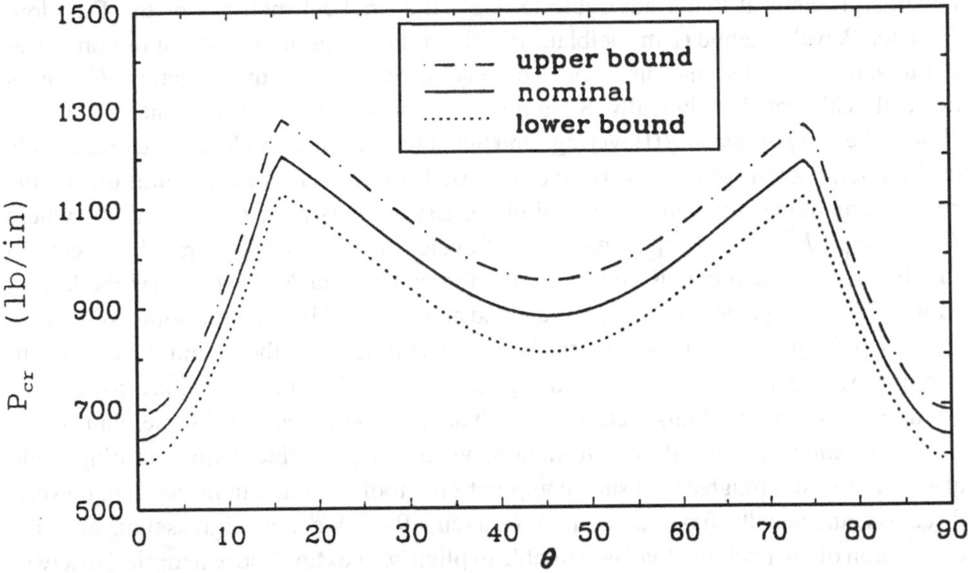

Figure 5.6 Uncertainty in buckling load of a set of 5-ply laminated shells (Case 2).

5.1.5 Numerical Analysis by Non-Linear Programming

The preceding problem can also be treated as a non-linear programming problem with bounds that are mathematically stated as follows:

$$\begin{array}{llll} \text{Find} & \min F(E_1, E_2, E_3, E_4) & \text{or} & \max F(E_1, E_2, E_3, E_4) \\ \text{Subject to} & E_i^L \leq E_i \leq E_i^U, & \text{for} & i = 1, 2, 3, 4 \end{array} \tag{5.56}$$

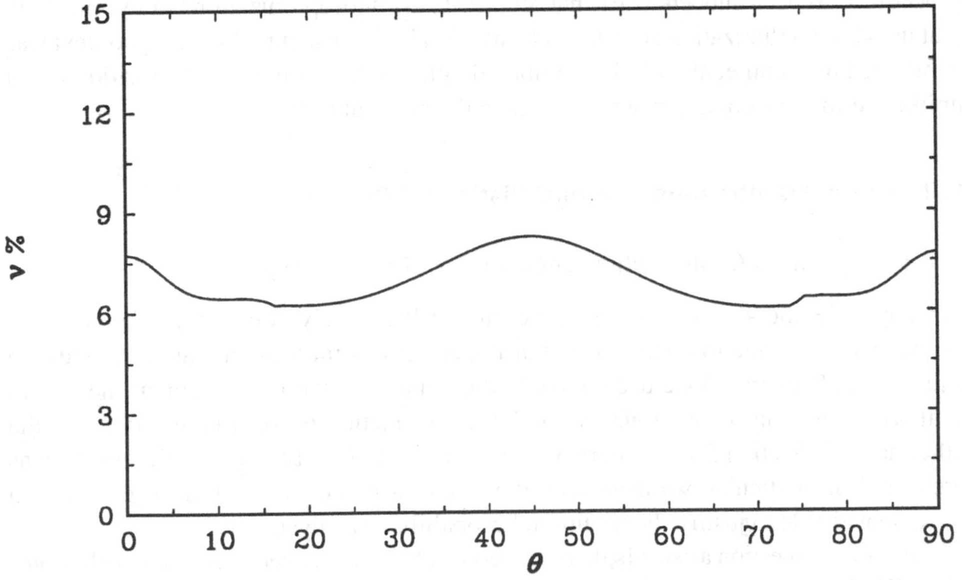

Figure 5.7 Effect of uncertainty in elastic moduli on the buckling load of a set of 5-ply laminated shells (Case 2).

For this problem, it is not advisable to apply the gradient methods or the Davidon-Fletcher-Powell method (Himmelblau, 1972), which are used most often in non-linear optimization, because the directional derivative of the objective function F cannot be easily calculated analytically. So, a direct search should be implemented. Here we choose the *complex method* (Beveridge and Schechter, 1970), which is based exclusively on function comparison; no derivatives are used. In the search for a minimum F, the complex method starts with $2n$ ($n = 4$ in our case) points $E^{(1)}, E^{(2)}, \ldots, E^{(2n)}$, where $E^{(i)} = \{E_1^{(i)}, E_2^{(i)}, E_3^{(i)}, E_4^{(i)}\}$. At each search cycle, a new point is generated by a certain rule in terms of the previous $2n$ points, and the worst point $E^{(j)}$, which has the largest value of F among these $2n$ points, is rejected and replaced by the new point. Whenever the new point generated is beyond the bound, it will be set to the bound. Progress will continue with repeated rejection and regeneration until some criteria are met. For a complete description of this method, consult several monographs (Himmelblau, 1972; Beveridge and Schechter, 1970). Nowadays, performing non-linear programming could also be realized through use of such computational tools as gradient projection, feasible direction, and penalty-function methods (Kirsch, 1981). We must digress that, since the expression of the buckling load is available explicitly, one can choose to optimize a two-term or three-term Taylor expression of the buckling load about the nominal values of elastic moduli subject to non-linear quadratic constraint function (Li et al., 1996). The constraint function, given in Equation (5.17), represents the equation of the ellipsoid of minimum volume that encloses the rectangular parallelepiped representing the original inequality constraints. Van den Nieuwendijk (1997) confirmed the present approach numerically and extended it.

It appears to be remarkable that the present treatment exhibits a basic philosophical difference from the classical optimization studies. In classical optimum design of structures, one looks to *maximize* the buckling loads; here we look for the *least favorable scenarios* (i.e., we determine, for design purposes, minimum buckling loads). This procedure has been dubbed by Elishakoff (1991) as "anti-optimization." A hybrid study that uses both optimization and anti-optimization in the structural buckling context was conducted by Adali et al. (1994). Combined optimization and anti-optimization is not unlike the folk wisdom, that advises "Make the best out of the worst."

5.2 Critical Contrasting of Probabilistic and Convex Analyses

5.2.1 Is There a Contradiction Between Two Methodologies?

In Chapters 3 and 4 we extensively used probabilistic analysis to predict the behavior of the structures due to uncertain initial imperfections modeled as random fields. In addition, in Section 4.6 we used a stochastic version of the finite element method to deal with randomly varying elastic moduli as a function of axial coordinate. On the other hand, in Section 5.1, an alternative, set-theoretical analysis of elastic moduli was presented; in particular, we dealt with the often-encountered situation when the data were unavailable to justify the traditional probabilistic analysis.

A natural question arises: Is there any contradiction between these two methodologies? We feel that the answer to this question is negative. If sufficient data are available, one can utilize probabilistic method; however, if the data are not available, we refrain from recommending, as it is often done, to "invent data out of nowhere," and to perform yet another academic study on the effect of random material properties on the buckling of a structure. In these circumstances, one should utilize non-probabilistic methods of dealing with uncertainty, such as convex modeling. For the detailed survey of the developments in this new alternative to probabilistic modeling, consult the essays of Ben-Haim (1994) and Elishakoff (1995).

The probabilistic analyst in actuality claims, "Give me the joint probability densities of random variables involved, and I will calculate the reliability of the structure through sophisticated numerical methods." This reminds us of the well-known statement by Archimedes: "Give me a firm spot on which to stand, and I will move the earth." It is clear that this analogy is not perfect. However, the resemblance of these two statements is clear. In this connection, the following quotation of Blekhman, Mishkis, and Panovko (1983) appears to be instructive:

> Significantly, the weakness of numerous works on stochastic models sometimes ruling out any application lies in the choice of statistically hypotheses, especially of assumptions regarding the probabilistic features of the given quantities and functions. These features are often regarded as fully known (like an assumption of a normal distribution with known parameters), or as capable of determination. In real situations, it mostly turns out that the needed information is lacking.

Moreover, as was shown by Ben-Haim and Elishakoff (1990) in the series of problems, Elishakoff and Hasofer (1992), Neal, Mathews, and Vangel (1992) and Elishakoff (1995), even small errors in probabilistic data may lead to large errors in estimating probabilities of failure. It should be borne in mind that (Wentzel, 1988)

> it is infrequent that the theory is viewed to be a sort of magic wand yielding information from nothing, i.e., from total ignorance. Those who think so are under a misapprehension since probability theory is used but to transform data on observed phenomena to infer the behavior of those which cannot be observed.

In some circumstances, as Wentzel (1988) notes, utilization of probabilistic method is questionable from the very start.

> A little comment is here in order on the difference between uncertainty and randomness. As will be recalled, probability theory refers the term "random event" to the events which recur and, more important, have a property of statistical stability. The latter implies that similar trials with random outcome tend to assume a stable distribution upon manifold repetition. The frequencies of the events tend then to the respective probabilities, and the sample means to the mathematical expectations. . . .
>
> There is however, nonstochastic uncertainty. . . . The factors . . . are as ever unknown beforehand, but additionally there is no point in speaking of, or trying to evaluate their distributions or other probability characteristics.

As Freudenthal (1956) digresses, "Ignorance of the cause of variation does not make such variation random." One of the present writers (Elishakoff, 1995) notes that "Uncertainty and randomness are not reciprocal."

Under these circumstances it makes sense not to abandon the alternatives to probabilistic methods, namely the theory of fuzzy sets (Zadeh, 1965) and anti-optimization, also known in the literature as the unknown-but-bounded uncertainty approach, guaranteed performance approach, or convex modeling of uncertainty, whose simplest form – interval analysis – was known for decades. First hints of the latter method appeared a long time ago (Bulgakov, 1940, 1946). The idea that bounding techniques, rather than the probabilistic methods, may be preferable in some circumstances has reappeared in works by Drenick (1968, 1970), Shinozuka (1970), Schweppe (1968, 1973), and Chernousko (1981), although in different engineering contexts. However, it was not until the monographs of Schweppe (1973), dealing with control theory, and Ben-Haim (1985), dealing with applications in nuclear engineering, that the non-probabilistic methods started to develop intensively. The convex modeling in the applied mechanics context was developed in monographs of Ben-Haim and Elishakoff (1990), Chernousko (1994), and Elishakoff et al. (1994). For extensive discussion on the convex models of uncertainty, consult with the review article of Ben-Haim (1994) and the essay of Elishakoff (1995).

Several additional quotations appear to be in order. Freudenthal (1961), one of the main architects of the modern probabilistic theory of structures, pinpoints the difficulties associated with probabilistic methods:

> when dealing with probabilities a clear distinction should be made between conditions arising in design of inexpensive mass products on which the probability figures are derived by statistical interpretation of actual observations or measurements (since a sufficiently large number of observations are actually obtainable), and conditions arising in design of structures of complex systems. In the latter, probability figures are used simply as a scale or measure of reliability that permits the comparison of alternative designs. The figures can never be checked by observations or measurements since they are obtained by extrapolations so far beyond any possible range of observation that such extrapolation can no longer be based on statistical arguments but could only be justified by relevant physical reasoning. Under these conditions the absolute probability figures have no real significance.

Bolotin's (1961) reasoning resonates well with the latter one:

> small probabilities of failure, if they are correctly found, still retain their importance as some objective characteristics of possibility of random events taking place. They become sensible when comparing them with each other, allowing contrast of the risk of failure of different structures or of the same structures in different working conditions. . . .

In the context of the probabilistic modeling, once the probability of failure of an ensemble of structures is evaluated, it should be compared with acceptable probability of failure. Grandori (1991) addresses a problem of assigning the acceptable probability of failure:

> the probabilistic approach to structural safety is today a well established paradigm. All overwhelming part of research effort, in fact, has been and still is devoted to estimating failure probabilities. By contrast, only sporadic research deals with the problem of

> choosing an acceptable risk of failure. . . . It is true that the adaptation of a probabilistic approach is in any case progress. . . . However, the concept of structural safety will not leave the "realm of metaphysics" unless we devise a method for justifying the choice of risk acceptability levels.

In this context, a natural question arises: Is there a possibility to directly contrast the probabilistic and the non-probabilistic methodologies? This question was posed to us by Crandall (personal communication, 1990). The reply to this question is affirmative. It was given in the recent study by Elishakoff, Cai, and Starnes (1994b). In order to conduct the probabilistic analysis, the authors of the preceding study first have "pretended" that the experimental information was sufficient to justify the traditional probabilistic analysis. Namely, the Fourier coefficients of the initial imperfections were treated as having a truncated normal distribution. Then the assumption of sufficiency of available information was abandoned, and the preceding Fourier coefficients were assumed to be uncertain but non-random: They were assumed to belong to a multi-dimensional box. The results of this comparison follow.

5.2.2 Deterministic Analysis of a Model Structure for a Specified Initial Imperfection

Consider buckling of an initial imperfection-sensitive structure-column on a non-linear elastic foundation. The system is illustrated in Figure 5.8, and the governing equation is

$$EI\frac{d^4w}{dx^4} + P\frac{d^2w}{dx^2} + K_1w - K_3w^3 = -P\frac{d^2\bar{w}}{dx^2} \tag{5.57}$$

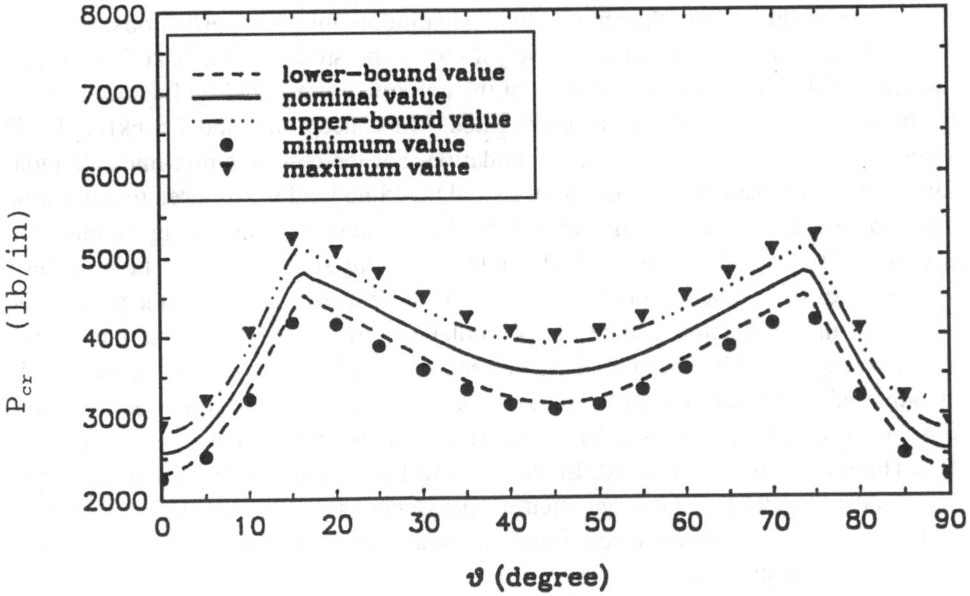

Figure 5.8 Comparison of results for the axial buckling load from the anti-optimization method and numerical non-linear programming.

subject to the boundary conditions

$$w = \frac{d^2 w}{dx^2} = 0, \qquad \text{at} \quad x = 0 \quad \text{and} \quad x = l \tag{5.58}$$

where $\bar{w}$ is the initial deflection, w is the additional deflection due to the axial load P, and EI is the bending rigidity; both K_1 and K_3 are positive constants, representing, respectively, the linear and nonlinear spring constants of the foundation.

If the initial deflection $\bar{w}$ is given as a deterministic function of x, then the Boobnov-Galerkin method can be applied to solve Equation (5.57) approximately. By expressing both the initial deflection $\bar{w}$ and additional deflection w as Fourier series, Equation (5.57) leads to a set of algebraic equations for the unknown Fourier coefficients of the additional deflection. If the foundation is linear, every term in the Fourier series represents a "separate" mode, different modes are uncoupled, and the Fourier coefficients (i.e., the magnitudes of the modes in the additional deflection) can be evaluated from the set of equations. However, in the case of a non-linear foundation, different modes are coupled, and the set of algebraic equations is an infinite hierarchy. To obtain an approximate solution for the additional deflection, this infinite set has to be truncated, and only a finite number of the "most important" modes should be taken into consideration. Based on the Galerkin method, two different numerical schemes (Fraser, 1965; Fraser and Budiansky, 1969; Elishakoff, 1979a) were developed to calculate the buckling load for the non-linear column with a given initial deflection. Fraser (1965) carried out Monte Carlo simulations – a general numerical method to deal with stochastic problems – to calculate the reliability of the column. In Fraser's simulation procedure, a single-mode Boobnov-Galerkin approximation was used to evaluate the buckling load in each sample. To significantly improve the accuracy, a multi-mode Boobnov-Galerkin approximation was used by Elishakoff (1979a), in conjunction with a development of a special procedure to simulate the multi-mode random initial imperfections.

In the present section, which closely follows the study by Elishakoff, Cai, and Starnes (1994a), the buckling problem of the column, represented by Equation (5.57) and boundary conditions (5.58) is investigated. The concept of "modal buckling load" is defined for the column on a linear foundation. For the case of a non-linear foundation, a criterion based on the concept of modal buckling load is proposed to determine which modes are the most significant and should be retained in calculating the buckling load by the Boobnov-Galerkin method. For the reliability study, a random field model – Fourier series with truncated normally distributed coefficients – is suggested to describe a random initial deflection. Monte Carlo simulations are performed to obtain the probability density of the buckling load and the reliability of the column. In the frequently encountered case where the sufficient knowledge about the initial imperfection is absent for substantiation of the stochastic analysis, an alternative non-stochastic approach (Ben-Haim and Elishakoff, 1990; Elishakoff and Ben-Haim, 1990; Elishakoff, 1991) is applied to model the initial geometric imperfections and to obtain the minimum buckling load. The objective is to critically contrast the results from the stochastic and non-stochastic approaches.

First, consider a perfect column on a linear elastic foundation, namely, $K_3 = 0$ and $\bar{w} = 0$ in Equation (5.57). Substituting the non-trivial solution $w = \sin(m\pi x/1)$ into

Equation (5.57), we obtain

$$P(m) = \frac{EI}{l^2}\left(\pi^2 m^2 + \frac{K_1 l^4}{EI\pi^2 m^2}\right) \tag{5.59}$$

When the axial load reaches the minimum value of $P(m), m = 1, 2, \ldots,$ buckling occurs. Namely, the buckling load P_* is determined by

$$P_* = P(m_*) = \min[P(m), m = 1, 2, \ldots] \tag{5.60}$$

where m_* is such an integer that $P(m_*)$ is the minimum of $P(m)$. From (5.59) and (5.60),

$$m_* = \begin{cases} [d], & \text{if} \quad P([d]) < P([d+1]) \\ [d+1], & \text{if} \quad P([d]) > P([d+1]) \end{cases} \tag{5.61}$$

where $d = 1/\pi(K_1/EI)^{1/4}$, and $[d]$ denotes the integer part of d. The buckling load is then

$$P_* = P_{cl} = \frac{EI}{l^2}\left(\pi^2 m_*^2 + \frac{K_1 l^4}{EI\pi^2 m_*^2}\right) \tag{5.62}$$

Equations (5.56) and (5.62) indicate that the buckling occurs in the m_*th mode once the external load reaches the buckling load P_{cl}, which depends on the system parameters.

Now, define the following non-dimensional quantities

$$u = \frac{w}{\Delta}, \qquad \bar{u} = \frac{\bar{w}}{\Delta}, \qquad \eta = \frac{x}{\Delta}, \qquad \alpha = \frac{P}{P_{cl}}$$
$$k_1 = \frac{K_1 l^4}{EI}, \qquad k_3 = \frac{K_3 \Delta^2 l^4}{EI}, \qquad \gamma = \pi^2 m_* + \frac{k_1}{\pi^2 m_*^2} \tag{5.63}$$

where Δ is the radius of gyration of the cross section. Equation (5.57) and boundary conditions (5.58) can be transformed to their non-dimensional forms, respectively,

$$\frac{d^4 u}{d\eta^4} + \alpha\gamma \frac{d^2 u}{d\eta^2} + k_1 u - k_3 u^3 = -\alpha\gamma \frac{d^2 \bar{u}}{d\eta^2} \tag{5.64}$$

and

$$u = \frac{d^2 u}{d\eta^2} = 0, \qquad \text{at} \quad \eta = 0 \quad \text{and} \quad \eta = 1 \tag{5.65}$$

In what follows, we only discuss the problem in non-dimensional form. The qualification phrase "non-dimensional" will be omitted for convenience; for example, u and $\bar{u}$ are called the initial and additional deflections, α is the external axial load, and so on.

The initial deflection can be expanded as

$$\bar{u}(\eta) = \sum_{m=1}^{\infty} \bar{\xi}_m \sin(m\pi\eta) \tag{5.66}$$

We seek the solution of Equation (5.64) also in the form of Fourier series

$$u(\eta) = \sum_{m=1}^{\infty} \xi_m \sin(m\pi\eta) \tag{5.67}$$

Substitution of Equations (5.66) and (5.67) into (5.65) results in (Fraser, 1965; Fraser and Budiansky, 1969)

$$\alpha_m \xi_m - \alpha(\xi_m + \bar{\xi}_m) - \frac{s}{8} \frac{m_*^2}{m^2} I_m = 0 \tag{5.68}$$

where m_* has been given in Equation (5.61),

$$s = \frac{2k_3}{k_1 + \xi^4 m_*^4}, \qquad \alpha_m = \frac{\pi^2 m^2 + k_1/(\pi^2 m^2)}{\pi^2 m_*^2 + k_1/(\pi^2 m_*^2)} \tag{5.69}$$

and

$$I_m = \sum_{p=1}^{\infty} \sum_{q=1}^{\infty} \sum_{r=1}^{\infty} \xi_p \xi_q \xi_r \times \left[\delta_{p+q,r+m} - \delta_{|p-q|,r+m} - \delta_{p+q,|r-m|} + \delta_{|p-q|,|r-m|} + \delta_{p,q}\delta_{r,m}\right] \tag{5.70}$$

in which $\delta_{p,q}$ is the Kronecker delta.

If the elastic foundation is linear, $s = 0$, and Equation (5.68) is reduced to

$$\xi_m = \frac{\alpha}{\alpha_m - \alpha} \bar{\xi}_m \tag{5.71}$$

which indicates that the mth mode in the initial deflection can only induce the mth mode in the additional deflection. The buckling load is usually defined as one that causes the additional deflection to become unbounded (Ziegler, 1969). If the initial deflection contains the m_*th mode, namely $\bar{\xi}_{m_*} \neq 0$, it can be seen from Equation (5.71) that the buckling occurs always in the m_* mode and the buckling load $\alpha_{cl} = \alpha_{m_*} = \min[\alpha_m, m = 1, 2, \ldots] = 1$, which is the same as that of the column without initial imperfection. However, if the initial deflection contains only the kth ($k \neq m_*$) mode, the additional deflection becomes unbounded not at unity load, but at α_k. Thus, we call α_m the mth *modal buckling load*. It is seen from Equation (5.69) that

$$\alpha_m \begin{cases} >1, & \text{if} \quad m \neq m_* \\ =1, & \text{if} \quad m = m_* \end{cases} \tag{5.72}$$

The m_*th mode is called the classical critical mode. In general, an initial deflection contains all the modes, including the m_*th mode. For a non-linear column with initial deflection and under an axial force, an explicit expression for the additional deflection is not obtainable, and a numerical algorithm has to be used to calculate the additional deflection. Depending on the non-linearity, the system may exhibit initial imperfection sensitivity. Fraser and Budiansky (1969) defined the buckling load α^* as

$$\left[\frac{d\alpha}{dF}\right]_{\alpha=\alpha^*} = 0 \tag{5.73}$$

where

$$F = F(u, \bar{u}) = \int_0^1 \left[\frac{1}{2}\left(\frac{du}{d\eta}\right)^2 + \frac{du}{d\eta}\frac{d\bar{u}}{d\eta}\right] d\eta \tag{5.74}$$

representing the end-shortening of the column. According to this definition, the buckling load α_* represents a maximum load the structure can sustain (i.e., the buckling load is defined as a limit load). If the initial and additional deflections are of the forms (5.66) and (5.67), respectively, the end-shortening can be obtained by substituting Equations (5.66) and (5.67) into (5.74) as follows:

$$F = \frac{\pi^2}{4} \sum_{m=1}^{\infty} m^2 \xi_m (\xi_m + 2\bar{\xi}_m) \tag{5.75}$$

Equation (5.73) cannot be solved either analytically or numerically since Equations (5.66)–(5.68) and (5.75) have an infinite number of terms. Approximate solutions were obtained by truncating these equations and retaining the "most important modes" (Fraser and Budiansky, 1969; Elishakoff, 1979a). It is obvious that for "weak" non-linearity (i.e., $k_3 \ll k_1$), the most significant contribution to the buckling load is expected to come from the m_*th mode, which is the critical mode for the corresponding linear case. In this case, an explicit relation between the buckling load and the initial deflection has been obtained (Fraser, 1965). However, if the non-linearity is not small and/or a high accuracy for the approximate solution is required, more modes must be taken into consideration. Thus, a criterion for determining which mode is more important is desirable.

Let us first consider a specific case where $k_1 = 16\pi^4$. It can be easily determined that $m_* = 2$ and that $\alpha_1 = 2.125$, $\alpha_2 = 1$, $\alpha_3 = 1.347$, $\alpha_4 = 2.125$, and $\alpha_5 = 3.205$. This specific example will be investigated in numerical calculations throughout this study. Figures 5.9(a)–(c) show the buckling load for the column with $k_3 = 0.1k_1$ computed from Equations (5.70)–(5.72) and (5.79) by retaining the first four modes. In all these figures, the magnitude of the second mode of the imperfection is taken as a variable for the horizontal coordinate. The buckling load curves in Figure 5.9(a) for different values of ξ_1 and in Figure 5.9(c) for different values of ξ_4 remain very close, whereas those in Figure 5.9(b) for different values of ξ_3 are, although not very close, still not significantly separated. The closeness of these curves indicates that neither ξ_1, ξ_3, nor ξ_4 has a profound influence on the buckling load. However, all these curves have sharp slopes, especially near the origin ($\xi_2 = 0$), which indicates a drastic change of the buckling load with an increasing ξ_2. Therefore, it can be concluded that the second mode plays a dominant role, the third mode also has significant contributions to the buckling load, but less important than the second mode, and the first and fourth modes are even less important. It is noticed that these results "correlate" well with the fact that $1 = \alpha_2 < \alpha_3 < \alpha_1 = \alpha_4$, namely, the closer the modal buckling load is to unity, the more significantly the corresponding mode contributes to the accurate evaluation of the buckling load. For the simplest approximation, only the critical mode needs to be considered. However, if more accurate results are required, the magnitude of the modal bucking load can be considered as a criterion to decide whether or not the corresponding mode should be included in calculating the buckling load by the Galerkin method.

To further confirm this, we return to the preceding specific case of $k_1 = 16\pi^2$, but with the non-linearity k_3 as a variable. The approximate buckling loads are calculated for

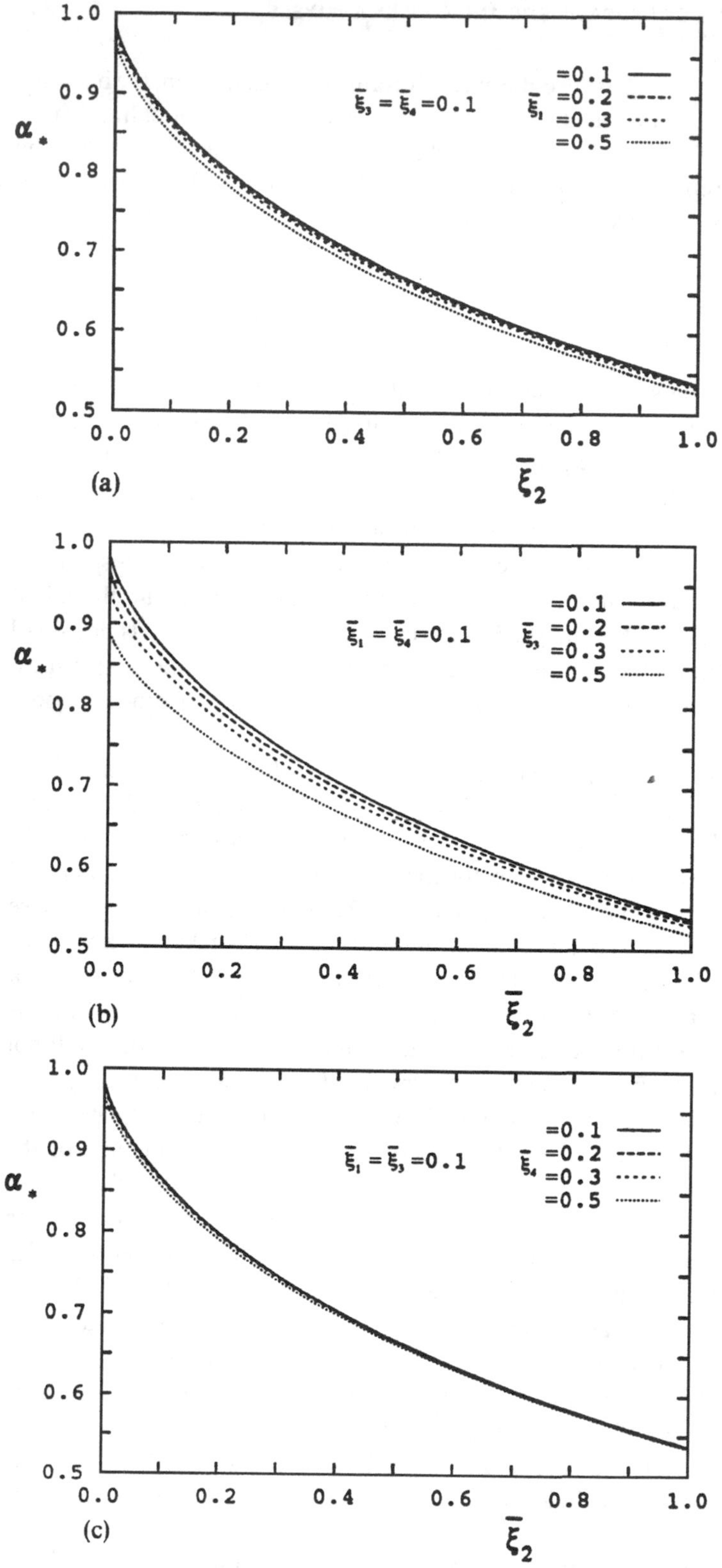

Figure 5.9 Buckling load α versus $\bar{\xi}_2$: (a) different $\bar{\xi}_1$; (b) different $\bar{\xi}_3$; (c) different $\bar{\xi}_4$ (after Elishakoff, Cai, and Starnes, 1994; Copyright © Elsevier Science Ltd., reprinted with permission).

two different initial deflections – (a) $\xi_1 = \xi_2 = \xi_3 = \xi_4 = \xi_5 = 0.1$, (b) $\xi_1 = \xi_3 = \xi_4 = \xi_5 = 0.3$, $\xi_2 = 0.1$ – and are depicted in Figures 5.10(a) and (b). Different combinations of modes are selected to carry out numerical computations. The second mode is included in all cases because it is the critical mode. It can be observed from Figures 5.10(a) and (b) that:

1. If the initial deflection is small and the non-linearity is weak, one mode approximation may give acceptable results.
2. Within two mode approximation, these two modes should be taken as the second mode and the third mode. This can be clearly seen from Figure 5.10(b) and coincides with the fact that α_1 and α_2 are less than the other modal buckling loads.
3. If the first, second, and third modes are included in the computations, the accuracy of the results is indeed improved, but not significantly because the fourth mode is almost as important as the first mode since $\alpha_1 = \alpha_4$. Thus, neglect of the fourth mode may lead to a considerable error.
4. Four-mode approximation gives very accurate results, and the contribution of the fifth mode is negligible because $\alpha_5 = 3.125$ is "far away" form unity.

The preceding analysis confirms that the modal buckling load reflects the importance of the corresponding mode to the buckling load and can serve as a criterion for selecting the modes involved in calculating the buckling load. In order to improve accuracy for the approximate results, the mode with a lower modal buckling load should be taken into account successively.

It is noted from Equation (5.71) that if the critical mode is a high mode (i.e., m_* is large), the α_m value for m near m_* is close to unity; thus, the mth mode is almost as important as the critical mode. In this case, a large number of modes are needed to obtain a highly accurate result.

5.2.3 Probabilistic Analysis

The probabilistic analysis in this section is based on Elishakoff et al. (1994b). In some cases, we can obtain statistical properties of a initial deflection by measurements or experience. In these cases, the initial deflection should be treated as a random field. Every sample in this random field can still be described by a Fourier series, namely, Equation (5.66). However, different samples have different coefficients in their respective Fourier series. Thus, a random initial deflection can be described by the Fourier expansion (5.66) with random coefficients $\bar{\xi}_m = (m = 1, 2, \ldots)$. To proceed with the reliability analysis, a knowledge of the joint probability distribution for random variables $\bar{\xi}_m$ is necessary. The normal distribution is a popular choice due to its simplicity in analysis, but it may not be appropriate for practical cases because the initial deflection can be visualized as limited in a certain range. (Indeed, quality control will discard "very imperfect" columns.) To improve this model, we propose a *truncated* normal distribution for each random variable $\bar{\xi}_m$ as follows:

$$p(\bar{\xi}_m) = \begin{cases} c_m \exp\left(-\dfrac{\bar{\xi}_m^2}{b_m^2}\right), & |\bar{\xi}_m| \le A_m \\ 0, & |\bar{\xi}_m| > A_m \end{cases} \tag{5.76}$$

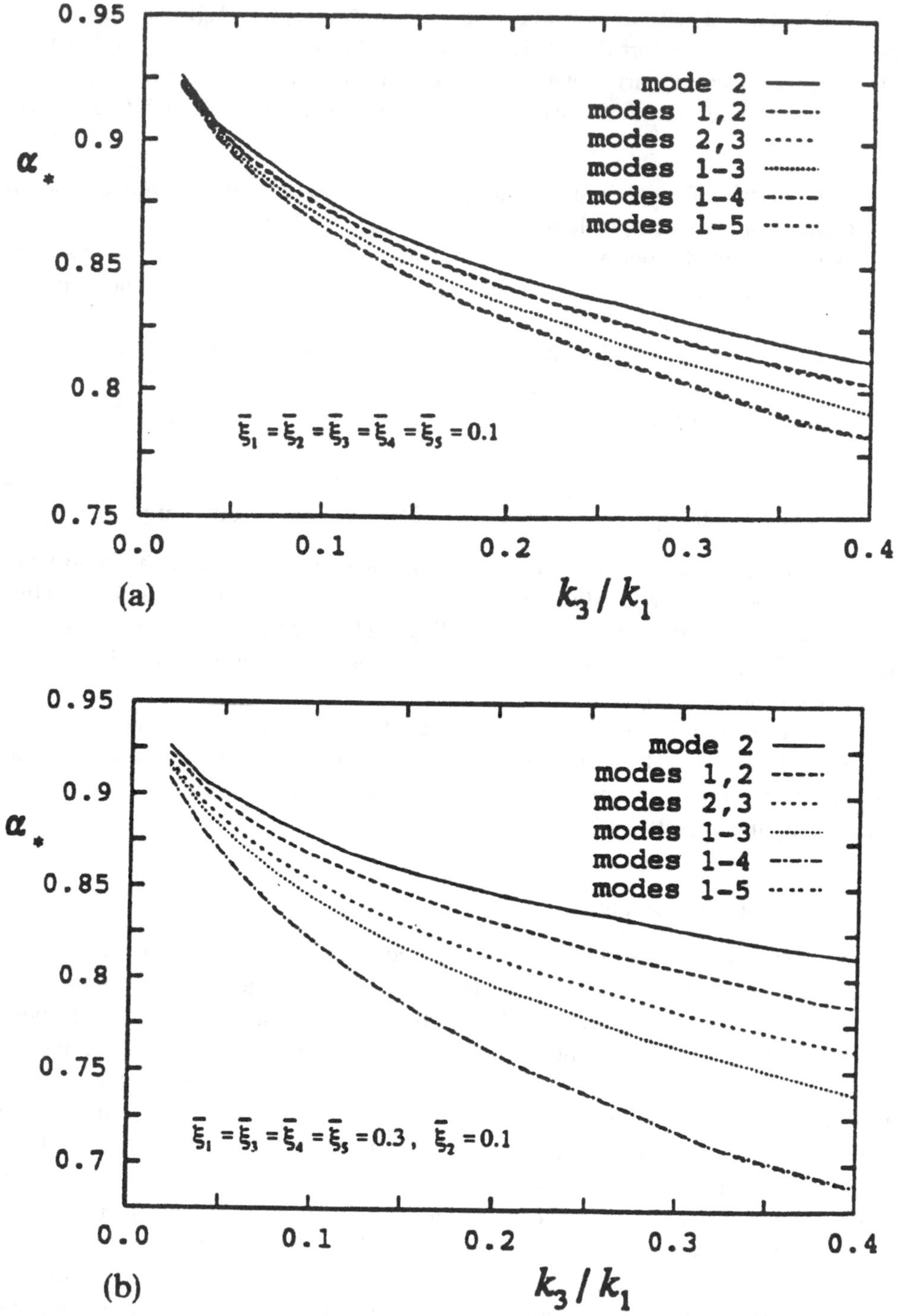

Figure 5.10 Buckling load α computed by using different combination of modes versus nonlinearity k_3/k_1: (a) $\bar{\xi}_1 = \bar{\xi}_2 = \bar{\xi}_3 = \bar{\xi}_4 = \bar{\xi}_5 = 0.1$; (b) $\bar{\xi}_1 = \bar{\xi}_3 = \bar{\xi}_4 = \bar{\xi}_5 = 0.3$, $\bar{\xi}_2 = 0.1$ (after Elishakoff, Cai, and Starnes, 1994; Copyright © Elsevier Science Ltd., reprinted with permission).

where $p(\bar{\xi}_m)$ is the probability density of $\bar{\xi}_m$, each A_m is a maximum possible value for the random variable ξ_m, b_m are parameters, and the normalization constants c_m are derived from

$$c_m = \frac{1}{2b_m \operatorname{erf}\left[\frac{A_m}{b_m}\right]} \tag{5.77}$$

in which the error function erf$(\cdot)$ is defined as

$$\operatorname{erf}(x) = \int_0^x e^{-t^2} dt \tag{5.78}$$

Also, we assume that the random coefficients $\bar{\xi}_m$ are independent for different m. The probability density functions of the random coefficients are shown in Figure 5.11 with $\bar{\xi}_m$, A_m, and b_m replaced by $\bar{\xi}$, A, and b for simplicity. With a given A, the probability density depends exclusively on b. A large b corresponds to a large deviation of $\bar{\xi}$. When $b^2 \gg A^2$, $\bar{\xi}$ is nearly uniformly distributed, as shown by the case of $b = 1$ in Figure 5.11.

The auto-correlation function and the mean square value of the initial deflection are, respectively,

$$R_{\bar{u}}(\eta_1, \eta_2) = \sum_{m=1}^{\infty} E\left[\bar{\xi}_m^2\right] \sin(m\pi\eta_1)\sin(m\pi\eta_2) \tag{5.79}$$

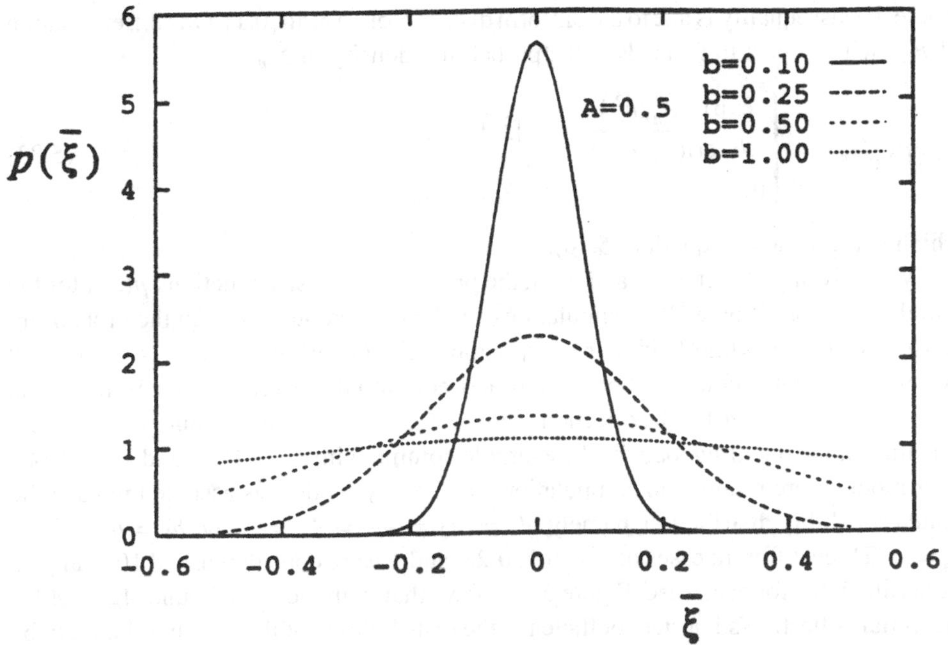

Figure 5.11 Probability density function for a truncated normally distributed random variable (after Elishakoff, Cai, and Starnes, 1994; Copyright © Elsevier Science Ltd., reprinted with permission).

and

$$E(\bar{u}^2(\eta)] = \sum_{m=1}^{\infty} E[\bar{\xi}_m^2] \sin^2(m\pi\eta) \tag{5.80}$$

Equations (5.79) and (5.80) show that the random field of the initial deflection is inhomogeneous. From the probability density function (5.77) and the auto-correlation function (5.79), it can be seen that the model is quite general and feasible for describing a real initial deflection.

Another advantage of the random field model is its simplicity to be simulated. The realization of $\bar{\xi}_m$, denoted by $(\bar{\Sigma}_m)_k$, $k = 1, 2, \ldots$, can be generated by

$$(\bar{\Sigma}_m)_k = b_m \mathrm{erf}^{-1}\left[(2\delta_k - 1)\mathrm{erf}\left(\frac{A_m}{b_m}\right)\right] \tag{5.81}$$

where δ_k, $k = 1, 2, \ldots$, are independent random numbers uniformly distributed in [0,1]. In fact, the probability distribution function for $\bar{\Sigma}_m$ is

$$\begin{aligned}
F_{\bar{\Sigma}_m}(\bar{\xi}_m) &= \mathrm{Prob}\,[\bar{\Sigma}_m \leq \bar{\xi}_m] \\
&= \mathrm{Prob}\{b_m \mathrm{erf}^{-1}[(2\delta_k - 1)\mathrm{erf}(A_m/b_m)] \leq \bar{\xi}_m\} \\
&= \mathrm{Prob}\left[\delta_k \leq \frac{1}{2} + \frac{\mathrm{erf}(\bar{\xi}_m/b_m)}{2\,\mathrm{erf}(A_m/b_m)}\right] \\
&= \frac{1}{2} + \frac{\mathrm{erf}(\bar{\xi}_m/b_m)}{2\,\mathrm{erf}(A_m/b_m)}
\end{aligned} \tag{5.82}$$

where the last equality is due to the uniform distribution of δ_k in [0,1]. One differentiation of $F_{\bar{\Sigma}_m}$ with respect to $\bar{\xi}_m$ leads to the probability density of $\bar{\Sigma}_m$

$$p_{\bar{\Sigma}_m}(\bar{\xi}_m) = \begin{cases} \dfrac{\exp\left(-\bar{\xi}_m^2/b_m^2\right)}{2b_m \mathrm{erf}(A_m/b_m)}, & |\bar{\xi}_m| \leq A_m \\ 0, & |\bar{\xi}_m| > A_m \end{cases} \tag{5.83}$$

which is the same as Equation (5.76).

With given parameter A_m and b_m in the probability density functions $p(\bar{\xi}_m)$ for the initial deflection, Monte Carlo simulations can be carried out to obtain the probability density for the buckling load. For every sample, the Galerkin method can be applied by retaining finite modes in both initial and additional deflections according to the criterion proposed in the last section. Figure 5.12 shows the computed probability densities of the buckling load for the example column with $k_1 = 16\pi^4$ and $k_2 = 0.1k_1$. Four modes were retained in computations, and every mode was assumed to have the same probability distribution, namely, $A_1 = A_2 = A_3 = A_4 = A$ and $b_1 = b_2 = b_3 = b_4 = b$. Three different cases of $b = 0.1$, 0.25, and 1 were considered, and 10^5 samples were calculated for each case. Figure 5.12 shows that with the same bound $A_m = A$ for the initial deflection's Fourier coefficients, the distributions of the buckling load can be significantly different, depending on the distributions of the initial deflections. For a larger deviation of the initial deflection (the case $b = 1$), the probability that a smaller buckling load occurs increases.

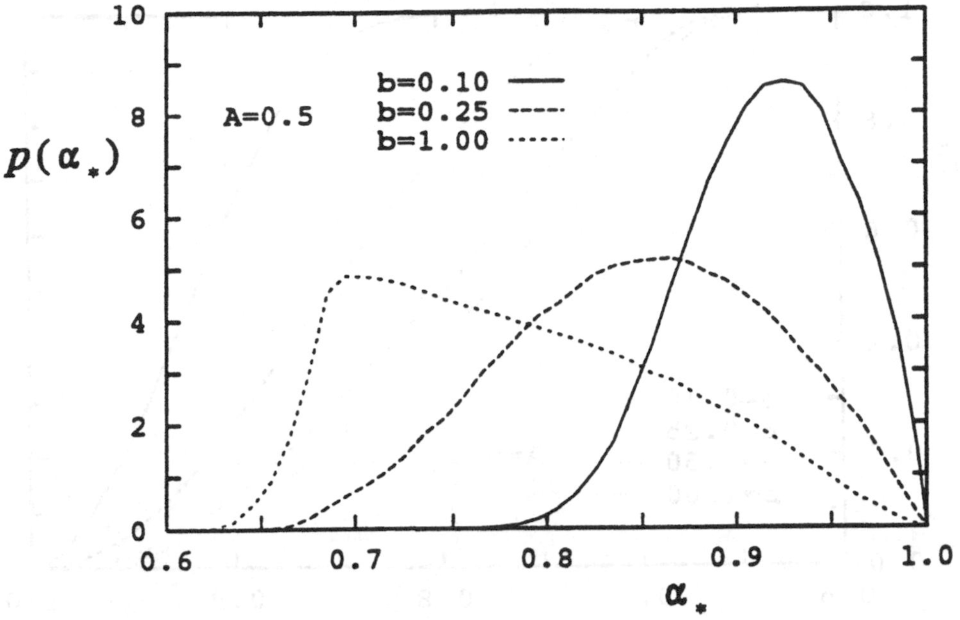

Figure 5.12 Probability density function of the buckling load (after Elishakoff, Cai, and Starnes, 1994; Copyright © Elsevier Science Ltd., reprinted with permission).

The reliability function for the column with a random initial imperfection, subject to a prespecified axial load α, is defined as (Bolotin, 1958)

$$R(\alpha) = \text{Prob}[\alpha_* \geq \alpha] \tag{5.84}$$

where α_* is the random buckling load. Figure 5.13 depicts the reliability functions for the preceding example column with $A = 0.5$ and four different values of b. As expected, the reliability for a given load α significantly depends on the parameter b, which indicates the deviation of the initial deflection from the nominal value. If we design a column based on the stochastic approach, the value of the load corresponding to R equal to a codified required reliability r is the maximum value for the admissible axial load. The latter load should be used as a design load for an ensemble of columns on non-linear elastic foundation with stochastic imperfection.

5.2.4 Non-Stochastic, Anti-Optimization Analysis

The non-stochastic, anti-optimization analysis in this section is based on Elishakoff et al. (1994b). In some cases, it is even difficult to estimate the probability distribution of the initial imperfection. In these circumstances, the stochastic approach is not applicable and a non-probabilistic model for the initial imperfection must be adopted. If we still expand the initial deflection as a Fourier series (5.66), then a simple non-probabilistic model for the initial deflection is its Fourier expansion coefficients vary in a hyper-cuboid set:

$$Z(A)\colon |\xi_m| \leq A_m \qquad (A_m \geq 0, m = 1, 2, \ldots) \tag{5.85}$$

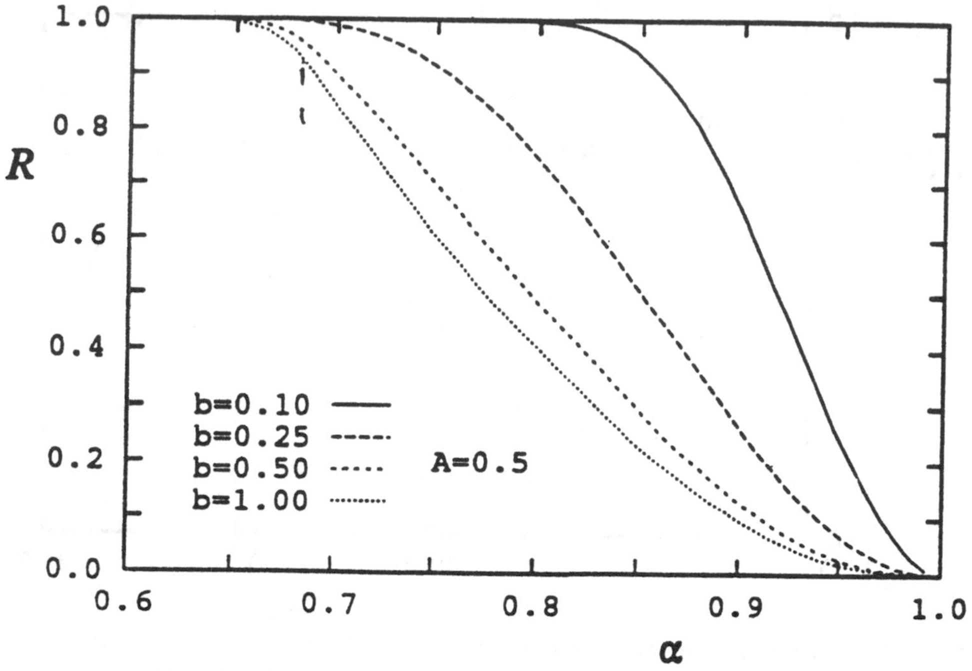

Figure 5.13 Reliability function versus actual axial load (after Elishakoff, Cai, and Starnes, 1994; Copyright © Elsevier Science Ltd., reprinted with permission).

where $A = \{A_1, A_2, \ldots\}$ is a constant vector. Our objective now is to find the minimum, least favorable limit load for all possible initial deflection $\bar{\xi} = \{\bar{\xi}_1, \bar{\xi}_2, \ldots\}$ belonging to the set $Z(A)$. If we design a column based on the non-stochastic approach, the minimum buckling load is the maximum value for the admissible axial load. Thus, we arrive at the alternative way of determining the admissible axial load, which can be applied to an ensemble of columns with bounded Fourier coefficients.

As discussed before, the buckling load is a function of $\bar{\xi}$ and can be expressed formally as

$$\alpha_* = \psi(\bar{\xi}) \tag{5.86}$$

where the function form ψ in (5.86) is implicit and non-linear; in fact, it represents a procedure of numerical evaluation of the buckling load as described previously. Then, the problem to find the minimum buckling load becomes an extreme value problem

$$\min_{\bar{\xi}\in Z(A)} (\alpha_*) = \min_{\bar{\xi}\in Z(A)} \psi(\bar{\xi}) \tag{5.87}$$

According to the method of Lagrange multipliers, we construct an auxiliary function

$$G(\bar{\xi}, A, x) = \psi(\bar{\xi}) + \sum_{m=1}^{N} \lambda_m (\bar{\xi}_m^2 - A_m^2 + x_m^2) \tag{5.88}$$

where N is the number of the most significant modes involved in computations, $x = (x_1, x_2, \ldots, x_N)$ is an auxiliary vector variable, and λ_m $(m = 1, 2, \ldots, N)$ are Lagrange multipliers. By letting $\partial G/\partial \bar{\xi}_m = 0$ and $\partial G/\partial x_m = 0$, we obtain

$$\frac{\partial \psi}{\partial \bar{\xi}_m} + 2\lambda_m \bar{\xi}_m = 0, \qquad m = 1, 2, 3, \ldots, N \tag{5.89}$$

$$\lambda_m x_m = 0, \qquad m = 1, 2, 3, \ldots, N \tag{5.90}$$

which, combined with

$$\bar{\xi}_m^2 - A_m^2 + x_m^2 = 0, \qquad m = 1, 2, \ldots, N \tag{5.91}$$

constitute a set of non-linear algebraic equations for unknown $\bar{\xi}_m$, x_m, and λ_m. If derivatives $\partial\psi/\partial\bar{\xi}_m$ exist in the entire set $Z(A)$, the extreme values of the function $\psi(\bar{\boldsymbol{\xi}})$ are reached at some of the solution points of the set of equations. For example, if $N = 4$, the solutions for the set of Equations (5.89)–(5.91) are

$$\begin{array}{lllll}
\text{(i)} & \bar{\xi}_1 = \pm A_1, & \bar{\xi}_2 = \pm A_2, & \bar{\xi}_3 = \pm A_3, & \bar{\xi}_4 = \pm A_4; \\
\text{(ii)} & \bar{\xi}_i = \pm A_i, & \bar{\xi}_j = \pm A_j, & \bar{\xi}_k = \pm A_k, & \partial\psi/\partial\bar{\xi}_l = 0; \\
\text{(iii)} & \bar{\xi}_i = \pm A_i, & \bar{\xi}_j = \pm A_j, & \partial\psi/\partial\bar{\xi}_k = 0, & \partial\psi/\partial\bar{\xi}_l = 0; \\
\text{(iv)} & \bar{\xi}_i = \pm A_i, & \partial\psi/\partial\bar{\xi}_j = 0, & \partial\psi/\partial\bar{\xi}_k = 0, & \partial\psi/\partial\bar{\xi}_l = 0; \\
\text{(v)} & \partial\psi/\partial\bar{\xi}_i = 0, & \partial\psi/\partial\bar{\xi}_j = 0, & \partial\psi/\partial\bar{\xi}_k = 0, & \partial\psi/\partial\bar{\xi}_l = 0;
\end{array} \tag{5.92}$$

where $i, j, k, l = 1, 2, 3, 4$ and $i \neq j \neq k \neq l$. However, $\partial\psi/\partial\bar{\xi}_m$ may not exist at some points; for example, $\partial\psi/\partial\bar{\xi}_2$ does not exist at $\bar{\xi}_2 = 0$ in the example column, as known from the single-mode approximation (Elishakoff, 1979) and the numerical results shown in Figures 5.9(a)–(c). Then, besides the solution points of Equations (5.89)–(5.91), the points at which any of the derivatives $\partial\psi/\partial\bar{\xi}_m$ does not exist have to be included as candidate points to seek the extreme values of the function $\psi(\bar{\boldsymbol{\xi}})$.

Let us consider the example column: Because of the symmetry of the problem, function ψ is an even function of $\bar{\xi}_m$. Thus, at $\bar{\xi}_m = 0$, either $\partial\psi/\partial\bar{\xi}_m$ does not exist or $\partial\psi/\partial\bar{\xi}_m = 0$. Therefore, if four modes are retained in computations, the following points:

$$\begin{array}{lllll}
\text{(i)} & \bar{\xi}_1 = \pm A_1, & \bar{\xi}_2 = \pm A_2, & \bar{\xi}_3 = \pm A_3, & \bar{\xi}_4 = \pm A_4; \\
\text{(ii)} & \bar{\xi}_i = \pm A_i, & \bar{\xi}_j = \pm A_j, & \bar{\xi}_k = \pm A_k, & \bar{\xi}_l = 0; \\
\text{(iii)} & \bar{\xi}_i = \pm A_i, & \bar{\xi}_j = \pm A_j, & \bar{\xi}_k = 0, & \bar{\xi}_l = 0; \\
\text{(iv)} & \bar{\xi}_i = \pm A_i, & \bar{\xi}_j = 0, & \bar{\xi}_k = 0, & \bar{\xi}_l = 0; \\
\text{(v)} & \bar{\xi}_i = 0, & \bar{\xi}_j = 0, & \bar{\xi}_k = 0, & \bar{\xi}_l = 0;
\end{array} \tag{5.93}$$

contain all the solution points for the set of Equations (5.89)–(5.91), as well as the points at which any of the derivatives $\partial\psi/\partial\bar{\xi}_m$ do not exist. Therefore, the points in (5.93) are the candidate points at which the ψ function may reach extreme value. Among these, the point (v), (i.e., the origin) corresponds to the case where the initial imperfection is absent. At this point, the function ψ obtains its maximum value for an imperfection sensitive structure. The minimum buckling load can be found by comparing the values of ψ at those points listed in (i) to (iv). Note that points in (i)–(iv) correspond, respectively, to vertices, middle points of side, center of planes, and centers of three-dimensional projected cuboids for the four-dimensional cuboid.

Figures 5.14(a) and (b) show the minimum buckling load for the example column with the initial deflection bound A_j = constant = A as a variable. Also, the admissible loads corresponding to different values of reliability are calculated from the stochastic approach and depicted in the same figures. For the case of $b = 1$ in Figure 5.14(b), namely, large deviation of the initial deflection, the minimum buckling load and the admissible load corresponding to different required reliability levels $r = 0.9, 0.99$, and 0.999 do not exhibit much difference regardless of the magnitude of the boundary A. Therefore, design can be made based on the non-stochastic approach because it is much simpler than the stochastic one. The same situation can also be found for the case of $b = 0.1$ and small A ($A < 0.2$ in the present case), shown in Figure 5.14(a). However, if the deviation of the initial deflection is small and the boundary for the initial deflection is large [$b = 0.1$ and $A > 0.2$ in the present case, shown in Figure 5.14(a)], the admissible value for the axial load obtained from the stochastic approach may be well above the minimum buckling load. This implies that the use of the non-probabilistic method when the sufficient probabilistic information is available may not be advisable; the latter technique may lead to a conservative design.

One of the non-stochastic models of uncertainty, namely, the Fourier coefficients varying within a hyper-cuboid set, is adopted for the initial deflection in this study, and the minimum buckling load is computed for this model. The comparison of the results with those obtained from the stochastic approach indicates that the design based on the non-stochastic approach is acceptable for a initial deflection with a large deviation but may be conservative for that with a small deviation.

It is remarkable that in some circumstances both approaches, although being of cardinally different nature, may yield close values for the design axial loads. If probabilistic information is unavailable, one should not propose a probabilistic model, based on an arbitrary assumption on the distribution of the Fourier coefficients. Rather, in such circumstances one should use the non-stochastic approach to uncertainty. Only when the full probabilistic information is available and the initial deflection's Fourier coefficients have a relatively small deviation, will use of the non-stochastic approach be unadvisable; at this time, purely stochastic analysis should be conducted. However, even when probabilistic information is available to substantiate the probabilistic analysis, if the distribution of the initial imperfections is "close" to uniform, one may prefer a simpler non-stochastic, convex analysis, because it yields admissible axial loads comparable to the results of the stochastic approach. Naturally, when the probabilistic information is unavailable, the probabilistic methods cannot be utilized. Use of the probabilistic modeling in such circumstances would be equivalent to treating it as a "magic wand," producing information, as Wentzel (1988) notes, from a void. In the case when scarce information is available on initial imperfections, application of non-stochastic methods appears to be a natural alternative to the probabilistic methods.

Thus we conclude that there are classes of problems where solely probabilistic or solely non-probabilistic convex modeling can be conducted to describe the uncertain quantities involved. It is also remarkable that there is a region of parameter variations, where either probabilistic and convex modeling can be applied. In this case the non-probabilistic convex modeling appears to be superior to that of the probabilistic approach because the former is numerically and conceptually easier than the latter.

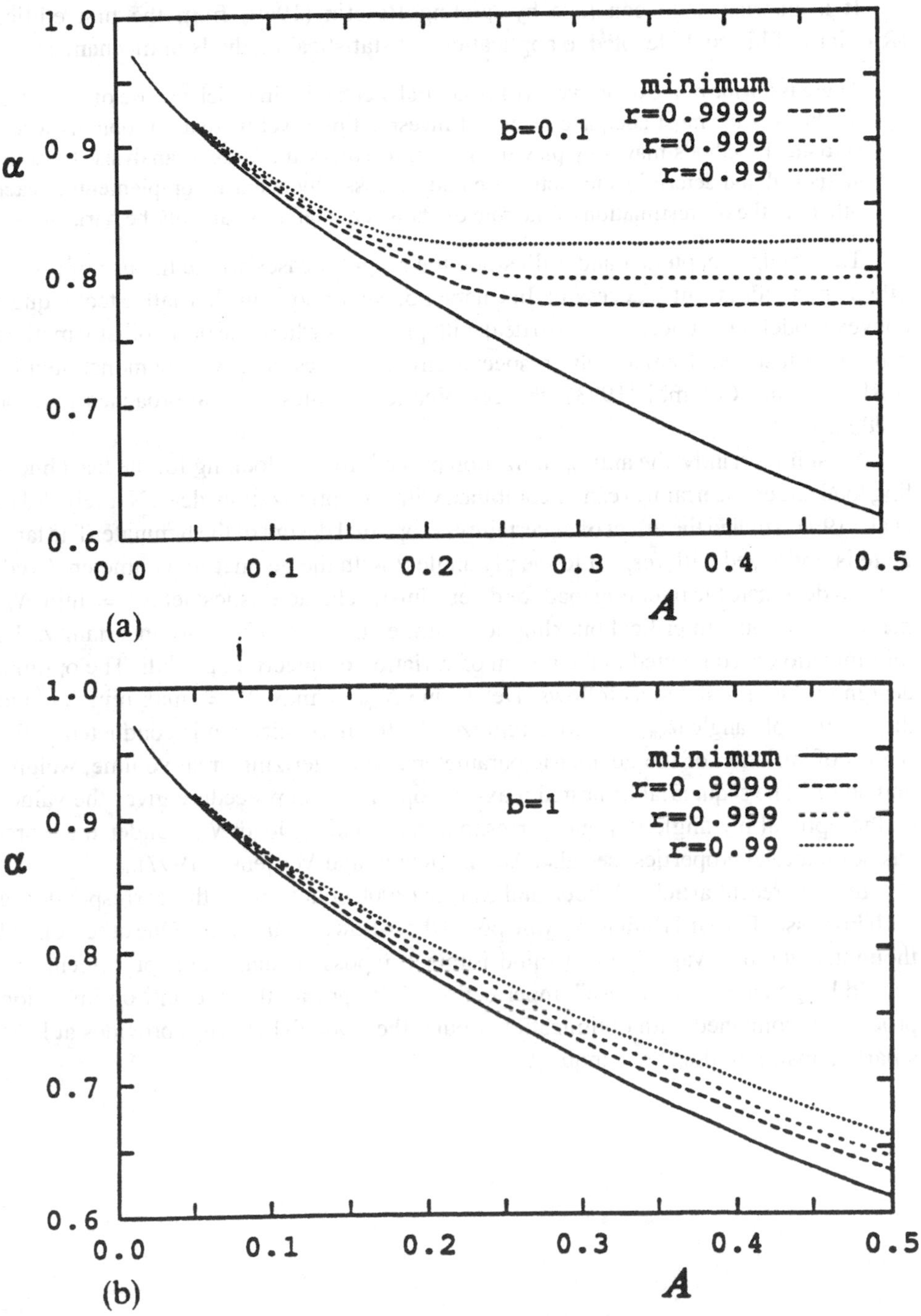

Figure 5.14 Comparison of admissible axial loads computed from stochastic and non-stochastic, anti-optimization approaches: (a) $b = 0.1$ and (b) $b = 1$ (after Elishakoff, Cai, and Starnes, 1994; Copyright © Elsevier Science Ltd., reprinted with permission).

It is instructive to conclude by quoting Bolotin (1969) from the first edition (Russian) of his book devoted to applications of statistical methods in mechanics:

> There is a wide range of problems of structural mechanics in which the use of statistical methods is the most adequate means of investigation – yet there are problems where statistical methods may only play the role of auxiliary methods of analysis. Here the statistical and deterministic methods could successfully coexist, complementing each other ... the overestimation of the role of statistical methods can only be harmful.

To sum up, Chapters 3 and 4 illustrate the range of cases where the use of probabilistic methods is quite successful. In Chapter 5, we propose an alternative technique – convex modeling of uncertainty – to deal with problems where the probabilistic method cannot be justified. Remarkably in special circumstances, as it was demonstrated by Elishakoff and Colombi (1993), the combination of these two approaches may be needed.

Most importantly, the anti-optimization procedure (i.e., looking for the least buckling load under uncertainty) can be combined with the optimization ideas. Namely, Adali et al. (1994) studied the effect of uncertainty on optimal design of the laminate. The laminate is optimized with respect to the ply angle θ with the geometric parameters fixed. Let N_{cl} designate the buckling load for deterministic characteristics; let $N^* = \min N_{cl}$ denote the minimum critical buckling load of the structure subject to uncertainty; the minimization is conducted in the region of variation of uncertain moduli. The optimal design problem is stated as follows: Determine $N_{\text{opt}} = \max N^* = \max \min N_{cl}$ and the optimal ply angle θ_{opt} so as to maximize N^*; the maximization is conducted in the region of variation of the geometric parameters, characterizing area, volume, weight, and so on. The sequential or nested max-min optimization procedure gives the values of the optimal ply angle θ_{opt} and corresponding buckling load $N_{\max}$ under the worst case of material properties (see also Adali, Richter, and Verijenko, 1997).

In their recent article, Arbocz and Singer (2000) quote from the correspondence with Professor Bernard Budiansky who posed the following question: "One other related thought, that is only vaguely in my mind, is this: Is it possible that a more prominent role should be given to "worst-case" imperfections?" It appears that the anti-optimization procedure, combined with optimization around the "worst" behavior, provides at least a partial answer to the above inquiry.

CHAPTER SIX

Application of the Godunov-Conte Shooting Method to Buckling Analysis

The purpose of computing is insight, not numbers. . . . Usually the first question to ask is "What are we going to do with the answers?"

R. W. Hamming

There is safety in numbers.

Euripides

I have no satisfaction in formulas unless I feel their numerical magnitude.

Lord Kelvin

This chapter does not give a general overview of numerical methods utilized in the buckling of linear and non-linear structure; rather, it addresses a specific numerical method, namely the Godunov-Conte method, which is often used in the buckling analysis. The Godunov-Conte, which avoids the loss of accuracy resulting from the numerical treatment often associated with stability and vibration analysis of elastic bodies, consists of parallel integration of the set of k homogeneous equations under the Kronecker-delta initial conditions, which are orthogonal (k being the number of "missing" conditions); after each step, subject to Conte's test, the set of solutions is reorthogonalized by the Gram-Schmidt procedure and integration is continued. The procedure prevents flattening of the base solutions, which otherwise become numerically dependent.

6.1 Introductory Remarks

Consider the eigenvalue problem defined by the set of $n = 2k$ linear ordinary differential equations

$$\dot{y} = A(t, \lambda)y \tag{6.1}$$

where $y(t, \lambda)$ is a $(n \times 1)$-vector with components $y_1(t, \lambda), y_2(t, \lambda), \ldots, y_n(t, \lambda)$; $A(t, \lambda)$ is an $n \times n$ matrix whose i, j-element is $a_{ij}(t, \lambda)$, λ being the sought eigenvalue; the initial conditions are

$$y_i(t_0) = 0 \qquad i = 1, 2, \ldots, k \tag{6.2}$$

The terminal conditions

$$y_{k+m}(t_f) = 0 \qquad (m = 1, 2, \ldots, k) \tag{6.3}$$

This problem can be solved by the following simple procedure, known in literature as the Goodman-Lance method (1956), or the method of complementary functions. Let $u^{(j)}(t, \lambda)$ $(j = 1, 2, \ldots, n; t_0 \le t \le t_f)$ be the set of n linearly independent solutions under the initial conditions

$$u_i^{(j)}(t_\delta, \lambda) = \delta_{ij} \qquad (i, j = 1, 2, \ldots, n) \tag{6.4}$$

where δ_{ij} is Kronecker's delta. The general solution of (6.1) is then found in a linear combination

$$y_i(t, \lambda) = \sum_{j=1}^{n} b_j u_i^{(j)}(t, \lambda) \tag{6.5}$$

where the b_j constants are to be determined by the boundary conditions. By virtue of (5.4)

$$y_i(t_0, \lambda) = b_i = 0, \qquad i = 1, 2, \ldots, k \tag{6.6}$$

$$y_{k+i}(t_0, \lambda) = b_{k+i}, \qquad i = 1, 2, \ldots, k \tag{6.7}$$

In other words, the constants $b_{k+i}, i = 1, 2, \ldots, k$, are equivalent to "missing" initial conditions. Using (5.3)

$$y_{k+i}(t_f, \lambda) = \sum_{\alpha=1}^{k} b_{k+\alpha}\, u_{k+i}^{(i)}(t_f, \lambda) = 0 \qquad (i = 1, 2, \ldots, k) \tag{6.8}$$

The homogeneous system (6.8) has a solution only if the determinant of the coefficient matrix $[f(\lambda)]$ vanishes

$$\det[f(\lambda)] = 0 \tag{6.9}$$

where

$$[f(\lambda)] = \begin{bmatrix} u_{(k+1)}^{(k+1)} & (t_f, \lambda) & u_{k+1}^{(k+2)} & (t_f, \lambda) & \cdots & u_{k+1}^{(n)} & (t_f, \lambda) \\ u_{k+2}^{(k+1)} & (t_f, \lambda) & u_{k+2}^{(k+2)} & (t_f, \lambda) & \cdots & u_{k+2}^{(n)} & (t_f, \lambda) \\ & & & \vdots & & & \\ u_n^{(k+1)} & (t_f, \lambda) & u_n^{(k+2)} & (t_f, \lambda) & \cdots & u_n^{(n)} & (t_f, \lambda) \end{bmatrix} \tag{6.10}$$

The procedure thus consists of determining a sequence of values λ until (6.9) is satisfied to a specified accuracy. Any root-finder technique may be used to arrive at the successive approximations of the eigenvalue λ^*, for which the eigenfunctions are readily obtainable. Let the rank of $[f(\lambda^*)]$ be $k - 1$. Then the system for determining the bs is (Grigolyuk et al., 1971)

$$\sum_{\alpha=1}^{k} b_{k+\alpha}\, u_{k+i}^{(i)}(t_f, \lambda^*) = -b_{k+r}\, u_{k+r}^{(i)}(t_f, \lambda^*), \qquad i, \alpha \neq r, \qquad b_{k+r} = b^0 \tag{6.11}$$

where b^0 is any constant, and the eigenfunctions are found by substituting $b_{k+1}, b_{k+2}, \ldots,$ $k_{k+r-1}, b_{k+r}, b_{k+r+1}, \ldots, b_n$ in (6.8); since the system is homogeneous, they are determined only up to a multiplicative constant, b^0. This procedure, although mathematically exact, may often lead to very poor or even completely incorrect results when applied in numerical form. In some cases, the vectors $u^{(k+1)}, u^{(k+2)}, \ldots, u^{(n)}$, which form the columns of $[f(\lambda)]$, become numerically dependent as t increases, irrespective of the integration procedure used. For example, Grigolyuk et al. (1971) found that, in stability problems of cylindrical shells, this situation occurs when the non-dimensional parameter Z satisfies the inequality

$$Z = L(1-\nu^2)^{0.25}(RH)^{-0.5} > 10 \tag{6.12}$$

where L, R and H are, respectively, the length, radius, and thickness of the shell, and ν is Poisson's ratio of the shell material. The matrix $[f(\lambda)]$ turns out to be poorly conditioned, the eigenvalues ν^* are inaccurate, and the expansion (6.8) leads to poor numerical results.

A method avoiding loss of accuracy was proposed by Godunov (1961), whereby the matrix $[f(t,\lambda)]$ of base solutions remains orthogonal throughout. This is achieved by integrating the set of k homogeneous equations in parallel under the Kronecker-delta initial conditions (which are orthogonal), with the set of solutions reorthogonalized after each integration step by the Gram-Schmidt procedure. The method was successfully applied by Grigolyuk and Lipovtsev (1975) and Valishvili (1975) in stability problems of axisymmetric shells, and by Novichkov and Indenbaum (1972) in non-homogeneous two-point boundary value problems (specifically, axisymmetric deformation of a circular cylinder made of low-modulus material and enclosed in a plastic shell subjected to internal pressure and a thermal effect). The method was subsequently modified and adapted for more practical use by Conte (1966), with the angle between the relevant pairs of vectors $u^{(k+1)}(t,\lambda), u^{(k+2)}(t,\lambda), \ldots, u^{(n)}(t,\lambda)$, for the points t_i at which the solution is computed – serving as criterion of the need for reorthogonalization. If the least angle is less than a specified tolerance β, the solution vectors are to be orthogonalized; otherwise, we proceed with the next step. Conte's test reads:

$$\min_{i,j} \cos^{-1} \frac{\left|\left(u^{(i,q-1)}(t), u^{(j,q-1)}(t)\right)\right|}{\left\|u^{(i,q-1)}(t)\right\|^2 \left\|u^{(j,q-1)}(t)\right\|^2} < \beta, \qquad i \neq j \tag{6.13}$$

$$\left\|u^{(i,q-1)}(t)\right\|^2 = \left(u^{(i,q-1)}(t), u^{(i,q-1)}(t)\right) \tag{6.14}$$

$$\left\|u^{(j,q-1)}(t)\right\|^2 = \left(u^{(j,q-1)}(t), u^{(j,q-1)}(t)\right) \tag{6.15}$$

the parentheses denoting an inner product. Here $u^{(m,q)}(t)$ is a $(k \times 1)$ vector solution of Equations (6.1)–(6.4) – m denoting the index counter for the number of the set of linear conditions $m = 1, 2, \ldots, k$, and q that of the point at which the vectors were last orthogonalized. For $q = 0$, $u^{(m,q)}(t_0) \equiv u^{(m)}(t_0)$ coincides with Kronecker-delta initial conditions in Equation (6.4); $u^{(m,q)}(t)$ hold for the interval $t^{(q)} \leq t \leq t^{(q+1)}$. In principle, the tolerance β can lie anywhere between 0° and 90°; at the lower bound reorthogonalization is unnecessary altogether, whereas at the upper one it is in order for every integration step. Usually, some numerical experience is needed for finding a suitable value of β for a given problem.

6.2 Brief Outline of Godunov-Conte Method As Applied to Eigenvalue Problems

The method consists in solving the equations

$$\dot{u} = A(t, \lambda)u \tag{6.16}$$

k times, under k sets of initial conditions

$$u_i^{(m)}(t_0) = \delta_{i,m+k} \qquad (m = 1, 2, \ldots, k) \tag{6.17}$$

where $u_i^{(m)}(t, \lambda)$ is the solution for the ith component of u with λ given for the mth integration of the homogeneous equations. The general solution, which reads

$$y_i(t, \lambda) = \sum_{j=1}^{k} b_j u_i^{(j)}(t, \lambda), \qquad i = 1, 2, \ldots, n \tag{6.18}$$

automatically satisfies (6.2). We subdivide the interval (t_0, t_f) into γ sub-intervals t_i, $i = 0, 1, \ldots, \gamma - 1$, and use any standard integration method to obtain the base solutions $u^{(1)}, u^{(2)}, \ldots, u^{(k)}$ at the mesh points, orthonormalizing them wherever they fall below Conte's criterion. Next, we define $U^{(q)}(t, \lambda)$ – an $(n \times k)$ matrix whose elements are the solutions $u^{(m,q)}(t, \lambda)(m = 1, 2, \ldots, k)$, which were last orthonormalized at $t^{(q)}$ and whose columns are the vectors $u^{(1,q)}(t, \lambda), u^{(2,q)}(t, \lambda), \ldots, u^{(k,q)}(t, \lambda)$, q again the index counter for the number of orthonormalizations, $0 \leq q \leq Q$:

$$U^{(q)}(t, \lambda) = \begin{bmatrix} u_1^{(1,q)}(t,\lambda) & u_1^{(2,q)}(t,\lambda) & \cdots & u_1^{(k,q)}(t,\lambda) \\ u_2^{(1,q)}(t,\lambda) & u_2^{(2,q)}(t,\lambda) & \cdots & u_2^{(k,q)}(t,\lambda) \\ \vdots & & & \\ u_n^{(1,q)}(t,\lambda) & u_n(t,\lambda) & \cdots & u_n^{(k,q)}(t,\lambda) \end{bmatrix} \tag{6.19}$$

and $u_i^{(m,0)}(0, \lambda) = u_i^{(m)}(0, \lambda) = \delta_{i,m+k}$. At $t^{(q)}$ (which need not be evenly spaced), we construct for the set of k linearly independent vectors $u^{(m,q-1)}(t^{(q)}, \lambda)$, $m = 1, 2, \ldots, k$ the set of orthonormal vectors $u^{(m,q)}(t^{(q)}, \lambda)$, which reads in matrix form

$$u^{(q)}\left(t^{(q)}, \lambda\right) = u^{(q-1)}\left(t^{(q)}\right) P^{(q)} \tag{6.20}$$

where $P^{(q)}$ is non-singular and upper triangular with elements P_{ij}

$$P_{ij} = \begin{cases} \dfrac{1}{w_{jj}}, & i = j \\ 0, & i > j \\ -\dfrac{\left(z^{(i)}, u^{(j)}\right)}{w_{ii}\, w_{jj}}, & i < j \end{cases} \tag{6.21}$$

and

$$w_{11} = \left(u^{(1)}, u^{(1)}\right)^{1/2}, \qquad z^{(1)} = u^{(1)}/w_{11}$$
$$\eta^{(2)} = u^{(20)} - \left(u^{(2)}\right), \qquad z^{(1)}z^{(1)}$$

$$w_{22} = \left(\eta^{(2)}, \eta^{(2)}\right)^{1/2}, \qquad z^{(2)} = \eta^{(2)}/w_{22} \tag{6.22}$$

$$\eta^{(j)} = u^{(j)} = \sum_{s=1}^{j-1} \left(u^{(j)}, z^{(s)}\right) z^{(s)}$$

$$w_{jj} = \left(\eta^{(j)}, \eta^{(j)^{1/2}}\right), \qquad z^{(j)} = \eta^{(j)}/w_{jj}, \qquad j = 1, 2, \ldots, k$$

The solution at $t^{(q)} = t_f$ is

$$y(t_f, \lambda) = U^{(Q)}(t_f, \lambda) b^{(Q)} \tag{6.23}$$

where $b^{(Q)}$ is a $(k \times 1)$ vector of constants with components $b_1^{(Q)}, b_2^{(Q)}, \ldots, b_n^{(Q)}$. The eigenvalue λ^* derives from the singularity condition of the matrix $U^{(Q)}(t_f, \lambda)$.

6.3 Application to Buckling of Polar Orthotropic Annular Plate

Several studies of buckling of homogeneous circular plates under radial compression are reported in the literature. The axisymmetric buckling problem was solved by Woinowsky-Krieger (1958) with the aid of Bessel functions. Later, Pandalai and Patel (1965) used power-series expansion in the same problem. The axisymmetric buckling problem was solved by Mossakowsky (1960) and Mossakowsky and Borsuk (1960) using Frobenius's method with the aid of hypergeometric functions. Swarmidas and Kunukkasseril (1973) examined the axisymmetric and first asymmetric buckling of a circular orthotropic plate, formulating the displacement of the buckled plate with the aid of Bessel and Lommel functions. Kaplyevatsky (1975) presented a solution for specific values of plate orthotropy measure by a generalization of Frobenius's method. Approximate values of the axisymmetric buckling loads, at different edge conditions, in a plate under uniform compression were obtained by Vijayakumar and Joga Rao (1971) using the Rayleigh-Ritz method, and by Uthgennant and Brand (1970) using a finite-difference technique and the Vianello-Stodola iterative procedure. The approach of Vijayakumar and Joga Rao (1971) was extended by Ramaiah and Vijayakumar (1975) for an annular plate under uniform internal pressure (in which case it always buckles axisymmetrically).

The axisymmetric and general asymmetric buckling problem of annular orthotropic plates, under arbitrarily specified internal and external pressures and with an arbitrary orthotropy measure, is still unsolved analytically. In what follows, we report results obtained by the Godunov-Conte method as described earlier.

The governing buckling equation in terms of the transverse deflection w and in-plane radial and circumferential forces N_2 and N_θ, respectively, reads:

$$\begin{aligned} D_r\left(\frac{\partial^4 w}{\partial r^4} + \frac{2}{r}\frac{\partial^3 w}{\partial r^3}\right) + D_\theta\left(-\frac{1}{r^2}\frac{\partial^2 w}{\partial r^2} + \frac{1}{r^3}\frac{\partial w}{\partial r} + \frac{2}{r^4}\frac{\partial^2 w}{\partial \theta^2} + \frac{1}{r^4}\frac{\partial^4 w}{\partial \theta^4}\right) \\ + 2D_{r_\theta}\left(\frac{1}{r^2}\frac{2^4 w}{2r^2\partial\theta^2} - \frac{1}{r^3}\frac{\partial^3 w}{\partial r \partial \theta^3} + \frac{1}{r^4}\frac{\partial^2 w}{\partial \theta^2}\right) \\ = N_r\frac{\partial^2 w}{\partial r^2} + N_\theta\left(\frac{1}{r}\frac{\partial w}{\partial r} + \frac{1}{r^2}\frac{\partial^2 w}{\partial \theta^2}\right) \end{aligned} \tag{6.24}$$

where

$$D_r = \frac{E_r h^3}{12(1-\nu_r\nu_\theta)}, \qquad D_\theta = \frac{E_\theta h^3}{12(1-\nu_r\nu_\theta)}, \qquad D_r = D_{r_\theta}\nu_\theta + \frac{G_{r_\theta}h^3}{12} \tag{6.25}$$

The plane stress-strain relations for the orthotropic plate material are

$$\begin{bmatrix} \sigma,_r \\ \sigma,_\theta \\ \tau_{r_\theta} \end{bmatrix} = \begin{bmatrix} E_r & E_{r_\theta} & 0 \\ E_{\theta_r} & E_\theta & 0 \\ 0 & 0 & G_{r_\theta} \end{bmatrix} \begin{bmatrix} \epsilon_r \\ \epsilon_\theta \\ \gamma_{r_\theta} \end{bmatrix} \tag{6.26}$$

where E_r, $E_{r_\theta}(=E_{\theta_r})$, E_θ, and G_{r_θ} are elastic stiffness moduli. In the isotropic case, $E_r = E_\theta = E/(1-\nu^2)$, $E_{r_\theta} = \nu E/(1-\nu^2)$, $G_{r_\theta} = G = E/2(1+\nu)$, $D_r = D_\theta = Eh^3/12(1-\nu^2)$ where E is Young's modulus, ν is Poisson's ratio, and D is the flexural rigidity.

Using the equilibrium equation for in-plane forces in the axisymmetric radial in-plane prebuckling state

$$\frac{d}{dr}(rN_r) - N_\theta = 0 \tag{6.27}$$

The following differential equation is obtained:

$$\begin{aligned} &D_r\left(\frac{\partial^4 w}{\partial r^4} + \frac{2}{r}\frac{\partial^3 w}{\partial r^3}\right) + D_\theta\left(-\frac{1}{r^2}\frac{\partial^2 w}{\partial r^2} + \frac{1}{r^3}\frac{\partial w}{\partial r} + \frac{2}{r^4}\partial^2 w \partial\theta^2 + \frac{1}{r^4}\frac{2^4 w}{\partial\theta^4}\right) \\ &+ 2D_{r_\theta}\left(\frac{1}{r^2}\frac{2^4 w}{\partial r^2\partial\theta^2} - \frac{1}{r^3}\frac{\partial^3 w}{\partial r\partial\theta^3} + \frac{1}{r^4}\frac{\partial^2 w}{\partial\theta^2}\right) - \frac{1}{r}\frac{\partial}{\partial r}\left(rN_r\frac{\partial w}{\partial r}\right) \\ &- \frac{\partial}{\partial r}(rN_r)\frac{1}{r^2}\frac{\partial^2 w}{\partial\theta^2} = 0 \end{aligned} \tag{6.28}$$

where N_r and N_θ are given by the plane-stress elasticity solution

$$N_r = \frac{\psi}{r}, \qquad N_\theta = \frac{d\psi}{d\theta} \tag{6.29}$$

$$\psi = Ar^k + Br^{-k} \tag{6.30}$$

ψ being the force resultant function, with

$$A = \left(-N_0 r_i^{-k-1} + N_i r_0^{-k-1}\right)R^{-1} \tag{6.31}$$

$$B = \left(N_0 r_i^{k-1} - N_i r_0^{k-1}\right)R^{-1} \tag{6.32}$$

$$R = r_i^{-k-1} r_0^{k-1}\left[1 - \left[\frac{r_i}{r_0}\right]^{2k}\right]. \tag{6.33}$$

$k^2 = E_r/E_\theta = D_r/D_\theta$ is an orthotropy measure, $N_0 = p_0^h$, $N_i = p_i^h$, p_i and p_0 denote uniform in-plane pressures at the inner and outer edges, respectively.

Now, assuming a deflection function such as

$$w(r,\theta) = W(r)\cos(n\theta) \tag{6.34}$$

where n is the number of nodal diameters, Equation (6.27) reduces to an ordinary differential equation with variable coefficients:

$$D_r\left(\frac{d^4W}{dr^4}+\frac{2}{r^3}\frac{d^3W}{dr^3}\right)+D_\theta\left(-\frac{1}{r^2}\frac{d^2W}{dr^2}+\frac{1}{r^3}\frac{dW}{dr}-\frac{2n^2}{r^4}W+\frac{n^4}{r^4}W\right)$$
$$-2D_{r\theta}\left(\frac{n^2}{r^2}\frac{d^2W}{dr^2}+\frac{n^3}{r^3}\frac{dW}{dr}+\frac{n^2}{r^2}W\right)-\frac{1}{r}\left(rN_r\frac{dW}{dr}\right)+\frac{n^2}{r^2}\frac{d}{dr}(rN_r)W=0 \tag{6.35}$$

For further convenience and generality, Equation (5.34) can be rewritten in non-dimensional form, namely

$$\rho^4W'''' + 2\rho^3W''' - \mu + (\rho^2W'' - \rho W') + \kappa W$$
$$+\lambda[\alpha\rho^{1+k}(\rho^2W'' + k\rho W' - n^2kW) + B\rho^{1-k}(\rho^2W'' - k\rho W' + n^2kW)] = 0 \tag{6.36}$$

where

$$\rho = \frac{r}{r_0}, \qquad \eta + \frac{r_i}{r_0}, \qquad \xi = \frac{P_i}{p_0}, \qquad k_1^2 = \frac{D_r\theta}{D_r}, \qquad \lambda = \frac{P_0b^2h}{D_r},$$
$$\mu = k^2 + 2k_1^2n^2, \qquad \kappa = n^2(k^2n^2 - 2k^2 - 2k_1^2) \tag{6.37}$$
$$\alpha = \frac{1-\xi\eta^{1+k}}{1-\eta^{2k}}, \qquad \beta = \frac{\eta^{1+k}(\xi - \eta^{k-1})}{1-\eta^{2k}}$$

r_i, r_0, and h are inside and outside radii and thickness, respectively; k_1 is a supplementary orthotropy measure, $\rho = 1$ and $\rho = \eta$ represent the outer and inner edges, respectively; λ is the non-dimensional critical buckling load for a specified load ratio ξ.

Equation (5.35) is supplemented by two conditions at each boundary point, which read in general form

$$F_\gamma(W) = G_\gamma(W) = 0, \qquad \gamma = r_1, r_0 \tag{6.38}$$

F_γ and G_γ being linear functionals.

Numerical results by the Godunov-Conte method were obtained, the lowest critical loads were calculated with the radius ratio η as parameter. As in the study of isotropic plates by Yamaki (1958), it turned out that the assumption of symmetrical loading often leads to a stability overestimate.

The plates under consideration were clamped at the inner and outer edges, with functionals F_γ and G_γ, as follows:

$$F_{r_i}(\cdots) = (\cdots)|_{r=r_1}, \qquad G_{r_i}(\cdots) = \left.\frac{\partial(\cdots)}{\partial r}\right|_{r=r_1} \tag{6.39}$$

$$F_{r_0}(\cdots) = (\cdots)|_{r=r_0}, \qquad G_{r_0}(\cdots) = \left.\frac{\partial(\cdots)}{\partial r}\right|_{r=r_0} \tag{6.40}$$

Curve 1 in Figure 6.1 presents the buckling behavior of the plate under uniform compression ($\xi = 1$) for different values of k, and curve 2 refers to axisymmetric buckling loads ($n = 0$) found by the Rayleigh-Ritz method. The nodal diameters n, which correspond

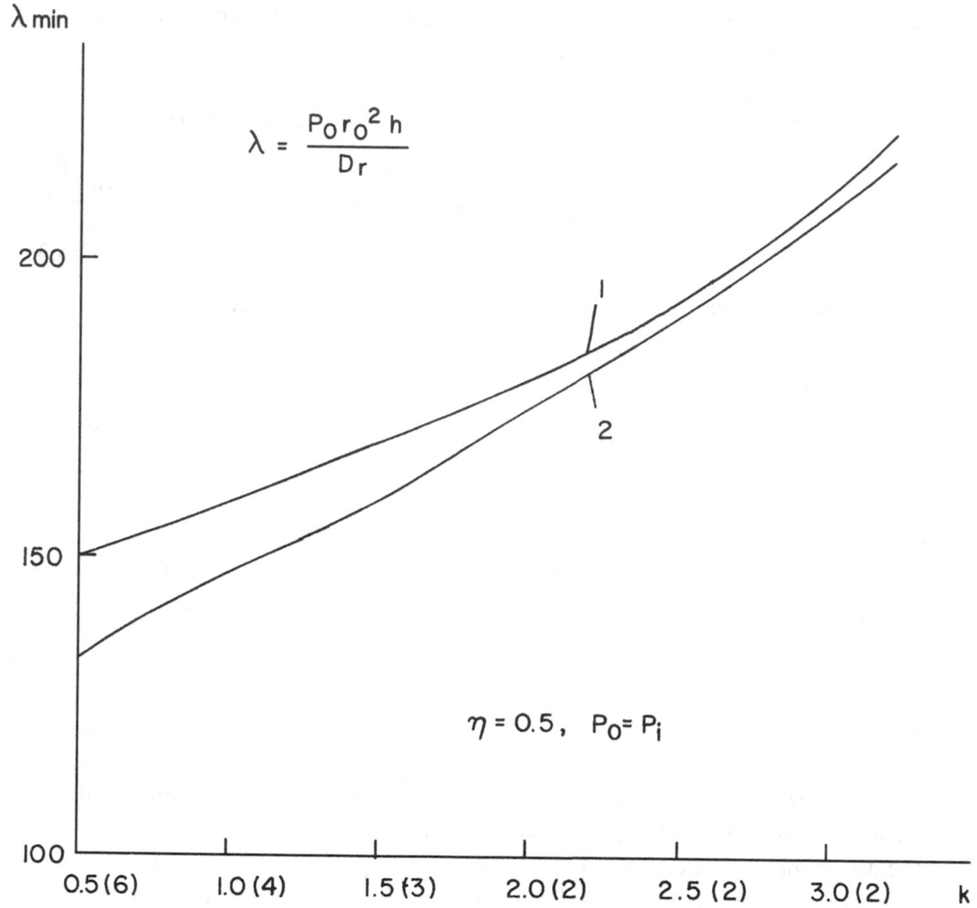

Figure 6.1 Buckling load parameter $\lambda = p_0 b^2 h / D_r$ as a function of orthotropy measure k; both edges clamped and $p_0 = p_i$. Curve 1, asymmetric buckling loads (Vijayakumar and Joga Rao, 1971); curve 2, asymmetric buckling loads.

Table 6.1. Least values of λ and corresponding values of n and N (plate under uniform compression)

	η	0.1	0.2	0.3	0.4	0.5	0.6	0.7	0.8
$k = 0.5$	$\sqrt{\lambda_{\min}}$	6.013	6.997	8.127	9.574	11.540	14.478	19.340	29.047
$(\beta = 5°)$	n	2	2	3	5	6	8	12	19
	N	6	0	1	1	6	2	20	
$k = 1$	$\sqrt{\lambda_{\min}}$	6.710	7.502	8.637	10.099	12.155	15.200	20.282	30.437
$(\beta = 14°)$	n	1	2	2	3	4	5	7	11
	N	6	5	7	7	2	25	45	
$k = 1.5$	$\sqrt{\lambda_{\min}}$	7.622	8.251	9.231	10.650	12.617	15.655	20.733	30.995
$(\beta = 5°)$	n	1	2	2	2	3	4	5	8
	N	3	2	1	0	0	1	2	

Table 6.2. Least values of λ for the plate under internal pressure, $\beta = 5°$

	$n = 0.1$	$n = 0.2$	$n = 0.3$	$n = 0.4$	$n = 0.5$	$n = 0.6$	$n = 0.7$	$n = 0.8$
$k = 0.5$	5.1737	14.109	30.039	58.869	113.63	227.034	499.13	1354.6
$k = 1$	7.6725	17.261	33.705	62.968	118.10	231.85	504.26	1360.0
	7.678[a]	17.26[a]	33.705[a]	62.972[a]	118.10[a]	231.86[a]	504.32[a]	1360.2[a]
$k = 2$	20.016	30.992	48.994	79,668	136.18	251.21	524.83	1381.7

[a] Results from Ramaiah and Vijayakumar (1974).

Table 6.3. Least values of λ and corresponding values of n for the plate under external pressure, $\beta = 5°$

	η	0.1	0.2	0.3	0.4	0.5	0.6	0.7	0.8
$k = 0.5$	λ_{min}	35.625	45.655	56.007	68.266	84.309	107.99	146.75	223.84
	n	2	3	5	6	9	12	18	30
$k = 1$	λ_{min}	44.515	55.684	70.241	89.101	113.80	150.01	208.64	324.36
	λ_{min}	44[a]	55[a]	69.5[a]	88.5[a]	114.5[a]			
	n	2	2	3	4	6	8	12	20
$k = 2$	λ_{min}	75.310	82.847	99.975	124.50	161.47	219.98	314.57	505.18
	n	1	2	2	3	4	6	8	14

[a] Results from Ramaiah and Vijayakumar (1975).

Table 6.4. Least values of λ and corresponding values of n and N (isotropic plate under internal tension, $\beta = 5°$)

	η	0.1	0.2	0.3	0.4	0.5	0.6	0.7	0.8
$k = 0.5$	λ_{min}	7.7591	14.808	23.628	34.850	50.586	73.767	112.39	189.78
	n	4	5	6	8	11	14	20	30
	N	21	22	16	17	19	31	28	41
$k = 2$	λ_{min}	66.37	75.413	98.866	128.21	166.66	225.86	322.16	512.70
	n	3	3	4	5	6	8	11	16
	N	18	19	17	18	20	8	7	7

Table 6.5. Least values of λ and corresponding values of n and N for isotropic plate under internal tension, $\beta = 5°$

η	1/8	1/7	1/6	1/5	1/4.5	1/2	1/3.5	1/3	0.4	0.5	0.6	0.7	0.8
λ_{min}	19.433	22.371	24.935	27.980	30.643	34.688	39.704	46.826	58.779	82.775	118.09	176.33	291.36
λ_{min}	19.5[a]	22.0[a]	25.7[a]	29.0[a]	32.0[a]	36.6[a]	42.5[a]	51.5[a]					
n	3	3	4	4	4	4	5	5	6	8	8	14	22
N	18	20	21	17	14	15	15	14	19	20	20	22	25

[a] Results from Lipovtsev (1970).

to the least values of the non-dimensional buckling load λ, are given in parentheses. Table 6.1 lists the least values of λ for different radius ratios η; N denotes the number of reorthogonalizations required. The tolerance β was chosen as 14°. [Note that the results for the isotropic plates are practically coincident with those of Yamaki (1958).] Table 6.2 lists the least values of λ for a plate under uniform internal pressure ($\xi \to \infty$), λ being defined in this case as

$$\lambda = p_i r_i^2 h / D_r \tag{6.41}$$

for different values of η and k (here, $n = 0$). Those for the isotropic case ($k = 1$) are compared with those of Ramaiah and Vijayakumar (1975). The percentage is of the order of 0.1%.

Table 6.3 present the results for a plate under uniform external pressure; those for the isotropic case are compared with those of Ramaiah and Vijayakumar (1974). Table 6.4 lists the non-dimensional buckling loads

$$\lambda = p_i r_i^2 h / D_r \tag{6.42}$$

for a plate under internal tension, as a function of k and η; Table 6.5 presents a comparison with Lipovtsev's (1970) results for the isotropic case, obtained by the double-sweep method (see Gelfand and Lokutsievski, 1964; Biderman, 1967; Grigolyuk and Lipovtsev, 1975; Godunov, 1971).

To sum up, in this chapter, application of the Godunov-Conte method to eigenvalue problems was demonstrated. Loss of accuracy was avoided by orthonormalizing the base solutions at a point where the least angle between the relevant pair of base vectors at each computation point is less than a specified tolerance. The method was applied to the buckling of a polar orthotropic plate. Numerous examples were evaluated, showing the effect of the orthotropy measure and radius ratio.

CHAPTER SEVEN

Application of Computerized Symbolic Algebra in Buckling Analysis

For the public at large and even for most scientists, numerical calculation and scientific calculation have become synonymous. ... However, numerical calculation does not rule out algebraic calculation. ... And, the power of computers does not solve everything.

J. H. Daventport, Y. Siret, and E. Tournier

There are ... problems in mechanics for which standard numerical procedures are available but which could be solved more elegantly and accurately using analytical methods if the formula work could be overcome.

J. Jensen and F. Niordson

Computer algebra systems can solve many problems more quickly than a human being. In our experience it is not unusual for a computer system to solve a problem which has been taxing a capable mathematician for several months, in a few minutes. One wonders how many tractable problems remain unsolved or have been forgotten about simply because they are making excessive demands on a researcher's time and sanity!

C. Wooff and D. Hodgkinson

In this chapter, we deal with the analytic, symbolic computation for buckling analysis. Symbolic algebra can significantly reduce the tedium of analytic computation and simultaneously increases its reliability. It has great impact on scientific computation, as more and more analytically minded researchers are using computers, which have been traditionally associated with "number crunching." Analytic work can now be extended as far as possible, and the numerical side of the analysis can be "delayed." Calculations with arbitrary arithmetic precision along with the automatic generation of computer codes have opened up more possibilities of computer use. In this chapter, we use a classical problem – buckling of non-isotropic plate – to illustrate the application of symbolic algebra, a neo-classical analytic-numerical tool.

7.1 Introductory Remarks

With the advent of high-speed computers, the scope of the structural analysis in all of its manifestations, including buckling, experienced a dramatic impact. As a result of this quiet "computer revolution," the analysis of even the most complex structures became tractable in many circumstances. Parallel with this increase in the capacity for solving large numbers of problems was a shift in emphasis so that classic analytic techniques became somewhat neglected, and engineers and researchers started to favor purely numerical analysis. A new discipline *computational mechanics* rightfully experienced huge developments. "Thus with the exception of relatively simple closed-form solutions, there is an increasing tendency to assume indiscriminately numerical methods are to be preferred even when long-established analytical techniques (whether exact or approximate) are readily available" (Pavlovic and Sapountzakis, 1986). Analytically minded researchers often associate this trend with "number crunching syndrome." According to Beltzer (1990a), researchers were even discussing the dilemma, "join the computers or fight them." Some researchers associated results of the numerical analyses as those that were "untouched by human mind." Human intellect, however, is getting around this dilemma by endowing computers with the capacity of performing analytic operations. Noor and Andersen (1979) stress that "... symbolic computation can provide an important link between analysis on the one hand, and numerical calculations on the other." In Beltzer's words (1990a),

> surprisingly enough, the same digital computers now help to restore a delicate equilibrium between analytic and numerical methods, which promises to be particularly fruitful. Yet, the arising field of the so-called "computer algebra" or "symbolic computation" may be useful in overcoming the difficulties due to limited speed the numerical calculations can be performed with. As noted in Davenport et al. (1988), even CRAY still needs much more than 48 hours to fairly forecast the weather 48 hours ahead [present numbers are unavailable to us].

According to Kleiber (1990), the coupled numeric-symbolic computation serves the purpose of "humanizing" structural mechanics computations. Although this may appear to be an over-dramatization of the state of affairs, it at least indicates the possibility that we can exercise more control over computation when computers are equipped with symbolic algebra software.

From the start of the electronic calculation, the main use of computers has been to conduct numerical calculations. Computers, or calculating machines, as they were initially known, deal effectively with numerical calculations in modern engineering and sciences. Yet, the very idea that computers would perform algebraic calculation apparently dates back to nearly a century and a half ago, first suggested by Lady Lovelace. According to Wooff and Hodgkinson (1987), Lady Lovelace was a patron of Charles Babbage, who is usually credited with the development of the first computer in the world. Her idea was very timely. Need for huge algebraic computations always existed. Famous large calculations of the nineteenth century include a large proportion of algebraic manipulation. The most widely known is Le Verrier's calculation of the orbit of Neptune, which started from the disturbances of the orbit of Uranus and which led to the discovery of Neptune. As Davenport, Siret, and Tournier (1988) mention, the most impressive calculation with pencil and paper was also in the field of astronomy: Delaunay took 10 years to calculate the orbit of the moon, and another 10 years to check the results of his calculations. The result was not numerical because it consisted primarily of a formula, which by itself occupied all the 128 pages of Chapter 4 of his book.

The first software to realize the idea of Lady Lovelace was developed in the 1950s; two Masters theses were written almost simultaneously – one at Temple University by Kahrimanian (1953) and the other one at M.I.T. by Nolan (1953) – and both were written on the computer implemention of analytic differentiation. About 10 years later, the first general-purpose systems appeared for algebraic calculations, namely, ALPAK (a forerunner of ALTRAN) and FORMAC. Since then, the development of systems and algorithms for processing formulas on computers has been extremely rapid. Presently, there are a number of general as well as more specialized languages on a broad variety of computers. Here are a few of the most renowned: DERIVE, MACSYMA, MAPLE, MATHEMATICA, muMATH, REDUCE, and SCRATCHPAD. Even handheld calculators with symbolic algebra capability are presently available (Patton, 1987). Some of these software are discussed in the books by Rayna (1987), Davenport et al. (1988), and Wolfram (1996). As Davenport et al. (1988) mention, the name of this discipline has long hesitated between "symbolic and algebraic calculation," or "symbolic and algebraic manipulations," and finally settled down as "computer algebra" in English and "Calcul Formel," abbreviated as CALSYF, in French. At the same time, societies were formed to bring together the research workers and the users of this new discipline: SIGSAM (Special Interest Group in Symbolic and Algebraic Manipulations) is the worldwide group of ACM (Association for Computing Machinery). SIGSAM organizes congresses SYMSAC and EUROSAM and publishes the bulletin entitled *SIGSAM*. The European group is called SAME (Symbolic and Algebraic Manipulation in Europe) and organizes congresses call EUROCAM. For French research workers, there is a body of the CNRS (Centre national de la recherche scientifique), the GRECO (Groupe de REcherches COordonées) of computer algebra. Since 1985, a specialized *Journal of Symbolic Computations* has been published by Academic Press. Presently, computer algebra in essence became an able and obedient "butler" – a reliable assistant to researchers and engineers.

In applied mechanics, computer algebra was apparently pioneered by Jensen and Niordson (1977), Pedersen (1977), Noor and Andersen (1979), Korncoff and Fenves (1978), Hussain and Noble (1983), and others. The state of the art of applications of computer algebra in applied mechanics is reflected in books by Rand (1984), Rand and Ambuster (1987), Klimov and Rudenko (1989), Noor, Elishakoff, and Hulbert (1990), Beltzer (l990a,b), and others. Extensive literature on applications of symbolic algebra in vibrations and buckling exists today. Crespo da Silva and Hodges (1986) used it in rotorcraft dynamics; Nagabhushnan, Gaonkar, and Reddy (1981) developed special analytic-numerical software for helicopter dynamics; Elishakoff, Hettema, and Wilson (1989) applied it in deterministic vibration problems; and Rehak et al. (1987) and Elishakoff and Hettema (1988) developed special symbolic manipulation techniques for random vibration analysis of structures; Klimov (1990) developed widely utilized symbolic algebra for solving non-linear dynamics problems; optimal design of laminated shells was performed by Verijenko et al. (1994); Yoakimidis (1994) used symbolic algebra for inverse design problems.

Applications to buckling problems can be found in the papers by Rizzi and Tatone (1985a, 1985b), Elishakoff and Tang (1988), and Elishakoff and Pletner (1990). This chapter represents an extended version of the study by Elishakoff and Tang (1988). Before illustrating the application of the symbolic algebra to a particular buckling problem, we will briefly review the main points of MATHEMATICA, one of the popular systems in this field.

7.2 Brief Review of MATHEMATICA®

Among the numerous systems for symbolic computation, MATHEMATICA® appears to be a newcomer. It was first announced in 1988, and Version 2 was released in 1991. MATHEMATICA is now available for a wide range of computers, including IBM-compatible PCs, Macintosh computers, UNIX workstations, and some mainframes. It is designed to handle analytical calculations as well as those numerical calculations that are usually carried out in FORTRAN.t Remarkably, in numerical computations, MATHEMATICA can handle numbers of any precision, which is usually impossible in traditional numerical analysis due to the hardware limitation in digit length. Besides, most special functions of mathematical physics, for example, Bessel functions and hypergeometrical functions, as well as linear algebra operations are built into MATHEMATICA. In addition, MATHEMATICA is a complete graphics programming language and has an extensive graphics capability for visualizing results of calculations. In fact, MATHEMATICA's popularity over several other symbolic languages rests in its graphics capability.

We now present some simple examples of MATHEMATICA. MATHEMATICA can be used in the interactive or batch mode. When MATHEMATICA is started, which is typically done by typing the command *math* at an operating system, the prompt *In*[*1*]: = pops up, signifying that it is ready for the user's input. The user can type the input and then press the RETURN key. MATHEMATICA will process the input and print out the result, starting with *Out*[*1*]: = . Before we start our first example, here are a few important points pertinent to the use of MATHEMATICA:

- Arguments of functions are given in square brackets such as in Sqrt[5], Sin[x].
- Names of built-in functions have their first letters capitalized; for example, Integrate[f(x), x] and DSolve [*eqns*, y[x], x].
- Multiplication can be represented by a space, though * could also be used. For example, a b means a times b.
- Powers are denoted by ^; for example, y^5 denotes y^5.

Differentiation

D[f, x]	Partial derivative, $\dfrac{\partial f}{\partial x}$
D[f, $x_1, x_2, \ldots, x_n$]	Multiple derivative, $\dfrac{\partial}{\partial x_1}\dfrac{\partial}{\partial x_2}\cdots\dfrac{\partial}{\partial x_n}f$
D[f, {x, n}]	Repeated derivative, $\dfrac{\partial^n f}{\partial x^n}$

Example 1:

In[*1*]: = D[Sin[4x],{x, 2}]

Out[*1*]: = −(16 Sin[4x])

Integration

Integrate[f, x]	Indefinite integral, $\int f dx$
Integrate[f, {x, a, b}]	Definite integral, $\int_a^b f dx$
integrate[f, {x, a, b}, {y, c, d}]	Multiple integral, $\int_a^b dx \int_c^d f dy$

Example 2:

In[*2*]: = Integrate[Sin[x]^3 + y^2, {x, 0, 1}, {y, 0, x}]

$$Out[2]: = \frac{3 - 27\,\mathrm{Cos}[1] + 3\,\mathrm{Cos}[3] + 27\,\mathrm{Sin}[1] - \mathrm{Sin}[3]}{36}$$

Algebraic Equations

Solve[*lhs* == *rhs*, x] — Solve an equation for x (*lhs* and *rhs* stand for the left-hand side and the right-hand side, respectively)

Solve[{lhs_1 == rhs_1, lhs_2 = rhs_2, ...}, {x, y, ...}] — Solve a set of simultaneous equations for $x, y, \ldots$

Eliminate[{lhs_1 == rhs_1, lhs_2 = rhs_2, ...}, {x, ...}] — Eliminate $x, \ldots$ in a set of simultaneous equations

Example 3:

In[*3*]: = Solve [ax + b == c, x]

$$Out[3]: = \left\{\left\{x -> -\left(\frac{b-c}{a}\right)\right\}\right\}$$

Example 4:

In[*4*]: = Solve [x^2 + y^2 == 1, x + y == 1}, {x, y}]

Out[*4*]: = {{x − > 0, y − > 1}, {x − > 1, y − > 0}}

Differential Equations

DSolve[*eqn*, y[x], x] — Solve a differential equation for $y(x)$ with x as the independent variable

Example 5:

In[5]: = DSolve[{y′[x] == ay[x], y[0] == 1}, y[x], x]

Out[5]: = {{y[x] − > E^{ax}}}

Operations with Polynomials

Expand[*expr*]	Multiply out expression *expr*
Expand[*expr*, Trig − > True]	Expand out trigonometric functions such as writing $\sin(x)^2$ in terms of $\sin(2x)$ and so on
Factor[*expr*]	Reduce *expr* to a product of factors
Collect[*expr*, x]	Collect terms in *expr* with like powers of *x*
Coefficient[*expr*, *form*]	Collect the coefficient of *form* in *expr*
Simplify[*expr*]	Try to obtain a simplest form of *expr*

Example 6:

In[6]: = Expand[(x^2 + y + 1)^2]

Out[6]: = $1 + 2x^2 + x^4 + 2y + 2x^2y + y^2$

In[7]: = Collect[%, y]   (% means the last result generated)

Out[7]: = $1 + 2x^2 + x^4 + (2 + 2x^2)y + y^2$

In[8]: = Coefficient[%, x^2y]

Out[8]: = 2

In[9]: = Simplify[% %]   (% % means the next-to-last result generated; in general, % % ⋯ % (*k* times) means the *k*th previous result)

Out[9]: = $(1 + x^2 + y)^2$

Substitution

expr/.x − > *value*	Replace *x* by *value* in the expression *expr*
expr/.{x − > *xval*, y − > *yval*}	Perform several replacements in an expression

Example 7:

In[10]: = x^2 + y/. x − > z + y

Out[10]: = y + (y + z)^2

Arbitrary-Precision Calculations

expr//N or N[*expr*]	Approximate numerical value of *expr*
N[*expr*, *n*]	Numerical value of *expr* with *n*-digit precision

Example 8:

In[*11*]: = N[Sqrt[7], 30]

Out[*11*]: = 2.64575131106459059050161575364

These are just a few of MATHEMATICA's capabilities. The interested reader may consult the comprehensive texts on MATHEMATICA by Wolfram (1996), Abell and Braselton (1992), Bahder (1995), and Höft and Höft (1998).

7.3 Buckling of Polar Orthotropic Circular Plates on Elastic Foundation by Computerized Symbolic Algebra

7.3.1 Results Reported in the Literature

In this section, we discuss the application of the *symbolic algebra* to buckling problems. It appears instructive to demonstrate its usefulness on specific examples. Since we used circular plates to demonstrate the Godunov-Conte method in Section 6.1, we will also utilize circular plates to elucidate the power of symbolic algebra. This will be done in conjunction with description and use of Rayleigh's method presently referred to as the method of *optimized parameters*.

In a number of interesting papers, Schmidt (1981, 1982, 1983) revived the quite forgotten version of Rayleigh's method, suggested by Rayleigh. He gave several applications of Rayleigh's method to the buckling of isotropic circular plates clamped or simply supported at their circumference. The essence of the method is an introduction of the undetermined power coefficient as one of the parameters describing the trial function. The desired eigenvalue then becomes a function of this undetermined power. Because the Rayleigh method provides the upper bound of the eigenvalues (buckling loads or natural frequencies), the presence of the additional "degree of freedom" – the non-integer power coefficient – allows one to minimize the eigenvalue with respect to this coefficient. Thus, the "best" value of the eigenvalue is found within the class of trial functions created by the undetermined power coefficient. The detailed description of Rayleigh's method with the new applications of this wonderful idea is given by Bert (1987).

Since we will focus our attention on buckling of circular plates, we will describe only the appropriate results reported in the literature. For circular isotropic plates, the Rayleigh quotient reads

$$N_r = \frac{D\int_0^R \left[\left(\frac{d^2w}{dr^2}\right)^2 + \left(\frac{1}{r}\frac{dw}{dr}\right)^2 + \frac{2\nu}{r}\frac{dw}{dr}\frac{d^2w}{dr^2}\right] r\,dr}{\int_0^R \left(\frac{dw}{dr}\right)^2 r\,dr} \tag{7.1}$$

where N_r is the radial compressive load per unit length, D is the flexural rigidity, R is the radius, w is the displacement, ν is Poisson's ratio, and r is the radial coordinate. In non-dimensional form, (6.41) becomes

$$\bar{N}_r = \frac{N_r R^2}{D} = \frac{\int_0^1 [(w_{,\rho\rho})^2 + (w_{,\rho}/\rho)^2 + 2\nu(w_{,\rho}/\rho)(w_{,\rho\rho})]\rho\,d\rho}{\int_0^1 (w_{,\rho})^2 \rho\,d\rho} \tag{7.2}$$

where $\rho = r/R$ and $w_{,\rho} = dw/d\rho$. For the clamped plates, instead of using the trial functions in either the form $\rho - \rho^2$ or $\rho - \rho^3$ utilized in the literature (and yielding the approximate buckling loads $\bar{N}_r = 15$ or $\bar{N}_r = 16$, respectively), Schmidt (1982) suggested a trial function

$$w_{,\rho} = \rho - \rho^n \tag{7.3}$$

where n is an undetermined power coefficient.

Using MATHEMATICA, we write the following program segment for calculating the buckling load:

wp =	ro − ro^n [Equation (7.3) with wp − > $w_{,\rho}$, ro − > ρ]
fden =	wp^2 ro [fden represents the integrand in the denominator of Equation (7.2)]
fnum =	(D[wp, ro]^2 + (wp/ro)^2 + 2 nu wp/ro D[wp, ro]) ro [fnum represents the integrand in the numerator of Equation (7.2); nu − > ν]
nr =	Integrate[fnum, {ro, 0, 1}]/Integrate[fden, {ro, 0, 1}] (nr − > $\bar{N}_r$)

The resulting expression for $\bar{N}_r$ is

$$\bar{N}_r = \frac{2(n+1)(n+3)}{n} \tag{7.4}$$

Minimizing $\bar{N}_r$ with respect to n, or requiring $\bar{N}_{r,n} = 0$, yields $n^2 = 3$, with buckling load

$$\bar{N}_r = 4(2 + \sqrt{3}) \simeq 14.928 \tag{7.5}$$

which is 1.66% higher than the exact value 14.682.

For the clamped plate on the elastic foundation, Bert (1987) applied the following buckling shape:

$$w = (1 - \rho^n)^2 \tag{7.6}$$

which satisfies the boundary conditions $w(1) = 0$, $w'(1) = 0$ and the regularity conditions

$$w_{,\rho}(0) = 0, \qquad w_{,\rho\rho}(0) < \infty \tag{7.7}$$

provided that $n > 1$. For the plate without elastic foundations, the buckling load equals

$$\bar{N}_{r,o} = 6n^3\left(\frac{1}{n-1} - \frac{8}{3n-2} + \frac{4}{3n-1}\right) \tag{7.8}$$

The minimal value is attained at $n = 1.722$ and is $\bar{N}_{r,o} = 15.161$ or 3.25% higher than the exact value. For the plate on elastic foundation with modulus q,

$$\bar{N} = \bar{N}_o + \frac{qR^4}{D}\frac{3}{2n}\left(1 - \frac{8}{n+2} + \frac{6}{n+1} - \frac{8}{2n+2} + \frac{1}{2n+1}\right) \tag{7.9}$$

and the minimal value occurs at $n = 1.63$ with $\bar{N} = 29.2$ for $qR^4/D = 100$. This turned out to be 5.07% higher than the exact value found by Kline and Hancock (1965).

In this section, we consider the buckling of clamped and simply supported polar orthotropic plates on elastic foundations.

7.3.2 Statement of the Problem

Several studies of buckling of polar orthotropic circular plates under radial compression are reported in the literature. The axisymmetric buckling problem was solved by Woinowsky-Krieger (1958) with the aid of Bessel functions. Later, Pandalai and Patel (1965) used a power-series expansion for the same problem. The axisymmetric buckling problem was treated by Mossakowsky (1960) and Mossakowsky and Borsuk (1960) by the Frobenius method with the aid of hypergeometric functions. We are focusing here on using the Rayleigh quotient, with an objective of obtaining the accurate results within the variable parameter approach. For the polar orthotropic circular plate, the Rayleigh quotient is obtained from the requirement

$$\frac{d}{dC}(V + T) = 0 \tag{7.10}$$

where the trial function is represented for the axisymmetric buckling as

$$W(r, \theta) = Cw(r), \tag{7.11}$$

and

$$V = \pi \int_0^R \left[D_r \left(\frac{d^2 W}{dr^2} \right)^2 + 2D_1 \left(\frac{d^2 w}{dr^2} \right) \left(\frac{1}{r} \frac{dW}{dr} \right)^2 + D_\theta \left(\frac{1}{r} \frac{dW}{dr} \right)^2 + qW^2 \right] r\, dr \tag{7.12}$$

$$T = \pi h \int_0^R \sigma_r \left(\frac{dW}{dr} \right)^2 r\, dr \tag{7.13}$$

where D_r, $D_1 = \nu_\theta D_r$, and D_θ are the flexural rigidities of the plate, ν_θ is Poisson's ratio, σ_r is the normal stress that equals (Woinowsky-Krieger, 1958)

$$\sigma_r = -N_r(r/R)^{k-1} \tag{7.14}$$

where N_r is the radial compression, k is the orthotropy coefficient

$$k = \sqrt{E_\theta / E_r} = \sqrt{D_\theta / D_r} \tag{7.15}$$

where E_r and E_θ are Young's moduli in radial and circumferential directions, respectively, and q is the elastic foundation modulus. Using the non-dimensional coordinate $\rho = r/R$, we get for the buckling load the Rayleigh quotient

$$\bar{N} = \frac{N_r R^2}{D_r} = \frac{\int_0^1 [(w_{,\rho\rho})^2 + 2\nu_\theta (w_{,\rho\rho})(w_{,\rho}/\rho) + k^2 (w_{,\rho}/\rho)^2 + Qw^2] \rho\, d\rho}{\int_0^1 \rho^k (w_{,\rho})^2\, d\rho} \tag{7.16}$$

where

$$Q = qR^4/D_r \tag{7.17}$$

The problem involves evaluating this quotient so as to yield accurate estimates of buckling loads of plates with or without elastic foundations, under different boundary conditions.

7.3.3 Buckling of Clamped Polar Orthotropic Plates

We first employ the variable power method. We use the trial function

$$w_{,\rho} = (1 - \rho^{n+1})^2 \tag{7.18}$$

[with $n + 1$ instead of n in (7.6) for numerical convenience]. For a buckling load, at fixed non-dimensional stiffness of the elastic foundations, $Q = 100$, we get

$$\bar{N} = I/J \tag{7.19}$$

$$\begin{aligned} I = {} & 288n^{13} + n^{12}(312k + 4200) + n^{11}(252k^2 + 4360k + 27{,}236) \\ & + n^{10}(168k^3 + 3468k^2 + 26{,}854k + 125{,}654) \\ & + n^9(54k^4 + 2254k^3 + 20{,}977k^2 + 119{,}818k + 125{,}654) \\ & + n^8(6k^5 + 693k^4 + 13{,}205k^3 + 81{,}839k^2 + 368{,}273k + 886{,}080) \\ & + n^7(73k^5 + 3840k^4 + 45{,}481k^3 + 212{,}123k^2 + 736{,}814k + 1{,}296{,}693) \\ & + n^6(378k^5 + 12{,}087k^4 + 100{,}585k^3 + 361{,}745k^2 + 959{,}417k + 1{,}289{,}712) \\ & + n^5(1091k^5 + 23{,}862k^4 + 147{,}447k^3 + 406{,}559k^2 + 819{,}022k + 873{,}523) \\ & + n^4(1924k^5 + 30{,}687k^4 + 144{,}685k^3 + 301{,}061k^2 + 454{,}435k + 396{,}520) \\ & + n^3(2127k^5 + 25{,}740k^4 + 94{,}039k^3 + 144{,}745k^2 + 157{,}666k + 115{,}371) \\ & + n^2(1442k^5 + 13{,}593k^4 + 38{,}947k^3 + 43{,}347k^2 + 31{,}029k + 19{,}434) \\ & + n(549k^5 + 4104k^4 + 9339k^3 + 7344k^2 + 2640k + 1440) \\ & + 90k^5 + 540k^4 + 990k^3 + 540k^2 \end{aligned} \tag{7.20}$$

$$\begin{aligned} J = {} & 4n(36n^{10} + 468n^9 + 2639n^8 + 8509n^7 + 17{,}377n^6 + 23{,}473n^5 \\ & + 21{,}211n^4 + 12{,}631n^3 + 4727n^2 + 999n + 90) \end{aligned} \tag{7.21}$$

The requirement of $\bar{N}_{,n} = 0$ yields a twenty-third-order polynomial equation ($k = 1$) for n:

$$\begin{aligned} & 864n^{23} + 23{,}616n^{22} + 302{,}640n^{21} + 2{,}385{,}756n^{20} + 13{,}003{,}362n^{19} \\ & + 52{,}546{,}379n^{18} + 165{,}046{,}136n^{17} + 416{,}094{,}782n^{16} + 858{,}351{,}464n^{15} \\ & + 1{,}459{,}534{,}754n^{14} + 2{,}038{,}352{,}060n^{13} + 2{,}304{,}580{,}458n^{12} \\ & + 2{,}048{,}146{,}644n^{11} + 1{,}342{,}033{,}680n^{10} + 528{,}482{,}612n^9 - 40{,}547{,}266n^8 \\ & - 248{,}719{,}048n^7 - 214{,}576{,}018n^6 - 113{,}350{,}828n^5 - 41{,}363{,}230n^4 \\ & - 10{,}511{,}286n^3 - 1{,}780{,}011n^2 - 179{,}820n - 8100 = 0 \end{aligned} \tag{7.22}$$

For the plate without an elastic foundation, the expressions for I and J are much simpler:

$$I = k^5 + k^4(9n+6) + k^3(28n^2+35n+11) + k^2(42n^3+67n^2+35n+6) + k(52n^4+94n^3+53n^2+11n) + 48n^5+116n^4+104n^3+41n^2+6n \tag{7.23}$$

$$J = 4n(6n^2+5n+1) \tag{7.24}$$

The equation for $\bar{N}_{,n} = 0$ for $k = 1$ yields an equation

$$24n^7 + 72n^6 + 78n^5 + 23n^4 - 28n^3 - 30n^2 - 10n - 1 = 0 \tag{7.25}$$

Simpler equations are obtained if one uses (7.3) as the trial function:

$$w_{,\rho} = \rho - \rho^n \tag{7.26}$$

for the plate without elastic foundation. The buckling load becomes

$$I = n^3(6+2k) + n^2(2k^3+9k^2+14k+15) + n(3k^4+15k^3+21k^3+11k+6) + (k^5+6k^4+11k^3+6k^2) \tag{7.27}$$

$$J = 4n(n+1) \tag{7.28}$$

The minimization requirement is

$$n^4(2k+6) + n^3(4k+12) + n^2(-3k^4-13k^3-12k^2+3k+9) + n(-2k^5-12k^4-22k^3-12k^2) - k^5-6k^4-11k^3-6k^2 = 0 \tag{7.29}$$

For $k = 1$, we have the following roots:

$$n_1 = -1.00078, \quad n_2 = -\sqrt{3}, \quad n_3 = -0.999218, \quad n_4 = \sqrt{3} \tag{7.30}$$

Only the last root is meaningful, with the buckling load 14.928 reported in Section 7.3.2.

For $k = 2$, we get

$$n_1 = 0.695678, \quad n_2 = -1.88, \quad n_3 = -4, \quad n_4 = 4.57915 \tag{7.31}$$

The last root has a significance and yields

$$\bar{N} = 43.7843 \tag{7.32}$$

Finally, for $k = 3$,

$$n_1 = -0.6259, \quad n_2 = -2.63, \quad n_3 = -6.78, \quad n_4 = 8.04277 \tag{7.33}$$

with

$$\bar{N} = 92.0826 \tag{7.34}$$

associates with $n = n_4$.

The undetermined power variant of Rayleigh's method in this problem appears to be relatively complex. For the trial function (7.18), it leads to the necessity to solve a polynomial of the twenty-third degree, whereas the trial function in (7.26) leads to one of the fourth degree. Under these circumstances, it appears that this undetermined power version may not always be preferable to the predetermined power variant of it. Indeed, substituting $n = 2$ in (7.26), or using $w_{j\rho} = \rho - \rho^2$, leads to $\bar{N} = 15$, which is also 2.166% higher than the exact value; the minimization procedure led to $N = 14.928$. The actual "gain" may appear illusory. This is especially true if one recalls that Rayleigh's method was designed, as well as applied, to get quick estimates.

These considerations naturally lead us to question whether it is possible to apply another version of the Rayleigh method to eliminate these disadvantages while retaining all the advantages of the original non-integer-power method. Here we use another version of the variable parameter method. It is equivalent to using the two-term Rayleigh-Ritz method (Schmidt, 1982).

To illustrate this undetermined multiplier method, consider first the plate without elastic foundation. Instead of the conventionally used trial functions, $\rho - \rho^2$ or $\rho - \rho^3$ for the plate's shape $w_{,\rho}$, Schmidt (1982) used a variable power $\rho - \rho^n$. Here we use a variable multiplier to form the following trail function:

$$w_{,\rho} = (\rho - \rho^2) + n(\rho - \rho^3) \tag{7.35}$$

The expressions for I and J for the buckling load become

$$\begin{aligned} I = {} & n^2(10k^7 + 250k^6 + 2480k^5 + 12{,}500k^4 + 34{,}890k^3 + 60{,}450k^2 \\ & + 82{,}620k + 75{,}600) + n(14k^7 + 350k^6 + 3464k^5 + 17{,}300k^4 \\ & + 46{,}886k^3 + 75{,}230k^2 + 93{,}636k + 85{,}680) + (5k^7 + 125k^6 \\ & + 1235k^5 + 6125k^4 + 16{,}220k^3 + 24{,}350k^2 + 27{,}540k) \end{aligned} \tag{7.36}$$

$$J = 120[n^2(4k^2 + 40k + 96) + n(4k^2 + 46k + 126) + (k^2 + 13k + 42)] \tag{7.37}$$

As we observe, the buckling load is given as a ratio of two quadratic expressions in terms of the undetermined power. Therefore, the minimization requirement $\bar{N}_{,n} = 0$ is also a quadratic equation:

$$\begin{aligned} & n^2(-8k^9 - 250k^8 - 3260k^7 - 22{,}890k^6 - 92{,}524k^5 - 207{,}710k^4 \\ & \quad - 188{,}740k^3 + 204{,}690k^2 + 735{,}732k + 650{,}160) \\ & \quad + n(-10k^9 - 320k^8 - 4270k^7 - 30{,}660k^6 - 126{,}890k^5 - 295{,}180k^4 \\ & \quad - 307{,}430k^3 + 148{,}560k^2 + 801{,}000k + 756{,}000) \\ & \quad + (-3k^9 - 99k^8 - 1359k^7 - 10{,}014k^6 - 42{,}483k^5 - 101{,}961k^4 \\ & \quad - 116{,}571k^3 + 13{,}434k^2 + 208{,}656k + 211{,}680) = 0 \end{aligned} \tag{7.38}$$

with roots

$$n_{1,2} = \frac{-5k^4 - 35k^3 - 35k^2 - 5k + 150 \pm A}{2(4k^4 + 25k^3 + 25k^2 - 5k - 129)} \tag{7.39}$$

where

$$A = [k^8 + 8k^7 + 33k^6 + 58k^5 + 111k^4 + 32k^3 + 7k^2 - 18k + 828]^{1/2} \tag{7.40}$$

Upon substituting $n_{1,2}$ into Equations (7.36) and (7.37), we obtain the buckling loads for specific values of the orthotropy coefficients. For $k = 1$, we obtain for $n = n_1$

$$\bar{N} = \frac{7}{6}(29 - \sqrt{265}) \simeq 14.841 \tag{7.41}$$

which is 1.08% higher than the exact value. For $k = 1.5$,

$$\bar{N} = \frac{143}{1024}(459 - 5\sqrt{2085}) \simeq 27.118 \tag{7.42}$$

For $k = 2$, $\bar{N} = 42$. For $k = 2.5$, we get

$$\bar{N} = \frac{221}{3072}(3409 - \sqrt{6{,}602{,}773}) \simeq 60.388 \tag{7.43}$$

The buckling load equals $\bar{N} = 84$ for $k = 3$. The values of the buckling loads reported here virtually coincide with the values read from the paper by Woinowsky-Krieger (1958, Figure 3), who obtained an exact solution in terms of Bessel functions. Moreover, the comparison with the results obtained through the use of the undetermined power method reveals that the variable parameter method yields both more straightforward (without recourse to high-degree polynomial equations) and more reliable results (the furnished upper bounds are lower).

Consider now the polar orthotropic plate on elastic foundation. Rayleigh's quotient, given in (7.16), contains also a term w^2 so that we approximate the displacement rather than the slope:

$$w = (1 - 3\rho^2 + 2\rho^3) + n(1 - 4\rho^3 + 3\rho^4) \tag{7.44}$$

Equation (7.16) can again be put as a ratio I/J with

$$\begin{aligned} I = {} & n^2(1008k^7 + 25{,}200k^6 + 56k^5Q + 253{,}008k^5 + 1400k^4Q + 1{,}335{,}600k^4 \\ & + 13{,}720k^3Q + 4{,}257{,}792k^3 + 65{,}800k^2Q + 9{,}646{,}560k^2 + 154{,}224kQ \\ & + 16{,}656{,}192k + 141{,}120Q + 15{,}240{,}960) + n(2016k^7 + 50{,}400k^6 \\ & + 89k^5Q + 500{,}976k^5 + 2225k^4Q + 2{,}545{,}200k^4 + 21{,}805k^3Q \\ & + 7{,}280{,}784k^3 + 104{,}575k^2Q + 13{,}371{,}120k^2 + 245{,}106kQ \\ & + 19{,}432{,}224k + 224{,}280Q + 17{,}781{,}120) + (1260k^7 + 31{,}500k^6 \\ & + 36k^5Q + 311{,}220k^5 + 900k^4Q + 1{,}543{,}500k^4 + 8820k^3Q \\ & + 4{,}087{,}440k^3 + 42{,}300k^2Q + 6{,}136{,}200k^2 + 99{,}144kQ \\ & + 6{,}940{,}080k + 90{,}720Q + 6{,}350{,}400) \end{aligned} \tag{7.45}$$

$$J = 30{,}240[n^2(4k^2 + 28k + 48) + n(4k^2 + 40k + 84) + (k^2 + 13k + 42)] \tag{7.46}$$

Again, to find the minimal value of the buckling load, we should solve a quadratic equation

$$\begin{aligned}
& n^2(-4032k^9 - 116{,}928k^8 - 132k^7Q - 1{,}407{,}168k^7 - 3552k^6Q \\
& \quad - 9{,}047{,}808k^6 - 38{,}208k^5Q - 32{,}727{,}744k^5 - 206{,}040k^4Q \\
& \quad - 58{,}427{,}712k^4 - 553{,}788k^3Q + 8{,}543{,}808k^3 - 519{,}048k^2Q \\
& \quad + 280{,}482{,}048k^2 + 554{,}688kQ + 578{,}140{,}416k + 210{,}325{,}248k \\
& \quad + 1{,}088{,}640Q + 426{,}746{,}880) + n(-8064k^9 - 245{,}952k^8 - 176k^7Q \\
& \quad - 3{,}128{,}832k^7 - 4960k^6Q - 21{,}434{,}112k^6 - 55{,}872k^5Q \\
& \quad - 84{,}518{,}784k^5 - 312{,}800k^4Q - 183{,}976{,}128k^4 - 836{,}944k^3Q \\
& \quad - 149{,}764{,}608k^3 - 519{,}360k^2Q + 245{,}331{,}072k^2 + 2{,}025{,}792kQ \\
& \quad + 773{,}515{,}008k + 3{,}144{,}960Q + 670{,}602{,}240) + (-3024k^9 \\
& \quad - 99{,}792k^8 - 55k^7Q - 1{,}369{,}872k^7 - 1658k^6Q - 10{,}094{,}112k^6 \\
& \quad - 19{,}836k^5Q - 42{,}822{,}864k^5 - 116{,}110k^4Q - 102{,}776{,}688k^4 \\
& \quad - 309{,}065k^3Q - 117{,}503{,}568k^3 - 79{,}032k^2Q + 13{,}541{,}472k^2 \\
& \quad + 1{,}253{,}196kQ + 210{,}325{,}248 + 1{,}799{,}280Q + 213{,}373{,}440) = 0
\end{aligned} \tag{7.47}$$

with roots

$$\begin{aligned}
n_{1,2} = {} & [-2016k^4 - 11{,}088k^3 - 44k^2Q - 11{,}088k^2 - 140k^2Q + 4032k \\
& + 312Q + 66{,}528 \pm B][6(336k^4 + 1344k^3 + 11k^2Q + 1344k^2 \\
& + 21kQ - 3696k - 36Q - 14{,}112)]^{-1}
\end{aligned} \tag{7.48}$$

where

$$B = [14(53Q^2 - 22{,}752Q + 4{,}808{,}160)]^{1/2} \tag{7.49}$$

Roots $n_{1,2}$ being substituted into $\bar{N} = I/J$ yield the minimal buckling loads for any combination of the orthotropy coefficient k and the non-dimensional stiffness Q to the elastic foundation. Consider a number of specific cases. For an isotropic plate, we get

$$\begin{aligned}
\bar{N} = {} & [C(1525Q^2 - 139{,}104Q - 18{,}797{,}184) \\
& + (41{,}552Q^3 \pm 122{,}602{,}016Q^2 + 1{,}522{,}063{,}872Q + 474{,}969{,}277{,}440] \\
& \times \{144[68CQ - 15{,}372C + 1855Q^2 \pm 796{,}320Q + 168{,}285{,}600]\}^{-1}
\end{aligned} \tag{7.50}$$

where

$$C = [14(53Q^2 - 22{,}752Q + 4{,}808{,}160)]^{1/2} \tag{7.51}$$

For a plate without elastic foundation, $Q = 0$ and

$$\bar{N} = \frac{14(1855 + 37\sqrt{265})}{1325 + 61\sqrt{265}} \simeq 14.841 \tag{7.52}$$

which coincides with Equation (7.41), which was derived by using the trial function $w_{,\rho} = (\rho - \rho^2) + m(\rho - \rho^3)$. The coincidence of the buckling loads is understandable due to the proportionality of the constituents of the trial function

$$(1 - 3\rho^2 + 2\rho^3)_{,\rho} \sim \rho - \rho^2 \tag{7.53}$$
$$(1 - 4\rho^3 + 3\rho^4)_{,\rho} \sim \rho - \rho^3 \tag{7.54}$$

For an isotropic plate with elastic foundation, without non-dimensional stiffness $Q = 100$,

$$\bar{N} = \frac{1}{108}(4654 - \sqrt{2{,}680{,}090}) \simeq 27{,}934 \tag{7.55}$$

which is only 0.5% higher than the value reported by Kline and Hancock (1965). Interestingly, the present approach yields a lower value than an approximate method utilized by Dinnik (1955). For the isotropic clamped plate, he suggested using the formula

$$\bar{N} = \frac{PR^2}{D} = \alpha_1^2 + \frac{2Q}{\alpha_1^2} \tag{7.56}$$

where $\alpha_1 = 3.8317$ is the first root of the equation $J_1(\sqrt{P/DR}) = 0$, with $J_1(x)$ being the Bessel function of the first order. For $Q = 0$ Dinnik's approximation coincides with the exact result, but for $Q = 100$ Dinnik's formula yields $\bar{N} = 28.3041$.

For orthotropy coefficient $k = 2$, we arrive at

$$\begin{aligned}\bar{N} = {} & [D(881Q^2 - 710{,}640Q - 152{,}409{,}600) + 7728Q^3 \pm 10{,}039{,}680Q^2 \\ & + 4{,}343{,}673{,}600Q - 1{,}152{,}216{,}576{,}000] \\ & \times [360(9DQ - 10{,}080D + 92Q^2 \pm 136{,}060Q + 76{,}204{,}800)]^{-1}\end{aligned} \tag{7.57}$$

where

$$D = [3(23Q^2 - 34{,}020Q + 19{,}051{,}200)]^{1/2} \tag{7.58}$$

For the specific value $Q = 100$, we obtain

$$\bar{N} = \frac{1}{18}(2553 - 4\sqrt{119{,}094}) \simeq 65.144 \tag{7.59}$$

When $k = 3$, we get

$$\begin{aligned}N = {} & [E(43Q^2 - 93{,}120Q - 27{,}820{,}800) + 56Q^3 \pm 11{,}760Q^2 \\ & + 254{,}016{,}000Q + 79{,}543{,}872{,}000][24(2EQ - 14{,}880E + 7Q^2 \\ & \pm 3{,}360Q + 32{,}659{,}200)]^{-1}\end{aligned} \tag{7.60}$$

where

$$E = (Q^2 - 480Q + 4{,}665{,}600)^{1/2} \tag{7.61}$$

For $Q = 100$, the buckling load is

$$\bar{N} = \frac{1}{6}(2942 - 21\sqrt{11{,}569}) \simeq 113{,}876 \tag{7.62}$$

It also appears instructive to study how the choice of the trial function influences the results of the buckling loads. Bert (1987) used the trial function $w = (1 - \rho^n)^2$, within the undetermined power version of the Rayleigh method, for the isotropic plate on elastic foundation. We will utilize trial functions of this class with predetermined powers, but with an undetermined multiplier,

$$w = (1 - \rho^2)^2 + n(1 - \rho^3)^2 \tag{7.63}$$

for the orthotropic plates on elastic foundation. For I and J, we get, respectively, (for $Q = 100$)

$$\begin{aligned} I = n^2(&8019k^{10} + 425{,}007K^9 + 9{,}789{,}741k^8 + 129{,}854{,}583k^7 \\ &+ 1{,}115{,}908{,}713k^6 + 6{,}698{,}662{,}173k^5 + 29{,}784{,}804{,}759k^4 \\ &+ 100{,}676{,}192{,}517k^3 + 247{,}223{,}400{,}750k^2 + 382{,}186{,}910{,}520k \\ &+ 269{,}168{,}961{,}600) + n(15{,}246k^{10} + 808{,}038k^9 + 18{,}526{,}805k^8 \\ &+ 242{,}337{,}205k^7 + 2{,}018{,}314{,}298k^6 + 11{,}422{,}984{,}804k^5 \\ &+ 46{,}426{,}816{,}205k^4 + 141{,}823{,}272{,}745k^3 + 322{,}949{,}067{,}606k^2 \\ &+ 483{,}513{,}979{,}608k + 340{,}532{,}216{,}640) + (9240k^{10} + 489{,}720k^9 \\ &+ 11{,}187{,}330k^8 + 144{,}696{,}090k^7 + 1{,}173{,}812{,}640k^6 \\ &+ 6{,}295{,}078{,}020k^5 + 23{,}257{,}835{,}370k^4 + 62{,}234{,}428{,}410k^3 \\ &+ 125{,}371{,}558{,}620k^2 + 176{,}747{,}099{,}760k + 124{,}480{,}540{,}800) \end{aligned} \tag{7.64}$$

$$\begin{aligned} J = n^2(&81k^5 + 2349k^4 + 26{,}325k^3 + 142{,}155k^2 + 368{,}874k + 367{,}416) \\ &+ n(72k^5 + 2412k^4 + 30{,}996k^3 + 190{,}548k^2 + 557{,}892k + 617{,}760) \\ &+ (16k^5 + 608k^4 + 9008k^3 + 6486k^2 + 226{,}176k + 304{,}128) \end{aligned} \tag{7.65}$$

The minimization requirement yields (for $Q = 100$)

$$\begin{aligned} n^2(&-657{,}558k^{15} - 51{,}321{,}600k^{14} - 1{,}821{,}568{,}257k^{13} - 38{,}923{,}767{,}378k^{12} \\ &- 558{,}486{,}724{,}554k^{11} - 5{,}676{,}827{,}194{,}026k^{10} - 41{,}947{,}587{,}724{,}428k^9 \\ &- 226{,}518{,}671{,}150{,}406k^8 - 875{,}831{,}248{,}231{,}164k^7 \\ &- 2{,}239{,}968{,}412{,}170{,}126k^6 - 2{,}605{,}843{,}279{,}603{,}827k^5 \\ &+ 5{,}359{,}615{,}554{,}667{,}824k^4 + 32{,}351{,}700{,}422{,}843{,}676k^3 \\ &+ 71{,}812{,}607{,}632{,}193{,}952k^2 + 83{,}002{,}742{,}955{,}266{,}112k \\ &+ 41{,}164{,}832{,}809{,}013{,}760) + n(-1{,}240{,}272k^{15} - 99{,}392{,}832k^{14} \\ &- 3{,}624{,}987{,}492k^{13} - 79{,}652{,}738{,}664k^{12} - 1{,}176{,}254{,}942{,}760k^{11} \\ &- 12{,}323{,}476{,}885{,}464k^{10} - 94{,}159{,}522{,}997{,}280k^9 \\ &- 529{,}897{,}561{,}143{,}624k^8 - 2{,}179{,}562{,}582{,}640{,}600k^7 \\ &- 6{,}313{,}972{,}266{,}390{,}024k^6 - 11{,}393{,}884{,}883{,}247{,}228k^5 \end{aligned}$$

$$
\begin{aligned}
&- 5{,}596{,}409{,}574{,}852{,}912k^4 + 32{,}469{,}254{,}939{,}758{,}032k^3 \\
&+ 100{,}263{,}966{,}172{,}139{,}520k^2 + 132{,}511{,}904{,}737{,}977{,}600k \\
&+ 72{,}251{,}351{,}149{,}824{,}000) + (-421{,}344k^{15} - 35{,}348{,}544k^{14} \\
&- 1{,}348{,}043{,}488k^{13} - 30{,}932{,}567{,}512k^{12} - 476{,}369{,}683{,}504k^{11} \\
&- 5{,}198{,}129{,}429{,}856k^{10} - 41{,}333{,}185{,}703{,}040k^9 \\
&- 242{,}248{,}300{,}830{,}096k^8 - 1{,}043{,}053{,}957{,}383{,}680k^7 \\
&- 3{,}222{,}489{,}305{,}035{,}616k^6 - 6{,}661{,}690{,}903{,}240{,}736k^5 \\
&- 7{,}048{,}828{,}900{,}788{,}312k^4 + 4{,}678{,}986{,}305{,}136{,}528k^3 \\
&+ 29{,}890{,}548{,}464{,}078{,}016k^2 + 45{,}436{,}368{,}005{,}259{,}264k \\
&+ 26{,}666{,}283{,}097{,}681{,}920)
\end{aligned}
\tag{7.66}
$$

with roots

$$
\begin{aligned}
n_{1,2} = {} & [-34{,}452k^7 - 934{,}956k^6 - 9{,}661{,}221k^5 - 47{,}659{,}833k^4 - 112{,}317{,}597k^3 \\
& - 59{,}691{,}411k^2 + 416{,}475{,}270k + 1{,}005{,}582{,}600 + 2F] \\
& \times [9(8{,}118k^7 + 203{,}346k^6 + 1{,}937{,}087k^5 + 8{,}824{,}507k^4 \\
& + 18{,}042{,}461k^3 - 9{,}824{,}949k^2 151{,}882{,}146k - 254{,}633{,}544)]^{-1}
\end{aligned}
\tag{7.67}
$$

where

$$
\begin{aligned}
F = {} & (331{,}822{,}656k^{14} + 16{,}583{,}919{,}264k^{13} + 367{,}934{,}944{,}608k^{12} \\
& + 4{,}787{,}539{,}400{,}556k^{11} + 40{,}663{,}247{,}789{,}641k^{10} \\
& + 237{,}527{,}161{,}809{,}208k^9 + 979{,}989{,}247{,}011{,}847k^8 \\
& + 2{,}877{,}371{,}177{,}782{,}438k^7 + 5{,}808{,}079{,}295{,}573{,}071k^6 \\
& + 6{,}446{,}393{,}327{,}811{,}628k^5 - 2{,}663{,}017{,}967{,}877{,}591k^4 \\
& - 19{,}081{,}572{,}469{,}289{,}766k^3 - 3{,}269{,}554{,}711{,}073{,}064k^2 \\
& + 88{,}717{,}428{,}179{,}238{,}672k + 160{,}660{,}980{,}513{,}392{,}832)
\end{aligned}
\tag{7.68}
$$

We consider specific cases to get more insight. For the isotropic case without elastic foundation, we get

$$
\bar{N} = \frac{1}{263}(9{,}961 - \sqrt{37{,}187{,}185}) \simeq 14.6877 \tag{7.69}
$$

which coincides with the exact solution within four significant digits. This is probably the approximate result that is closest to the exact solution, using the trial functions, reported here. For an isotropic plate on elastic foundation with $Q = 100$, we have

$$
\bar{N} = \frac{1}{52{,}074}(2{,}465{,}703 - 7\sqrt{21{,}325{,}774{,}110}) \simeq 27.7196 \tag{7.70}
$$

For the plate with orthotropy coefficient $k = 2$ and without elastic foundation, we arrive at

$$
\bar{N} = \frac{693}{415{,}787}(60{,}331 - 4\sqrt{73{,}778{,}091}) \simeq 43.2902 \tag{7.71}
$$

Table 7.1. Values of buckling loads for clamped plates without elastic foundation

	$k = 1$	$k = 2$	$k = 3$
$w_{,\rho} = \rho - \rho^n$	14.928[a]	43.7843	92.0826
$w_{,\rho} = (\rho - \rho^2) + n(\rho - \rho^3)$	14.841	42	84
$w = (1 - 3\rho^2 - 2\rho^3) + n(1 - 4\rho^3 + 3\rho^4)$	14.841	42	84
$w = (1 - \rho^2)^2 + n(1 - \rho^3)^2$	14.6877	43.2902	82.8093

[a] Result reported by Schmidt (1982).

which is higher than the value $\bar{N} = 42$ obtained with Equation (7.35) as the trial function. For $Q = 100$, the result is $\bar{N} = 67.5251$. For $k = 3$, $Q = 0$,

$$\bar{N} = \frac{3}{563}(54{,}523 - 9\sqrt{18{,}760{,}889}) \simeq 82{,}8093 \tag{7.72}$$

which is lower than the value $\bar{N} = 84$ obtained with (7.35). For $k = 3$, $Q = 100$,

$$\bar{N} = \frac{1}{6756}(2{,}136{,}503 - 9\sqrt{23{,}807{,}753{,}219}) \simeq 110.7120 \tag{7.73}$$

The lesson to be learned by "trying" various trial functions in the same problem is that the degree of accuracy depends on the orthotrophy coefficient k: Whereas for one particular value of k the specified trial function may be "better," the same trial function may perform "worse" for another value of k. Generally, however, the undetermined multiplier (or, in other words, the Rayleigh-Ritz) method is much easier to implement than its undetermined power counterpart (Table 7.1).

7.3.4 Buckling of Simply Supported Polar Orthotropic Plates

For a simply supported isotropic plate, Schmidt used the trial function

$$w_{,\rho} = \frac{n + \nu}{1 + \nu}\rho - \rho^n \tag{7.74}$$

satisfying both the boundary conditions at the circumference and the regularity conditions at the center; ν is the Poisson's ratio. The minimum occurs at $n = 2.5$ for which $\bar{N} = 4.20$, agreeing precisely with the exact value reported by Timoshenko and Gere (1961) to three significant figures.

We use the following simple function:

$$\psi = 1 - \rho^3 + n(\rho^2 - \rho^3) \tag{7.75}$$

which satisfies the geometric boundary condition at $\rho = 1$ – namely $w(1) = 0$ – as well as the regularity conditions at the center.

$$\begin{aligned} I = {} & n^2(k^5 + 12k^4 + 4k^3\nu_\theta + 59k^3 + 48k^2\nu_\theta + 204k^2 + 188k\nu_\theta + 564k \\ & + 240\nu_\theta + 720) + n(2k^5 + 24k^4 + 24k^3\nu_\theta + 134k^3 + 288k^2\nu_\theta + 600k^2 \\ & + 1{,}128k\nu_\theta + 1880k + 1440\nu_\theta + 2400) + (9k^5 + 108k^4 + 36k^3\nu_\theta + 459k^3 \\ & + 432k^2\nu_\theta + 972k^2 + 1692k\nu_\theta + 1692k + 2160\nu_\theta + 2160) \end{aligned} \tag{7.76}$$

$$J = 4[n^2(k^2 + 3k + 8) + n(6k^2 + 30k + 36) + (9k^2 + 63k + 108)] \tag{7.77}$$

For $\nu_\theta = 0.3$, the minimization requirement leads to the quadratic

$$\begin{aligned} &n^2(5k^7 + 90k^6 + 660k^5 + 2808k^4 + 8793k^3 + 20{,}310k^2 + 24{,}814k + 7320) \\ &\quad + n(90k^6 + 1350k^5 + 9198k^4 + 41{,}454k^3 + 128{,}952k^2 \\ &\quad + 227{,}196k + 157{,}680) + (-45k^7 - 720k^6 - 4230k^5 - 8658k^4 \\ &\quad + 19{,}125k^3 + 146{,}106k^2 + 318{,}222k + 255{,}960) = 0 \end{aligned} \tag{7.78}$$

with roots

$$\begin{aligned} n_{1,2} &= [3(-15k^3 - 45k^2 - 288k - 438 \pm G)] \\ &\quad \times [5k^4 + 30k^3 + 65k^2 + 318k + 122]^{-1} \end{aligned} \tag{7.79}$$

where

$$\begin{aligned} G &= (25k^8 + 250k^7 + 1225k^6 + 3000k^5 + 8400k^4 + 11{,}520k^3 + 21{,}620k^2 \\ &\quad + 74{,}960k + 134{,}016)^{1/2} \end{aligned} \tag{7.80}$$

For the case of the isotropic plate, we arrive at, for $n = n_2$,

$$\bar{N} = \frac{101 - \sqrt{7081}}{4} \simeq 4.213 \tag{7.81}$$

For the orthotropy coefficient $k = 2$, for $n = n_2$,

$$\bar{N} = \frac{3}{5}(98 - \sqrt{5579}) \simeq 13.984 \tag{7.82}$$

Whereas for $k = 3$, we get for $n = n_2$,

$$\bar{N} = \frac{7}{30}(609 - \sqrt{237{,}801}) \simeq 28.135 \tag{7.83}$$

For the plate on elastic foundation, to the numerator J should be added the following increment:

$$\begin{aligned} \Delta J &= Q[5k^3n^2 + 54k^3n + 189k^3 + 60k^2n^2 + 648k^2n + 2268k^2 + 235kn^2 \\ &\quad + 2538kn + 8883k + 300n^2 + 3240n + 11{,}340] \end{aligned} \tag{7.84}$$

Accordingly, the minimization condition changes too. For $Q = 100$, the minimal values of $\bar{N}$ occur at

$$\begin{aligned} n_{1,2} &= [3(-105k^3 - 115k^2 - 1666k - 1716 \pm H)] \\ &\quad \times [35k^4 + 210k^3 + 355k^2 + 2176k - 196]^{-1} \end{aligned} \tag{7.85}$$

where

$$\begin{aligned} H &= (1225k^8 + 12{,}250k^7 + 62{,}125k^6 + 189{,}00k^5 + 554{,}600k^4 + 521{,}480k^3 \\ &\quad + 1{,}064{,}380k^2 - 429{,}760k + 3{,}506{,}784)^{1/2} \end{aligned} \tag{7.86}$$

Let us list some results. For $k = 1$ we have, for $n = n_1$,

$$\bar{N} = \frac{1}{84}(2{,}971 - \sqrt{1{,}370{,}521}) \simeq 21.432 \tag{7.87}$$

which is very close to the extrapolated value 20 read from the curve in the paper by Kline and Hancock (1965). Our result is also in good agreement with the approximate formula by Dinnik (1955) for isotropic plates,

$$\bar{N} = \alpha^2 + \frac{Q}{\alpha^2}\frac{\alpha^2 - 2 + 2\nu^2}{\alpha^2 - 1 + \nu^2} \tag{7.88}$$

where $\alpha = 2.047$ and which is the first root of the transcendental equation

$$\sqrt{\frac{P}{D}}RJ_0\left(\sqrt{\frac{P}{D}}R\right) - (1-\nu)J_1\left(\sqrt{\frac{P}{D}}R\right) = 0 \tag{7.89}$$

for the buckling of the plate without elastic foundation, ν being the Poisson ratio.

For $Q = 100$, $\nu = 0.3$, Dinnik's formula yields

$$\bar{N} = 20.7799 \tag{7.90}$$

so that our result is 3.13% higher than Dinnik's approximation. Since Dinnik's value is also an upper bound, this means that the percentagewise error of (7.87) in comparison to the exact solution is at least 3.13%. We are unaware of results in the literature for the polar orthotropic plate.

Let us list some results; at $k = 2$,

$$\bar{N} = \frac{3}{35}(861 - 13\sqrt{1309}) \simeq 33.485 \tag{7.91}$$

for $k = 3$, for $n = n_1$,

$$\bar{N} = \frac{1}{15}(2469 - 49\sqrt{1281}) \simeq 47.682 \tag{7.92}$$

The values of buckling loads are listed in Table 7.2.

Note that Makushin (1962) suggested an analogous method for the buckling of isotropic circular plates. He utilized the beam deflection functions stemming from two different loads, T_1 and T_2, which resulted in deflections

$$w_1 = T_1\varphi_1, \qquad w_2 = T_2\varphi_2 \tag{7.93}$$

Table 7.2. Values of buckling loads for clamped plates with elastic foundation

	$k = 1$	$k = 2$	$k = 3$
$w = (1-\rho^n)^2$	29.2[5]	—	—
$w = (1 - 3\rho^2 + 2\rho^3) + n(1 - 4\rho^3 + 3\rho^4)$	27.934	65.144	113.879
$w = (1-\rho^2)^2 + n(1-\rho^3)^2$	27.7196	67.5251	110.7120

The Makushin method combined these two functions to construct the trial function

$$w = \varphi_1 + \frac{T_2}{T_1}\varphi_2 = \varphi_1 + n\varphi_2 \tag{7.94}$$

where n is an undetermined multiplier. He then solved the problem of the buckling of a clamped circular plate with trial functions

$$w = (2+n) - (6+2n)\rho^2 + 4\rho^3 + n\rho^4 \tag{7.95}$$

or

$$w = (10+3n) - 5(4+n)\rho^2 + 10\rho^3 + 2n\rho^5 \tag{7.96}$$

For the isotropic plate, which is simply supported, he used

$$w = (8+5n) - (12+6n)\rho^2 + 4\rho^3 + n\rho^4 \tag{7.97}$$

or

$$w = (25+9n) - 3(10+n)\rho^2 + 5\rho^4 + n\rho^5 \tag{7.98}$$

Equations (7.93)–(7.98) are based on the following observation. For the axisymmetric buckling modes, the shape $w_{,\rho}$ vanishes. Accordingly, Makushin (1962) used the functions that satisfy the condition $w_{,\rho} = 0$ at $\rho = 0$ and the appropriate boundary condition at the circumference. For clamped plates, functions in (7.93) correspond to the displacements due to uniformly distribute (T_1) and to concentrate force at the center (T_2), respectively; note that the slope $w_{,\rho}$ coincides with the assumption we used in (7.35), from other considerations. Therefore, our numerical result for the *isotropic* plate coincides with that of Makushin (Table 7.2).

7.3.5 Buckling of Free Polar Orthotropic Plates

We employ the following trial function:

$$w = \rho^2 + n\rho^3 \tag{7.99}$$

The results for I and J read

$$\begin{aligned} I = {} & n^2(9k^5 + 108k^4 + 36k^3\nu_\theta + 459k^3 + 432k^2\nu_\theta + 972k^2 + 1692k\nu_\theta \\ & + 1692k + 2160\nu_\theta + 2160) + n(16k^5 + 192k^4 + 48k^3\nu_\theta + 784k^3 \\ & + 576k^2\nu_\theta + 1344k^2 + 2256k\nu_\theta + 1504k + 2880\nu_\theta + 1920) \\ & + (8k^5 + 96k^4 + 16k^3\nu_\theta + 384k^3 + 192k^2\nu_\theta + 576k^2 + 752k\nu_\theta \\ & + 376k + 960\nu_\theta + 480) \end{aligned} \tag{7.100}$$

$$I = 4[n^2(9k^2 + 63k + 108) + n(12k^2 + 96k + 180) + (4k^2 + 36k + 80)] \tag{7.101}$$

The minimization requirement is also a quadratic equation

$$\begin{aligned} & n^2(-9k^7 - 144k^6 - 846k^5 + 108k^4\nu_\theta - 1764k^4 + 1620k^3\nu_\theta + 3339k^3 \\ & \quad + 8964k^2\nu_\theta + 26{,}532k^2 + 21{,}708k\nu_\theta + 57{,}132k + 19{,}440\nu_\theta + 45{,}360) \\ & \quad + n(-18k^7 - 306k^6 - 1962k^5 + 144k^4\nu_\theta - 5346k^4 + 2304k^3\nu_\theta - 1332k^3 \\ & \quad + 13{,}680k^2\nu_\theta + 28{,}548k^2 + 35{,}712k\nu_\theta + 71{,}136k + 34{,}560\nu_\theta + 60{,}480) \\ & \quad + (-8k^7 - 144k^6 - 984k^5 + 48k^4\nu_\theta - 3024k^4 + 816k^3\nu_\theta - 2952k^3 \\ & \quad + 5136k^2\nu_\theta + 5952k^3 + 14{,}160k\nu_\theta + 18{,}920k + 14{,}400\nu_\theta + 16{,}800) = 0 \end{aligned} \tag{7.102}$$

with roots

$$\begin{aligned} n_{1,2} = {} & [3k^4 - 15k^3 - 6k^2 + 24k\nu_\theta + 66k + 96\nu_\theta + 168 \pm L] \\ & \times [3(k^4 + 4k^3 - k^2 - 12k\nu_\theta - 40k - 36\nu_\theta - 84)]^{-1} \end{aligned} \tag{7.103}$$

where

$$\begin{aligned} L = {} & (k^8 + 10k^7 + 45k^6 + 120k^5 + 336k^4 - 144k^3\nu_\theta + 504k^3 - 144k^2\nu_\theta \\ & + 908k^2 + 288k\nu_\theta + 2912k + 576\nu_\theta^2 + 2016\nu_\theta + 4707)^{1/2} \end{aligned} \tag{7.104}$$

Substitution of specific values of $\nu_\theta = 0.3$ and $k = 1, 2, 3$ leads to the conclusion that the buckling loads of the free plate coincide with those of the simply supported plates. Indeed, in the absence of an elastic foundation, the trial function itself does not enter in the analysis but rather in the slope.

The first derivative of the trial function used for the simply supported cases can be represented as

$$[1 - \rho^3 + m(\rho^2 - \rho^3)]_{,\rho} = m\left[2\rho - \frac{(m+1)}{m}3\rho^2\right] \tag{7.105}$$

Then, with

$$-(m+1)/m \equiv M \tag{7.106}$$

the simply supported plate's slope coincides with that of the free plate.

As we have seen, within trial functions' choice, the free and simply supported plates share the same buckling load when they are unsupported by the elastic foundation. Note that the computerized symbolic algebra proved extremely useful to obtain closed-form solutions in buckling of structures for columns (Elishakoff and Rollot, 1999) and plates (Elishakoff, 2000).

Bibliography

He who is able to write a book and does not write it is as one who has lost a child.
Nahman of Bratslav

Abell, M. L., and Braselton, J. P., *The Mathematica Handbook*, AP Professional, Boston, 1992.

Aboudi, J., Cederbaum, G., and Elishakoff, I., Dynamic stability analysis of viscoelastic plates by Lyapunov exponents, *Journal of Sound and Vibration*, Vol. 139, 459–68, 1990.

Abramovich, H., and Elishakoff, I., Application of the Krein's method for determination of natural frequencies of periodically supported beam based on simplified Bresse-Timoshenko equations, *Acta Mechanica*, Vol. 66, 39–59, 1987.

Abramovich, H., Singer, J., and Yaffe, R., Imperfection characteristics of stiffened shells-Group 1, *TAE Report 406*, Technion, Israel Institute of Technology, Department of Aeronautical Engineering, Haifa, 1981.

Adali, S., Elishakoff, I., Richter, A., and Verijenko, V. E., Optimal design of symmetric angle-ply laminates for maximum buckling load with scatter in material properties, *Paper AIAA-94-4365-CP, A Collection of Technical Papers, 5th AIAA/USAF/NASA/ISSMO Symposium on Multidisciplinary Analysis and Optimization* (J. Sobieski, ed.), Part 2, pp. 1041–5, 1994.

Adali, S., Richter, A., and Verijenko, V. E., Minimum weight design of symmetric angle-ply laminates under multiple uncertain loads, *Structural Optimization*, Vol. 9, 89–95, 1995.

Adali, S., Richter, A., and Verijenko, V. E., Minimum weight design of symmetric angle-ply laminates with incomplete information on initial imperfections, *Journal of Applied Mechanics*, Vol. 64, 90–7, 1997.

Almroth, B. O., Influence of edge conditions on the stability of axially compressed cylindrical shells, *AIAA Journal*, Vol. 4, 134–40, 1966.

Almroth, B. O., and Brogan, F. A., The STAGS computer code, *NASA CR-2950*, 1978.

Almroth, B. O., and Rankin, C. R., Imperfection sensitivity of cylindrical shells, *Advances in Engineering Mechanics*, 1071–4, 1983.

Almorth, B. O., and Starnes, J. H., Jr., The computer in shell stability analysis, *Journal of the Engineering Mechanics Division*, Vol. 101, 873–88, 1974.

Almroth, B. O., Brogan, F. A., Miller E., Zelle, F., and Peterson, M. T., Collapse analysis or shells of general shape, *User's Manual for the STAGS-A Computer Code*, Air Force Flight Dynamics Laboratory, Wright-Patterson AFB, AFFDL-TR-71-8, 1973.

Amazigo, J. C., Buckling under axial compression of long cylindrical shells with random axisymmetric imperfections, *Quarterly of Applied Mathematics*, Vol. 26, 537–66, 1969.

Amazigo, J. C., Buckling of stochastically imperfect columns on non-linear elastic foundations, *Quarterly of Applied Mathematics*, Vol. 28, 403–9, 1971.

Amazigo, J. C., Asymptotic analysis of the buckling of externally pressurized cylinders with random imperfections, *Quarterly of Applied Mathematics*, Vol. 31, 429–42, 1972.

Amazigo, J. C., Dynamic buckling of structures with random imperfections, *Stochastic Problems in Mechanics* (H. H. E. Leipholz, ed.), University of Waterloo Press, Waterloo, pp. 234–54, 1974.

Amazigo, J. C., Buckling of stochastically imperfect structures, *Buckling of Structures* (B. Budiansky, ed.), Springer Verlag, Berlin, pp. 172–82, 1976.

Amazigo, J. C., and Budiansky, B., Asymptotic formulas for the buckling stresses of axially compressed cylinders with localized or random axisymmetric imperfections, *Journal of Applied Mechanics*, Vol. 39, 179–84, 1972.

Amazigo, J. C., Budiansky, B., and Carrier, G. F., Asymptotic analysis of the buckling of imperfect columns on nonlinear elastic foundation, *International Journal of Solids and Structures*, Vol. 6, 1341–56, 1970.

Ambartsumyan, S. A., *Theory of Anistropic Shells*, Technomic Publishers, Lancaster, 1970.

Ambartsumyan, S. A., *General Theory of Anisotropic Shells*, "Nauka" Publishing House, Moscow, 1974 (in Russian).

Ambartsumyan, S. A., Some problems of the theory of shells from composite materials, *Uspekhi Mekhaniki*, Vol. 6, Nos. 3–4, 1983 (in Russian).

Ambartsumyan, S. A., *Theory of Anisotropic Plates: Strength, Stability and Vibration*, Hemisphere Publishing, New York, 1991.

Anderson, P. W., Absence of diffusion in certain random lattices, *Physical Review*, Vol. 109, 1492–1505, 1958.

Aoki, T., and Fukumoto, Y., On scatter in buckling strength of steel columns – Effect of residual stress distributions, *Proceedings of JSCE*, No. 201, 31–41, 1972 (in Japanese).

Arbocz, J., The effect of initial imperfections in shell stability, *Thin Shell Structures: Theory, Experiment and Design* (Y. C. Fung and E. E. Sechler, eds.), Prentice-Hall, Englewood Cliffs, NJ, pp. 205–45, 1974.

Arbocz, J., Present and future of shell stability analysis, *Zeitschrift für Flugwissenschaften und Weltraumforschung*, Vol. 5, 335–48, 1981.

Arbocz, J., The imperfection data bank: A means to obtain realistic buckling loads, in *Buckling of Shells* (E. Ramm, ed.), Springer Verlag, Berlin, pp. 535–67, 1982a.

Arbocz, J., Shell stability analysis: theory and practice, *Collapse: Buckling of Structures in Theory and Practice* (J. M. T. Thompson and G. W. Hunt, eds.), Cambridge University Press, Cambridge, pp. 43–74, 1982b.

Arbocz, J., About the state of the art of shell design, *Memorandum M-629*, Faculty of Aerospace Engineering, Delft University of Technology, Delft, the Netherlands, 1990.

Arbocz J., Towards an improved design procedure for buckling critical structures, *Buckling of Shell Structures on Land, in the Sea and in the Air* (J. F. Jullien, ed.), Elsevier Applied Science, London, pp. 270–6, 1991.

Arbocz, J., Future directions and challenges in shell stability, 38th AIAA/ASME/ASCE/AHS/ASC Structures, *Structural Dynamics, and Materials Conference and Exhibit*, AIAA paper 97-1077, pp. 1949–62, 1997.

Arbocz, J., and Abramovich, H., The initial imperfection data bank at the Delft University of Technology, Part 1, *Report LR-290*, Delft University of Technology, Department of Aerospace Engineering, Delft, the Netherlands, 1979.

Arbocz, J., and Babcock, Jr., C. D., Experimental investigation of the effect of general imperfections on the buckling of cylindrical shells, *NASA CR-1163*, 1968.

Arbocz, J., and Babcock, Jr., C. D., A multi-mode analysis for calculating buckling loads of imperfect cylindrical shells, *GALCIT Report SM-74-4*, California Institute of Technology, Pasadena, 1974.

Arbocz, J., and Babcock, Jr., C. D., Prediction of buckling loads based on experimentally measured initial imperfections, *Buckling of Structures* (B. Budiansky, ed.), Springer Verlag, Berlin, 1976.

Arbocz, J., and Babcock, Jr., C. D., Utilization of STAGS to determine knockdown factors from measured initial imperfections, *Report LR-275*, Delft University of Technology, Department of Aerospace Engineering, Delft, the Netherlands, 1978.

Arbocz, J., and Babcock, Jr., C. D., Computerized stability analysis using measured initial imperfections, *12th Congress of International Council of Aeronautical Science*, Munich, pp. 688–701, 1980a.

Arbocz, J., and Babcock, Jr., C. D., The buckling analysis of imperfection sensitive structures, *NASA CR-3310*, 1980b.

Arbocz, J., and Hol, J. M. A. M., ANILISA – Computational module for Koiter's imperfection theory, *Report LR-582*, Delft University of Technology, Faculty of Aerospace Engineering, Delft, the Netherlands, 1989.

Arbocz, J., and Hol, J. M. A. M., Koiter's stability theory in a computer aided engineering environment, *International Journal of Solids and Structures*, Vol. 26, 945–73, 1990a.

Arbocz, J., and Hol, J. M. A. M., Recent developments in shell stability analysis, *Report LR-633*, Delft University of Technology, Faculty of Aerospace Engineering, Delft, the Netherlands, 1990b.

Arbocz, J., and Hol, J. M. A. M., Collapse of axially compressed cylindrical shells with random imperfections, *AIAA Journal*, Vol. 29, 2247–56, 1991 (also reprinted in *Thin-Walled Structures*, Vol. 23, 131–58, 1995).

Arbocz, J., and Singer, J., Professor Bernard Budiansky's contributions to buckling and postbuckling of shell structures, AIAA Paper 2000-1322, 41st AIAA/ASME/ASCE/AHS/ASC Structures, Structural Dynamics and Materials Conference and Exhibit, 3–6 April 2000, Atlanta, Ga.

Arbocz, J., and Williams, J. G., Imperfection surveys on a 10 ft. diameter shell structure, *AIAA Journal*, Vol. 15, 949–56, 1976.

Arbocz, J., Poiter-Ferry, M., Singer, J., and Tvergaard, V., *Buckling and Postbuckling*, Springer Verlag, Berlin, 1987.

Arbocz, J., and Singer, J., Professor Bernard Budiansky's contributions to buckling and postbuckling of shell structures, (preprint), 41st AIAA/ASME/ASCE/AHS/ASC Structures, Structural Dynamics, and Materials Conference and Exhibit, Atlanta, GA, 2000.

Arbocz, J., Starnes, J., and Nemeth, M., A comparison of probabilistic and lower bound methods for predicting the response of buckling-sensitive structures, Paper No. AIAA-2000-1382, AIAA Press, Washington, D. C., 2000.

Ari-Gur, J., and Elishakoff, I., Influence of the shear deformation on the buckling of a structure with overhang, *Journal of Sound and Vibration*, Vol. 139, 165–9, 1990.

Ariaratnam, S. T., and Xie, W. C., Wave localization in randomly disordered nearly periodic long continuous beams, *Journal of Sound and Vibration*, Vol. 181, 7–22, 1995.

Ariaratnam, S. T., and Xie, W. C., Buckling node localization in randomly disordered continuous beams using a simplified model, *Chaos, Solitons and Fractals*, Vol. 7, 1127–44, 1996.

Arnold, L., Crauel, H., and Eckmann, J. P. (eds.), *Lyapunov Exponents*, Springer Verlag, Berlin, 1991.

Aston, P. J., Analysis and computation of symmetry-breaking bifurcation and scaling laws using group theoretic methods, *SIAM Journal of Mathematical Analysis*, Vol. 22, 181–212, 1991.

Augusti, G., Probabilistic treatments of column buckling problems, *Stochastic Problems in Mechanics*, University of Waterloo Press, Waterloo, pp. 255–74, 1974.

Augusti, G., and Baratta, A., Teoria probabilistica della resistenza delle aste compresse, *Construzioni Metalliche*, No. 1, 1–15, 1971 (in Italian).

Augusti, G., and Baratta, A., Limit analysis of structures with stochastic strength variations, *Journal of Structural Mechanics*, Vol. 1, 43–62, 1972a.

Augusti, G., and Baratta, A., Probabilistic theory of slender columns, *Proceedings of 13th Polish Solid Mechanics Conference*, Warsaw, 1972b.

Augusti, G., and Baratta, A., Reliability of slender columns: Comparison of different approximations, *Buckling of Structures* (B. Budiansky, ed.), Springer Verlag, Berlin, pp. 183–98, 1976.

Augusti, G., Baratta, A., and Casciati, F., *Probabilistic Methods in Structural Engineering*, Chapman and Hall, London, 1984.

Ayyub, B. M., and Chia, C. Y., Generalized conditional expectation for structural reliability assessment, *Structural Safety*, Vol. 11, 131–46, 1992.

Ayyub, B. M., and Haldar, A., Improved simulation techniques as structural reliability models, *Proceedings, International Conference on Structural Safety and Relibility*, Kobe, Japan, pp. I. 17–I. 26, 1985.

Ayyub, B. M., and Lai, K. L., Selective sampling in simulation-based reliability assessment, *International Journal of Pressure Vessels and Piping*, Vol. 46, 229–49, 1991.

Ayyub, B. M., and McCuen, R. H., Simulation-based reliability methods, *Probabilistic Structural Mechanics Handbook* (C. R. Sundararajan, ed.), Chapman and Hall, New York, pp. 53–69, 1995.

Babcock, Jr., C. D., Shell stability, *Journal of Applied Mechanics*, Vol. 50, 935–40, 1983.

Babcock, Jr., C. D., and Sechler, E. E., The effect of initial imperfections on the buckling stress of cylindrical shells, *NASA TN D-1510*, pp. 135–42, 1962.

Badala, A. Marinetti, A., and Obiveto, G., The collapse load of pin-jointed networks of randomly imperfect rods, *Engineering Structures*, Vol. 2, 9–14, 1980.

Bahder, Th. B., *Mathematica for Scientists and Engineers*, Addison-Wesley, Reading, MA, 1995.

Baker, E. H., Cappelli, A. P., Kovalevsky, L., Rish, F. L., and Verette, R. M., Shell analysis manual, *NASA CR-512*, 1968.

Ball, J. M., Extensible beam stability theory, *Journal of Differential Equations*, Vol. 14, 399–418, 1974.

Bažant, Z. P., and Cedolin, L., *Stability of Structures*, Oxford University Press, New York, 1991.

Bažant, Z., Structural stability, *International Journal of Solids and Structures*, Vol. 37, 55–67, 2000.

Bellini, P. X., and Chulya, A., An improved automatic incremental algorithm for the efficient solution of non-linear finite element equations, *Computers and Structures*, Vol. 26, 99–110, 1987.

Beltzer, A. I., Engineering analysis via symbolic computation - A breakthrough, *Applied Mechanics Reviews*, Vol. 43, 119–23, 1990a.

Beltzer, A. I., *Variational and Finite Element Methods – A Symbolic Computation Approach*, Springer Verlag, Berlin, 1990b.

Benaroya, H., and Rehak, M., Finite element methods in probabilistic structural analysis: A selective review, *Applied Mechanics Reviews*, Vol. 41, 201–13, 1988.

Benedetti, A., On the reliability analysis of elastic-plastic struts with initial imperfections, *IASS Symposium – 91*, Vol. III, pp. 123–30, 1991.

Benedetti, A., On the probabilistic stability analysis of tubular struts with initial imperfections, *Structural Safety and Reliability* (G. I. Schuëller, M. Shinozuka, and J. T. P. Yao, eds.), Balkema Publishers, Rotterdam, Vol. 2, pp. 745–52, 1994.

Ben-Haim, Y., *The Assay of Spatially Random Material*, Kluwer, Dortrecht, 1985.

Ben-Haim, Y., Convex models for uncertainty in radial pulse buckling of shells, *Journal of Applied Mechanics*, Vol. 60, 683–8, 1993.

Ben-Haim, Y., Convex models of uncertainty: Applications and implications, *Erkenntnis*, Vol. 41, 139–56, 1994.

Ben-Haim, Y., On convex models of uncertainty for small initial imperfections, *Zeitschrift für angewandte Mathematik und Mechanik*, Vol. 75, 901–8, 1995.

Ben-Haim, Y., *Robust Reliability in the Mechanical Sciences*, Springer-Verlag, Berlin, 1996.

Ben-Haim, Y., and Elishakoff, I., Non-probabilistic models of uncertainty in the nonlinear buckling of shells with general imperfections: Theoretical estimates of the knockdown factor, *Journal of Applied Mechanics*, Vol. 56, 403–10, 1989a.

Ben-Haim, Y., and Elishakoff, I., Dynamics and failure of structures based on the unknown-but-bounded imperfection model, *Recent Advances in Impact of Engineering Structures* (D. Hui and N. Jones, eds.), AMD – Vol. 105, AD – Vol. 17, pp. 89–95, 1989b.

Ben-Haim, Y., and Elishakoff, I., *Convex Models of Uncertainty in Applied Mechanics*, Elsevier Science Publishers, Amsterdam, 1990.

Bernard, M. C., and Bogdanoff, J. L., Buckling of columns with random initial displacements, *Journal of Engineering Mechanics*, Vol. 97, 755–71, 1971.

Bernard, P., and Fogli, M., Utilization des méthodes de Monte-Carlo en sécurité structurale, *Technical Report*, Université de Clermont II, Laboratoire de Génie Civil, 1984 (in French).

Bert, C. W., Analysis of shells, *Composite Materials, Vol. 7, Structural Design and Analysis, Part 1* (C. C. Chamis, ed.), Academic Press, New York, pp. 205–58, 1975.

Bert, C. W., Applications of the Rayleigh technique to problems of bars, beams, columns, membranes, and plates, *Journal of Sound and Vibration*, Vol. 119, 317–26, 1987.

Bert, C. W., Techniques for estimating buckling loads, *Handbook of Civil Engineering* (P. N. Cheremisinoff, ed.), Technomic Publishers, Lancaster, 1987.

Beveridge, G., and Schechter, R., *Optimization: Theory and Practice*, McGraw Hill, New York, pp. 453–6, 1970.

Biderman, V. L., Application of the double-sweep method to problems in structural mechanics, *Mekhanika Tverdogo Tela*, Vol. 5, 62–6, 1967 (in Russian).

Bielewicz, E., Gorski, J., Schmidt, R., and Walukiewicz, H., Random fields in the limit analysis of elastic-plastic shell structures, *Computers and Structures*, Vol. 51, 267–75, 1994.

Birkemoe, P. C., Stability: Directions in experimental research, *Engineering Structures*, Vol. 18, 807–11, 1996.

Bjaerum, D., Finite element formulation and solutions algorithms for buckling and collapse analysis of thin shells, PhD Thesis, Division of Structural Engineering, University of Trondheim, 1992.

Bjorhovde, R., Deterministic and probabilistic approaches to the strength of steel columns, PhD Dissertation, Department of Civil Engineering, Lehigh University, 1972.

Bljuger, F., Probabilistic analysis of reinforced concrete columns and walls in buckling, *ACI Structural Journal*, 124–31, 1987.

Bleich, F., *Buckling Strength of Metal Structures*, McGraw-Hill, New York, 1952.

Blekhman, I. I., Myshkis, A. D., and Panovko, Ya. G., *Mechanics and Applied Mathematics: Logics and Specifics of Applications of Mathematics*, "Nauka" Publishers, Moscow, pp. 205–9, 1983 (in Russian).

Bogdanovich, A. E., *Nonlinear Problems of Dynamics of Cylindrical Composite Shells*, "Zinatne" Publishers, Riga, 1987 (in Russian).

Bogdanovich, A. E., and Yushanov, S. P., Analysis of buckling of cylindrical shells with a random field of initial imprefections under axial dynamic compression, *Mechanics of Composite Materials*, Vol. 17, 552–60, 1981.

Bogdanovich, A. E., and Yushanov, S. P., About reliability analysis of anisotropic shells via the probability of rare out-crossing of vector random field through the limit surface, *Mechanics of Composite Materials*, No. 1, 80–9, 1983a (in Russian).

Bogdanovich, A. E., and Yushanov, S. P., Computing the reliability of anisotropic shells from the probability of infrequent excursions of a random vector field beyond the limiting surface, *Mechanics of Composite Materials*, Vol. 19, 67–75, 1983b (in Russian).

Bolotin, V. V., Statistical methods in the nonlinear theory of elastic shells, *Izvestija Academii Nauk SSSR, Otdelenie Tekhnicheskikh Nauk*, No. 3, 1958 (English translations, *NASA TTF-85*, 1–16, 1962).

Bolotin, V. V., *Statistical Methods in Structural Mechanics*, Grosstroiizdat, Moscow, 1961 (in Russian).

Bolotin, V. V., *Non-Conservative Problems of Theory of Elastic Stability*, Pergaman Press, London, 1964.

Bolotin, V. V., Application of methods of the theory of probability in the theory of plates and shells, *Theory of Shells and Plates* (S. M. Durgaryan, ed.), Israel Program for Scientific Translation, Jerusalem, 1966a.

Bolotin, V. V., Theory of layered medium with random initial imperfections, *Mechanics of Polymers*, Vol. 1, 11–19, 1966b (in Russian).

Bolotin, V. V., Statistical aspects in the theory of structural stability, *Dynamic Stability of Structures* (G. Herrmann, ed.), Pergamon Press, New York, pp. 67–81, 1967.

Bolotin, V. V., *Statistical Methods in Structural Mechanics*, Holden-Day, San Francisco, 1969.

Bolotin, V. V., Stochastic edge-effects in subcritical deformations of elastic shells, *Mechanics of Solids*, Vol. 5, 82–6, 1970.

Bolotin, V. V., Reliability theory and stochastic stability, *Stability* (H. H. E. Leipholz, ed.), University of Waterloo Press, Waterloo, pp. 385–442, 1972.

Bolotin, V. V., and Makarov, B. P., On correlation of displacements in thin elastic shells with random initial imperfections, *Reliability Problems in Structural Mechanics* (V. V. Bolotin and A. Cyras, eds.), "RINTIP" Publishers, Vilnius, pp. 163–8, 1968a (in Russian).

Bolotin, V. V., and Makarov, B. P., Correlation theory of pre-critical deformations in thin elastic shells, *Prikladnaya Matematika i Mekhanika*, Vol. 32, 428–34, 1968b (in Russian).

Bolotin, V. V., and Novichkov, Yu. N., *Mechanics of Multilayered Structures*, "Mashinostroenie" Publishers, Moscow, 1981 (in Russian).

Bonello, M., Reliability assessment and design of stiffened compression flanges, PhD Thesis, Department of Civil Engineering, Imperial College, University of London, 1992.

Bonello, M. A., Chrystanthopoulos, M. K., and Dowling, P. J., Probabilistic strength modeling of unstiffened plates under axial compression, *Proceedings of the 10th Offshore Mechanics and Arctic Engineering Symposium*, ASME, Norway, 1991.

Booton, M., Buckling of imperfect anisotropic cylinders under combined loading, *Report No. 203*, Institute for Aerospace Studies, University of Toronto, 1976.

Borst, R. de, Kusters, G., Nauta, P., and Witte, F. de, DIANA, a complex but flexible finite element system, *Finite Element Systems* (C. A. Brebbia, ed.), 3rd ed., Springer Verlag, Berlin, 1983.

Boyce, W. E., Buckling of a column with random initial displacements, *Journal of Aerospace Sciences*, Vol. 28, 308–20, 1961.

Brabré, R., Stabilität gleichmässig gedrückter Rechteckplatten mit Längs- oder Quersteifen, *Ingenieur Archiv*, Vol. 8, 117–50, 1937 (in German).

Brasil, R. M., and Hawwa, M. A., The localization of buckling modes in nearly periodic trusses, *Computers and Structures*, Vol. 56, 927–32, 1995.

Brenner, C. E., Stochastic finite element method, *Report No. 35-91*, Institute of Engineering Mechanics, University of Innsbruck, Austria, 1991.

Brillouin, L., *Wave Propagation in Periodic Structures*, Dover, New York, 1953.

Britvec, S. J., *Stability and Optimization of Flexible Space Structures*, Birkhauser, Boston, 1995.

Brown, W. S., *ALTRAN Users Manual*, Bell Laboratories, Murray Hill, NJ, 1973.

Brush, D. D., and Almroth, B. O., *Buckling of Bars, Plates and Shells*, McGraw-Hill, New York, 1975.

Budiansky, B., Theory of buckling and postbuckling behavior of elastic structures, *Advances in Applied Mechanics* (C. Yih, ed.), Academic Press, New York, pp. 1–65, 1974.

Budiansky, B., and Hutchinson, J. W., Dynamic buckling of imperfection-sensitivity structures, *Proceedings of the Eleventh International Congress of Applied Mechanics* (H. Goertler, ed.), Munich, pp. 636–51, 1964.

Budiansky, B., and Hutchinson, J. W., A survey of some buckling problems, *AIAA Journal*, Vol. 4, 1505–10, 1966.

Budiansky, B., and Hutchinson, J. W., Buckling of circular cylindrical shells under axial compression, *Contributions to the Theory of Aircraft Structures*, The Van Der Neut Anniversary Volume, Delft University Press, Delft, pp. 239–59, 1972.

Budiansky, B., and Hutchinson, J. W., Buckling: Progress and challenge, *Trends in Solid Mechanics* (J. F. Besseling and A. M. A. Van der Heijden, eds.), Proceedings of the Symposium Dedicated to the 65th Birthday of W. T. Koiter, Sijthoff and Noordhoff, Delft, pp. 93–116, 1979.

Budiansky, B., and Sanders, Y. L., Jr., On the "best" first order linear theory, *Progress in Applied Mechanics*, Macmillan, New York, pp. 129–40, 1963.

Bulgakov, B. V., Fehleranhaeufung bei Kreiselapparaten, *Ingenieur-Archiv*, Vol. 11, 461–9, 1940 (in German).

Bulgakov, B. V., On the accumulation of disturbances in linear systems with constant coefficients, *Doklady Akademii Nauk SSR*, Vol. LI, No. 5, 339–42, 1946 (in Russian).

Bulson, P. S., *The Stability of Flat Plates*, American Elsevier, New York, 1969.

Bushnell, D., Stress, stability, and vibration of complex branched shells of revolution: Analysis and user's manual for BOSOR 4, *NASA CR-2116*, 1972.

Bushnell, D., BOSOR-54 – Program for buckling of elastic-plastic complex shells of revolution including large deflections and creep, *Computers and Structures*, Vol. 6, 221–39, 1976.

Bushnell, D., Buckling of shells – Pitfall for designers, *AIAA Journal*, Vol. 11, 1183–226, 1981.

Bushnell, D., PANDA – Interactive program for minimum weight design of stiffened cylindrical panels and shells, *Computers and Structures*, Vol. 16, 167–85, 1983.

Bushnell, D., *Computerized Buckling Analysis of Shells*, Martinus Nijhoff Publishers, Dortrecht, 1985.

Cai, G. Q., and Lin, Y. K., Localization of wave propagation in disordered periodic structures, *AIAA Journal*, Vol. 29, 450–6, 1991.

Calladine, C. R., *Theory of Shell Structures*, Cambridge University Press, London, 1983.

Cambou, B., Application of first-order uncertainty analysis in the finite element method in linear elasticity, *Proceedings of the Second International Conference on Applied Statistics and Probability in Soil and Structural Engineering*, Aachen, pp. 67–87, 1975.

Capsoni, A., and Poggi, C., The role of symbolic algebra in the initial post-buckling and imperfection sensitivity analysis of axially compressed composite cylindrical shells, *Technical Report*, Politecnico di Milano, Dip. di Ingegneria Strutturale, Milan, Italy, 1990.

Card, M. F., The sensitivity of buckling of axially compressed fiber-reinforced cylindrical shells to small geometric imperfections, *PhD Dissertation*, Department of Engineering Science and Mechanics, Virginia Polytechnic Institute, 1969.

Carstensen, C., and Wagner, W., Detecting symmetry-breaking bifurcation points of symmetric structures, *International Journal for Numerical Methods in Engineering*, Vol. 36, 3019–39, 1993.

Čausević, M., Prikaz uzroka opadanju vrijednosti kriticnoy opterecenja za neke staticke sisteme, *Tehnika*, Vol. 36, No. 7–8, 1087–91, 1981 (in Serbian).

Cederbaum, G., and Arbocz, J., Reliability of axially compressed cylindrical shells, Delft University of Technology, *Report LR-767*, Faculty of Aerospace Engineering, Delft, the Netherlands, 1994.

Cederbaum, G., and Arbocz, J., Reliability of shells via Koiter formulation, *Thin-Walled Structures*, Vol. 24, 173–87, 1996a.

Cederbaum, G., and Arbocz, J., On the reliability of imperfection-sensitive long isotropic cylindrical shells, *Structural Safety*, Vol. 18, 1–9, 1996b.

Cederbaum, G., and Arbocz, J., Reliability of imperfection-sensitive composite shells via the Koiter Cohen criterion, *Reliability Engineering and System Safety*, Vol. 56, 257–63, 1997.

Cederbaum, G., Aboudi, J., and Elishakoff, I., Dynamic stability of viscoelastic laminated plates via the Lyapunov exponents, *International Journal of Solids and Structures*, Vol. 28, 317–27, 1991.

Champneys, A. R., van der Heijden, G. H. M., and Thompson, J. M. T., Spatially complex localization after one-twist-per-wave equilibria in twisted circular rods with initial curvature, *Philosophical Transactions of the Royal Society of London, Series A – Mathematical, Physical and Engineering Science*, Vol. 355, 2151–74, 1997.

Cheriachukin, V. V., Application of the Monte Carlo method to some statistical problems of stability and reliability, *Reliability Problems in Structural Mechanics* (V. V. Bolotin and A. Cyras, eds.), Vilnius, Rintip, pp. 79–85, 1968 (in Russian).

Chernousko, F. L., Optimal guaranteed estimates of indeterminacies with the aid of ellipsoids, *Tekhnicheskaya Kibernetika*, pp. 1–9, 1981 (in Russian).

Chernousko, F. L., *State Estimation for Dynamic Systems*, CRC Press, Boca Raton, 1994.

Chetaev, N. G., *The Stability of Motion*, Pergamon Press, New York, 1961.

Cho, H.-N., Lee, S.-J., Choi, Y.-M., and Shin, J.-C., Buckling reliability of laminated cylindrical shells, *Structural Safety and Reliability* (G. I. Schuëller, M. Shinozuka, and Y. T. P. Yao, eds.), Balkema Publishers, Rotterdam, Vol. 2, pp. 753–60, 1994.

Choi, S.-T., and Liao, Y. L., Nonlinear forced vibration of rectangular plates with random geometric imperfections, *Structural Safety and Reliability*, (G. I. Schuëller, M. Shinozuka, and J. T. P. Yao, eds.), Balkema, Rotterdam, pp. 101–8, 1994.

Chryssanthopoulos, M. K., Probabilistic buckling analysis of plates and shells, *Thin-Walled Structures*, Vol. 30, 135–57, 1997.

Chryssanthopoulos, M. K., and Poggi, C., Stochastic imperfection modelling in shell buckling studies, *Thin-Walled Structures*, Vol. 23, 179–200, 1995a.

Chryssanthopoulos, M. K., and Poggi, C., Probabilistic imperfection sensitivity analysis of axially compressed composite cylinders, *Engineering Structures*, Vol. 17, 398–406, 1995b.

Chryssanthopoulos, M. K., Baker, M. J., and Dowling, P. J., Statistical analysis of imperfections in stiffened cylinders, *Journal of Structural Engineering*, Vol. 117, 1979–97, 1991a.

Chryssanthopoulos, M. K., Baker, M. J., and Dowling, P. J., Imperfection modelling for buckling analysis of stiffened cylinders, *Journal of Structural Engineering*, Vol. 117, 1998–2017, 1991b.

Chryssanthopoulos, M. K., Giavotto, V., and Poggi, C., Statistical imperfection models for buckling analysis of composite shells, *Buckling of Shell Structures on Land, in the Sea and in the Air* (J. F. Jullien, ed.), Elsevier Applied Science, London, pp. 43–52, 1991.

Chryssanthopoulos, M. K., Giavotto, V., and Poggi, C., Characterization of manufacturing effects for buckling-sensitive composite cylinders, *Composites Manufacturing*, Vol. 6, 93–101, 1995.

Chung, B. T., Random-parameter analysis of stability of inelastic structures, PhD Thesis, State University of New York, 1962.

Chung, B. T., and Lee, G. T., Buckling strength of column based on random parameters, *Journal of the Structure Division*, Vol. 97, 1927–45, 1971.

Ciavarella, M., Modelli probabilistici di imperfezioni geometriche per lo studio dell instabilita di gusci cilindrici in materiale composito, Diploma Thesis, Department of Aerospace Engineering, Politecnico di Milano, 1991 (in Italian).

Como, M., and Grimaldi, A., Stability, buckling and post-buckling of elastic structures, *Meccanica*, Vol. 4, 254–68, 1975.

Combescure, A., Gusik, G., and Jullien, J. F., Influence of thickness imperfections on the buckling of cylinders under external pressure. *Fourth Int'l Coll. on Computation of Shell & Spatial Structures*, Chania-Crete, Greece, pp. 62–63, June 4–7, 2000.

Como, M., and Grimaldi, A., *Theory of Stability of Continuous Elastic Structures*, CRC Press, Boca Raton, 1995.

Conte, S. D., The numerical solution of linear boundary value problems, *SIAM Review*, Vol. 8, 309–21, 1966.

Cornell, C. A., Bounds on the reliability of structural systems, *Journal of Structural Division*, Vol. 93, 171–200, 1967.

Cornell, C. A., Probability based structural code, *ACI Journal*, 66, 974–85, 1969.

Crandall, S. H., *Engineering Analysis, A Review of Numerical Analysis Procedures*, McGraw-Hill, New York, pp. 297–8, 1956.

Crandall, S. H., Personal communication to I. E., 1990.

Crespo da Silva, M. R. M., and Hodges, D. H., Dynamic modeling of rotating beams undergoing flexure in two directions and torsion, with application to helicopter rotor blades, *Computers and Mathematics*, 1986.

Crisfield, M. A., A fast incremental/iterative procedure that handles "snap-through," *Computers and Structures*, Vol. 13, 387–99, 1979.

Crisfield, M. A., An arc-length method including line searches and accelerations, *International Journal of Numerical Methods in Engineering*, Vol. 19, 1269–89, 1983.

Damil, N., and Potier-Ferry, M., A new method to compute perturbed bifurcations: Application to the buckling of imperfect elastic structures, *International Journal of Engineering Science*, Vol. 28, 943–57, 1990.

Das, P. K., Frieze, P. A., and Faulkner, D., Structural reliability modelling of stiffened components of floating structures, *Structural Safety*, Vol. 12, 1984.

Day, W. B., Buckling of a column with nonlinear restraints and random initial displacement, *Journal of Applied Mechanics*, Vol. 47, 204–5, 1980.

Davenport, J., Siret, N., and Tournier, E., *Computer Algebra*, Academic Press, London, 1988.

Davidenko, D. F., On one new method of numerical solution of the systems of non-linear equations, *Doklady Akademii Nauk SSSR*, Vol. 88, pp. 601–2, 1953 (in Russian).

Davister, M., and Britvec, S., Evaluation of the most unstable dynamic buckling mode of pin jointed space lattices, *Third International Conference on Space Structures*, pp. 468–73, 1984.

Davletkhanova, A. D., Stability of multi-part column and discrete supports with random charateristics, *Stroitelnayia Mekhanika i Raschet Sooruzhenii*, No. 6, 48–53, 1980.

De Jong, R. G., Localization and transmission loss in ribbed plates, *Proceedings of the NOISE – CON '94, Progress in Noise Control for Industry* (J. M. Cuschieri, S. A. L. Glegg, and D. M. Yeager, eds.), Fort Lauderdale, pp. 603–8, 1994.

Dellnitz, M., and Werner, B., Computational methods for bifurcation problems with symmetries, with special attention to steady state and Hopf bifurcation points, *Journal of Computer Applications in Mathematics*, Vol. 26, 97–123, 1989.

Deodatis, G., and Shinozuka, M., Bounds on response variability of stochastic systems, *Journal of Engineering Mechanics*, Vol. 115, 2543–63, 1989.

Der Kiureghian, A., and Zhang, J., Space-variant finite element reliability analysis, *Computer Methods in Applied Mechanics and Engineering*, Vol. 168, 173–83, 1999.

Dimentberg, M. F., Some problems of stability of shells subjected to random excitations, *Problems of Stability in Structural Mechanics*, "Stroiizdat" Publishers, Moscow, pp. 217–22, 1965 (in Russian).

Diniz, S. M. C., Reliability evaluation of high strength concrete columns, PhD Thesis, Department of Civil Engineering, University of Colorado at Boulder, 1994.

Diniz, S. M. C., and Frangopol, D. M., Reliability basis for high-strength concrete columns, *Journal of Structural Engineering*, Vol. 123, 1375–81, 1997.

Diniz, S. M. C., and Frangopol, D. M., Reliability of high-strength concrete columns, *Materials for the New Millennium* (K. P. Chong, ed.), ASCE Press, New York, Vol. 1, pp. 213–22, 1996.

Dinnik, A. N., *Selected Papers*, Vol. 2, Academy of Ukrainian SSR Publishing House, Kiev, 1955 (in Russian).

Doedel, E., AVTO: Software for continuation and bifurcation problems in ordinary differential equations, California Institute of Technology, 1986.

Donnell, L. H., *Beams, Plates and Shells*, McGraw-Hill, New York, 1976.

Donnell, L. H., and Wan, C. C., Effect of imperfections on the buckling of thin cylinders and columns under axial compression, *Journal of Applied Mechanics*, Vol. 17, 73–83, 1950.

Drenick, R. F., Functional analysis of effects of earthquakes, *2nd Joint United States-Japan Seminar on Applied Stochastics*, Washington, DC, Sept. 19–24, 1968.

Drenick, R. F., Model-free design of aseismic structures, *Journal of Engineering Mechanics Division*, Vol. 96, 483–93, 1970.

Dugundji, J., Nonlinear problem of aeroelasticity, *Computational Nonlinear Mechanics in Aerospace Engineering* (S. N. Atluri, ed.), AIAA Press, Washington, D.C., pp. 127–55, 1992.

Duprat, F., Pinglot, M., and Lorrain, M., Structural reliability analysis of reinforced concrete columns: A comparison of methods to assess the probability of buckling failure, *Materiaux et Constructions, Materials and Structures*, Vol. 29, 485–93, 1998 (in French).

Edlund, B. L. O., Computer simulation for structures, *Publication S73-6*, Chalmers University of Technology, Göteborg, 1973.

Edlund, B. L. O., and Leopoldson, U. L. C., Monte Carlo simulation of the strength of steel structure: Part 1; Method and basic data, *Publication S71-3*, Chalmers University of Technology, Department of Structural Engineering, 1971a.

Edlund, B. L. O., and Leopoldson, U. L. C., Monte Carlo simulation of the strength of steel structure: Part 2; Rolled I- and H-beam, statistical variation of the load carrying capacity, *Publication S71-5*, Chalmers University of Technology, Department of Structural Engineering, 1971b.

Edlund, B. L. O., and Leopoldson, U. L. C., Simulation of strength of axially loaded cylindrical shells, CTH Institute für Konstruktonsteknik, Stål-och Träbyggnad, Inst. Strifter S 73:7, Göteborg, 1974.

Edlund, B. L. O., and Leopoldson, U. L. C., Computer simulation of the scatter in steel member strength, *Computers and Structures*, Vol. 5, 209–24, 1975.

Eggwertz, S., and Palmberg, B., Structural safety of axially loaded cylindrical shells, *Technical Report FFA-TN-1985-50*, Aeronautical Research Institute of Sweden, Bromma, Sweden, 1985.

Elishakoff, I., Axial impact buckling of a column with random initial imperfection, *Journal of Applied Mechanics*, Vol. 45, 361–5, 1978a.

Elishakoff, I., Impact buckling of thin bar via Monte Carlo method, *Journal of Applied Mechanics*, Vol. 45, 586–90, 1978b.

Elishakoff, I., Buckling of a stochastically imperfect finite column on a nonlinear elastic foundation – A reliability study, *Journal of Applied Mechanics*, Vol. 46, 411–16, 1979a.

Elishakoff, I., Simulation of space-random fields for solution of stochastic boundary-value problems, *Journal of Acoustical Society of America*, Vol. 65, 399–403, 1979b.

Elishakoff, I., Remarks on the static and dynamic imperfection-sensitive of nonsymmetric structures, *Journal Applied Mechanics*, Vol. 47, 111–15, 1980a.

Elishakoff, I., Hoff's problem in a probabilistic setting, *Journal of Applied Mechanics*, Vol. 47, 403–8, 1980b.

Elishakoff, I., How to introduce initial-imperfection sensitivity concept into design, *Collapse: The Buckling of Structures of Theory and Practice* (J. M. T. Thompson, and G. W. Hunt, eds.), pp. 345–57, 1983a.

Elishakoff, I., How to introduce initial-imperfection sensitivity concept into design 2, M. Stein Volume, NASA CP, 1998, 206280.

Elishakoff, I., *Probabilistic Methods in the Theory of Structures*, John Wiley, New York, pp. 433–68, 1983b (2nd ed., Dover, Mineola, 1999).

Elishakoff, I., Reliability approach to the initial imperfection sensitivity, *Acta Mechanica*, Vol. 55, 151–70, 1985.

Elishakoff, I., A remark on the adjustable parameter version of Rayleigh's method, *Journal of Sound and Vibration*, Vol. 118, 163–5, 1987.

Elishakoff, I., Simulation of an initial imperfection data bank, Part I: Isotropic shells with general imperfections, *TAE Report 500*, Technion–Israel Institute of Technology, Department of Aeronautical Engineering, Haifa, 1982. Also in *Buckling of Structures – Theory and Experiment* (I. Elishakoff, J. Arbocz, C. D. Babcock, Jr., and A. Libai, eds.), Elsevier Science Publishers, Armsterdam, pp. 195–210, 1988.

Elishakoff, I., An idea of the uncertainty triangle, *Shock and Vibration Digest*, Vol. 22, 1, 1990.

Elishakoff, I., Convex versus probabilistic modelling of uncertainty in structural dynamics, *Structural Dynamics – Recent Advances* (M. Petyt, H. F. Wolfe, and C. Mei, eds.), Elsevier, London, pp. 3–21, 1991.

Elishakoff, I., Essay on uncertainties in elastic and viscoelastic structures: From A. M. Freudenthal's criticisms to modern convex modeling, *Computers and Structures*, Vol. 56, 871–96, 1995.

Elishakoff, I., How to introduce the imperfection sensitivity concept into design 2, *237–267 NASA/CP-1998-206280*, 1998.

Elishakoff, I., *Unusual Closed-form Solutions for Eigenvalues of Inhomogeneous Structures*, 2000 (in press).

Elishakoff, I., and Arbocz, J., Reliability of axially compressed cylindrical shells with random axisymmetric imperfections, *International Journal of Solids and Structures*, Vol. 18, 563–85, 1982.

Elishakoff, I., and Arbocz, J., Reliability of axially compressed cylindrical shells with general non-symmetric imperfections, *Journal of Applied Mechanics*, Vol. 52, 122–8, 1985.

Elishakoff, I., and Ben-Haim, Y., Dynamics of a thin cylindrical shell under impact with limited deterministic information on initial imperfections, *Journal of Structural Safety*, Vol. 8, 103–12, 1990.

Elishakoff, I., and Charmats, M., Godunov-Conte method for solution of engineering problems and its applications, *Journal of Applied Mechanics*, Vol. 44, 776–9, 1977.

Elishakoff, I., and Colombi, P., Combination of probabilistic and convex models of uncertainty when scarce knowledge is present on acoustic excitation parameters, *Computer Methods in Applied Mechanics and Engineering*, Vol. 104, 187–209, 1993.

Elishakoff, I., and Hasofer, A. M., On the accuracy of Hasofer-Lind reliability index, *Structural Safety and Reliability* (I. Konishi, A. H-S. Ang, and M. Shinozuka, eds.), *Proceedings of ICOSSAR'85, the 4th International Conference on Structural Safety and Reliability*, Kobe, Japan, Vol. 1, pp. 1. 229–1.239, 1985.

Elishakoff, I. and Hasofer, A. M., Effect of human error on reliability of structures, *Proceedings, 33rd AIAA/ASME/ASCE/AMS/ASC Structures, Structural Dynamics and Materials Conference*, AIAA Press, Washington, DC, pp. 3222–37, 1992.

Elishakoff, I., and Hettema, Ch. D., Nonstationary random vibration with REDUCE, *Probabilistic Methods in Civil Engineering* (P. D. Spanos, ed.), ASCE, New York, pp. 523–32, 1988.

Elishakoff, I., Hettema, Ch. D., and Wilson, E. L., Direct superposition of Wilson trial functions by computerized symbolic algebra, *Acta Mechanica*, Vol. 74, 69–79, 1989.

Elishakoff, I., and Pletner, B., Computerized symbolic algebraic evaluation of buckling loads by Snitko's method, *Symbolic Computations and Their Impact on Mechanics* (A. K. Noor, I. Elishakoff, and G. Hubbert, eds.), ASME Press, New York, pp. 175–88, 1990.

Elishakoff, I., and Ren, I. J., The bird's eye view on finite element method for stochastic structures, *Computer Methods in Applied Mechanics and Engineering*, Vol. 168, 51–61, 1999.

Elishakoff, I., and Rollot, New Closed-form solutions for buckling of variable stiffness column by Mathematica, *Journal of Sound and Vibration*, Vol. 224, 172–182, 1999.

Elishakoff, I., and Tang, J., Buckling of polar orthotropic circular plates on elastic foundation by computerized symbolic algebra, *Computer Methods in Applied Mechanics and Engineering*, Vol. 68, 229–47, 1988.

Elishakoff, I., Arbocz, J., Babcock, Jr., C. D., and Libai, A. (eds.), *Buckling of Structures – Theory and Experiment*, Elsevier Science Publishers, Amsterdam, 1988.

Elishakoff, I., Cai, G. Q., and Starnes, J. H., Jr., Probabilistic and convex models of uncertainty in buckling of structures, *Structural Safety and Reliability*, (G. I. Schuëller, M. Shinozuka, and J. T. P. Yao, eds.), Balkema Publishers, Rotterdam, Vol. 2, pp. 761–6, 1994a.

Elishakoff, I., Cai, G. Q., and Starnes, J. H., Jr., Nonlinear buckling of a column with initial imperfection via stochastic and non-stochastic, convex models, *International Journal of Nonlinear Mechanics*, Vol. 29, 71–82, 1994b.

Elishakoff, I., Gana Shvili, Y., and Givoli, D., Treatment of uncertain imperfections as a convex optimization problems, *Proceedings of the Sixth International Conference on Applications of Stochastics and Probability in Civil Engineering* (L. Esteva and S. E. Ruiz, eds.), Mexico, Vol. 1, pp. 150–7, 1991.

Elishakoff, I., Li, Y. W., and Starnes, Jr., J. H., The combined effect of the thickness variation and axisymmetric initial imperfection on the buckling of the isotropic cylindrical shell under axial compression, *Preliminary Report*, Florida Atlantic University, Department of Mechanical Engineering, Boca Raton, May 1992.

Elishakoff, I., Li, Y. W., and Starnes, Jr., J. H., A deterministic method to predict the effect of unknown-but-bounded uncertain elastic moduli on the buckling of composite structures, *Computer Methods in Applied Mechanics and Engineering*, Vol. 111, pp. 157–67, 1993.

Elishakoff, I., Li, Y. W., and Starnes, Jr., J. H., Buckling mode localization in elastic plates due to misplacement in the stiffener location, *Chaos, Solitons and Fractals*, Vol. 5, 1517–31, 1995.

Elishakoff I., Li, Y. W., and Starnes, J. H., Jr., Imperfection sensitivity due to the elastic moduli in the Roorda-Koiter frame, *Chaos, Solitons and Fractals*, Vol. 7, 1179–86, 1996.

Elishakoff, I., Lin, Y. K., and Zhu, L. P., *Probabilistic and Convex Modeling of Acoustically Excited Structures*, Elsevier Science Publishers, Amsterdam, 1994.

Elishakoff, I., Van Manen, S., Vermeulen, P. G., and Arbocz, J., First-order second-moment analysis of the buckling of shells with random initial imperfections, *AIAA Journal*, Vol. 25, 1113–17, 1987.

Ellingwood, B., Statistical analysis of RC beam-column interaction, *Journal of Structural Division*, Vol. 103, 1377–88, 1977.

El Naschie, M. S., Exact asymptotic solution for the initial post-buckling of a strut on a linear elastic foundation, *Zeitschrift für angewandte Mathematik und Mekhanik*, Vol. 54, 677–83, 1974.

El Naschie, M. S., Local postbuckling of compressed cylindrical shells, *Proceedings of the Institution of Civil Engineers*, London, Part 2, Vol. 59, 523–5, 1975a.

El Naschie, M. S., Localized diamond shaped buckling patterns of axially compressed cylindrical shells, *AIAA Journal*, Vol. 13, 837–8, 1975b.

El Naschie, M. S., Nonlinear isometric bifurcation and shell buckling, *Zeitschrift für angewandte Mathematik und Mechanik*, Vol. 57, 293–6, 1977.

El Naschie, M. S., *Stress, Stability and Chaos in Structural Engineering: an Energy Approach*, McGraw-Hill, London, 1990.

Englestad, S. P., and Reddy, Y. N., A probabilistic postbuckling analysis of composite shells, *Computational Stochastic Mechanics* (P. D. Spanos and C. A. Brebbia, eds.), Elsevier Applied Science, London, pp. 839–50, 1991.

Ernst, L. J., A finite element approach to shell problems, *Research Report WTHD-114*, Delft University of Technology, Department of Mechanical Engineering, Delft, the Netherlands, 1979.

Farshad, M., *Stability of Structures*, Elsevier Science Publishers, Amsterdam, 1994.

Fedorov, E. I., About probabilistic analysis of beam-column, *Problems of Reliability in Structural Design* (S. A. Timoshenko, ed.), Sverdlovsk, pp. 255–61, 1972 (in Russian).

Fersht, R. S., Buckling of cylindrical shells with random imperfections, PhD Thesis, California Institute of Technology, 1968a.

Fersht, R. S., Almost sure stability of long cylindrical shells with random imperfections, *NASA CR-1161*, 1968b.

Fersht, R. S., Buckling of cylindrical shells with random imperfections, *Thin Shell Structures: Theory, Experiment and Design* (Y. C. Fung and E. E. Sechler, eds.), Prentice Hall, Englewood Cliffs, NJ, pp. 325–41, 1974.

Floris, C., and Mazzucchelli, A., Reliability assessment of RC column under stochastic stress, *Journal of Structural Engineering*, Vol. 117, 3274–92, 1991.

Folz, B., and Foschi, R. O., Stochastic stiffness properties and buckling of laminated wood columns, *Structural Safety and Reliability* (G. I. Schuëller, M. Shinozuka, and J. T. P. Yao, eds.), Balkema, Rotterdam, pp. 577–89, 1994.

Frangopol, D. M., Ide, Y., Spacone, E., and Iwaki, I., A new look at reliability of reinforced concrete columns, *Structural Safety*, Vol. 18, 123–50, 1996.

Fraser, W. B., Buckling of a structure with random imperfections, PhD Thesis, Division of Engineering and Applied Physics, Harvard University, 1965.

Fraser, W. B., and Budiansky, B., The buckling of a column with random initial deflections, *Journal Applied Mechanics*, Vol. 36, 232–40, 1969.

Freudenthal, A. M., Safety and probability of structural failure, *Transactions of the ASME*, Vol. 121, 1337–75, 1956.

Freudenthal, A. M., Fatigue sensitivity and reliability of mechanical systems, especially aircraft structures, *WADD Technical Report 61–53*, Wright-Patterson AFB, Dayton, OH, 1961.

Frieze, P. A., Das, P. K., and Faulkner, D., Partial safety factors for stringer stiffened cylinders under extreme compressive loads, *Proceedings, Second International Symposium on Practical Design in Shipbuilding*, Tokyo, pp. 475–82, 1983.

Fukumoto, Y., and Itoh, Y., Statistical study of experiments on welded beams, *Journal of Structural Division*, Vol. 107, 89–102, 1981.

Fukumoto, Y., and Itoh, Y., Basic compressive strength of steel plates from test data, *Transactions of Japan Society of Civil Engineers*, Vol. 344/I-1, 129–39, 1984.

Fujimoto, M., and Iwata, M., A study on buckling strength of steel columns by the application of probabilistic method – Consideration of random initial imperfection, *Translation of the Architectural Institute of Japan*, No. 218, 17–25, 1974 (in Japanese).

Fukimoto, Y., Itoh, Y., and Kubo, M., Strength variation of laterally unsupported beams, *Journal of Structural Division*, Vol. 106, 165–81, 1979.

Fujimoto, M., Iwata, M., and Nakatani, F., A study on buckling strength of steel columns by the application of probabilistic method – Weak axis buckling of H-section members with random residual stress, random yield stress and initial deflection, *Transactions of the Architectual Institute of Japan*, No. 229, 53–61, 1975 (in Japan).

Fukimoto, Y., Kajita, N., and Aoki, T., Evaluation of column curves based on probabilistic concept, *Preliminary Report on Second International Colloquium on Stability*, Tokyo, 1976.

Furstenberg, H., Noncommuting random products, *Transactions of the American Mathematical Society*, Vol. 108, 377–428, 1963.

Galambos, T. V., Reliability of axially loaded columns, *Engineering Structures*, Vol. 5, 73–8, 1983.

Galambos, T. V., *Guide to Stability Design Criteria for Metal Structures*, John Wiley and Sons, New York, 1988.

Galoussis, E., and Vasilas, V., On the investigation of the stochastic non-linear snap-through phenomenon in structures, *Zeitschrift für angewandte Mathematik und Mechanik*, Vol. 61, 40–1, 1981.

Gáspár, Z. S., Probability of the instability of imperfect structures, *Zeitschrift für angewandte Mathematik und Mechanik*, Vol. 63, T48–49, 1983.

Gatermann, K., and Hohmann, A., Symbolic exploitation of symmetry in numerical pathfollowing, *Impact of Computational Science and Engineering*, Vol. 3, 330–65, 1991.

Gauthier, R. W., Maximum statistical strength of steel columns, MS Thesis, Department of Civil Enginering, MIT, 1971.

Geier, B., and Rohwer, K., On the analysis of the buckling behavior of laminated composite plates and shells, *International Journal of Numerical Methods in Engineering*, Vol. 27, 403–27, 1989.

Geier, B., Klein, H., and Zimmerman, R., Buckling tests with axially compressed unstiffened cylindrical shells made from CFRP, *Buckling of Shells Structures on Land, in the Sea and in the Air* (J. F. Jullien, ed.), Elsevier Applied Science, London, pp. 498–507, 1991.

Gelfand, M., and Lokutsievski, O. V., The double sweep method for solution of difference equations, *Theory of Difference Schemes* (S. K. Godunov and V. S. Ryabenkii, eds.), North Holland, Amsterdam, pp. 239–62, 1964.

Ghanem, R. G., and Spanos, P. D., *Stochastic Finite Elements: A Spectral Approach*, Springer, New York, 1991.

Ghanem, R. G., and Spanos P. D., A spectral formulation of stochastic finite elements, *Probabilistic Methods for Structural Design* (C. Guedes Soares, ed.), Kluwer, Dordrecht, pp. 289–312, 1997.

Giavotto, V., Poggi, C., and Chrissanthopoulus, M., Buckling of imperfect composite cylindrical shells under compression and torsion, *Annual Forum of the American Helicopter Society*, Atlanta, Paper 43, 1991.

Godunov, S. K., On the numerical solution of boundary value problems for systems of linear ordinary differential equations, *Uspekhi Matematicheskikh Nauk*, Vol. 16, 171–4, 1961 (in Russian).

Godunov, S. K., *Equations in Mathematical Physics*, "Nauka" Publishing House, Moscow, pp. 394–403, 1971 (in Russian).

Godunov, S. K., and Ryabenkii, V. S., *Theory of Difference Schemes*, North Holland, Amsterdam, 1964.

Goggin, P. R., The elastic constants of carbon-fiber composites, *Journal of Materials Science*, Vol. 8, 233–44, 1973.

Goncherenko, V. M., Towards determination of probability of stability loss of the shell, *Izvestiya Akademii Nauk SSSR, Otdelenie Tekhnicheskykh Nauk, Mekhanika i Mashinostoenie*, No. 1, 159–60, 1962a (in Russian).

Goncharenko, V. M., Investigation of probability of snapping of a long cylindrical panel under random pressure, *Prikladnaya Matematika i Mekhanika*, Vol. 26, 740–4, 1926b (in Russian).

Goncharenko, V. M., On one aspect of the statistical method in the theory of stability of shells, *Theory of Plates and Shells*, Academy of Ukraine Press, Kiev, pp. 368–71, 1962c (in Russian).

Goncharenko, V. M., Application of Markov processes in the statistical theory of stability of shells, *Ukrainskii Matematicheskii Zhurnal*, Vol. 14, 198–202, 1962d (in Russian).

Goncharenko, V. M., Snap-through of a panel in presence of random excitation, *Theory of Shells and Plates*, Academy of Armenia Press, Erevan, pp. 383–90, 1964 (in Russian).

Goncharenko, V. M., About dynamic problems of statistical theory of stability of elastic systems, *Problems of Stability in Structural Mechanics*, "Stroiizdat" Publishers, Moscow, pp. 210–16, 1965 (in Russian).

Goodman, T. R., and Lance, G. N., The numerical solution of two-point boundary value problems, *Mathematical Tables and Other Aids of Computation*, Vol. 10, 82–6, 1956.

Goodman, R. A., and Movatt, G. A., Allowance for imperfections in ship structural design, *Journal of Strain Analysis*, Vol. 12, 153–62, 1977.

Grandori, G., Paradigms and falsification in earthquake engineering, *Meccanica*, Vol., 26, 17–21, 1991.

Greenberg, J. B., and Stavsky, Y., Buckling and vibration of orthotropic composite cylindrical shells, *Acta Mechanica*, Vol. 36, 15–29, 1980.

Greenberg, J. B., and Stavsky, Y., Stability and vibration of compressed, aelotropic, composite cylindrical shells, *Journal of Applied Mechanics*, Vol. 49, 843–8, 1982.

Greenberg, J. B., and Stavsky, Y., Buckling of composite orthotropic cylindrical shells under non-uniform axial loads, *Composite Structures*, Vol. 30, 399–406, 1995.

Greene, W. H., Effects of random member length errors on the accuracy and internal loads of truss antennas, *Proceedings of AIAA/ASME/ASCE/AHS 24th Structures, Structural Dynamics and Materials Conference*, Lake Tahoe, pp. 697–704, 1983.

Grigolyuk, E. I., and Kabanov, V. V., *Stability of Shells*, "Nauka" Publishers, Moscow, 1978a (in Russian).

Grigolyuk, E. I., and Kabanov, V. V., *Stability of Circular Cylindrical Shells*, "VINITI" Publishers, Moscow, 1978b (in Russian).

Grigolyuk, E. I., and Lipovtsev, Yu. V., On the solution of a class of eigenvalue problems in thin shells of revolution, *Problems of Mechanics of Deformed Solids*, Volume Dedicated to 60th Anniversary of V. V. Novozhilov, "Sudostroenie" Publishing House, Leningrad, pp. 129–41, 1975 (in Russian).

Grigolyuk, E. I., Maltsev, V. P., Myachenkov, V. I., and Frolov, A. N., On a solution method for stability and vibration problems in shells of revolution, *Mekhanika Tverdogo Tela*, Vol. 1, 9–19, 1971, (in Russian).

Grigorenko, Yu. M., *Isotropic and Anisotropic Rotational Shells of Varying Stiffness*, "Naukova Dumka" Publisher, Kiev, 1970 (in Russian).

Grigoriu, M., *Applied Non-Gaussian Processes*, PTR Prentice Hall, Englewood Cliffs, NJ, 1995.

Grove, T., and Didriksen, T., Buckling experiments on large axial stiffened and one ring stiffened cylindrical shells, *Report 76-432*, Det norske Veritas, 1976.

Groves, S. E., and Highsmith, A. L. (eds.), *Compression Response of Composite Structures*, ASTM, New York, 1994.

Guedes Soares, C., Uncertainty modeling in plate buckling, *Structural Safety*, Vol. 5, 17–34, 1988.

Guedes Soares, C., Probabilistic modelling of the strength of flat compression members, *Probabilistic Methods for Structural Design* (C. Guedes Soares, ed.), Kluwer Academic Publishers, Dortrecht, pp. 113–40, 1997.

Guedes Soares, C., and Silva, A. G., Reliability of unstiffened plate elements under in-plane combined loading, *Proceedings 10th Offshore Mechanics and Arctic Engineering Symposium*, ASME, Norway, 1991.

Guliaev, V. I., Bazhenov, V. A., and Gozuliak, E. I., *Stability of Nonlinear Mechanical Systems*, Lvov University Press, Lvov, 1982 (in Russian).

Guliaev, V. I., Bazhenov, V. A., and Lizunov, P. P., *Non-classical Theory of Shells and Its Applications to Solution of Engineering Problems*, Lvov University Press, Lvov, 1978 (in Russian).

Gusic, G., Combescure, A. and Jullien, J. F., Influence of circumferential thickness variations on buckling of cylindrical shells under external pressure, *Computers and Structures*, 2000 (in press).

Guz, A. N., *Stability of Three Dimensional Deformable Bodies*, "Naukova Dumka" Publishers, Kiev, 1971 (in Russian).

Guz, A. N., and Babich, I. Yu., *Three Dimensional Theory of Stability of Beams, Plates and Shells*, "Vischa Shkola" Publishers, Kiev, 1980 (in Russian).

Haines, W., Hierarchy methods for random vibrations of elastic strings and beams, *Journal of Engineering Mathematics*, Vol. 1, 293–305, 1967.

Hammersley, J. M., and Handscomb, D. G., *Monte Carlo Method*, Methuen, London, 1964.

Handbook of Structural Stability, Column Research Committee of Japan, Corona, Tokyo, 1971.

Hangai, Y., and Kawamata, S., Analysis of geometrically nonlinear and stability problems by static perturbation method, *Report of the Institute of Industrial Science*, The University of Tokyo, Vol. 22, No. 5 (Serial No. 143), 1973.

Hansen, J. S., Buckling of imperfection-sensitive structures: A probabilistic approch, PhD Dissertation, Department of Civil Engineering, University of Waterloo, 1973.

Hansen, J. S., Influence of general imperfections in axially loaded cylindrical shells, *International Journal of Solids Structures*, Vol. 11, 1223–33, 1975.

Hansen, J. S., General random imperfection in the buckling of axially loaded cylindrical shells, *AIAA Journal*, Vol. 15, 1250–6, 1977.

Hansen, J. S., and Roorda, J., Reliability of imperfection sensitive structures, *Stochastic Problems in Mechanics*, (S. T. Ariaratnam and H. H. E. Leipholz, eds.), University of Waterloo Press, Waterloo, pp. 229–42, 1973.

Hansen, J. S., and Roorda, J., On a probabilistic stability theory for imperfection sensitive structures, *International Journal of Solids and Structures*, Vol. 10, 341–59, 1974.

Harris, L., Suer, H. S., Skene, W. T., and Benjamin, R. J., The stability of thin-walled unstiffened circular cylinders under axial compression including the effects of internal pressure, *Journal of the Aeronautical Sciences*, 587–96, 1957.

Hart, D. K., Rutherford, S. E., and Wickham, A. H. S., Structural reliability analysis of stiffened plates, *Transactions of the Royal Institution of Naval Architects*, Vol. 128, 293–310, 1986.

Hart-Smith, L. J., Buckling of thin cylindrical shells under uniform axial compression, *International Journal of Mechanical Sciences*, Vol. 12, 299–313, 1970.

Hasofer, A. M., and Lind, N. C., Exact and invariant second-moment code format, *Journal of the Engineering Mechanics Division*, Vol. 100, 111–21, 1974.

Hawranek, R., and Rackwitz, P., Reliability calibration for steel columns, *Bulletin d'Information No. 112*, Comité Européen du Beton, pp. 125–57, 1976.

Healey, T. J., A group theoretic approach to computational bifurcation problems with symmetry, *Computational Methods in Applied Mechanics and Engineering*, Vol. 67, 257–95, 1988.

Herrmann, G. (ed.), *Dynamic Stability of Structures*, Pergamon Press, New York, 1967.

Hilton, H. H., Yi, S., and Danyluk, M. J., Stochastic delamination buckling of elastic and viscoelastic columns, *Computational Stochastic Mechanics* (P. D. Spanos, ed.), Balkema, Rotterdam, pp. 687–96, 1995.

Himmelblau, D. M., *Applied Nonlinear Programming*, McGraw-Hill, New York, pp. 177–8, 1972.

Hirano, Y., Buckling of angle-ply laminated circular cylindrical shells, *Journal of Applied Mechanics*, Vol. 46, 233–4, 1979.

Hirano, Y., Optimization of laminated composite shells for axial buckling, *Transactions of the Japan Society for Aeronautical and Space Science*, Vol. 26, 154–62, 1983.

Hodges, C. H., Confinement of vibration by structural irregularity, *Journal of Sound and Vibration*, Vol. 82, 441–4, 1982.

Hodges, C. H., and Woodhouse, J., Vibration isolation from irregularity in a nearly periodic structure: Theory and measurements, *Journal of the Acoustical Society of America*, Vol. 74, 894–905, 1983.

Hoff, N. J., The accuracy of Donnell's equations, *Journal of Applied Mechanics*, Vol. 23, 329–34, 1955.

Hoff, N. J., The perplexing behavior of thin circular cylindrical shells in axial compression, *Israel Journal of Technology*, Vol. 4, 1–28, 1966.

Höft, H. F. W., and Höft, M., *Computing with Mathematica®*, Academic Press, San Diego, 1998.

Hoh, Y., Usami, T., and Fukumoto, Y., Experimental and numeral analysis and database on structural stability, *Engineering Structures*, Vol. 18, 812–20, 1996.

Horne, M. R., and Merchant, W., *Stability of Frames*, Pregamon Press, Oxford, 1965.

Horton, W. H., and Durham, S. C., Imperfection, a main contributor to scatter in experimental values of buckling load, *International Journal of Solids and Structures*, Vol. 1, 59–72, 1965.

Hoshiya, M., and Shah, H. C., Free vibration of a stochastic beam column, *Journal of Engineering Mechanics*, Vol. 97, 1239–55, 1971.

Hoshiya, M., and Spence, S. T., Reliability of a single flexible column with three spring supports. *Proceeding of JSCE*, No. 183, 1970.

Hour, K. Y., Damage characterization of a random-oriented fiber reinforced composite, *Mechanics of Composite Materials–Nonlinear Effects* (M. W. Hyer, ed.), AMD-Vol. 159, ASME Press, New York, pp. 303–18, 1993.

Hu, H. T., Buckling analysis of fiber composite laminate plates with material nonlinearity, *Finite Element Analysis and Design*, Vol. 19, 169–80, 1995.

Hunt, G. W., Elastic stability in structural mechanics and applied mechanics, *Collapse: The Buckling of Structures in Theory and Practice* (J. M. T. Thompson and G. W. Hunt, eds.), Cambridge University Press, London, pp. 123–47, 1983.

Hunt, G. W., Hidden asymmetries of elastic and plastic bifurcation, *Applied Mechanics Reviews*, Vol. 39, 1165–86, 1986.

Hunt, G. W., and Wadee, M. K., Comparative Lagrangian formulations for localized buckling, *Proceedings of the Royal Society of London*, Vol. A 434, 485–502, 1991.

Hunt, G. W., and Wadee, M. A., Localization and mode interaction in sandwich structures, *Proceedings of the Royal Society of London, Series A–Mathematical, Physical and Engineering Sciences*, Vol. 454, 1197–216, 1998.

Hunt, G. W., Bolt, H. M., and Thompson, J. M. T., Structural localization phenomena and the dynamical phase-space analogy, *Proceedings of the Royal Society of London*, Vol. A 425, 245–67, 1986.

Hussain, M. A., and Noble, B., Applications of MACSYMA to calculations in applied mechanics, *Report No. 83CRD054*, General Electric, Schenectady, New York, March 1983.

Hutchinson, J. W., Plastic buckling, *Advances in Applied Mechanics* (C. S. Yih, ed.), Academic Press, New York, pp. 67–143, 1974.

Hutchinson, J. W., and Amazigo, J. C., Imperfection-sensitivity of eccentrically stiffened cylindrical shells, *AIAA Journal*, 392–401, 1967.

Hutchinson, J. W., and Frauenthal, J. C., Elastic postbuckling behavior of stiffened and barreled cylindrical shells, *Journal of Applied Mechanics*, Vol. 37, 784–90, 1969.

Hutchinson, J. W., and Koiter, W. T., Postbuckling theory. *Applied Mechanics Reviews*, Vol. 23, 1353–66, 1970.

Ibrahim, A., Buckling mode localization in rib-stiffened plates with misplaced stiffness, *Master of Applied Science Thesis*, Department of Civil Engineering, University of Waterloo, 1997.

Ikeda, K., and Murota, K., Computation of critical initial imperfections of truss structures, *Journal of Engineering Mechanics*, Vol. 116, 2101–17, 1990a.

Ikeda, K., and Murota, K., Critical initial imperfection of structures, *International Journal of Solids and Structures*, Vol. 26, 865–86, 1990b.

Ikeda, K., and Murota, K., Bifurcation analysis of symmetric structures using block diagonalization, *Computational Methods in Applied Mechanics and Engineering*, Vol. 86, 215–43, 1991a.

Ikeda, K., and Murota, K., Random initial imperfections of structures, *International Journal of Solids Structures*, Vol. 28, 1003–21, 1991b.

Ikeda, K., and Murota, K., Statistics of normally distributed initial imperfections, *International Journal of Solids Structures*, Vol. 30, 2445–67, 1993.

Ikeda, K., Chida, T., and Yanagisawa, E., Imperfection sensitive strength variation of soil specimens, *Journal of Mechanics of Physics of Solids*, Vol. 45, 293–315, 1997.

Ikeda K., Murota, K., and Elishakoff. I., Reliability of structures subject to normally distributed initial imperfections, *Computers and Structures*, Vol. 59, 463–9, 1995.

Ikeda, K., Murota, K., and Fuji, H., Bifurcation hierarchy of symmetric structure, *International Journal of Solids and Structures*, Vol. 27, 1551–73, 1991.

Ilin, V. A., Ovcharov, N. P., and Sukalho, A. A., Statistical diagnosis of critical forces during the investigation of stability of imperfect cylindrical shells, *Izvestiya Akademii Nauk SSSR, Mekhanika Tverdogo Tela*, No. 5, 191–4, 1975 (in Russian).

Imbert K., The effect of imperfections on the buckling of cylindrical shells, Aeronautical Engineer Thesis, California Institute of Technology, Pasadena, 1971.

Itoh, Y., and Fukimoto, Y., Stochastic evaluation of compressive strength of unstiffened plate components, *4th International Colloquium on Stability of Plate and Shell Structures*, Belgium, 1987.

Itoh, Y., Kubo, M., and Fukimoto, Y., Statistical experiment of strength of laterally unsupported rolled H-beams, *Japan Society of Civil Engineers*, Annual Conference, pp. 579–80, 1978.

Itoh, Y., Usami, T., and Fukumoto, Y., Experimental and numerical analysis database on structural stability, *Engineering Structures*, Vol. 18, 812–20, 1996.

Ivanov, L. D., and Rousev, S. H., Statistical estimation of reduction coefficient of ship's hull plates with initial deflections, *The Naval Architect*, No. 4, 158–60, 1979.

Iwatsubo, T., Tomita, K., and Kawai, R., Evaluation of unstable probability of rotors having stiffness errors of asymmetry, *Bulletin of the JSME*, Vol. 25, No. 206, 1299–1305, 1982.

Jaquet, J., Essais de flambement et exploitation statistique, *Construction Métallique*, No. 3, 1970 (in French).

Jasquot, R. G., Nonstationary random column buckling problem, *Journal of the Engineering Mechanics Division*, Vol. 98, 1173–82, 1972.

Jensen, J., and Niordson, F., Symbolic and algebraic manipulation languages and their application in mechanics, *Structural Mechanics Software Series*, Vol. 1, 541–76, 1977.

Jeong, G. D., Critical buckling load analysis of uncertain column, *Probabilistic Mechanics and Structural and Geotechnical Reliability*, 6th Specialty Conference, pp. 563–6, 1992.

Jin, W-L., and Luz, E., Improving importance sampling method in structural reliability, *Nuclear Engineering Design*, Vol. 147, 393–401, 1994.

Johns, K. C., Some statistical aspects of coupled buckling structures, *Buckling of Structures* (B. Budiansky, ed.), Springer, Berlin, pp. 199–207, 1974.

Johns, K. C., Strength statistics for thin shell structures, *Proceedings of the Fifth Canadian Congress of Applied Mechanics*, Fredericton, pp. 169–70, 1975.

Jones, R. M., *Mechanics of Composite Materials*, Hemisphere Publishing Company, Fredericton, pp. 45–54, 1975.

Julien, J. F. (ed.), *Buckling of Shell Structures on the Land, in the Sea, and in the Air*, Elsevier Applied Science, London, 1991.

Kabanov, V. V., Investigation of stability of shells by methods of finite difference, *Izvestiya Akademii Nauk SSSR, Mekhanika Tverdogo Tela*, No. 1, 24–9, 1971 (in Russian).

Kabanov, V. V., *Stability of Non-homogeneous Cylindrical Shells*, "Mashinostroenie" Publishers, Moscow, 1982 (in Russian).

Kac, M., and Siegert, A. J. F., An explicit representation of stationary Gaussian process, Ann. Mat. Stat, Vol. 18, 438–42, 1947.

Kahrimanian, H. G., Analytic differentiation by a digital computer, MS Thesis, Temple University, Philadelphia, PA, 1953.

Kaplyevatsky, Y., Generalization of Frobenius' method for stability analysis of orthotropic annular plates, *Acta Mechanica*, Vol. 22, 295–307, 1975.

Karadeniz, H., Van Manen, S., and Vrouwenvelder, A., Probabilistic reliability analysis for the fatigue limit state of gravity and jacket type structures, *Proceedings of the Third International Conference – BOSS*, McGraw-Hill, London, 1982.

Karhunen, K., Ueber lineare methoden in der wahrscheinlichkeitsrechnung, *Amer. Acad. Sci., Fennicade, Ser. A, I*, Vol. 37, 3–79, 1947 (in German); English translation: RAND Corp., Santa Monica, (A, Report T-131, Aug. 1960).

Von Karman, T., and Tsien, H. S., The buckling of thin cylindrical shells under axial compression, *Journal of Aerospace Sciences*, Vol. 8, 303–12, 1941.

Kasagi, A., Interactive buckling of ring stiffened composite cylindrical shells, D. Sc. Thesis, Washington University, 1994.

Kasagi, A., and Sridharan, S., Postbuckling analysis of layered composite using p-version finite strips, *International Journal of Numerical Methods in Engineering*, Vol. 33, 2091–107, 1992.

Kasagi, A., and Sridharan, S., Buckling and postbuckling analysis of thick composite cylindrical shells under hydrostatic pressure using axisymmetric solid elements, *Composites Engineering*, Vol. 3, 467–87, 1993.

Kasagi, A., and Sridharan, S., Imperfection sensitivity of layered composite cylinders, *Journal of Engineering Mechanics*, Vol. 121, 810–18, 1995.

Keener, J. P., Buckling imperfection sensitivity of columns and spherical caps, *Quarterly Journal of Applied Mathematics*, Vol. 32, 173–99, 1974.

Keller, J. B., Stochastic equations and wave propagation in random media, *Proceedings of Symposia on Applied Mathematics*, XVI, Providence, American Mathematical Society, RI, pp. 145–70, 1964.

Keller, J. B., and Antman, S., *Bifurcation Theory and Non-Linear Eigenvalue Problems*, W. A. Benjamin, New York, 1969.

Kendall, K., and Stuart, A., *The Advanced Theory of Statistics*, Vol. 2, Charles Griffin & Co. Ltd., London, 1973.

Kennedy, D., and Williams, F. W., More efficient use of determinants to solve transcendental structural eigenvalue problems reliably, *Computers and Structures*, Vol. 41, 973–9, 1991.

Khot, N. S., On the influence of initial geometric imperfections on the buckling and postbuckling behavior of fiber-reinforced cylindrical shells under uniform axial compression, *AFFDL-TR-68-136*, 1968.

Khot, N. S., Buckling and postbuckling behavior of composite cylindrical shells under axial compression, *AIAA Journal*, Vol. 8, 229–35, 1970a.

Khot, N. S., Postbuckling behavior of geometrically imperfect composite cylindrical shells under axial compression, *AIAA Journal*, Vol. 8, 579–81, 1970b.

Khot, N. S., and Venkayya, V. B., Effect of fiber orientation on initial postbuckling behavior and imperfection sensitivity of composite cylindrical shells, *AFFDL-TR-79-125*, 1970.

Kirsch, U., *Optimum Structural Design*, McGraw-Hill, New York, 1981.

Kisliakov, S., An investigation of buckling danger of bars, columns and other structural elements as a result of random overloadings or small deviations from the standard of materials, *Annuaire des Ecoles Techniques Supérieures, Mechanique Appliqué*, Vol. VI, Livre, 1, 43–58, 1971 (in Bulgarian).

Kissel, G. I., Localization in disordered periodic structures, *Proceedings of 28th AIAA/ASME/ASCE/AHS Structures, Structural Dynamics and Materials Conference*, AIAA Paper 87–0819, Monterey, 1987.

Kissel, G. I., Localization on disordered periodic structures, PhD Thesis, Department of Aeronautics and Astronautics, MIT, 1988.

Kissel, G. I., Localization factor for multichannel disorder system, *Physical Reviews*, Vol. 44, 1008–14, 1991.

Kleiber, M., Coupled numeric-symbolic computations in structural mechanics, *Discretization Methods in Structural Mechanics* (G. Kuhin and H. Mang, eds.), Springer, Berlin, 317–26, 1989.

Kleiber, M., and Hien, T. D., *Stochastic Finite Element Method*, Wiley, New York, 1993.

Klimanov, V. I., and Timashev, S. A., *Nonlinear Problems of Stiffened Shells*, Academy of Sciences, Ural Scientific Center, Sverdlovsk, 1985 (in Russian).

Klimov, D. M., Computer algebra in mechanics, *General and Applied Mechanics* (A. Yu. Ishlinsky et al., eds.), Hemisphere, New York, pp. 163–78, 1990.

Klimov, D. M., and Rudenko, V. M., *Computer Algebra in Applied Mechanics*, "Nauka" Publishing House, Moscow, 1989 (in Russian).

Kline, L. V., and Hancock, J. O., Buckling of circular plate on elastic foundation, *Journal of Engineering and Industry*, Vol. 87B, 323–4, 1965.

Klompé, A. W. H., Initial imperfection survey on a cylindrical shell (C. S. E.) at the Fokker B. V., *Report LR-495*, Delft University of Technology, Department of Aerospace Engineering, Delft, the Netherlands, 1986.

Klompé, A. W. H., UNIVIMP–A universal instrument for the survey of initial imperfections of thin-walled shells, *Report LR-570*, Delft University of Technology, Department of Aerospace Engineering, Delft, the Netherlands, 1988.

Klompé, A. W. H., The initial imperfection data bank at the Delft University of Technology, Part IV, *Report LR-569*, Delft University of Technology, Department of Aerospace Engineering, Delft, the Netherlands, 1989.

Klompé, A. W. H., and Den Reyer, P. C., The initial imperfection data bank at the Delft University of Technology, Part III, *Report LR-568*, Delft University of Technology, Department of Aerospace Engineering, Delft, the Netherlands, 1989.

Kmiecik, M., Jastrzeboski, T., and Kuzniar, J., Statics of ship plating distortions, *Marine Structures*, Vol. 8, No. 2, 119–32, 1995.

Kogiso, N., Reliability analysis and reliability-based optimization of composite laminated plate for buckling, PhD Thesis, Department of Aerospace Engineering, Osaka Prefecture University, Sakai, Osaka, Japan, 1997.

Kogiso, N., Shao, S., and Murotsu, Y., Reliability-based optimum design of symmetric laminated plate subject to buckling, *Structural Optimization*, Vol. 14, 184–92, 1997.

Kogiso, N., Shao, S., and Murotsu, Y., Effect of correlation on reliability-based design of composite plate for buckling, *AIAA Journal*, Vol. 36, 1706–13, 1998.

Koiter, W. T., On the stability of elastic equilibrium, PhD Dissertation, Delft University, 1945 (English translations: a, NASA TT-F-10, 833, 1967; b, AFFDL-TR-70-25, 1970).

Koiter, W. T., The effect of axisymmetric imperfections on the buckling of cylindrical shells under axial compression, *Proceedings of Royal Netherlands Academy of Sciences*, Amsterdam, Series 13, Vol. 66, 265–79, 1963.

Koiter, W. T., The energy criterion of stability for continuous elastic bodies, I, II, *Proceedings, Koninklijke Nederlandse Akademie van Wetenschappen, Series B*, Vol. 68, 178–202, 1965.

Koiter, W. T., Purpose and achievements of research in elastic stability, *Proceedings of 4th Technical Conference, Society of Engineering Science*, North Carolina State University, Raleigh, NC, pp. 197–218, 1966a.

Koiter, W. T., General equations of elastic stability for thin shells. *Proceedings of the Symposium on the Theory of Shells*, Houston, pp. 187–227, 1966b.

Koiter, W. T., Postbuckling analysis of single two-bar frames, *Recent Progress in Applied Mechanics*, (B. Broberg et al., eds.), Almquist and Wiksell, Göteborg, 1967.

Koiter, W. T., A basic open problem in the theory of elastic stability, *Proceedings, IUTAM/IMU Symposium on Application of Methods of Functional Analysis to Problems of Mechanics*, Marseille, Part XXIX, 1975.

Koiter, W. T., Current trends in the theory of buckling, *Buckling of Structures* (B. Budiansky, ed.), Springer Verlag, Berlin, pp. 1–16, 1976.

Koiter, W. T., Buckling of cylindrical shells under axial compression and external pressure, *Thin Shell Theory. New Trends and Applications* (W. Olszak, ed.), CISM Courses and Lectures No. 240, Springer Verlag, Vienna, pp. 77–87, 1980.

Koiter, W. T., Elastic Stability, 28th Ludwig Prandtl Memorial Lecture, *Zeitschrift für Flugwissenschaften und Weltraumforschung*, Vol. 9, 205–10, 1985.

Koiter, W. T., Elishakoff, I., Li, Y. W., and Starnes, J. H., Buckling of an axially compressed cylindrical shell of variable thickness, *International Journal of Solids and Structures*, Vol. 31, pp. 797–805, 1994a.

Koiter, W. T., Elishakoff, I., Li, Y. W., and Starnes, J. H., Buckling of an axially compressed imperfect cylindrical perfect shell of variable thickness, *Proceedings of 35th AIAA ASME/ASCE/AHS/ASC Structures, Structural Dynamics and Materials Conference*, Hilton Head, SC, AIAA Press, Washington, DC, pp. 277–289, 1994b.

Kondubhatla, S. R. V., Stability of elastic-plastic columns with random geometric imperfections, *SECTAM, Southeastern Theoretical and Applied Mechanics Conference*, Deerfield Beach, FL, 1998.

Korncoff, A. R., and Fenves, S. J., Symbolic generation of finite element stiffness matrices, *Computers and Structures*, Vol. 10, 119–24, 1978.

Kotoguchi, H., Leonard, J. W., and Shiomi, H., Statistical evaluation of steel beam-column resistance, *Engineering Structures*, Vol. 3, 573–88, 1985.

Kowal, Z., and Raducki, K., Interaction of random critical load of spatial truss joint loaded at the joint and on the bars, *Third International Conference on Space Structures* (H. Nooshin, ed.), Elsevier Applied Science Publishers, Amsterdam, pp. 606–12, 1984.

Köylüoğlu, H. U., Nielsen, S. R. K., and Cakmak, A. A., Uncertain buckling load and reliability of columns with uncertain properties, Department of Building Technology and Structural Engineering, *Paper No. 141*, Aalborg University, Denmark, 1995.

Krätzig, W. B., and Oñate, E. (eds.), *Static and Dynamic Stability of Shells*, Springer Verlag, Berlin, 1990.

Krätzig, W. B., Basar, Y., and Wittek, U., Nonlinear behaviour and elastic stability of shells: Theoretical concepts, numerical computations and results, *Buckling of Shells: Proceedings of a State-of-the-Art Colloquium* (E. Ramm, ed.), Springer Verlag, Berlin, pp. 19–57, 1982.

Krein, M. G., Vibration theory of multi-span beams, *Vestnik Inzhenerov i Tekhnikov*, Vol. 4, 142–5, 1933 (in Russian).

Kweon, J. H., and Hong, C. S., Improved arc-length method for postbuckling analysis of composite cylindrical panels, *Computers and Structures*, Vol. 53, 541–9, 1994.

La Tegola, A., Il carico di collasso delle strutture con imperfezioni random, *Giornale del Genio Ciorle*, 1976 (in Italian).

La Tegola, A., Simulation methods for the analysis of pin-jointed structures with randomly imperfect members, *Second International Symposium on Innovative Numerical Analysis in Applied Engineering Science*, Ecole Polytechnique de Montreal, 1980.

Lagace, P. A., Jensen, D. W., and Finch, D. C., Buckling of unsymmetric composite laminates, *Composite Structures*, Vol. 5, 101–23, 1986.

Langley, R. C., Forced response of 1D periodic structures: Vibration localization by damping, *Journal of Sound and Vibration*, Vol. 178, 411–28, 1994.

Law, A., and Kelton, W., *Simulation Modeling and Analysis*, McGraw-Hill, New York 1982.

Lebedev, N. N., *Special Functions and Their Applications*, Dover Publications, New York, 1972.

Leipholz, H. H. E., *Stability of Elastic Systems*, Sijthoff and Noordhoff, Alphen san den Rijn, 1980.

Leissa, A. W., Buckling of laminated composite plates and shell panels, *AFWAL-TR-85 3069*, AF Wright Aeronautical Laboratory, Dayton, OH, 1985.

Leizerakh, V. M., Statistical analysis of random imperfections in cylindrical shells with the aid of the computer, *Proceedings of the Moscow Power Engineering Institute, Subsection of Dynamics and Strength of Machines* (V. V. Bolotin, ed.), MEI Press, Moscow, pp. 45–56, 1969 (in Russian).

Leizerakh, V. M., and Makarov, B. P., Statistical correlation analysis of deformation of compressed shell with initial imperfections, *Theory of Plates and Shells, Proceedings of the 8th All-Union Conference*, Rostov-on-Don, pp. 320–324, 1971 (in Russian).

Lenz, J. C., Reliability based design rules for column buckling, PhD Thesis, Department of Civil Engineering, Washington University, 1972.

Lenz, J., Ravindra, M. K., and Galambos, T. V., Reliability based rules for column buckling, *Computers and Structures*, Vol. 3, 573–88, 1973.

Li, Y. W., Prediction of natural frequency variability due to uncertainty in material properties, *Proceedings of the NOISE-CON'94, Progress in Noise Control for Industry* (J. M. Cuschieri, S. A. L. Glegg, and D. M. Yeager, eds.), Fort Lauderdale, pp. 917–22, 1994.

Li, Y. W., and Gan, H. L., Evaluation of the probability of failure for engineering structures by a modified simulation method, *Journal of Shanghai Institute of Railway Technology*, Vol. 13, 29–39, 1992 (in Chinese).

Li, Y. W., Elishakoff, I., and Starnes, J. H., Jr., Axial buckling of composite cylindrical shells with periodic thickness variation, *Buckling and Postbuckling of Composite Structures* (A. K. Noor, ed.), ASME Press, pp. 95–114, 1994 (see also *Computers and Structures*, Vol. 56, pp. 65–74, 1995).

Li, Y. W., Elishakoff, I., and Starnes, J. H., Jr., Buckling mode localization in a multi-span periodic structure with a disorder in a single span, *Chaos, Solitons and Fractals*, Vol. 5, 955–69, 1995a.

Li, Y. W., Elishakoff, I., and Starnes, J. H., Jr., Effect of thickness variation and initial imperfection on buckling of composite cylindrical shells, *Computers and Structures*, Vol. 56, 65–74, 1995b.

Li, Y. W., Elishakoff, I., Starnes, J. H., Jr., and Bushnell, D., Effect of the thickness variation and initial imperfection on buckling of composite cylindrical shells: Asymptotic analysis and numerical results by BOSOR 4 and PANDA2, *International Journal of Solids and Structures*, Vol. 34, 3755–67, 1997.

Li, Y. W., Elishakoff, I., Starnes, J. H., Jr., and Shinozuka, M., Nonlinear buckling of a structure with random imperfection and random axial compression by a conditional simulation technique, *Computers and Structures*, Vol. 56, 59–64, 1995.

Li, Y. W., Elishakoff, I., Starnes, J. H., Jr., and Shinozuka, M., Prediction of natural frequency and buckling load variability by convex modelling, *Fields Institute Communications*, Vol. 9, 139–54, 1996.

Libai, A., and Durban, A., A method of approximate stability analysis and its application to circular cylindrical shells under circumferentially varying edge loads, *Journal of Applied Mechanics*, Vol. 40, 971–6, 1973.

Librescu, L., and Lin, W., Classical versus non-classical postbuckling behaviour of laminated composite panels under complex loading conditions, *Report*, Department of Engineering Science and Mechanics, Virginia Polytechnic Institute and State University, 1993.

Liew, K. M., and Wang, C. M., Elastic buckling of rectangular plates with curved internal supports, *Journal of Structural Engineering*, Vol. 118, 1480–93, 1990.

Lim, K. B., and Junkins, J. L., Probability of stability: New measures of stability robustness for linear dynamical systems, *The Journal of the Astronautical Sciences*, Vol. 35, 383–97, 1987.

Lin, S. C., and Kam, T. Y., Buckling analysis of imperfect frames, using a stochastic finite element method, *Computers and Structures*, Vol. 42, 895–901, 1992.

Lin, S. C., Kam, T. Y., and Chu, K. H., Evaluation of buckling and first-ply failure, probabilities of composite laminates, *International Journal of Solids and Structures*, Vol. 35, 1395–410, 1998.

Lin, Y. K., and McDaniel, T. J., Dynamics of beam type periodic structures, *Journal of Engineering for Industry*, Vol. 93, 1133–41, 1969.

Lindberg, H. E., Random imperfection for dynamic pulse buckling, *Journal of Engineering Mechanics*, Vol. 114, 1144–65, 1990.

Lindberg, H. E., Dynamic response and buckling failure measures for structures with bounded and random imperfections, *Journal of Applied Mechanics*, Vol. 58, 1092–5, 1991.

Lindberg, H. E., An evaluation for convex modeling for multimode dynamic buckling, *Journal of Applied Mechanics*, Vol. 59, 923–36, 1992a.

Lindberg, H. E., Convex models for uncertain imperfection control in multimode dynamic buckling, *Journal of Applied Mechanics*, Vol. 59, 937–45, 1992b.

Lipovskii, D. E., Altukher, G. M., Koz, V. M., Nazarov, V. A., Todchuk, V. A., and Shun, V. M., Statistical estimation of the effect of random disturbances on the stability of stiffened shells based on the experimented investigations, *Analysis of Spacewise Structures*, Vol. 17, "Stroiizdat" Publishers, Moscow, pp. 32–44, 1977 (in Russian).

Lipovtsev, Y. V., Features of application of the double-sweep method to stability problems in shells and plates, *Mekhanika Tverdogo Tela*, Vol. 3, 43–9, 1970 (in Russian).

Litle, W. A., Reliability of shell buckling predictions based on experimental analysis of plastic models, ScD Thesis, Department of Civil Engineering, MIT, 1963.

Liu, J. K., and Zhao, L. C., Mode localization and eigenvalue loci veering phenomena in disordered aircraft structures, *Proceedings of 12th International Modal Analysis Conference*, Honolulu, Vol. 1, pp. 374–8, 1994.

Liu, W. K., Belytschko T., and Mani, A., Random field finite elements, *International Journal of Numerical Methods in Engineering*, Vol. 23, 1831–45, 1984.

Lockhart, D., Dynamic buckling of damped imperfect columns on a nonlinear foundation, *Quarterly of Applied Mathematics*, Vol. 36, 49–55, 1978.

Lockhart, D., and Amazigo, J. C., Dynamic buckling of externally pressurized imperfect cylindrical shells, *Journal of Applied Mechanics*, Vol. 42, 316–20, 1975.

Loeve, M., Functions aleatoires du second ordre, Supplement to P. Levy, *Processus Stochastic et Mouvement Brownian*, Gauthier Villars, Paris, 1948 (in French).

Lombardi, M., and Haftka, R. T., Anti-optimization technique for structural design under load uncertainties, *Computer Methods in Applied Mechanics and Engineering*, Vol. 157, 19–31, 1998.

Loo, T. T., An extension of Donnell's equations for circular cylindrical shell, *Journal of Aeronautical Science*, Vol. 24, 390–1, 1957.

Lorenz, R., Achsensymmetrische Verzerrung in dünnwandigen Hohlzylindern, *Zeitschrift der Vereines Deutscher Ingenieure*, Vol. 52, 1706–13, 1908 (in German).

Lorenz, R., Die nichtachsensymmetrische Knickung dünnwanger Hohlzylinder, *Physikalische Zeitschrift*, Vol. 13, 241–60, 1911 (in German).

Lottati, I., and Elishakoff, I., Influence of shear deformation and rotary inertia on flutter of a cantilevered beam-exact and symbolic computerized solutions, *Refined Dynamical Theories in Beams, Plates and Shells and Their Applications* (I. Elishakoff and H. Irretier, eds.), Springer Verlag, Berlin, pp. 261–73, 1987.

Lukoshevichus, R. S., Rikards, R. B., and Teters, G. A., Probabilistic analysis of stability and minimization of mass of the cylindrical shell made of composite materials with random geometric imperfections, *Mekhanika Polimerov*, No. 1, 80–9, 1977 (in Russian).

Lust, S. D., Friedmann, P. P., and Bendiksen, O. O., Free and forced response of multi-span beams and multi-bay trusses with localized modes, *Journal of Sound and Vibration*, Vol. 180, 313–32, 1995.

Maidanik, G., and Dickey, J., Localization and delocalization in one-dimensional dynamic systems, Proceedings of the IEEE Ultrasonic Symposium, Vol. 1, pp. 507–14, 1996.

Maimon, G., and Libai, A., Dynamics and failure of cylindrical shells subjected to axial impact, *AIAA Journal*, Vol. 15, 1624–30, 1977.

Mak, C. K. K., Probabilistic buckling behavior of arches with random imperfections, *Recent Advances in Engineering Science*, Vol. 7, 443–50, 1970.

Mak, C. K. K., and Kelsey, S., Statistical aspects in the analysis of structures with random imperfections, *Statics and Probability in Civil Engineering* (P. Lumb, ed.), Hong Kong University Press, Hong Kong, pp. 569–84, 1971.

Mak, C. K. K., and Kelsey, S., Probabilistic buckling behavior of structure with random lack-of-fit and geometric imperfections, *Proceedings of the 2nd International Conference on Structural Mechanics in Reactor Technology*, Berlin, Vol. V – Part M, Paper M7/10, 1973.

Makar, P. F., Application of an implicit linear statistical analysis of the estimation of the resistance of a reinforced concrete beam-column, *Miscellaneous Paper N-78-7*, Defense Nuclear Agency, 1978.

Makarov, B. P., Application of a statistical method for analysis of experimental data on shell stability, *Izvestiya Akademii Nauk, SSSR, Otdelenie Tekhnicheskikh Nauk, Mekhanika i Mashinostroenie*, No. 1, 157–8, 1962a (in Russian).

Makarov, B. P., Application of statistical method for analysis of nonlinear problems of shell stability, *Theory of Plates and Shells, Proceedings of the All-Union Conference*, Kiev, pp. 363–7, 1962b (in Russian).

Makarov, B. P., Analysis of nonlinear shell stability problems with the aid of a statistical method, *Inzhenernyi Zhurnal*, Vol. 3, 100–106, 1963 (in Russian).

Makarov, B. P., Statistical analysis of stability of imperfect cylindrical shells, *Proceedings of 7th All-Union Conference on the Theory of Plates and Shells*, Dnepropetrovsk, pp. 387–91, 1969 (in Russian).

Makarov, B. P., Statistical analysis of nonideal cylindrical shells, *Izvestija Akademii Nauk SSSR, Mekhanika Tverdogo Tela*, Vol. 5, 97–104, 1970 (in Russian).

Makarov, B. P., Statistical analysis of deformation of imperfect cylindrical shells, *Strength Analysis*, Vol. 15, "Mashinostroenie" Publishing House, Moscow, 240–56, 1971 (in Russian).

Makarov, B. P., Postcritical deformations of non-ideal spherical shells, *Problems of Reliability in Structural Design*, Sverdlovsk, pp. 126–131, 1972 (in Russian).

Makarov, B. P., and Chechenev, N. A., About snap-through of thin elastic panels under random impulsive loads, *Raschety na Prochnost*, Vol. 11, "Mashinostroenie" Publishing House, Moscow, pp. 378–84, 1965 (in Russian).

Makarov, B. P., and Leizerakh, V. M., Towards stability of cylindrical shells with random initial imperfections, *Proceedings of the Moscow Power Engineering Institute* (V. V. Bolotin, ed.), Moscow, pp. 35–45, 1969 (in Russian).

Makarov, B. P., and Leizerakh, V. M., On stability of nonideal cylindrical shell in axial compression, *Proceedings of the Moscow Power Engineering Institute, Dynamics and Strength of Machines* (V. V. Bolotin, ed.), Vol. 74, 24–9, 1970 (in Russian).

Makarov, B. P., Leizerakh, V. M., and Sudakova, N. I., Statistical investigation of initial imperfections of cylindrical shells, *Problems of Reliability in Structural Mechanics* (V. V. Bolotin and A. Cyras, eds.), "RINTIP" Publishing House, Vilnius, 1968 (in Russian).

Makushin, V. M., Effective application of an energy method of investigation of elastic stability of columns and plates, *Raschety na Prochnost'* (*Strength Calculations) 8*, "Mashinostroenie" Publishing House, Moscow, pp. 225–52, 1962.

Manevich, A. I., *Stability and Optimum Design of Stiffened Shells*, "Visha Shkola" Publishers, Kiev, 1975 (in Russian).

Manevich, L. I., and Prokopalo, E. F., On statistical properties of load carrying capacity of smooth cylindrical shells, *Problems of Reliability in Structural Mechanics* (V. V. Bolotin and A. Cyras, eds.), "RINTIP" Publishing House, Vilnius, 1968.

Manevich, L. I., Milzin, A. M., Mossakvskii, V. I., Probkpalo, E. F., Smelii, G. N., and Sotnikov, D. I., Experimental investigation of stability of unstiffened cylindrical shells of various scales in axial compression, *Izvestiya Akademii Nauk SSSR, Mckhanika Tverdoygo Tela*, No. 5, 1968 (in Russian).

Massey, F. J., Jr., The Kolmogorov-Smirnov test for goodness of fit, *Journal of American Statistical Assessment*, Vol. 46, 68–78, 1951.

Matthies, H. G., and Bucher, Ch., Finite elements for stochastic media problems, *Computer Methods in Applied Mechanics and Engineering*, Vol. 168, 3–17, 1999.

McDaniel, T. J., The deflection of columns with random centroidal loci, M. Sci. Thesis, Department of Aeronautical and Astronautical Engineering, University of Illinois, Urbana, 1962.

McFarland, D., Smith, B., and Bernhart, W., *Analysis of Plates*, Spartan Books, New York, 1972.

Mead D. J., Vibration response and wave propagation in periodic structures, *Journal of Engineering for Industry*, Vol. 41, 467–86, 1971.

Mead, D. J., Free wave propagation in periodically supported infinite beams, *Journal of Sound and Vibration*, Vol. 11, 181–97, 1973.

Mester, S. S., and Benaroya, H., Periodic and near-periodic structures, *Shock and Vibration*, Vol. 2, 69–95, 1995.

Miles, J. W., Vibration of beams on many supports, *Journal of Engineering Mechanics*, Vol. 82, 1–9, 1956.

Miller, R. K., and Hedgepeth, J. M., The buckling of lattice columns with stochastic imperfections, *International Journal of Solids and Structures*, Vol. 15, 73–84, 1979.

Mirza, S. A., Probability based strength criterion for reinforced concrete slender columns, *ACI Structures Journal*, Vol. 84, 459–66, 1987.

Mirza, S. A., and MacGregor, J. G., Slenderness and strength reliability of reinforced concrete columns, *ACI Structures Journal*, Vol. 86, 428–38, 1989.

Mirza, S. A., and Skrabek, B. W., Reliability of short composite beam-column strength interaction, *Journal of Structural Engineering*, Vol. 117, 2320–1, 1991.

Mirza, S. A., Lee, P. M., and Morgan, D. L., ACI stability resistance factor for RC columns, *Journal of Structural Engineering*, Vol. 113, 1963–76, 1987.

Mlakar, P. F., Reliability based safety factors for concrete structures sliding stability, *Proceedings of the 26th Conference on the Design of Experiments in Army Research Development and Testing*, ARO Report 81-82, 1981.

Moghtaderi-Zadeh, M., and Madsen, H., Probabilistic analysis of buckling and collapse of plates in marine structures, *Proceedings 6th Offshore Mechanics and Arctic Engineering Symposium*, ASME, Houston, 1986.

Morandi, A. C., Computer aided reliability based design of externally pressurized vessels, PhD Thesis, Department of Naval Architecture and Ocean Engineering, University of Glasgow, 1994.

Morandi, A. C., Das, P. K., and Faulkner, D., Finite element analysis and reliability based design of externally pressurized ring stiffened cylinders, *The Royal Institution of Naval Architects*, pp. 1–18, W4, 1996.

Mork, K. J., Bjornsen, T., Venas, A., and Thorkildsen, F., Reliability-based calibration study for upheaval buckling of pipelines, *Journal of Offshore Mechanics and Arctic Engineering*, Vol. 119, 203–8, 1997.

Morley, L. S., An improvement of Donnell's approximation for thin walled circular cylinders, *Quarterly Journal of Mechanics and Applied Mathematics*, Vol. 12, 89–99, 1959.

Morris, N. F., Shell stability: The long road from theory to practice, *Engineering Structures*, Vol. 18, 801–6, 1996.

Mossakowsky, J., Buckling of circular plates with cylindrical orthotropy, *Archiwum Mechaniki Stosowanej*, Vol. 12, 583–96, 1960.

Mossakowsky, J., and Borsuk, K., Buckling and vibrations of circular plates with cylindrical orthotropy, *Applied Mechanics* (A. Rolla and W. T. Koiter, eds.), Elsevier, Amsterdam, pp. 266–9, 1960.

Movsisian, L. A., On stability of cylindrical shell with random initial stresses, *Reliability Problems in Structural Mechanics* (V. V. Bolotin and A. A. Cyras, eds.), "RINTIP" Publishing House, Vilnius, pp. 187–91, 1968 (in Russian).

Muggeridge, D. B., The effect of initial axisymmetric shape imperfections on the buckling behavior of circular cylindrical shells under axial compression, *UTIAS Report No. 148*, Institute for Aerospace Studies, University of Toronto, 1969.

Mura, T., and Koya, T., *Variational Methods in Mechanics*, Oxford University Press, New York, 1992.

Murota, K., and Ikeda, K., Critical imperfection of symmetric structures, *SIAM Journal of Applied Mathematics*, Vol. 51, 1222–54, 1991.

Murota, K., and Ikeda, K., On random imperfections for structures of regular-polygonal symmetry, *SIAM Journal of Applied Mathematics*, Vol. 52, 1780–803, 1992.

Murzewski, J. W., Random instability of elastic-plastic frames, *Structural Safety and Reliability* (A. H.-S. Ang, M. Shinozuka, and G. I. Schuëller, eds.), ASCE Press, New York, pp. 1011–14, 1990.

Murzewski, J. W., Overall instability of steel frames with random imperfections, *International Colloquium, East-European Session, Stability of Steel Structures* (M. Ivanji and B. Veröci, eds.), Bucharest, pp. II/221–8, 1990.

Mutoh, I., Kato, S., and Chiba, Y., Alternative lower bounds analysis of elastic thin shells of revolution, *Engineering Computations*, Vol. 13, 41–75, 1996.

Muzean, J. P., Fogli, M., and Lemaire, M., Buckling of structures in probabilistic context, *Third International Conference on Space Structures* (H. Nooshin, ed.), Elsevier Applied Science Publishers, London, pp. 486–91, 1984.

Nagabhushnan, J., Gaonkar, G. K., and Reddy, T. S. R., Automatic generation of equations for rotor-body systems with dynamic inflow for a priori ordering schemes, Paper No. 7, *7th European Rotorcraft Forum*, Garmish-Partenkirchen, Federal Republic of Germany, 1981.

Nakagiri, S., and Hisada, T., *An Introduction to Stochastic Finite Element Method: Analysis of Uncertain Structures*, Bai Fukan, Tokyo, 1985 (in Japanese).

Nakagiri, S., Takabatake, M., and Tani, S., Uncertain eigenvalue analysis of composite laminated plates by the stochastic finite element method, *Journal of Engineering for Industry*, Vol. 109, 9–12, 1987.

NASA, Buckling of thin-walled circular cylinders, *NASA SP 8007*, Aug. 1968.

Nash, W. A., Instability of thin shells, *Applied Mechanics Surveys* (H. N. Abramson, H. Liebowitz, J. M. Crowley, and S. Juhasz, eds.), Spartan Books, New York, pp. 339–56, 1966.

Nayfeh, A. H., and Hawwa, M., Buckling mode localization in multi-span columns, *Proceedings of 35th AIAA/ASME/ASCE/AHS/ASC Structures, Structural Dynamics, and Materials Conference*, April 18–20, Hilton Head, pp. 277–89, 1994a.

Nayfeh, A. H., and Hawwa, M. A., The use of mode localization in the passive control of structural buckling, *AIAA Journal*, Vol. 32, 2131–3, 1994b.

Neal, D. M., Mathews, W. T., and Vangel, M. G., Uncertainties in obtaining high reliability from stress-strength models, *Proceedings, Ninth DoD/NASA/FAA Conference on Fibrous Composites In Structural Design*, Lake Tahoe, pp. 503–21, 1992.

Nemeth, M. P., Importance of anisotropy on buckling of compression-loaded symmetric composite plates, *AIAA Journal*, Vol. 24, 1831–5, 1986.

Nemirovskii, B. Ya., Towards probabilistic analysis of a flexible shallow shell, *Stroitelnaya Mekhaniki i Raschet Sooruzhenii (Structural Mechanics and Analysis of Buildings)*, No. 3, 17–21, 1968 (in Russian).

Nemirovskii, B. Ya., Probabilistic design of flexible shallow shells under combined loading, *Stroitel'naya Mekhanika i Raschet Sooruzhenii*, No. 3, 38–42, 1969 (in Russian).

Newton, D. A., and Ayaru, B., Statistical methods in the design of structural timber columns, *The Structural Engineer*, Vol. 50, 191–5, 1972.

Niedenfuhr, F. W., Scatter of observed buckling loads of pressurized shells, *AIAA Journal*, Vol. 1, 1923–5, 1963.

Nieuwendijk van den, H. L., Preliminary study of a new design criterion for shells under axial compression, Memorandum M-805, Faculty of Aerospace Engineering, Delft University of Technology, July 1997.

Nikolaidis, E., Hughes, O., Ayyub, B. M., and White, G. J., Reliability analysis of stiffened panels, *Structures Congress XII* (N. C. Baker and B. J. Goodno, eds.), Vol. 2, ASCE Press, New York, pp. 1466–71, 1994.

Niordson, F. I., *Shell Theory*, North Holland, Amsterdam, 1985.

Nolan, J., Analytic differentiation on a digital computer, SM Thesis, MIT Cambridge, MA, 1953.

Noor, A. K., Survey of computer programs for solution of non-linear structural and solid mechanics problems, *Computers and Structures*, Vol. 13, 425–65, 1981.

Noor, A. K., Recent advances and applications of reduction methods, *Applied Mechanics Reviews*, Vol. 47, 125–46, 1994.

Noor, A. K., and Andersen, C. M., Computerized symbolic manipulation in structural mechanics-progress and potential, *Computers and Structures*, Vol. 10, 95–118, 1979.

Noor, A. K., and Peters, J. M., Recent advances in reduction methods for instability analysis of structures, *Computers and Structures*, Vol. 16, 67–80, 1983.

Noor, A. K., Andersen, C. M., and Peters, J. M., Reduced basis techniques for collapse analysis of shells, *AIAA Journal*, Vol. 19, 393–7, 1981.

Noor, A. K., Elishakoff, I., and Hulbert, G. (eds.), *Symbolic Computations and Their Impact on Mechanics*, ASME Press, New York, 1990.

Nordgren, R. P., On One-Dimensional Random Fields with Fixed End Values, *Probabilistic Engineering Mechanics*, Vol. 14, 301–10, 1999.

Nordgren, R. P., and Hussain, K. S., Probabilistic stability analysis of shallow arches, *Engineering Mechanics, Proceedings of the 11th Conference* (Y. K. Lin and T. C. Su, eds.), ASCE Press, New York, pp. 124–7, 1996.

Novichkov, Yu. N., and Indenbaum, V. M., Deformation of a circular cylinder made of low-modulus material and enclosed in a shell, *Proceedings of the Moscow Energy Institute, Dynamics and Strength of Machines*, Vol. 101, pp. 48–55, 1972 (in Russian).

Oceledec, Y. I., A multiplicative ergodic theorem. Lyapunov characteristic number for dynamical systems, *Transactions of the Moscow Mathematical Society*, Vol. 19, 197–231, 1968.

Oh, D. M., and Librescu, L., Free vibration and reliability of composite cantilevers featuring uncertain properties, *Reliability Engineering and System Safety*, Vol. 56, 265–72, 1997.

Ohsaki, M., and Nakanura, T., Optimum design with imperfection sensitivity coefficients for limit point loads, *Structural Optimization*, Vol. 8, 131–7, 1997.

Pai, S. S., and Chamis, C. C., Probabilistic progressive buckling of trusses, *Journal of Spacecraft and Rockets*, Vol. 31, 466–74, 1994.

Palassopoulos, G. V., A probabilistic approach to the buckling of thin cylindrical shells with general random imperfections: Solution of the corresponding deterministic problem, *Theory of Shells* (W. T. Koiter and G. K. Mikhailov, eds.), North Holland, Amsterdam, pp. 417–43, 1980.

Palassoupoulos, G. V., On the effect of stochastic imperfections on the buckling strength of certain structures, *Computational Stochastic Mechanics*, (P. D. Spanos and C. A. Brebbia, eds.), Elsevier Applied Science, London, pp. 211–22, 1991a.

Palassopoulos, G. V., Reliability-based design of imperfection sensitive structures, *Journal of Engineering Mechanics*, Vol. 117, 1220–40, 1991b.

Palassopoulos, G. V., Response variability of structures subjected to bifurcation buckling, *Journal of Engineering Mechanics*, Vol. 118, 1164–83, 1992.

Palassopoulos, G. V., A new approach to the buckling of imperfection-sensitive structures, *Journal of Engineering Mechanics*, Vol. 119, 850–69, 1993.

Palassopoulos, G. V., An improved approach to the reliability-based design of structures subjected to bifurcation buckling, *Structural Safety and Reliability* (G. I. Schuëller, M. Shinozuka, and J. T. P. Yao, eds.), Balkema Publishers, Rotterdam, Vol. 2, pp. 767–76, 1994.

Palassopoulos, G. V., Efficient finite element analysis of imperfection-sensitive frames and arches with stochastic imperfections, *Computational Stochastic Mechanics* (P. D. Spanos, ed.), Balkema Publishers, Rotterdam, pp. 461–8, 1995.

Palassopoulos, G. V., Buckling analysis and design of imperfection sensitive structures, *Uncertainty Modeling in Finite Element, Fatigue and Stability of Systems*, (A. Haldar, A. Guran, and B. M. Ayyub, eds.), World Scientific, Singapore, pp. 311–56, 1997.

Palassopoulos, G. V., and Shinozuka, M., On the elastic stability of thin shells, *Journal of Structural Mechanics*, Vol. 1, 439–49, 1973.

Pandalai, K. A. V., and Patel, S. A., Buckling of orthotropic circular plates, *Journal of Royal Aeronautical Society*, Vol. 69, 279–80, 1965.

Panovko, Ya. G., and Gubanova, I. I., *Stability and Oscillations of Elastic Systems, Paradoxes, Fallacies and New Concepts*, Consultants Bureau, New York, 1964.

Pantelides, C. P., Buckling of elastic columns using convex model of uncertain springs, *Journal of Engineering Mechanics*, Vol. 121, 837–44, 1995.

Parkus, H., Die Beulwahrscheinlichkeit einer Platte in einem transversalen magnetfeld, *Sitzungsberichten der Österreichischen Akademie der Wissenschaften*, Mathem.-naturw, Klasse, Abteilung II, 180, Bcl. 4. bis. 7, Heft, 185–91, 1971 (in German).

Patton, C. M., Symbolic computation for handheld calculators, *Hewlett-Packard Journal*, Aug., 21–25, 1987.

Pavelle, R. (ed.), *Applications of Computer Algebra*, Kluwer, Boston, 1985.

Pavlovic, M. N., and Sapountzakis, E. J., Computers and structures: Non-numerical applications, *Computers and Structures*, Vol. 24, 455–74, 1986.

Pedersen, P., On computer-aided analytic element analysis and the similarities of tetrahedron elements, *International Journal of Numerical Methods in Engineering*, Vol. 11, 611–22, 1977.

Peek, R., Worst shapes of imperfections for space trusses with multiple global and local buckling nodes, *International Journal of Solids and Structures*, Vol. 30, 2243–60, 1993.

Peek, R., and Triantafylidis, N., Worst shapes of imperfections for space trusses with many simultaneously buckling members, *International Journal of Solids and Structures*, Vol. 29, 2385–403, 1992.

Perov, V. A., Roev, V. A., Riashina, N. A., and Sharygin, I. A., About post-critical deformations of thin shells with random initial imperfections, *Proceedings of the Moscow Power Engineering Institute* (M. M. Orakhelashvili, ed.), pp. 86–95, 1970 (in Russian).

Perry, S. H., Statistical variation of buckling strength, PhD Thesis, Department of Civil and Mechanical Engineering, University College, London, 1966.

Pflüger, A., *Stabilitätsprobleme der Elastostatik*, Springer Verlag, Berlin, 1967 (in German).

Pflüger, W., Die Stabilität de Kreiszylinderschale, *Ingenieur-Archiev*, Vol. 3, 463–506, 1932 (in German).

Pi, H. N., Ariaratnam, S. T., and Lennox, W. C., First-passage time for the snap-through of a shell-type structure, *Journal of Sound and Vibration*, Vol. 14, 375–84, 1971.

Pierre, C., Weak and strong vibration localization in disordered structures: A statistical investigation, *Journal of Sound and Vibration*, Vol. 139, 111–32, 1990.

Pierre, C., and Chat, P. D., Strong mode localization in nearly periodic disordered structures, *AIAA Journal*, Vol. 27, 227–41, 1989.

Pierre, C., and Dowell, E. H., Localization of vibrations by structural irregularity, *Journal of Sound and Vibration*, Vol. 114, 549–64, 1987.

Pierre, C., and Plaut, R. H., Curve veering and mode localization in a buckling problem, *Journal of Applied Mathematics and Physics* (*ZAMP*), Vol. 40, 758–61, 1989.

Pignataro, M., Rizzi, N., and Luongo, A., *Stability, Bifurcation and Postbuckling Behavior of Elastic Structures*, Elsevier Science Publishers, Amsterdam, 1991.

Potier-Ferry, M., *Multiple Bifurcation, Symmetry and Secondary Bifurcation*, Pitman Press, London, 1981a.

Potier-Ferry, M., Critères de l'énergie en élasticité et viscoélasticité, *Le Flambement des Structures* (R. L'Hermite, ed.), Editions Bâtiment Travaux Publics, Paris, pp. 23–37, 1981b (in French).

Potier-Ferry, M., Foundations of elastic postbuckling theory, *Buckling and Post-Buckling* (J. Arbocz, M. Potier-Ferry, J. Singer, and V. Tvergaard, eds.), Springer Verlag, Berlin, pp. 1–82, 1987.

Power, T. L., and Kyriakides, S., Localization and propagation of instabilities on long shallow panels under external pressure, *Journal of Applied Mechanics*, Vol. 61, 755–63, 1994.

Pshenichnov, G. I., Statistical analysis of net-cylindrical shells, *Inzhenernyi Sbornik*, Vol. 27, 171–8, 1960 (in Russian).

Pu, Y., Das, P. K., and Faulkner, D., Ultimate compression strength and probabilistic analysis of stiffness plates, *Proceedings of the 15th Offshore Mechanics and Arctic Engineering Conference*, Vol. II, ASME Press, New York, pp. 151–7, 1996.

Qiria, V. S., Motion of the bodies in resisting media, *Proceedings of the Tbilisi State University*, Vol. 44, 1–20, 1951 (in Russian).

Qiu, Z. P., and Elishakoff, I., Antioptimization of structures with large uncertain-but-nonrandom parameters via interval analysis, *Computer Methods in Applied Mechanics and Engineering*, Vol. 152, 361–72, 1998.

Ramaiah, G. K., and Vijayakumar, K., Buckling of polar orthotropic annular plates under uniform internal pressure, *AIAA Journal*, Vol. 12, 1045–50, 1974.

Ramaiah, G. K., and Vijayakumar, K., Elastic stability of annular plates under uniform compressive forces along the outer edge, *AIAA Journal*, Vol. 13, 832–5, 1975.

Ramm, E. (ed.), *Buckling of Shells, Proceedings of the State-of-the-Art Colloquium*, Springer Verlag, Berlin, 1982a.

Ramm, E., The Riks/Wempner approach – an extension of the displacement control method in nonlinear analysis, *Recent Advances in Nonlinear Computational Mechanics* (E. Hinton, D. R. J. Owen, and C. Taylor, eds.), Pineridge Press, Swansea, pp. 63–86, 1982b.

Ramu, S. A., and Ganesan, R., Stability analysis of a stochastic column subjected to stochastically distributed loadings using the finite element method, *Finite Elements in Analysis and Design*, Vol. 11, 105–15, 1992.

Ramu, S. A., and Ganesan, R., Parametric stability of stochastic columns, *International Journal of Solids and Structures*, Vol. 30, 1339–54, 1993.

Ramu, S. A., and Ganesan, R., Response and stability of a stochastic beam-column using stochastic FEM, *Computers and Structures*, Vol. 54, 207–22, 1995.

Rand, R. H., *Computer Algebra in Applied Mathematics: An Introduction to MACSYMA*, Pitman, London, 1984.

Rand, R. H., and Armbruster, D., *Perturbation Methods, Bifurcation Theory and Computer Algebra*, Springer Verlag, Berlin, 1987.

Rankin, C. C., Riks, E., Starnes, Jr., J. H., and Waters, Jr. A., An experimental and numerical verification of the postbuckling behavior of a composite cylinder in compression, *NASA Contract Report*, to appear.

Rankin, C. S., Stehlin, P., and Brogan, F. A., Enhancements to the STAGS computer code, *NASA CR-4000*, 1986.

Rantis, T. D., and Taingjitham, S., Probability based stability analysis of a laminated composite plate under confined in-plane load, *CCMS-94-17*, Virginia Polytechnic Institute and State University, Center for Composite Materials and Structures, Blacksburg, VA, 1993.

Rayleigh, B., *The Theory of Sound*, Dover Publications, New York, 2nd ed., 1945 (1st ed., Macmillan, New York, 1894).

Rayna, G., *REDUCE: Software for Algebraic Computation*, Springer, Berlin, 1987.

Rehak, M., Dillaggio, F., Benaroya, H., and Elishakoff, I., Random vibrations with MACSYMA, *Computer Methods in Applied Mechanics and Engineering*, Vol. 61, 61–70, 1987.

Reissner, E., On postbuckling behavior and imperfection sensitivity of thin plates on a non-linear foundation, *Studies on Applied Mathematics*, Vol. XLIX, 45–7, 1970.

Reissner, E., A note on imperfection sensitivity of thin plates on a non-linear elastic foundation, *Instability of Continuous Systems* (H. H. E. Leipholz, ed.), Springer Verlag, Berlin, pp. 15–18, 1971.

Reznikov, B. S., Probability of failure of thermoelastic reinforced shells, *Strength of Materials*, Vol. 14, 1364–8, 1982 (in Russian).

Richardson, J. M., The application of truncated hierarchy techniques in the solution of stochastic linear differential equations, *Proceedings of Symposium in Applied Mechanics*, pp. 290–302, 1964.

Rikards, R. B., and Teters, G. A., *Stability of Shells Made of Composite Materials*, "Zinatne" Publishers, Riga, 1974 (in Russian).

Riks, E., An incremental approach to the solution of snapping and buckling problems, *International Journal of Solids and Structures*, Vol. 15, 529–51, 1979a.

Riks, E., Development of an improved analysis capability of the STAGS computer code with respect to the collapse behavior of shell structures, *Memorandum SC-79-054*, National Aerospace Laboratory NRL, the Netherlands, 1979b.

Riks, E., Some computational aspects of the stability analysis of nonlinear structures, *Computer Methods in Applied Mechanics and Engineering*, Vol. 47, 219–59, 1984.

Riks, E., Progress in collapse analysis, *Journal of Pressure Vessels Technology*, Vol. 109, 33–41, 1987.

Riks, E., Brogan, F. A., and Rankin, C. C., Aspects of the stability analysis of shells, *Static and Dynamic Stability of Shells* (W. B. Krätzig and E. Oñate, eds.), Springer Verlag, Berlin, 1988.

Rizzi, N., and Tatone, A., Symbolic manipulation in buckling and postbuckling analysis, *Computers and Structures*, Vol. 21, 691–700, 1985a.

Rizzi, N., and Tatone, A., Using symbolic computation in buckling analysis, *Journal of Symbolic Computation*, Vol. 1, 317–21, 1985b.

Roberts, S. M., and Shipman, J. S., *Two-Point Boundary Values Problems: Shooting Methods*, American Elsevier, New York, pp. 50–86, 1972.

Rokach, A. J., A statistical study of the steel columns, M.Sc. Thesis, *Report R0-60*, Department of Civil Engineering, MIT, 1970.

Romstad, K. M., Hutchinson, J. R., and Runge, K. H., Design parameter variation and structural response, *International Journal of Numerical Methods in Engineering*, Vol. 5, 337–49, 1973.

Roorda, J., Stability of structures with small imperfections, *Journal of Engineering Mechanics Division*, Vol. 91, 87–106, 1965.

Roorda, J., Some statistical aspects of the buckling of imperfection-sensitive structures, *Journal of Mechanics, Physics and Solids*, Vol. 17, 111–23, 1969.

Roorda, J., An experience in equilibrium and stability, *Technical No. 3*, University of Waterloo, Solid Mechanics Division, Waterloo, 1971.

Roorda, J., Buckling of shells; an old idea with a new twist, *Journal of Engineering Mechanics Division*, Vol. 98, 531–8, 1972a.

Roorda, J., Concepts in elastic structural stability, *Mechanics Today* (S. Nemat-Nasser, ed.), Pergamon Press, New York, Vol. 1, pp. 322–72, 1972b.

Roorda, Y., The random nature of column failure, *Journal of Structural Mechanics*, Vol. 3, 239–57, 1975.

Roorda, J., *Buckling of Elastic Structures*, University of Waterloo Press, Waterloo, 1980.

Roorda, J., and Chilver, A. M., Frame buckling: An illustrative of the perturbation technique, *International Journal of Non-Linear*, Vol. 5, 235–46, 1970.

Roorda, J., and Hansen, J. S., Random buckling behavior in axially loaded cylindrical shells with axisymmetric imperfections, *Journal of Spacecraft*, Vol. 9, 88–91, 1972.

Rubinstein, R., *Simulation and the Monte Carlo Method*, John Wiley, New York, 1981.

Ruiz, S. E., and Aguilar, J. C., Reliability of short and slender reinforce concrete columns, *Journal of Structural Engineering*, Vol. 120, 1850–65, 1994.

Rurlykina, V. N., and Minaev, V. A., About a statistical problem of stability of elastic beam, *Analysis of Strength, Stability and Vibrations of Elements of Engineering Structures*, Vononezh, pp. 114–17, 1981 (in Russian).

Rzadkowski, J. W., Random ultimate bearing capacity of elastic-plastic space trusses, *Third International Conference on Space structures* (H. Nooshin, ed.), Elsevier Applied Science Publishers, London, pp. 561–6, 1984.

Rzhanitsin, A. R., *Structural Design with Consideration of Plastic Properties of Materials*, "Stroivoenmorizdat," Moscow, 1949 (in Russian).

Rzhanitsin, A. R., *Statistical Method of Determination of Allowable Stresses for Beam-Columns*, Gosudarstvennoe Izdatelstvo Stroitelnoi Literatury, Moscow, 1951 (in Russian).

Rzhanitsin, A. R., Statistical stability of a compressed column, *Reliability Problems in Structural Mechanics* (V. V. Bolotin and A. Cyras, eds.), "RINTIP" Publishing House, Vilnius, pp. 192–8, 1968 (in Russian).

Rzhanitsin, A. R., and Visir, P. L., Stability of equilibrium and motion under conditions of random initial and boundary conditions, *Loadings and Reliability of Building Constructions Proceedings of ZNIISK*, Vol. 21, "Stroiizdat" Publishers, Moscow, 1973 (in Russian).

Sabag, M., Stavsky, Y., and Greenberg, J. B., Buckling of edge-damaged, cylindrical composite shells, *Journal of Applied Mechanics*, Vol. 56, 121–6, 1989.

Sadovsky, Z., Stochastic resistance of square plate under uniaxial compression, *Engineering Structures*, Vol. 19, 827–33, 1997.

Salim, S., Iyengar, N. G. R., and Yadav, D., Buckling of laminated plates with random material characteristics, *Applied Composite Materials*, Vol. 5, 1–9, 1998.

Salim, S., Yadav, D., and Iyengar, N. G. R., Analysis of composite plates with random material characteristics, *Mechanics Research Communications*, Vol. 20, 405–14, 1993.

Sanders, Jr., J. L., Nonlinear theories for thin shells, *Quarterly of Applied Mechanics*, Vol. 21, 21–36, 1963.

Sankar, T. S., and Ariaratnam, S. T., Snap buckling of shell-type structures under stochastic loading, *International Journal of Solids and Structures*, Vol. 7, 655–66, 1971.

Schenk, C. A., Schuëller, G. I. and Arbocz, J., On the analysis of cylindrical shells with random imperfections, *Fourth Int'l Coll. on Computation of Shell & Spatial Structures*, Chania-Crete, Greece, June 4–7, 2000.

Schenk, C. A., Schuëller, G. I. and Arbocz, J., Buckling analysis of cylindrical shells with random imperfections, *International Conference on Monte Carlo Simulation*, Monte Carlo, June 18–21, 2000.

Scheurkogel, A., and Elishakoff, I., On ergodicity assumption in an applied mechanics problem, *Journal of Applied Mechanics*, Vol. 52, 133–6, 1985.

Scheurkogel, A., Elishakoff, I., and Kalker, J., On the error that can be induced by an ergodicity assumption, *Journal of Applied Mechanics*, Vol. 48, 654–6, 1981.

Schmidt, R., A variant of the Rayleigh-Ritz method, *Industrial Mathematics*, Vol. 31, 37–46, 1981.

Schmidt, R., Estimation of buckling loads and other eigenvalues via a modification of the Rayleigh-Ritz method, *Journal of Applied Mechanics*, Vol. 49, 639–40, 1982.

Schmidt, R., Modification of Timoshenko's technique for estimating buckling loads, *Industrial Mathematics*, Vol. 33, 169–73, 1983.

Schmidt, H., and Krysik, R., Towards recommendations for shell stability design by means of numerically determined buckling loads, *Buckling of Shell Structures on Land, in the Sea and in the Air* (J. F. Jullien, ed.), Elsevier Applied Science, London, pp. 508–19, 1991.

Schweppe, F. C., Recursive state estimation: unknown but bounded errors and system inputs, *IEEE Transactions on Automatic Control*, Vol. AC-13, 22–8, 1968.

Schweppe, F. C., *Uncertain Dynamic Systems*, Prentice Hall, Englewood Cliffs, NJ, 1973.

Sebec, R. W. L., Imperfection surveys and data reduction of Ariane interstage I/II and II/III, Ir. Thesis, Department of Aerospace Engineering, Delft University of Technology, 1981.

Semeniuk, N. P., About initial post-buckling behavior of composite cylindrical shells in axial compression, *Prikladnaya Mekhanika*, Vol. 88, No. 5, 34–60, 1988 (in Russian).

Semeniuk, N. P., Formulas for analysis of stability, initial post-buckling behavior and imperfection sensitivity of layered composite cylindrical shells, *Prikladnaya Mekhanika*, Vol. 27, No. 7, 31–6, 1991a (in Russian).

Semeniuk, N. P., Towards analysis of stability, initial post-buckling behavior and imperfection-sensitivity of layered composite cylindrical shells, *Prikladnaya Mekhanika*, Vol. 27, No. 6, 74–80, 1991b (in Russian).

Semeniuk, N. P., and Zhukova, N. B., Towards solution of a problem on initial post-buckling behavior of stringer-stiffened cylindrical shells, *Prikladnaya Mekhanika*, Vol. 27, 81–8, 1991 (in Russian).

Semeniuk, N. P., and Zhukova, N. B., About stability and initial post-buckling behavior of stiffened cylindrical composite shells, *Prikladnaya Mekhanika*, Vol. 28, 48–53, 1992 (in Russian).

Semeniuk, N. P., and Zhukova, N. B., Towards the problem of interaction of modes of stability loss for imperfect composite cylindrical shells, *Prikladnaya Mekhanika*, Vol. 30, 69–75, 1994 (in Russian).

Sen-Gupta, G., Natural flexural wave and the normal modes of periodically supported beams and plates, *Journal of Sound and Vibration*, Vol. 13, 89–101, 1970.

Seward, L. R., *REDUCE User's Guide, for IBM 360 and Derivative Computers*, Version 3.2, Rand Publication CP83, 1985.

Shah, H. C., Regression analysis of reinforced concrete columns, *Proceedings of the ASCE*, Vol. 90, 1964.

Sheinman, I., Cylindrical buckling load of laminated columns, *Journal of Engineering Mechanics*, Vol. 115, 659–61, 1989.

Sheinman, I., and Simitses, G. J., Buckling of geometrically imperfect stiffened cylinders under axial compression, *AIAA Journal*, Vol. 15, 374–82, 1977.

Shilkrut, D. I., The influence of the paths of multi-parametrical conservative loading on the behaviour of a geometrically nonlinear deformable body, *Buckling of Structures: Theory and Experiment* (I. Elishakoff, J. Arbocz, C. D. Babcock, Jr., and A. Libai, eds.), Elsevier Science Publishers, Amsterdam, 1988.

Shilkrut, D. I., and Vyrlan, P. M., *Stability of Nonlinear Shells*, "Shtiinza" Publishing House, Kishinev, 1977 (in Russian).

Shinozuka, M., Maximum structural response of seismic excitations, *Journal of Engineering Mechanics Division*, Vol. 96, 729–38, 1970.

Shinozuka, M., Structural response variability, *Journal of Engineering Mechanics*, Vol. 113, 825–42, 1987.

Shinozuka, M., and Deodatis, G., Response variability of stochastic finite element systems, *Journal of Engineering Mechanics*, Vol. 114, 499–519, 1988.

Shinozuka, M., and Deodatis, G., Simulation of stochastic processes by spectral representation, *Applied Mechanics Reviews*, Vol. 44, 191–204, 1991.

Shinozuka, M., and Henry, L., Random vibration of a beam column, *Journal of the Engineering Mechanics Division*, Vol. 91, 123–43, 1965.

Shinozuka, M., and Yamazaki, F., Stochastic finite element analysis: An introduction, *Stochastic Structural Dynamics* (S. T. Ariaratnam, G. I. Schuëller, and I. Elishakoff, eds.), Elsevier, London, pp. 241–91, 1988.

Shiraki, W., and Takaoka, N., Reliability analysis of compression members with rectangular cross section, *Theoretical and Applied Mechanics*, Vol. 27, 247–58, 1979.

Shiraki, W., and Takaoka, N., Reliability analysis of compression members with nonstationary random initial deflection, *Proceedings of JSCE*, No. 297, 37–46, 1980 (in Japanese).

Shreider, Ya. A., *Method of Statistical Testing: Monte Carlo Method*, Elsevier, Amsterdam, 1964.

Shulga, S. A., Sudol, D. E., and Nishino, F., Influence of the mode of initial geometrical imperfections of the load-carrying capacity of cylindrical shells made of composite materials, *Thin-Walled Structures*, Vol. 14, 89–103, 1992.

Simitses, G. J., Buckling and postbuckling of imperfect cylindrical shells, *Applied Mechanics Reviews*, Vol. 39, 1517–24, 1986.

Simitses, G. J., and Anastasiadis, J. S., Shear deformable theories for cylindrical laminates equilibrium and buckling with applications, *AIAA Journal*, Vol. 30, 826–34, 1992.

Simitses, G. J., Shaw, D., and Sheinman, I., Stability of imperfect laminated cylinders: A comparison between theory and experiments, *AIAA Journal*, Vol. 23, 1086–92, 1985.

Singer, J., The influence of stiffener geometry and spacing on the buckling of axially compressed cylindrical and conical shells, *Theory of Thin Shells* (F. I. Niordson, ed.), Springer Verlag, Berlin, pp. 234–63, 1969.

Singer, J., Buckling experiments on shells-a review of recent developments, *Solid Mechanics Archives*, Vol. 7, 213–313, 1982.

Singer, J., and Rosen, A., The influence of boundary conditions of the buckling of stiffened cylindrical shells, *Buckling of Structures* (B. Budiansky, ed.), Springer Verlag, Berlin, pp. 227–50, 1976.

Singer, J., Abramovich, H., and Yaffe, R., Initial imperfection measurements of integrally stringer-stiffened cylindrical shells, *TAE Report 330*, Technion–Israel Institute of Technology, Department of Aeronautical Engineering, Haifa, Israel, 1978.

Singer, J., Abramrovich, H., and Yaffe, R., Initial imperfection measurements of stiffened shells and buckling predictions, *Israel Journal of Technology*, Vol. 17, 324–38, 1979.

Singer J., Arbocz, J., and Weller, T., *Buckling Experiments: Experimental Methods in Buckling of Thin-Walled Structures*, Wiley, New York, 1997.

Singh, M. P., Analysis and reliability-based design of columns with random parameters, *PhD Thesis*, Department of Civil Engineering, University of Illinois at Urbana, 1972.

Singh, M. P., and Ang, A. H.-S., Columns with random imperfections and associated bracing requirements, *Journal of Structural Mechanics*, Vol. 4, 161–80, 1976.

Slaughter, W. S., and Fleck, N. A., Microbuckling of fiber composites with random initial fiber waviness, *Journal of Mechanics of Physics and Solids*, Vol. 42, 1743–66, 1994.

Slezinger, I. N., and Barskaya, S. Ya., On influence of random form imperfections on stability of shallow double-curvative panel, *Dynamics and Strength of Machines*, Kharkov, pp. 96–104, 1977 (in Russian).

Sobel, L. H., Effect of boundary conditions on the stability of cylinders subjected to lateral and axial pressure, *AIAA Journal*, Vol. 2, 1437–40, 1964.

Southwell, R. V., On the collapse of tubes by external pressure, *Philosophical Magazine*, Part I, Vol. 25, 687–8, 1913; Part II, Vol. 26, 502–11, 1913; Part III, Vol. 29, 66–7, 1915.

Spencer, H. H., Are measurements of geometric imperfections of plates and shells useful?, *Experimental Mechanics*, Vol. 18, 107–11, 1978.

Stam, A. R., Stability of imperfect cylindrical shells with random properties, Paper AIAA-96-1462-CP, *37th AIAA/ASME/ASCE/AHS/ASC Structures, Structural Dynamics, and Materials Conference*, Salt Lake City, 1307–15, 1996.

Stam, A., Experimental Verification of Probabilistic Design Analysis, Paper No. AIAA-2000-1511, AIAA Press, Washington, D. C., 2000.

Stam, A. R., and Arbocz, J., Stability analysis of anisotropic cylindrical shells with random layer properties, *Reliability in Nonlinear Structural Mechanics*, (O. D. Ditlevsen and J. C. Mitteau, eds.), Euromech-372, University of Blaise Pascal, Clermon-Ferrand, pp. 77–84, 1997.

Starnes, Jr., J. H., and Haftka, R. T., Preliminary design of composite wings for buckling, strength and displacement constraints, *Journal of Aircraft*, Vol. 16, 564–70, 1979.

Starnes, Jr., J. H., Kright, Jr., N. F., and Rouse, M., Postbuckling behavior of selected flat stiffened graphite-epoxy panels loaded in compression, *AIAA Journal*, Vol. 23, 1236–46, 1985.

Stavsky, Y., and Friedland, S., Stability of heterogeneous orthotropic cylindrical shells in axial compression, *Israel Journal of Technology*, Vol. 7, 111–19, 1969.

Stavsky, Y., and Hoff, N. J., Mechanics of composite structures, *Composite Engineering Laminates* (A. G. H. Dietz, ed.), MIT Press, Cambridge, pp. 5–59, 1969.

Stein, M., Some recent advances in the investigation of shell buckling, *AIAA Journal*, Vol. 6, 2239–45, 1968.

Strating, J., Probabilistic theory and buckling curves, *Report 6-70-8, CEACM 8-1*, Stewin-Laboratorium, Delft University of Technology, Delft, the Netherlands, 1970.

Strating, J., and Vas, H., Computer simulation of the R.C.C.S. buckling curve using a Monte-Carlo method, *Heron*, Vol. 19, 1–25, 1973.

Streletskii, N. S., *Osnovy Statisticheskogo Ucheta Koeffizienta Zapasa Prochnosti Sooruzhenii* (*Basics of Statistical Analysis of Safety Factor of Structures*), "Stroiizdat" Publishers, Moscow, 1947 (in Russian).

Stroud, W. J., and Anderson, M. S., PASCO – Structural panel analysis and sizing code, capability and analytical foundations, *NASA TM-80181*, 1981.

Stroud, W. J., Davis, Jr., D. D., Maring, L. D., Krishnamurthy, T., and Elishakoff, I., Reliability of stiffened structural panels: Two examples, *Reliability Technology* (T. A. Cruse, ed.), ASME Press, New York, pp. 196–216, 1992.

Stroud, W. J., Krishnamurthy, T., Sykes, N. P., and Elishakoff, I., Effect of bow-type initial imperfection on the reliability of minimum-weight stiffened structural panels, *NASA TN-3263*, 1993.

Swamidas, A. S. J., and Kunukkasseril, V. X., Buckling of orthotropic circular plates, *AIAA Journal*, Vol. 11, 1633–36, 1973.

Takaoka, N., and Shiraki, W., Probabilistic design method of elastic compression members, *Reports of the Faculty of Engineering*, Tottori University, Vol. 10, No. 1, 140–52, 1979 (in Japanese).

Takaoka, N., Shiraki, W., and Fujiwara, T., Reliability analysis of H-shaped compression members with random initial imperfections, *Reports of the Faculty of Engineering*, Tottori University, Vol. 11, No. 1, 134–48, 1980 (in Japanese).

Talaslidis, D., Anwendung statistischer Methoden auf das Stabilitatsproblem "ungenauigkeitsemfindlicher" Schalen, *Konstruktver Ingenieurbau-Berichte* (E. H. Zerna, ed.), Vulkan Verlag, Essen, pp. 151–63, 1976 (in German).

Tasi, J., Effect of heterogeneity on the stability of cylindrical shells under axial compression, *AIAA Journal*, Vol. 4, 1058–62, 1966.

Tatsa, E., Tene, Y., and Baruch, M., Remarks on the application of Koiter's general post-buckling theory, *Acta Mechanics*, Vol. 22, 153–62, 1975.

Temple, S. G., and Bickley, B. G., *Rayleigh's Principle and Its Applications to Engineering,* Oxford University Press, New York, 1933 (reprinted by Dover, New York, 1956).

Tenerelli, D. J., and Horton, W. M., Theoretical and experimental study of the local buckling of ring-stiffened cylinders subject to axial compression, *Israel Journal of Technology*, Vol. 7, 181–94, 1969.

Tennyson, R. C., Buckling of laminated composite cylinders: a review, *Composites*, Vol. 6, 17–24, 1975.

Tennyson, R. C., Interaction of cylindrical shell buckling experiments with theory, *Theory of Shells*, North-Holland, Amsterdam, pp. 65–116, 1980.

Tennyson, R. C., Buckling of composite cylinders under axial compression, *Developments in Engineering Mechanics* (A. P. S. Salvadurai, ed.), Elsevier Science Publishers, Amsterdam, pp. 229–58, 1987.

Tennyson, R. C., and Muggeridge, D. B., Buckling of axisymmetric imperfect cylindrical shells under axial compression, *AIAA Journal*, Vol. 7, 2127–31, 1969.

Tennyson, R. C., and Muggeridge, D. B., Buckling of laminated anisotropic imperfect circular cylinders under axial compression, *AIAA 10th Aerospace Sciences Meeting*, AIAA Paper No. 72-139, 1972.

Tennyson, R. C., Chan, M. K., and Muggeridge, D. B., The effect of axisymmetric shape imperfections on the buckling of laminated anisotropic circular cylinders, *Canadian Aeronautics and Space, Transactions of the Institute*, Vol. 4, 131–8, 1971.

Tennyson, R. C., Muggeridge, D. B., and Caswell, R. D., Buckling of circular cylindrical shells having axisymmetric imperfections distributions, *AIAA Journal*, Vol. 9, 924–30, 1971.

Tewary, V. K., *Mechanics of Fiber Composites*, John Wiley and Sons, New York, pp. 108–21, 1978.

Thangjitham, S., and Rantis, T. D., Probability-based buckling instability analysis of a laminated composite plate, *Proceedings of 33th AIAA/ASME/ASCE/AHS/ASC Structures, Structural Dynamics and Materials Conference*, AIAA Press, New York, pp. 454–62, 1992.

Thoft-Christensen, P., and Baker, M. J., *Structural Reliability Theory and Its Applications*, Springer, Berlin, 1982.

Thompson, J. M. T., Toward a general statistical theory of imperfection-sensitivity in elastic post-buckling, *Journal of Mechanics, Physics and Solids*, Vol. 15, 413–17, 1967.

Thompson, J. M. T., and Hunt, G. W., *A General Theory of Elastic Stability*, Wiley, London, 1973.

Thomson, W. T., *Theory of Vibration with Applications*, 2nd ed., Prentice Hall, Englewood Cliffs, NJ, pp. 30–4, 1981.

Tichy, M., and Norlicek, M., Safety and eccentrically loaded reinforced concrete columns, *Journal of Structural Division*, Vol. 88, 1–10, 1962.

Timashev, S. A., Statistical investigation of stability of thin convex orthotropic shells, *Reliability Problems in Structural Mechanics* (V. V. Bolotin and A. Cyras, eds.), "RINTIP" Publishing House, Vilnius, pp. 157–64, 1968 (in Russian).

Timashev, S. A., Stochastic stability and reliability of stiffened shells, ScD Dissertation, Leningrad Civil Engineering Institute, 1975 (in Russian).

Timashev, S. A., and Kantor, S. L., Stochastic stability and reliability of stiffened shells, *International Conference on Light-Weight Spacewise Structures in Concentional and Seismic Regions*, Alma-Ata, pp. 140–5, 1977 (in Russian).

Timoshenko, S. P., Einige Stabilitätsprobleme der Elastizitätstheorie, *Zeitschrift für angewandte Mathematik und Physik*, Vol. 58, 337–85, 1910 (in German).

Timoshenko, S. P., and Gere, J. M., *Theory of Elastic Stability*, 2nd ed., McGraw-Hill, New York, pp. 94–8, 1961.

Trendafilova, I., and Ivanova, J., Loss of stability of thin, elastic, strongly convex shells of revolution with initial imperfections, subjected to uniform pressure. A probabilistic approach, *Thin-Walled Structures*, Vol. 23, 201–14, 1995.

Triantafylidis, N., and Peek, R., On stability and the worst imperfection shape in solids with nearby simultaneous eigenmodes, *International Journal of Solids and Structures*, Vol. 29, 2281–99, 1992.

Troshenkov, M. K., Application of variational methods in nonlinear stochastic problems of dynamics and statics of shells, PhD Dissertation, Novosibirsk Electrotechnic Institute, 1976 (in Russian).

Tsai, C. T., and Palazotto, A. N., A modified Riks approach to composite shell snapping using a high-order shear deformation theory, *Computers and Structures*, Vol. 35, 221–6, 1990.

Tvergaard, V., Buckling Behavior of Plate and Shell Structure, *Theoretical and Applied Mechanics* (W. T. Koiter, ed.), North-Holland, Amsterdam, pp. 233–47, 1976.

Tvergaard, V., and Needleman, A., On the localization of buckling pattern, *Journal of Applied Mechanics*, Vol. 50, 935–40, 1980.

Tvergaard, V., and Needleman, A., On the development of localized buckling patterns, *Collapse: The Buckling of Structures in Theory and Practice*, (J. M. T. Thompson and G. W. Hunt, eds.), Cambridge University Press, Cambridge, pp. 299–332, 1983.

Uthgenannt, E. B., and Brand, R. S., Buckling of orthotropic annular plates, *AIAA Journal*, Vol. 8, 2102–4, 1970.

Vainberg, M. M., and Trenogin, V. A., *Theory of Branching of Solution of Nonlinear Equations*, Noordhoff International Publishers, Leyden, 1974.

Valishvili, N. V., Stability of spherical shells under finite displacements, *Mechanics of Deformable Solids and Structures*, volume dedicated to 60th anniversary of Ju. N. Rabotnov, "Mashinostroenie" Publishing House, Moscow, pp. 92–8, 1975 (in Russian).

Van Eekelen, A. J., Computation module for the buckling and postbuckling behavior of a cylindrical shell with a two-mode imperfection, *Memorandum LR 773*, Delft University of Technology, Department of Aerospace Engineering, Delft, the Netherlands, 1994.

Van Slooten, R. A., and Soong, T. T., Buckling of a long, axial compressed, thin cylindrical shell with random initial imperfection, *Journal of Applied Mechanics*, Vol. 38, 1066–71, 1972.

Vanin, G. A., and Semeniuk, N. P., *Stability of Shells Made of Composite Materials with Imperfections*, "Naukova Dumka" Publishers, Kiev, 1978 (in Russian).

Vanin, G. A., Semeniuk, N. P., and Emelianov, R. F., *Stability of Shells of Composite Materials*, "Naukova Dumka" Publishers, Kiev, 1978 (in Russian).

Vanmarcke, E., Shinozuka, M., Nakagiri, S., Schuëller, G., and Grigoriu, M., Random fields and stochastic finite-elements, *Structural Safety*, Vol. 3, 143–66, 1986.

Vásárhelyi, A., Collapse load analysis of panel structures by stochastic programming, *Periodica Politechnica*, Vol. 25, No. 314, 243–50, 1981.

Vasiliev, V. V., *Mechanics of Composite Structures*, Taylor and Francis, Washington, DC, 1993.

Vassalos, D., Turan, O., and Pawlowski, M., Dynamic stability assessment of damaged passenger/ro-ro ships and proposal of rational survival criteria, *Marine Technology*, Vol. 34, 241–66, 1997.

Verduijn, W. D., and Elishakoff, I., A testing machine for statistical analysis of small imperfect shells, *Report LR-375*, Delft University of Technology, Department of Aerospace Engineering, Delft, the Netherlands, 1982.

Verijenko V. E., Adali, S., Summers, E. B., and Reiss, T., Optimal design of laminated shells with shear and normal deformation using symbolic Computation, *AIAA-94-4428-CP*, AIAA Press, pp. 1066–71, 1994.

Videc, B. P., and Sanders, J. L., Application of Khas' minskii limit theorem to the buckling problem of a column with random initial deflections, *Quarterly of Applied Mathematics*, Vol. 33, 422–8, 1976.

Vijayakumar, K., and Joga Rao, C. V., Buckling of polar orthotropic annular plates, *Journal of Engineering Mechanics Division*, Vol. EM3, 701–10, 1971.

Vinson, J. R., and Sierakowski, R. L., *The Behavior of Structures Composed of Composite Materials*, Martinus Nijhoff Publishers, Dortrecht, 1986.

Volkov, S. D., *Statistical Strength Theory*, Gordon and Breach, New York, 1962.

Volkov, S. D., and Stavrov, V. P., *Statistical Mechanics of Composite Materials*, Belorussian State University, Minsk, 1978 (in Russian).

Volkov, S. I., Branching of probability functionals' extremals and stability of nonlinear elastic column with random imperfections, *Technical Report*, Novosikirsk Electrotechnic Institute, Novosikirsk, 1978 (in Russian).

Volkov, S. I., An example of stochastic bifurcation in the theory of bending of non-ideal plates, *Prikladnaya Matematika i Mekhanika*, Vol. 45, 876–83, 1981 (in Russian).

Volmir, A. S., *Biegsame Platten und Schalen*, Verlag für Bauwsen, Berlin, 1962 (in German).

Volmir, A. S., *Stability of Elastic Systems*, Wright-Patterson Air Force Base, Dayton, OH, *FTD-MT 64-335*, 1964, and *NASA AD-628508*, 1965.

Volmir, A. S., and Kildibekov, I. G., Probabilistic characteristics of behavior of cylindrical shells subjected to acoustic loading, *Prikladnaya Mekhanika*, Vol. 1, 1–9, 1965 (in Russian).

Volmir, A. S., and Kildibekov, I. G., Probabilistic characteristics of behavior of shallow shells in an acoustic field, *Reliability Problems in Structural Mechanics* (V. V. Bolotin and A. A. Cyras, eds.), "RINTIP" Publishing House, Vilnius, pp. 169–74, 1968a (in Russian).

Volmir, A. S., and Kildibekov, I. G., Nonlinear vibrations and buckling of a shallow shell in acoustic field, *Reliability Problems in Structural Mechanics* (V. V. Bolotin and A. Cyras, eds.), pp. 169–74, 1968b (in Russian).

Volosovich, O. V., and Timashev, S. A., Initial imperfections and buckling modes of rectangular convex shells, *Theory of Shells and Plates, Proceedings of All-Union Conference*, "Nauka" Publishing House, Rostov-on-Don, pp. 254–8, 1973 (in Russian).

Von Slooten, R. A., and Soong, T. T., Buckling of a long, axially compressed, thin cylindrical shell with random initial imperfections, *Journal of Applied Mechanics*, 1066–71, 1972.

Vorovitch, I. I., Statistical method in the theory of stability of shells, *Prikladnaya Matematika i Mekhanika*, Vol. 23, 885–92, 1959 (in Russian).

Vorovich, I. I., Some problems of applications of statistical methods in the theory of stability of plates and shells, *Theory of Plates and Shells*, Academy of Armenia Press, Erevan, pp. 69–94, 1964 (in Russian).

Vorovich, I. I., Paths of developments of problem of stability in the theory of shells, *Actual Problems of Science*, Rostov University Press, Rostov-na-Donu, pp. 111–26, 1967 (in Russian).

Vorovich, I. I., and Minakova, N. I., Stability problem and numerical methods in the theory of spherical shells, *Mechanics of Solid Deformable Bodies*, Vol. 7, "VINITI" Publishers, pp. 1–86, 19732 (in Russian).

Vorovich, I. I., and Zipalova, V. F., Towards solution of nonlinear boundary-value problems of elasticity by transfer to the Cauchy problems, *Prikladnaya Matematika i Mekhanika*, Vol. 29, 894–901, 1965 (in Russian).

Wah, T., and Calcote, L. R., *Structural Analysis by Finite Difference Calculus*, Van Nostrand Reinhold, New York, 1970.

Wang, D., Zhang, S., and Yang, G., Improved Monte Carlo method and its application to structural reliability, *Acta Mechanica Solida Sinica*, Vol. 7, 256–64, 1994.

Wang, I-Y., Monte Carlo analysis of nonlinear vibration of rectangular plates with random geometric imperfections, *International Journal of Solids and Structures*, Vol. 26, 99–109, 1990.

Wang, L. R. L., and Rodriguez-Agrait, L., Statistical tests on structural models for the buckling behavior of spherical shells, *Statistics and Probability in Civil Engineering* (P. Lumb, ed.), Hong Kong University Press, Hong Kong, pp. 489–512, 1971.

Waszczyszyn, Z., Cichon, C., and Radwanska, M., *Stability of Structures by Finite Element Methods*, Elsevier Science Publishers, Amsterdam, 1994.

Weingarten, V. I., Morgan, E. J., and Seide, P., Elastic stability of thin-walled cylindrical and conical shells under axial compression, *AIAA Journal*, Vol. 3, 500–5, 1965.

Weingarten, V. I., Seide, P., and Peterson, J. P., Buckling of thin-walled structures, *NASA SP-8007*, 1968.

Wentzel, E. S., *Operations Research*, "Mir" Publishers, Moscow, 1988.

Whitney, J. M., *Structural Analysis of Laminated Anisotropic Plates*, Technomic Publishing Company, Lancaster, PA, pp. 151–6, 1987.

Williams, F. W., Development history of efficient algorithms for finding roots with certainty in dynamic stiffness methods, *Proceedings of Second ESA Workshop on Modal Representation of Flexible Structures by Continuum Methods*, ESA WPP-034 European Space Agency, Noordwijk, pp. 75–94, 1992.

Wirsching, P. H., and Yao, J. T. P., Random behavior of columns, *Journal of the Engineers Mechanics Division*, Vol. 97, 605–18, 1971.

Wittrick, W. H., and Williams, F. W., An algorithm for computing critical buckling loads of elastic structures, *Journal of Structural Mechanics*, Vol. 1, 497–518, 1973.

Wohlever, J. C., and Healey, T. J., A group theoretic approach to the golbal bifurcation analysis of an axially compressed cylindrical shell, *Computer Methods in Applied Mechanics and Engineering*, Vol. 122, 315–50, 1995.

Woinowsky-Krieger, S., Ueber die Beulsicherheit von Kreisplatten mit kreiszylindrischer Aelotropie, *Ingenieur Archiev*, Vol. 26, 129–31, 1958.

Wolfram, S., *The Mathematica*^{R} *Book*, 3rd ed., Cambridge University Press, 1996.

Wooff, G., and Hodgkinson, D., *muMATH: A Microcomputer Algebra System*, Academic Press, London, 1987.

Wriggers, P., and Simo, J. C., A general procedure for the direct computation of turning and bifurcation points, *International Journal of Numerical Methods in Engineering*, Vol. 30, 155–67, 1990.

Wu, C. H., Buckling of anisotropic circular cylindrical shells, PhD Thesis, Case Western Reserve University, 1971.

Wu, Y. T., and Burnside, D. M., Computational methods for probability of instability calculations, *Proceedings, AIAA/ASME/ASCE/AHS/ASC 31st SDM Conference, Structures, Structural Dynamics and Materials*, Long Beach, 1990.

Wu, Y. T., and Burnside, O. H., Computational methods for probability of instability calculations, *Proceedings of the 31st Structures, Structural Dynamics and Materials Conference*, AIAA Press, Washington, DC, pp. 1081–91, 1990.

Wunderlich, W., Rensch, H. J., and Obrecht, H., Analysis of elastic-plastic buckling and imperfection-sensitivity of shells of revolution, *Buckling of Shells: Proceedings of a State-of-the-Art Colloquium* (E. Ramm, ed.), Springer Verlag, Berlin, pp. 137–74, 1982.

Xie, W. C., Buckling mode localization in randomly disordered multispan continuous beam, *AIAA Journal*, Vol. 33, 1142–9, 1995.

Xie, W. C., Buckling mode localization in rib-stiffened plates with randomly misplaced stiffness, *Computers and Structures*, Vol. 67, 175–89, 1998.

Xie, W. C., Buckling mode localization in nonhomogeneous beams on elastic foundation, *Chaos, Solitons and Fractals*, Vol. 8, 411–31, 1997.

Xie, W. C., Buckling mode localization in rib-stiffened plates with randomly misplaced stiffener, *Computers and Structures*, Vol. 67, 175–89, 1998.

Xie, W. C., and Elishakoff, I., Buckling mode localization in rib-stiffened plates with misplaced stiffeners-Kantorovich approach, *Chaos, Solitons and Fractals*, Vol. 11 (10), 1559–1574, 2000.

Yadav, D., and Verma, N., Buckling of composite circular cylindrical shells with random material properties, *Composite Structures*, Vol. 37, 385–91, 1997.

Yaffe, R., Singer, J., and Abramovich, H., Further initial imperfection measurements of integrally stringer-stiffened cylindrical shells-series 2, *TAE Report 404*, Department of Aeronautical Engineering, Technion – Israel Institute of Technology, 1981.

Yamaki, N., Buckling of a thin annular plate under uniform compression, *Journal Applied Mechanics*, Vol. 25, 267–72, 1958.

Yamaki, N., *Elastic Stability of Circular Cylindrical Shells*, North-Holland, New York, 1984.

Yamaki, N., and Kodama, S., Postbuckling behavior of circular cylindrical shells under compression, *International Journal of Non-Linear Mechanics*, Vol. 11, 99–111, 1976.

Yeigh B. W., Imperfections and instabilities, *PhD Dissertation*, Department of Civil Engineering and Operations Research, Princeton University, 1995.

Yamazaki, F., and Shinozuka, M., Neumann expansion for stochastic finite element analysis, *Journal of Engineering Mechanics*, Vol. 114, 1335–54, 1988.

Yeigh B. W., Stochastic modeling of imperfections in beams, *Probabilistic Mechanics and Structural Reliability* (D. M. Frangopol and M. D. Grigoriu, eds.), ASCE Press, New York, pp. 676–9, 1996.

Yeigh, B. W., and Shinozuka, M., Uncertainty modeling in structural stability, *Uncertainty Modeling in Finite Element, Fatigue and Stability of Systems* (A. Haldar, A. Guran, and B. M. Ayyub, eds.), World Scientific, Singapore, pp. 215–60, 1997.

Yoakimidis, N. I., Symbolic computations for the solution of inverse/design problems with UAPLE, *Computers and Structures*, Vol. 53, 63–8, 1994.

Yushanov, S. P., Probabilistic model of layerwise collapse of a composite and analysis of reliability of layered cylindrical shells, *Mechanics of Composite Materials*, No. 4, 642–53, 1985 (in Russian).

Yushanov, S. P., and Bogdanovich, A. E., Method of analysis of reliability of layered cylindrical shells with allowance for scatter in the strength characteristics of the composite, *Mechanics of Composite Materials*, Vol. 22, 725–30, 1987 (in Russian).

Yusupov, A. K., Stability of compressed column during random rotation of the support, *Stroitelnaya Mekhanika i Raschet Sooruzhenii*, No. 1, 38–41, 1976 (in Russian).

Zadeh, L. A., Fuzzy sets, *Information and Control*, Vol. 8, 338–53, 1965.

Zhang, J., and Ellingwood, B., Orthogonal series expansions of random fields in reliability analysis, *Journal of Engineering Mechanics*, Vol. 120, 2660–77, 1994.

Zhang, J., and Ellingwood, B., Orthogonal series representations of random fields in stochastic FEA, *Computational Stochastic Mechanics* (P. D. Spanos, ed.), Balkema Publishers, Rotterdam, pp. 57–65, 1995a.

Zhang, J., and Ellingwood, B., Effects of uncertain material properties on structural stability, *Journal of Structural Engineering*, Vol. 121, 705–16, 1995b.

Zhang, J., and Ellingwood, B., SFE-Based structural stability analysis, *Application of Statistics and Probability* (M. Lemaire, J. L. Favre, and A. Mebarki, eds.), Balkema Publishers, Rotterdam, pp. 1041–6, 1995c.

Zhang, J., and Ellingwood, B., Error measure for reliability studies using reduced variable set, *Journal of Engineering Mechanics*, Vol. 121, 935–7, 1995d.

Zhu, L. P., Elishakoff, I., and Lin, Y. K., Free and forced vibrations of periodic multi-span beams, *Shock and Vibration*, Vol. 1, 217–32, 1994.

Zhu, L. P., Elishakoff I., and Starnes J. H., Jr., Derivation of multi-dimensional ellipsoidal convex model for experimental data, *Mathematical Computing and Modelling*, Vol. 24, 103–14, 1996.

Zhukova, N. B., and Semeniuk, N. P., Timoshenko-type theory in cubic approximation with application to the initial post-buckling behavior of composite shells, *Prikladnaya Mekhanika*, Vol. 28, No. 10, 41–6, 1992 (in Russian).

Ziegler, H., *Principles of Structural Stability*, Blaisdell, Boston, 1968; Birkhauser, Basel, 1977.

Zimmermann, R., Optimierung axial gedrückter CFK-Zylinderschalen, *VDI Report Nr. 207*, VDI-Verlag, Düsseldorf, 1992 (in German).

Zingales, M., and Elishakoff, I., Localization of the bending response in presence of axial load, *International Journal of Solids and Structures*, (to appear).

Zipalova, V. F., and Nenastieva, V. M., On application of Galerkin method in temperature-force problems in nonlinear stability of shells, *Izvestiya Akademii Nauk SSSR, Mekhanika Tverdogo Tela*, Vol. II, No. 1, 100–3, 1971 (in Russian).

Zucarro, G., Elishakoff, I., and Baratta, A., Antioptimization of earthquake excitation and response, *Mathematical Problems in Engineering*, Vol. 4, 1–19, 1998.

Zyczkowski, M., and Gajewski, A., Optimal structural design under stability constraints, in *Collapse: The Buckling of Structures in Theory and Practice*, (J. M. T. Thompson and G. W. Hunt, eds.), Cambridge University Press, Cambridge, pp. 299–332, 1983.

Author Index

Not too many intellectuals have the courage to admit they pick a book and look first for their name in the index.

I. Howe

Only those authors who are referenced in the text are listed. The reader may also consult the extensive bibliography of about 900 references. We tried to cover the subject as completely as we could, but some inadvertent omissions may have occurred.

Subject Index